An Introduction to Thermodynamics and Statistical Mechanics, Second Edition

This introductory textbook for standard undergraduate courses in thermodynamics has been completely rewritten to explore a greater number of topics more clearly and concisely. Starting with an overview of important quantum behaviors, the book teaches students how to calculate probabilities in order to provide a firm foundation for later chapters. It then introduces the ideas of "classical thermodynamics" – internal energy, interactions, entropy, and the fundamental second law. These ideas are explored both in general and as they are applied to more specific processes and interactions. The remainder of the book deals with "statistical mechanics" – the study of small systems interacting with huge reservoirs.

The changes in this Second Edition have been made as a result of more than 10 years of classroom testing and feedback from students. To help students review the important concepts and test their newly gained knowledge, each topic ends with a boxed summary of ideas and results. Every chapter has numerous homework problems, covering a broad range of difficulties. Answers are given to odd-numbered problems, and solutions to even-numbered problems are available to instructors at www.cambridge.org/9781107694927.

KEITH STOWE is a professor of physics at California Polytechnic State University and has worked there for 32 years. He has spent time at the University of Washington, Harvard, the University of North Carolina, and the University of Michigan. As well as having written the First Edition of *Introduction to Thermodynamics and Statistical Mechanics*, he has also written books on ocean science.

An Introduction to Thermodynamics and Statistical Mechanics

Second Edition

Keith Stowe
California Polytechnic State University

CAMBRIDGE
UNIVERSITY PRESS

CAMBRIDGE
UNIVERSITY PRESS

University Printing House, Cambridge CB2 8BS, United Kingdom

One Liberty Plaza, 20th Floor, New York, NY 10006, USA

477 Williamstown Road, Port Melbourne, VIC 3207, Australia

314-321, 3rd Floor, Plot 3, Splendor Forum, Jasola District Centre, New Delhi - 110025, India

79 Anson Road, #06-04/06, Singapore 079906

Cambridge University Press is part of the University of Cambridge.

It furthers the University's mission by disseminating knowledge in the pursuit of education, learning and research at the highest international levels of excellence.

www.cambridge.org
Information on this title: www.cambridge.org/9781107694927

First published, second edition 2007
First paperback edition 2013

A catalogue record for this publication is available from the British Library

ISBN 978-0-521-86557-9 Hardback
ISBN 978-1-107-69492-7 Paperback

Contents

Preface

Goals

The subject of thermodynamics was being developed on a postulatory basis long before we understood the nature or behavior of the elementary constituents of matter. As we became more familiar with these constituents, we were still slow to place our trust in the "new" field of quantum mechanics, which was telling us that their behaviors could be described correctly and accurately using probabilities and statistics.

The influence of this historical sequence has lingered in our traditional thermodynamics curriculum. Until recently, we continued to teach an introductory course using the more formal and abstract postulatory approach. Now, however, there is a growing feeling that the statistical approach is more effective. It demonstrates the firm physical and statistical basis of thermodynamics by showing how the properties of macroscopic systems are direct consequences of the behaviors of their elementary constituents. An added advantage of this approach is that it is easily extended to include some statistical mechanics in an introductory course. It gives the student a broader spectrum of skills as well as a better understanding of the physical bases.

This book is intended for use in the standard junior or senior undergraduate course in thermodynamics, and it assumes no previous knowledge of the subject. I try to introduce the subject as simply and succinctly as possible, with enough applications to indicate the relevance of the results but not so many as might risk losing the student in details. There are many advanced books of high quality that can help the interested student probe more deeply into the subject and its more specialized applications.

I try to tie everything straight to fundamental concepts, and I avoid "slick tricks" and the "pyramiding" of results. I remain focused on the basic ideas and physical causes, because I believe this will help students better understand, retain, and apply the tools and results that we develop.

Active learning

I think that real learning must be an *active* process. It is important for the student to apply new knowledge to specific problems as soon as possible. This should be a daily activity, and problems should be attempted while the knowledge is still fresh. A routine of frequent, timely, and short problem-solving sessions is far superior to a few infrequent problem-solving marathons. For this reason, at the end of each chapter the text includes a very large number of suggested homework problems, which are organized by section. Solutions to the odd-numbered problems are at the end of the book for instant feedback.

Active learning can also be encouraged by streamlining the more passive components. The sooner the student understands the text material, the sooner he or she can apply it. For this reason, I have put the topics in what I believe to be the most learning-efficient order, and I explain the concepts as simply and clearly as possible. Summaries are frequent and are included within the chapters wherever I think would be helpful to a first-time student wrestling with the concepts. They are also shaded for easy identification. Hopefully, this streamlining of the passive aspects might allow more time for active problem solving.

Changes in the second edition

The entire book has been rewritten. My primary objective for the second edition has been to explore more topics, more thoroughly, more clearly, and with fewer words. To accomplish this I have written more concisely, combined related topics, and reduced repetition. The result is a modest reduction in text, in spite of the broadened coverage of topics.

In addition I wanted to correct what I considered to be the two biggest problems with the first edition: the large number of uncorrected typos and an incomplete description of the chemical potential. A further objective was to increase the number and quality of homework problems that are available for the instructor or student to select from. These range in difficulty from warm-ups to challenges. In this edition the number of homework problems has nearly doubled, averaging around 40 per chapter. In addition, solutions (and occasional hints) to the odd-numbered problems are given at the back of the book. My experience with students at this level has been that solutions give quick and efficient feedback, encouraging those who are doing things correctly and helping to guide those who stumble.

The following list expands upon the more important new initiatives and features in this edition in order of their appearance, with the chapters and sections indicated in parentheses.

- Fluctuations in observables, such as energy, temperature, volume, number of particles, etc. (Sections 3A, 3C, 7C, 9B, 19A)
- Improved discussion and illustrations of the chemical potential (Sections 5C, 8A, 9E, 14A)
- The explicit dependence of the number of accessible states on the system's internal energy, volume, and number of particles (Chapter 6)
- Behaviors near absolute zero (Sections 9H, 24A, 24B)

- Entropy and the third law (Section 8D)
- A new chapter on interdependence among thermodynamic variables (Chapter 11)
- Thermal conduction, and the heat equation (Section 12E)
- A more extensive treatment of engines, including performance analysis (Section 13F), model cycles, a description of several of the more common internal combustion engines (Section 13H), and vapor cycles (Section 13I)
- A new chapter on diffusive interactions, including such topics as diffusive equilibrium, osmosis, chemical equilibrium, and phase transitions (Chapter 14)
- Properties of solutions (colligative properties, vapor pressure, osmosis, etc.) (Section 14B)
- Chemical equilibrium and reaction rates (Section 14C)
- A more thorough treatment of phase transitions (Section 14D)
- Binary mixtures, solubility gap, phase transitions in minerals and alloys, etc. (Section 14E)
- Conserved properties (Section 16E)
- Calculating the chemical potential for quantum systems (Section 19E)
- Chemical potential and internal energy for quantum gases (Section 20D)
- Entropy and adiabatic processes in photon gases (Section 21E)
- Thermal noise (Section 21F)
- Electrical properties of materials, including band structure, conductors, intrinsic and doped semiconductors, and p–n junctions (Chapter 23)
- Update of recent advances in cooling methods (Section 24A)
- Update of recent advances in Bose–Einstein condensation (Section 24B)
- Stellar collapse (Section 24C)

Organization

The book has been organized to give the instructor as much flexibility as possible. Some early chapters are essential for the understanding of later topics. Many chapters, however, could be skipped at a first reading or their order rearranged as the instructor sees fit. To help the instructor or student with these choices, I give the following summary followed by more detailed information.

Summary of organization

Part I   Introduction
Chapter 1   essential if the students have not yet had a course in quantum mechanics. Summarizes important quantum effects

Part II   Small systems
Chapter 2 and Chapter 3   insightful, but not needed for succeeding chapters

Part III   Energy and the first law
Chapter 4, Chapter 5 and Chapter 6   essential

Part IV   States and the second law
Chapter 6, Chapter 7 and Chapter 8   essential

Part V   Constraints
Chapter 9   essential
Chapter 10, Chapter 11, Chapter 12, Chapter 13 and Chapter 14   any order, and
any can be skipped

Part VI   Classical statistics
Chapter 15   essential
Chapter 16, Chapter 17 and Chapter 18   any order, and any can be skipped

Part VII   Quantum statistics
Chapter 19, Chapter 20 A, B   essential
Chapter 21, Chapter 22, Chapter 23 and Chapter 24   any order, and any can be
skipped

More details

Part I – Introduction   Chapter 1 is included for the benefit of those students
who have not yet had a course in quantum mechanics. It summarizes important
quantum effects that are used in examples throughout the book.

Part II – Small systems   Chapters 2 and 3 study systems with only a few elements. By studying small systems first the student develops both a better appreciation and also a better understanding of the powerful tools that we will need for
large systems in subsequent chapters. However, these two chapters are not *essential* for understanding the rest of the book and may be skipped if the instructor
wishes.

Part III – Energy and the first law   Chapters 4 and 5 are intended to give the
student an intuitive physical picture of what goes on within interacting systems on
a microscopic scale. Although the mathematical rigor comes later, this physical
understanding is essential to the rest of the book so these two chapters should not
be skipped.

Part IV – States and the second law   Chapters 6, 7, and 8 are the most important in the book. They develop the statistical basis for much of thermodynamics.

Part V – Constraints   Chapter 9 derives the universal consequences of the fundamental ideas of the preceding three chapters. So this chapter shows why things
must behave as they do, and why our "common sense" is what it is. Chapters 10–14

all describe the application of constraints to more specific systems. None of these topics is essential, although some models in Chapter 10 would be helpful in understanding examples used later in the book; if Chapters 11 and 12 are covered, they should be done in numerical order. Topics in these five chapters include equations of state and models, the choice and manipulation of variables, isobaric, isothermal, and adiabatic processes, reversibility, important nonequilibrium processes, engines, diffusion, solutions, chemical equilibrium, phase transitions, and binary mixtures.

Part VI – Classical statistics Chapter 15 develops the basis for both classical "Boltzmann" and quantum statistics. So even if you go straight to quantum statistics, this chapter should be covered first. Chapters 16, 17, and 18 are applications of classical statistics, each of which has no impact on any other material in the book. So they may be skipped or presented in any order with no effect on subsequent material.

Part VII – Quantum statistics Chapter 19 introduces quantum statistics, and the first two sections of Chapter 20 introduce quantum gases. These provide the underpinnings for the subsequent chapters and therefore must be covered first. The remaining four (Chapters 21–24) are each independent and may be skipped or presented in any order, as the instructor chooses.

Acknowledgments

I wish to thank my students for their ideas, encouragement, and corrections, and my colleagues Joe Boone and Rich Saenz for their careful scrutiny and thoughtful suggestions. I also appreciate the suggestions received from Professors Robert Dickerson and David Hafemeister (California Polytechnic State University), Albert Petschek (New Mexico Institute of Mining and Technology), Ralph Baierlein (Wesleyan University), Dan Wilkins (University of Nebraska, Omaha), Henry White (University of Missouri), and I apologize to the many whose names I forgot to record.

List of constants, conversions, and prefixes

Constants

acceleration of gravity	$g = 9.807$ m/s^2
Avogadro's number	$N_A = 6.022 \times 10^{23}$ particles/mole
Boltzmann's constant	$k = 1.381 \times 10^{-23}$ J/K $= 8.617 \times 10^{-5}$ eV/K
Coulomb constant	$1/4\pi\varepsilon_0 = 8.988 \times 10^9$ kg m^3/(s^2 C^2)
elementary unit of charge	$e = 1.602 \times 10^{-19}$ C
gas constant	$R = N_A k = 8.315$ J/(K mole)
	$= 0.08206$ liter atm/(K mole)
gravitational constant	$G = 6.673 \times 10^{-11}$ m^3/(kg s^2)
magnetons	
Bohr magneton	$\mu_B = 9.274 \times 10^{-24}$ J/T $= 5.788 \times 10^{-5}$ eV/T
nuclear magneton	$\mu_N = 5.051 \times 10^{-27}$ J/T $= 3.152 \times 10^{-8}$ eV/T
masses	
atomic mass unit	$u = 1.661 \times 10^{-27}$ kg
electron mass	$m_e = 9.109 \times 10^{-31}$ kg
neutron mass	$m_n = 1.675 \times 10^{-27}$ kg
proton mass	$m_p = 1.673 \times 10^{-27}$ kg
Planck's constant	$h = 6.626 \times 10^{-34}$ J s $= 4.136 \times 10^{-15}$ eV s
	$\hbar = h/2\pi = 1.055 \times 10^{-34}$ J s $= 6.582 \times 10^{-16}$ eV s
speed of light in vacuum	$c = 2.998 \times 10^8$ m/s
Stefan–Boltzmann constant	$\sigma = 5.671 \times 10^{-8}$ W/(m^2 K^4)

Conversions

1 Å $= 10^{-10}$ m

1 liter $= 10^{-3}$ m^3

1 atm $= 1.013 \times 10^5$ Pa

$\log_{10} x = 0.4343 \ln x$

$e^x = 10^{0.4343x}$

1 eV $= 1.602 \times 10^{-19}$ J

1 cal $= 4.184$ J $= 0.04129$ liter atm

1 T $= 1$ Wb/m$^2 = 10^4$ G

temperature (K) = temperature (°C) + 273.15 K

Prefixes

tera	T	10^{12}
giga	G	10^{9}
mega	M	10^{6}
kilo	k	10^{3}
centi	c	10^{-2}
milli	m	10^{-3}
micro	μ	10^{-6}
nano	n	10^{-9}
pico	p	10^{-12}
femto	f	10^{-15}

Part I
Setting the scene

Chapter 1
Introduction

Imagine you could shrink into the atomic world. On this small scale, motion is violent and chaotic. Atoms shake and dance wildly, and each carries an electron cloud that is a blur of motion. By contrast, the behavior of a very large number of atoms, such as a baseball or planet, is quite sedate. Their positions, motions, and properties change continuously yet predictably. How can the behavior of macroscopic systems be so predictable if their microscopic constituents are so unruly? Shouldn't there be some connection between the two?

Indeed, the behaviors of the individual microscopic elements are reflected in the properties of the system as a whole. In this course, we will learn how to make the translation, either way, between microscopic behaviors and macroscopic properties.

A The translation between microscopic and macroscopic behavior

A.1 The statistical tools

If you guess whether a flipped coin will land heads or tails, you have a 50% chance of being wrong. But for a very large number of flipped coins, you may safely

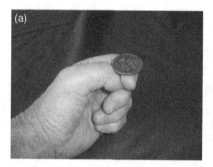

Figure 1.1 (a) If you know the probabilities for one single coin flip then you can predict the heads–tails distribution for a large number of them. Conversely, by observing the heads–tails distribution for a large number of flipped coins, you can infer the probabilities for any one of them. (b) What is the probability that a rolled dice will land with six dots up? If a large number of dice were rolled, roughly what fraction of them would land with six dots up?

assume that nearly half will land heads. Even though the individual elements are unruly, the behavior of a large system is predictable (Figure 1.1).

Your prediction could go the other way, too. From the behavior of the entire system, you might predict probabilities for the individual elements. For example, if you find that one sixth of a large number of rolled dice show sixes (i.e., six dots up), you can correctly infer that the probability for any one die to show a six is 1/6 (Figure 1.1b). When a system is composed of a large number of identical elements, you can use the observed behavior of an individual element to predict the properties of the whole system, or conversely, you can use the observed properties of the entire system to deduce the probable behaviors of the individual elements.

The study of this two-way translation between the behavior of the individual elements and the properties of the system as a whole is called statistical mechanics. One of the goals of this book is to give you the tools for making this translation, in either direction, for whatever system you wish.

A.2 Thermodynamics

The industrial revolution and the attendant proliferation in the use of engines gave a huge impetus to the study of thermodynamics, a name that obviously reflects the early interest in turning heat into motion. The study now encompasses all forms of work and energy and includes probing the relationships among system parameters, such as how pressure influences temperature, how energy is converted from one form to another, etc.

Considerable early progress was made with little or no knowledge of the atomic nature of matter. Now that we understand matter's elementary constituents better, the tools of thermodynamics and statistical mechanics help us improve our understanding of matter and macroscopic systems at a more fundamental level.

Summary of Section A

If a system is composed of many identical elements, the probable behaviors of an individual element may be used to predict the properties of the system as a whole or, conversely, the properties of the system as a whole may be used to infer the probable behaviors of an individual element. The study of the statistical techniques used to make this two-way translation between the microscopic and macroscopic behaviors of physical systems is called statistical mechanics. The study of interrelationships among macroscopic properties is called thermodynamics. Using statistical tools, we can relate the properties of a macroscopic system to the behaviors of its individual elements, and in this way obtain a better understanding of both.

B Quantum effects

When a large number of coins are flipped, it is easy to predict that nearly half will land heads up. With a little mathematical sophistication, you might even be able to calculate typical fluctuations or probabilities for various possible outcomes. You could do the same for a system of many rolled dice.

Like coins and dice, the microscopic constituents of physical systems also have only certain discrete states available to them, and we can analyze their behaviors with the same tools that we use for systems of coins or dice. We now describe a few of these important "quantized" properties, because we will be using them as examples in this course. You may wish to refer back to them when you arrive at the appropriate point later in the book.

B.1 Electrical charge

For reasons we do not yet understand, nature has provided electrical charge in fundamental units of 1.6×10^{-19} coulombs, a unit that we identify by e:

$$e = 1.602 \times 10^{-19} \, \text{C}.$$

We sometimes use collisions to study the small-scale structure of subatomic particles. No matter how powerful the collision or how many tiny fragments are

produced, the charge of each is always found to be an integral number of units of
the fundamental charge, e.[1]

B.2   Wave nature of particles

In the nineteenth century it was thought that energy could go from one point to
another by either of two distinct processes: the transport of matter or the propagation of waves. Until the 1860s, we thought waves could only propagate through
matter. Then the work of James Clerk Maxwell (1831–79) demonstrated that
electromagnetic radiation was also a type of wave, with oscillations in electric
and magnetic fields rather than in matter. These waves traveled at extremely high
speeds and through empty space. Experiments with appropriate diffraction gratings showed that electromagnetic radiation displays the same diffractive behavior
as waves that travel in material media, such as sound or ocean waves.

Then in the early twentieth century, experiments began to blur the distinction
between the two forms of energy transport. The photoelectric effect and Compton
scattering demonstrated that electromagnetic "waves" could behave like "particles." And other experiments showed that "particles" could behave like "waves:"
when directed onto appropriate diffraction gratings, beams of electrons or other
subatomic particles yielded diffraction patterns, just as waves do.

The wavelength λ for these particle–waves was found to be inversely proportional to the particle's momentum p; it is governed by the same equation used for
electromagnetic waves in the photoelectric effect and Compton scattering,

$$\lambda = \frac{h}{p} \qquad (h = 6.626 \times 10^{-34}\,\mathrm{J\,s}). \tag{1.1}$$

Equivalently, we can write a particle's momentum in terms of its wave number,
$k = 2\pi/\lambda$.

$$p = \frac{h}{\lambda} = \frac{h}{2\pi}\frac{2\pi}{\lambda} = \hbar k \qquad (\hbar = h/2\pi = 1.055 \times 10^{-34}\,\mathrm{J\,s}). \tag{1.2}$$

The constant of proportionality, h, is Planck's constant, and when divided by 2π
it is called "h-bar."

We do not know why particles behave as waves any more than we know
why electrical charge comes in fundamental units e. But they do, and we can
set up differential "wave equations" to describe any system of particles we like.
The solutions to these equations are called "wave functions," and they give us
the probabilities for various behaviors of the system. In the next few pages we
describe some of the important consequences.

---

[1] For quarks the fundamental unit would be $e/3$. But they bind together to form the observed elementary particles (protons, neutrons, mesons, etc.) only in ways such that the total electrical charge
is in units of e.

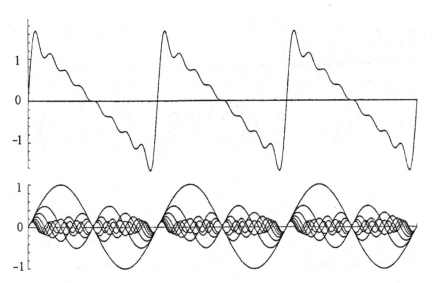

Figure 1.2 The superposition of the sine waves below yields the sawtooth wave above.

B.3   Uncertainty principle

Any function of the variable x on $(-\infty, \infty)$ can be written as a superposition of sine wave components of various wavelengths (Figure 1.2). These sine wave components may be either of the form $\sin kx$ and $\cos kx$, or e^{ikx}, and the technique used to determine the contributions of each component to any function, $f(x)$, is called Fourier analysis. In mathematical terms, any function $f(x)$ on $(-\infty, +\infty)$ can be written as

$$f(x) = \int_0^\infty [a(k)\sin kx + b(k)\cos kx]\,dk$$

or

$$f(x) = \int_{-\infty}^\infty c(k)e^{ikx}\,dk,$$

where the coefficients $a(k)$, $b(k)$, $c(k)$ are the "amplitudes" of the respective components.

We now investigate the behavior of a particle's wave function in the x dimension. Although a particle exists in a certain region of space, the sine wave components, e.g., $\sin kx$, extend forever. Consequently, if we are to construct a localized function from the superposition of infinitely long sine waves, the superposition must be such that the various components cancel each other out everywhere except for the appropriate small region (Figure 1.3).

To accomplish this cancellation requires an infinite number of sine wave components, but the bulk of the contributions come from those whose wave numbers k lie within some small region Δk. As we do the Fourier analysis of various functions, we find that the more localized the function is in x, the broader is the characteristic spread in the wave numbers k of the sine wave components.

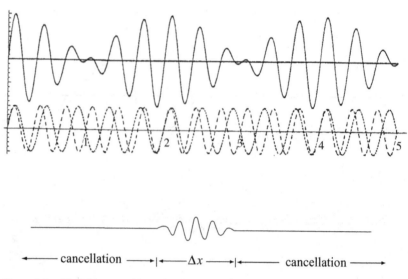

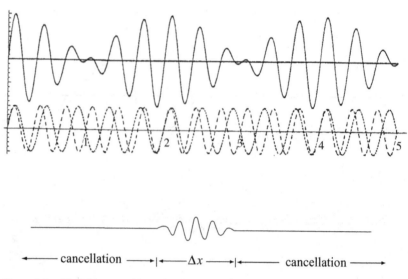
cancellation ⟶|⟵ Δx ⟶|⟵ cancellation ⟶

Figure 1.3 (Top) Superposition of two sine waves of nearly the same wavelengths (the broken and the dotted curves), resulting in beats (the solid curve). The closer the two wavelengths, the longer the beats. There is an inverse relationship. (Bottom) In a particle's wave function, the sine wave components must cancel each other out everywhere except for the appropriate localized region of space, Δx. To make a waveform that does *not* repeat requires the superposition of an infinite number of sine waves, but the same relationship applies: the spread in wavelengths is inversely related to the length of the beat. (The cancellation of the waves farther out requires the inclusion of waves with a *smaller* spread in wavelengths. So the wave numbers of these additional components are closer together and therefore lie within the range Δk of the "primary" wave number.)

In fact, the two are inversely related. If Δx represents the characteristic width of the particle's wave function and Δk the characteristic spread in the components' wave numbers, then

$$\Delta x \Delta k = 2\pi.$$

If we multiply both sides by $\hbar$ and use the relationship 1.2 between wave number and momentum for a particle, this becomes the uncertainty principle,

$$\Delta x \Delta p_x = h. \tag{1.3}$$

This surprising result[2] tells us that because particles behave like waves, they cannot be pinpointed. We cannot know exactly either where they are or where

---

[2] The uncertainty principle is written in many closely related forms. Many authors replace the equals sign by $\geq$, to indicate that the actual measurement may be less precise than the mathematics allows. Furthermore, the spread is a matter of probabilities, so its size reflects your confidence level (i.e., 50%, 75%, etc.). We use the conservative value h because it coincides with Nature's choice for the size of a quantum state, as originally discovered in the study of blackbody radiation.

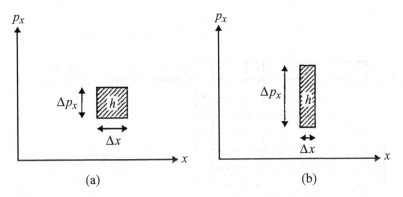

Figure 1.4 (a) According to classical physics, a particle could be located as a point in (x, p_x) space. That is, both its position and momentum could be specified exactly. In modern physics, however, the best we can do is to identify a particle as being somewhere within a box of area $\Delta x \Delta p_x = h$. (b) Because of the wave nature of particles, if we try to specify better the location of a particle in x-space, we lose accuracy in the determination of its momentum p_x. The area $\Delta x \Delta p_x$ of the minimal quantum box does not change.

they are going. If we try to locate a particle's coordinates in the two-dimensional space (x, p_x), we will not be able to specify either coordinate exactly. Instead, the best we can do is to say that its coordinates are somewhere within a rectangle of area $\Delta x \Delta p_x = h$ (Figure 1.4a). If we try to specify its position in x better then our uncertainty in p_x will increase, and vice versa; the area of the rectangle $\Delta x \Delta p_x$ remains the same (Figure 1.4b).

B.4 Quantum states and phase space

The position (x, y, z) and momentum (p_x, p_y, p_z) specify the coordinates of a particle in a six-dimensional "phase space." Although the uncertainty relation 1.3 applies to the two-dimensional phase space (x, p_x), identical relationships apply in the y and z dimensions. And by converting to angular measure, we get the same uncertainty principle for angular position and angular momentum. Thus we obtain

$$\Delta y \Delta p_y = h, \quad \Delta z \Delta p_z = h, \quad \Delta\theta \Delta L = h. \qquad (1.3', 1.3'', 1.3''')$$

We can multiply the three relationships 1.3, 1.3', 1.3'' together to get

$$\Delta x \Delta y \Delta z \Delta p_x \Delta p_y \Delta p_z = h^3,$$

which indicates that we cannot identify a particle's position and momentum coordinates in this six-dimensional phase space precisely. Rather, the best we can do is to say that they lie somewhere within a six-dimensional quantum "box" or "state" of volume $\Delta x \Delta y \Delta z \Delta p_x \Delta p_y \Delta p_z = h^3$.

Figure 1.5 The total
number of quantum
states accessible to a
particle whose
momentum is confined to
the range [p_x] and whose
position is confined to the
range [x] is equal to the
total accessible area in
phase space divided by
the area of a single
quantum state, [x][p_x]/h.

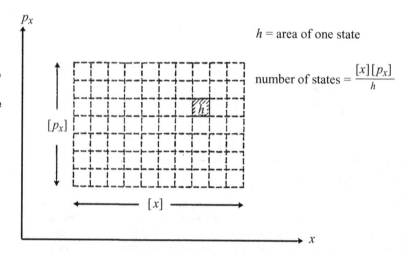

Figure 1.5 The total number of quantum states accessible to a particle whose momentum is confined to the range [p_x] and whose position is confined to the range [x] is equal to the total accessible area in phase space divided by the area of a single quantum state, [x][p_x]/h.

Consider a particle moving in the x dimension whose position and momentum coordinates lie within the ranges [x] and [p_x], respectively (Figure 1.5). The number of different quantum states that are available to this particle is equal to the total accessible area in two-dimensional phase space, [x][p_x], divided by the area of a single quantum state, $\Delta x \Delta p_x = h$. That is,

$$\text{number of accessible states} = \frac{\text{total area}}{\text{area of one state}} = \frac{[x][p_x]}{h}.$$

Extending this to motion in three dimensions we have

$$\text{number of accessible states} = \frac{V_r V_p}{h^3}, \tag{1.4}$$

where V_r and V_p are the accessible volumes in coordinate and momentum space, respectively. In particular, the number of quantum states available in the six-dimensional volume element $d^3 r\, d^3 p$ is given by

$$\text{number of accessible states} = \frac{d^3 r\, d^3 p}{h^3} = \frac{dx\,dy\,dz\,dp_x\,dp_y\,dp_z}{h^3}. \tag{1.5}$$

One important consequence of the relations 1.4 and 1.5 is that the number of quantum states included in any interval of any coordinate is directly proportional to the length of that interval. If ξ represents any of the phase-space coordinates (i.e., the position and momentum coordinates) then

$$\text{number of quantum states in the interval } d\xi \propto d\xi. \tag{1.6}$$

B.5 Density of states

Many calculations require a summation over all states accessible to a particle. Since quantum states normally occupy only a very small region of phase space and are very close together, it is often convenient to replace discrete summation by continuous integration, using the result 1.5:

$$\sum_{\text{states}} \rightarrow \int \frac{\mathrm{d}^3r\mathrm{d}^3p}{h^3}. \tag{1.7}$$

Sometimes the most difficult part of doing this integral is trying to determine the limits of integration. Interactions among particles may restrict the region of phase space accessible to them.

In ideal gases, particles have access to the entire container volume. Changing the sum over states to an integral over the volume and all momentum directions (i.e., the angles in $\mathrm{d}^3p = p^2\mathrm{d}p\sin\theta\,\mathrm{d}\theta\,\mathrm{d}\phi$) gives

$$\sum_{\text{states}} \rightarrow \int \frac{\mathrm{d}^3r\mathrm{d}^3p}{h^3} = \frac{4\pi V}{h^3} \int p^2\mathrm{d}p.$$

We can also write this as a distribution of states in the particle energy, ε. Energy and momentum are related by $\varepsilon = p^2/2m$ for massive nonrelativistic particles and by $\varepsilon = pc$ (c is the speed of travel) for massless particles such as electromagnetic waves (photons) or vibrations in solids (phonons). For these "gases" the sum over states becomes (homework)

$$\sum_{\text{states}} \rightarrow \int \frac{\mathrm{d}^3r\mathrm{d}^3p}{h^3} = \begin{cases} \dfrac{2\pi V(2m)^{3/2}}{h^3} \displaystyle\int \sqrt{\varepsilon}\,\mathrm{d}\varepsilon & \text{(nonrelativistic)}, \\[3mm] \dfrac{4\pi V}{h^3 c^3} \displaystyle\int \varepsilon^2\mathrm{d}\varepsilon & \text{(massless or relativistic)} \end{cases}$$

It is customary to write the summation over phase space as an integral over a function $g(\varepsilon)$:

$$\sum_{\text{states}} \rightarrow \int \frac{\mathrm{d}^3r\mathrm{d}^3p}{h^3} = \int g(\varepsilon)\,\mathrm{d}\varepsilon, \tag{1.8}$$

where $g(\varepsilon)$ is the number of accessible states per unit energy and is therefore called the "density of states." From the above case of an ideal gas, we see that the density of states for a system of noninteracting particles is given by

$$\begin{aligned} g(\varepsilon) &= \frac{2\pi V(2m)^{3/2}}{h^3}\sqrt{\varepsilon} & \text{(nonrelativistic gas)} \\[3mm] g(\varepsilon) &= \frac{4\pi V}{h^3 c^3}\varepsilon^2 & \text{(massless or relativistic gas)} \end{aligned} \tag{1.9}$$

For other systems, however, $g(\varepsilon)$ may be quite different (Figure 1.6). The density of states contains within it the constraints placed on the particles by their mutual interactions.

B.6 Angular momentum

Another surprising result of quantum mechanics is that the angular momentum of a particle or a system of particles can only have certain values; furthermore, a fundamental constraint (the uncertainty principle) prohibits us from knowing its

Figure 1.6 The solid line shows the actual density of states for the atomic vibrations in a mole of sodium metal. The broken line shows the density of states for the motion of the sodium atoms viewed as an ideal gas of massless phonons occupying the same volume (equation 1.9).

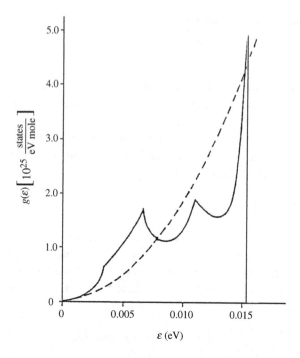

exact orientation in space. In fact, we can know only one of its three components at a time. It is customary to call the direction of the known component the z direction.

The angular momentum of the particles of a system comes from either or both of two sources. They may be traveling in an orbit and may have intrinsic spin as well. The total angular momentum **J** of a particle is the vector sum of that due to its orbit, **L**, and that due to its intrinsic spin, **S**:

$$\mathbf{J} = \mathbf{L} + \mathbf{S}.$$

The orbital angular momentum of a particle must have magnitude

$$|\mathbf{L}| = \sqrt{l(l+1)}\hbar, \qquad l = 0, 1, 2, \ldots, \tag{1.10}$$

where the integer l is called the "angular momentum quantum number." Its orientation is also restricted; the component along any chosen axis (usually called the z-axis) must be an integral multiple of $\hbar$ (Figure 1.7):

$$L_z = l_z \hbar, \qquad l_z = 0, \pm 1, \pm 2, \ldots, \pm l. \tag{1.11}$$

For example, if the particle is in an orbit with $l = 1$ then the total angular momentum has magnitude $\sqrt{2}\hbar$, and its z-component can have any of the values $(-1, 0, 1)\hbar$.

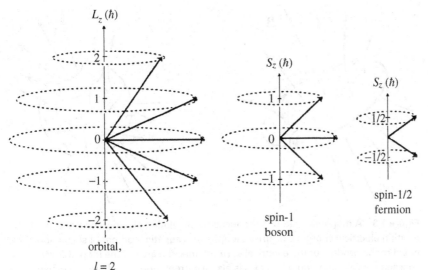

Figure 1.7 Illustration of the quantization of one component (here the z-component) of angular momentum, which can take the values $(0, \pm 1, \pm 2, \dots)\hbar$. The first illustration is for an $l=2$ orbit. Also shown are the possible spin angular momentum orientations for a spin-1 boson, and for a spin-1/2 fermion.

Similar constraints apply to the intrinsic spin angular momentum $\mathbf{S}$ of a particle, for which the magnitude and z-component are given by

$$|\mathbf{S}| = \sqrt{s(s+1)}\,\hbar, \tag{1.12}$$

$$S_z = s_z\,\hbar, \qquad s_z = -s, -s+1, \dots, +s, \tag{1.13}$$

but with one major difference. The spin quantum number s may be either integer or half integer. Those particles with integer spins are called "bosons," and those with half-integer spins are called "fermions."

For later reference, we summarize the constraints on the z-component of angular momentum as follows:

$$L_z = (0, \pm 1, \pm 2, \dots, \pm l)\,\hbar \tag{1.14}$$

and

$$
\begin{aligned}
S_z &= (0, \pm 1, \pm 2, \dots, \pm s)\,\hbar &&\text{(bosons)},\\
S_z &= (\pm 1/2, \pm 3/2, \dots, \pm s)\,\hbar &&\text{(fermions)}.
\end{aligned}
$$

We label particles by the value of their spin quantum number, s. For example, a spin-1 particle has $s = 1$. Its z-component can have the values $s_z = (-1, 0, 1)\,\hbar$. A spin-1/2 particle can have z-component $(-1/2, +1/2)\,\hbar$. We often say simply that it is "spin down" or "spin up," respectively. Protons, neutrons, and electrons are all spin-1/2 particles.

The quantum mechanical origin of these strange restrictions lies in the requirement that if either the particle or the laboratory is turned through a complete rotation around any axis, the observed situation will be the same as before the rotation. Because observables are related to the *square* of the wave function, the

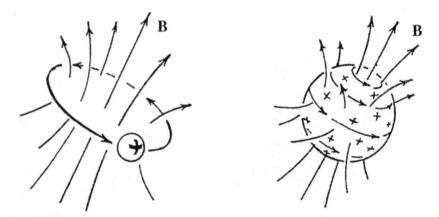

Figure 1.8 A magnetic field **B** is produced by circulating electrical charges. (Left) An orbiting electrical charge is a current loop. The magnetic moment of such a loop is equal to the product of the electrical current times the area of the loop. (Right) A charged particle spinning on its axis is also a current loop, and therefore it also produces a magnetic field. (The figures show positive charges.)

wave function must turn into either plus or minus itself under a rotation by 2π radians. Its sign remains unchanged if the angular momentum around the rotation axis is an integer multiple of $\hbar$ (i.e., for bosons) but changes if the angular momentum around the rotation axis is a half-integer multiple of $\hbar$ (i.e., for fermions). Because of this difference in sign under 2π rotations, bosons and fermions each obey a different type of quantum statistics, as we will see in a later chapter.

B.7 Magnetic moments

Moving charges create magnetic fields (Figure 1.8). For a particle in orbit, such as an electron orbiting the atomic nucleus, the magnetic moment μ is directly proportional to its angular momentum **L** (see Appendix A):

$$\mu = \left(\frac{q}{2m}\right)\mathbf{L},$$

where q is the charge of the particle and m is its mass. Since angular momenta are quantized, so are the magnetic moments:

$$\mu_z = \left(\frac{q}{2m}\right)L_z, \qquad \text{where} \quad L_z = (0, \pm1, \pm2, \ldots, \pm l)\hbar. \qquad (1.15)$$

For particle spin, the relationship between the magnetic moment and the spin angular momentum **S** is similar:

$$\mu = g\left(\frac{e}{2m}\right)\mathbf{S} \qquad \text{and} \qquad \mu_z = g\left(\frac{e}{2m}\right)S_z, \qquad (1.16)$$

where e is the fundamental unit of charge and g is called the "gyromagnetic ratio."

By comparing formulas 1.15 and 1.16, you might think that the factor g is simply the charge of the particle in units of e. But the derivation of equation 1.15

(Appendix A) assumes that the mass and charge have the same distribution, which is *not* true for the intrinsic angular momentum (i.e., spin) of quark-composite particles such as nucleons. Furthermore, in the area of particle spins our classical expectations are wrong anyhow. Measurements reveal that for particle spins:

$$g = -2.00 \quad \text{(electron)},$$
$$g = +5.58 \quad \text{(proton)},$$
$$g = -3.82 \quad \text{(neutron)}.$$

As equations 1.15 and 1.16 indicate, the magnetic moment of a particle is inversely proportional to its mass. Nucleons are nearly 2000 times more massive than electrons, so their contribution to atomic magnetism is normally nearly 2000 times smaller.

The interaction energy of a magnetic moment, μ with an external magnetic field $\mathbf{B}$ is $U = -\mu \cdot \mathbf{B}$. If we define the z direction to be that of the external magnetic field, then

$$U = -\mu_z B. \tag{1.17}$$

In general there are two contributions to the magnetic moment of a particle, one from its orbit and one from its spin. Both are quantized, so the interaction energy U can have only certain discrete values.

B.8 Bound states

Whenever a particle is confined, it may have only certain discrete energies. With the particle bouncing back and forth across the confinement, the superposition of waves going in both directions results in standing waves. Like waves on a string (Figure 1.9), standing waves of only certain wavelengths fit – hence only certain momenta, (1.2), and therefore certain energies, are allowed.

The particular spectrum of allowed energies depends on the type of confinement. Those allowed by a Coulomb potential are different from those of a harmonic oscillator or those of a particle held inside a box with rigid walls, for example. Narrower confinements require shorter wavelengths, which correspond to larger momenta, higher kinetic energies, and greater energy spacing between neighboring states.

The harmonic oscillator confinement is prominent in both the macroscopic and microscopic worlds. If you try to displace any system away from equilibrium, there will be a restoring force that tries to bring it back. (If not, it wouldn't have been in equilibrium in the first place!) For sufficiently small displacements, the restoring force is proportional to the displacement and in the opposite direction. That is,

$$F = -\kappa x,$$

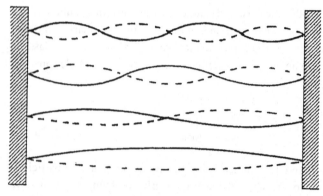

Figure 1.9 Particles behave as waves. A particle in a rigid confinement cannot leave, but must move back and forth across it. This generates standing waves, which must vanish at the boundaries because the particle cannot go beyond. Only certain wavelengths fit. Here are shown the four longest allowed wavelengths, corresponding to the four lowest momenta ($p = h/\lambda$), and hence the four lowest kinetic energies ($p^2/2m$).

where x is the displacement and κ is the constant of proportionality, sometimes called the "elastic" or "spring" constant. The corresponding potential energy is

$$U(x) = U_0 + \tfrac{1}{2}\kappa x^2$$

where U_0 is a constant.

When we solve the wave equation for the spectrum of energies (relative to U_0) allowed by this harmonic oscillator potential, we find that they are given by

$$E = \left(n + \tfrac{1}{2}\right)\hbar\omega, \qquad n = 0, 1, 2, \ldots \tag{1.18}$$

for a one-dimensional harmonic oscillator, and

$$E = \left(n + \tfrac{3}{2}\right)\hbar\omega, \qquad n = 0, 1, 2, \ldots \tag{1.19}$$

for a three-dimensional harmonic oscillator, where the angular frequency is given by

$$\omega = \sqrt{\frac{\kappa}{m}}.$$

Notice that the lowest possible energy (with $n = 0$) is not zero. In fact, no particle may ever have exactly zero kinetic energy, because then its momentum would be zero and its momentum would be fixed. That would violate the uncertainty principle, which dictates that we can never know the momentum and the position exactly. Consequently, even at absolute zero temperature, a particle must still be moving. This motion is sometimes called the "zero-point energy" or

"Zitterbewegung." We know neither in which direction it is going nor where it is in the confinement, so it still obeys the uncertainty principle.

C   Description of a state

We began this chapter with examples involving coins and dice. Each of these could have only a limited number of configurations or states: a coin has two and a die has six. Then we learned that important characteristics of the microscopic components of real physical systems also have discrete values, such as the electrical charge, the angular momentum, the magnetic moment and magnetic interaction energy, or the energy in a confinement.

In any particular problem there will be only one or two properties of the element of the system that would be relevant, so we can ignore all others. When dealing with flipped coins, for example, we wish to know their heads–tails configurations only. Their colors, compositions, designs, interactions with the table, etc. are irrelevant. Likewise, in studying the magnetic properties of a material we may wish to know the magnetic moment of the outer electrons only, and nothing else. Or, when studying a material's thermal properties, we may wish to know the vibrational states of the atoms and nothing else. Consequently, when we describe the "state" of a system, we will only give the properties that are relevant for the problem we are considering.

The state of a system is determined by the state of each element. For example, a system of three coins is identified by the heads–tails configuration of each. And the spin state of three distinguishable particles is identified by stating the spin orientation of each. When the system becomes large (10^{24} electrons, for example) the description of the system becomes hopelessly long. Fortunately, we can use statistical methods to describe these large systems; the larger the systems, the simpler and more useful these descriptions will be. In Chapter 2, we begin with small systems and then proceed to larger systems, to illustrate the development and utility of some of these statistical techniques.

Summary of Sections B and C

Many important properties of the microscopic elements of a system are quantized. One is electrical charge. Others are due to the wave nature of particles and include their position and momentum coordinates, angular momentum, magnetic moment, magnetic interaction energies, and the energies of any particles confined to a restricted region of space.

We normally restrict our description of the state of an element of a system to those few properties in which we are interested. The state of a system is determined by specifying the state of each of its elements. This is done statistically for large systems.

Problems

The answers to the first three problems are given here. After that, you will find the answers to the odd-numbered problems at the back of the book.

Section A

1. You flip one million identical coins and find that six of them end up standing on edge. What is the probability that the next flipped coin will end up on edge? (Answer: 6×10^{-6})

2. (a) If you deal one card from a well-shuffled deck of 52 playing cards, what is the probability that the card will be an ace? (Answer: 4/52, since there are four aces in a deck.)
   (b) Suppose that you deal one card from each of one million well-shuffled decks of 52 playing cards each. How many of the dealt cards would be aces? (Answer: 7.7×10^4)

3. Flip a coin twice. What percentage of the time did it land heads? Repeat this a few times, each time recording the percentage of the two flips that were heads. Now flip the coin 20 times, and record what percentage of the 20 flips were heads. Repeat. For which case (2 flips, or 20 flips) is the outcome generally closer to a 50–50 heads–tails distribution? If you flip 20 coins, why would it be unwise to bet on exactly 10 landing heads?

4. A certain puddle of water has 10^{25} identical water molecules. As the temperature of this puddle falls to $0\,°C$ and below, the puddle freezes, resulting in a considerable change in the thermodynamic properties of this system. What do you suppose happens to the individual molecules to cause this remarkable change?

5. List eight systems that have large numbers of identical elements.

6. In a certain city, there are 2 000 000 people and 600 000 autos. The average auto is driven 30 miles each day. If the average driver drives about 80 000 miles per accident, roughly how many auto accidents are there per day in this city?

7. Suppose that you flip a coin three times, and each time it lands tails. Many a gambler would be willing to bet better than even odds (e.g., 2 to 1, or 3 to 1) that the next time it will land heads, citing the "law of averages." Are these gamblers wise or foolish? Explain.

Section B

8. The density of liquid water is 10^3 kg/m^3. There are 6.022×10^{23} molecules in 18 grams of water. With this information, estimate the width of a molecule. Of an atom. Of an atomic nucleus, which is about 5×10^4 times narrower than an atom.

9. After combing your hair, you find your comb has a net charge of -1.92×10^{-18} C. How many extra electrons are on your comb?

10. What is the wavelength associated with an electron moving at a speed of 10^7 m/s? What is the wavelength associated with a proton moving at this speed? What is the wavelength of a 70 kg sprinter running at 10 m/s?

11. For waves incident on a diffraction grating, the diffraction formula is given by $2d \sin \theta = m\lambda$, where m is an integer, d is the grating spacing, and θ is the angle for constructive interference, measured from the direction of the incoming waves. Suppose that we use the arrangements of atoms in a crystal for our grating.
    (a) For first-order diffraction ($m = 1$) and a crystal lattice spacing of 0.2 nm, what wavelength would have constructive interference at an angle of $30°$ with the incoming direction?
    (b) What is the momentum of a particle with this wavelength?
    (c) At what speed would an electron be traveling in order to have this momentum? A proton?
    (d) What would be the energy in eV of an electron with this momentum? Of a proton? ($1\text{eV} = 1.6 \times 10^{-19}$ J.)
    (e) What would be the energy in eV of an x-ray of this wavelength? (For an electromagnetic wave $E = pc$, where c is the speed of light.)

12. Consider the superposition of two waves with wavelengths $\lambda_1 = 0.020$ nm and $\lambda_2 = 0.021$ nm, which produces beats (i.e., alternate regions of constructive and destructive interference).
    (a) What is the width of a beat?
    (b) What is the difference between the two wave numbers, $\Delta k = k_2 - k_1$?
    (c) What is the product $\Delta k \Delta x$, where Δx is the width of a beat?
    (d) Repeat the above for the two wavelengths $\lambda_1 = 5$ m, $\lambda_2 = 5.2$ m.

13. Suppose we know that a certain electron is somewhere in an atom, so that our uncertainty in the position of this electron is the width of the atom, $\Delta x = 0.1$ nm. What is our minimum uncertainty in the x-component of its momentum? In its x-component of velocity?

14. Consider a particle moving in one dimension. Estimate the number of quantum states available to that particle if:
    (a) It is confined to a region 10^{-4} m long and its momentum must lie between -10^{-24} and $+10^{-24}$ kg m/s;
    (b) it is an electron confined a region 10^{-9} m long with speed less than 10^7 m/s (i.e., the velocity is between $+10^7$ and -10^7 m/s).

15. Consider a proton moving in three dimensions, whose motion is confined to be within a nucleus (a sphere of radius 2×10^{-15} m) and whose momentum must have magnitude less than $p_0 = 3 \times 10^{-19}$ kg m/s. Roughly how many

quantum states are available to this proton? (Hint: The volume of a sphere of radius p_0 is $(4/3)\pi p_0^3$.)

16. A particle is confined within a rectangular box with dimensions 1 cm by 1 cm by 2 cm. In addition, it is known that the magnitude of its momentum is less than 3g cm/s. How many states are available to it? (Hint: In this problem, the available volume in momentum space is a sphere of radius 3g cm/s.)

17. In this problem you will estimate the lower limit to the kinetic energy of a nucleon in a nucleus. A typical nucleus is 8×10^{-15} m across.
    (a) What is the longest wavelength of a standing wave that fits inside this confinement?
    (b) What is the energy of a proton of this wavelength in MeV? (1 MeV $=$ 1.6×10^{-13} joules.)

18. (a) Using the technique of the problem above, estimate the typical kinetic energies of electrons in an atom. The atomic electron cloud is typically 10^{-10} m across. Express your answer in eV.
    (b) Roughly, what is our minimum uncertainty in the velocity of such an electron in any one direction?

19. Consider a particle in a box. By what factor does the number of accessible states increase if you:
    (a) double the height of the box,
    (b) double the width of the box,
    (c) double the magnitude of the maximum momentum allowed to the particle?

20. Starting with the replacement of the sum over states by an integral, $\sum_s \rightarrow \int d^3r\, d^3p / h^3 = \int g(\varepsilon)d\varepsilon$, derive the results 1.9 for the density of states $g(\varepsilon)$ for an ideal gas.

21. (a) Estimate the density of states accessible to an air molecule in a typical classroom. Assume that the classroom is 6 m by 8 m by 3 m and that the molecule's maximum energy is about 0.025 eV (4×10^{-21} joules) and its mass is 5.7×10^{-26} kg. Express your answer in states per joule and in states per eV.
    (b) If this air molecule were absorbed into a metallic crystal lattice which confined it so that it could move only approximately 10^{-11} m in each direction, what would be the density of states available to it, expressed in states per eV?

22. The total angular momentum of a particle is the sum of its spin and orbital angular momenta and is given by $J_{total} = [j(j+1)]^{1/2}\hbar$, where j is the

maximum z-component in units of $\hbar$. With this information, calculate the angles that a particle's angular momentum can make with the z-axis for:
(a) a spinless particle in an $l = 2$ orbit,
(b) a spin-1 boson by itself (in no orbit),
(c) a spin-1/2 fermion by itself.

23. A hydrogen atom is sometimes found in a state where the spins of the proton and the electron are parallel to each other (e.g., $s_z = +1/2$ for both), yet the atom's total angular momentum is zero. How is this possible?

24. Use the relationship 1.16 to estimate the magnetic moment of a spinning electron, given that an electron is a spin-1/2 particle. If an electron were placed in an external magnetic field of 1 tesla, what would be the two possible values of its magnetic interaction energy? (1 tesla $= 1$ weber/m$^2 = 1\,\text{J}\,\text{s}/(\text{C}\,\text{m}^2)$)

25. Repeat the above problem for a proton.

26. Estimate the number of quantum states available to an electron if all the volume and energy of the entire universe were available to it. The radius of the universe is about 2×10^{10} LY, and one LY is about 10^{16} m. The total energy in the universe, including converting all the mass to energy, is about 10^{70} J. The electron would be highly relativistic, so use $E = pc$.

27. Consider an electron in an $l = 1$ orbit, which is in a magnetic field of 0.4 T. Calculate the magnetic interaction energies for all possible orientations of its spin and orbital angular momenta (i.e., all l_z, s_z combinations).

28. The strength of the electrostatic force between two charges q_1 and q_2 separated by a distance r is given by $F = k_C q_1 q_2 / r^2$, where k_C is a constant given by $8.99 \times 10^9\,\text{N}\,\text{m}^2/\text{C}^2$.
(a) What is the electrostatic force between an electron and a proton separated by 0.05 nm, as is typical in an atom?
(b) If this same amount of force were due to a spring stretched by 0.05 nm, what would be the force constant for this spring? ($F = -\kappa x$, where κ is the force constant.)
(c) Suppose that an electron were connected to a proton by a spring with force constant equal to that which you calculated in part (b). What would be the angular frequency ($\omega^2 = \kappa/m$) for the electron's oscillations?
(d) What would be the separation between allowed energy levels, in eV?
(e) How does this compare with the 10.2 eV separation between the ground state and the first excited state in hydrogen?

29. According to our equation for a particle in a harmonic oscillator potential, the lowest possible energy is not zero. Explain this in terms of wavelengths of the standing waves in a confinement.

Section C

30. In how many different ways can a dime and a nickel, land when flipped? A dime, nickel, and quarter? How about 10^{24} different coins?

31. A certain fast-food restaurant advertises that its hamburger comes in over 10^{23} different ways. How many different yes–no choices (e.g., with or without ketchup, with or without pickles, etc.) would this require?

Part II
Small systems

Chapter 2
Statistics for small systems

As indicated in Chapter 1, we will begin our studies by considering "small systems" – those with relatively few elements. Small systems are important in many fields, such as microelectronics, thin films, surface coatings, and materials at low temperatures. The elements of small systems may be impurities in semiconductors, signal carriers, vortices in liquids, vibrational excitations in solids, elements in computer circuits, etc. We may wish to study some behavioral characteristic of a small population of plants or people or to analyze the results of a small number of identical experiments. Besides being important in their own right, the pedagogical reason for studying small, easily comprehensible systems first is that we gain better insight into the behaviors of larger systems and better appreciation for the statistical tools we must develop to study them.

The introduction to larger systems will begin in Chapter 4. Each macroscopic system contains a very large number of microscopic elements. A glass of water has more than 10^{24} identical water molecules, and the room you are in probably has over 10^{27} identical nitrogen molecules and one quarter that number of identical molecules of oxygen. The properties of large systems are very predictable, even though the behavior of any individual element is not (Figure 2.1). This predictability allows us to use rather elegant and streamlined statistical tools in analyzing them.

By contrast, the behaviors of smaller systems are more erratic and unpredictable, requiring the use of more detailed statistical tools. These tools become cumbersome when the number of elements in the system is large. But fortunately, this is the point where the simpler and more elegant methods for large systems become useful.

Figure 2.1 The behavior of a swarm of gnats is much more predictable than the behavior of just one or two. The larger the system, the more predictable its behavior.

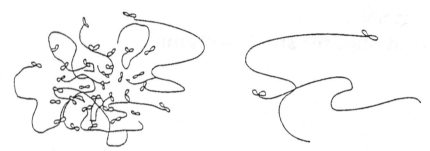

A Mean values

We now develop machinery to relate the properties of a system to the behaviors of the individual elements. To start our development, we imagine we have many identically prepared systems.[1] For example, if our system is a flipped coin then we have many of them, or if our system is two rolled dice then we have many pairs of rolled dice. Each system could be in any of several different possible configurations or "states." We let P_s indicate the probability that a system is in the state s.

Suppose that we are interested in some function f, which has the value f_s when the system is in state s. Then the average or mean value $\overline{f}$ is determined as follows:

$$\overline{f} = \sum_s f_s P_s. \tag{2.1}$$

Example 2.1 Our system is a single coin and the function f is the number of heads. That is, $f = 1$ for heads, and $f = 0$ for tails. What is the mean value of f if many coins are flipped?

The probabilities for heads and tails are $P_h = 1/2$, $P_t = 1/2$, so the mean value of f is

$$\overline{f} = f_h P_h + f_t P_t = 1 \times \tfrac{1}{2} + 0 \times \tfrac{1}{2} = \tfrac{1}{2}.$$

The average number of heads showing per coin is $1/2$.

Example 2.2 Suppose that now each system is a single rolled die and n indicates the number of dots showing upward. Suppose that f is the *square* of the number of dots showing upward ($f_n = n^2$). What is the mean value of f if large numbers of dice are rolled?

Each of the six faces has probability $1/6$ of facing upwards, so

$$\overline{f} = \sum_n n^2 P_n = 1^2 \times \tfrac{1}{6} + 2^2 \times \tfrac{1}{6} + 3^2 \times \tfrac{1}{6} + 4^2 \times \tfrac{1}{6}$$
$$+ 5^2 \times \tfrac{1}{6} + 6^2 \times \tfrac{1}{6} = \tfrac{91}{6}.$$

The mean value of $f = n^2$ is $91/6$, or about 15.2.

---

[1] A large set of identically prepared systems is called an ensemble.

If f and g are two functions that depend on the state of a system and c is a constant then

$$\overline{f + g} = \overline{f} + \overline{g}, \qquad (2.2)$$
$$\overline{cf} = c\overline{f}. \qquad (2.3)$$

These two relations follow directly from the definition of the mean value 2.1 (homework).

B Probabilities for various configurations

B.1 One criterion

We now calculate the probability for a system to be in each of its possible configurations or states. For example, what is the probability that three flipped coins land with two heads and one tails? Or what is the probability that 12 flipped coins land with five heads and seven tails?

In this type of problem, we first select the appropriate criterion for the individual elements. Then we let p represent the probability that the criterion is satisfied and q the probability that it is not. Examples that we will use in this chapter include the following.

- *Criterion: a flipped coin lands heads up.* The probability that this criterion is satisfied is $1/2$, and the probability that it is not satisfied (i.e., the coin lands tails up) is also $1/2$. Therefore

$$p = 1/2, \qquad q = 1/2.$$

- *Criterion: a certain air molecule is in the front third of an otherwise empty room.* In this case,

$$p = 1/3, \qquad q = 2/3.$$

- *Criterion: a rolled dice lands with six dots up.* In this case,

$$p = 1/6, \qquad q = 5/6.$$

A correctly formulated criterion is either satisfied or not satisfied, so we can say with certainty that it must be one or the other:

$$\text{probability for one or the other} = p + q = 1.$$

Now suppose that a system has two identical elements, which we label 1 and 2. The possible configurations and probabilities for the two elements are given by writing

$$(p_1 + q_1)(p_2 + q_2) = 1 \times 1 = 1 = p_1 p_2 + p_1 q_2 + q_1 p_2 + q_1 q_2.$$

Here $p_1 p_2$ is the probability that both elements satisfy the criterion, $p_1 q_2$ is the probability that element 1 does and 2 does not, and so on. There are a total of four possible configurations, as indicated by the four terms on the right in the equation above, and each term is the probability for that particular configuration. The fact that the four terms add up to unity reflects the certainty that the system must be in one of the four configurations.

Example 2.3 We are interested in whether two rolled dice (labeled 1 and 2) both land with six dots up. What are the probabilities for the various possible configurations of the two dice?

The probability that either die lands with six dots up is $1/6$, and the probability that it does not is $5/6$:

$$p_1 = p_2 = 1/6, \qquad q_1 = q_2 = 5/6.$$

The probabilities for all possible configurations are again given by writing

$$(p_1 + q_1)(p_2 + q_2) = p_1 p_2 + p_1 q_2 + q_1 p_2 + q_1 q_2 = 1.$$

Accordingly, the probabilities for the four possible configurations of the two dice are:

- both show sixes, $\quad p_1 p_2 = \left(\frac{1}{6}\right)\left(\frac{1}{6}\right) = \left(\frac{1}{36}\right)$;
- die 1 shows six, but die 2 does not, $\quad p_1 q_2 = \left(\frac{1}{6}\right)\left(\frac{5}{6}\right) = \left(\frac{5}{36}\right)$;
- die 1 doesn't show six, but die 2 does, $\quad q_1 p_2 = \left(\frac{5}{6}\right)\left(\frac{1}{6}\right) = \left(\frac{5}{36}\right)$;
- neither shows six, $\quad q_1 q_2 = \left(\frac{5}{6}\right)\left(\frac{5}{6}\right) = \left(\frac{25}{36}\right)$.

If two elements have identical probabilities, such as two coins, two dice, or two air molecules in the room, we can write

$$p_1 = p_2 = p \qquad \text{and} \qquad q_1 = q_2 = q.$$

The probabilities for the various possible configurations are then given by

$$(p_1 + q_1)(p_2 + q_2) = (p + q)^2 = p^2 + 2pq + q^2 = 1.$$

The probabilities are p^2 that both elements satisfy the criterion, q^2 that neither does, and $2pq$ that one does and the other does not. The coefficient 2 in this last expression indicates that there are two ways in which this can happen:

- $p_1 q_2$, die 1 satisfies the criterion and die 2 doesn't, or
- $q_1 p_2$, die 2 satisfies the criterion and die 1 doesn't.

If we extend our analysis to systems of three elements, we find that the probabilities are given by writing

$$(p_1 + q_1)(p_2 + q_2)(p_3 + q_3) = (p + q)^3 = p^3 + 3p^2 q + 3pq^2 + q^3 = 1.$$

Accordingly, the probabilities of the various possible configurations are as follows:

- p^3, all three satisfy the criterion;
- $3p^2q$, two satisfy the criterion and one doesn't;
- $3pq^2$, one satisfies the criterion and two don't;
- q^3, none of the three satisfies the criterion.

Looking at the $3p^2q$ term, for example, the coefficient 3 indicates that there are three different configurations for which two elements satisfy the criterion and one doesn't. The following table lists these possibilities.

Elements that do satisfy the criterion	Element that does not
1, 2	3
1, 3	2
2, 3	1

Example 2.4 You flip three coins, labeled 1, 2, and 3. What are the three different ways in which they could land with two heads and one tail, and what is the probability of this happening?

The three different possibilities would be hht, hth, and thh. The probability for two heads and one tail would be

$$3p^2q = 3\left(\frac{1}{2}\right)^2\left(\frac{1}{2}\right) = \frac{3}{8}.$$

We can continue to expand the above development to systems of four elements, or five, or any number N. For a system of N elements, the probabilities for all the possible configurations are given by the binomial expansion:

$$(p+q)^N = \sum_{n=0}^{N} \frac{N!}{n!(N-n)!} p^n q^{N-n} = 1^N = 1.$$

The nth term in this expansion represents the probability $P_N(n)$ that n elements satisfy the criterion and the remaining $N-n$ elements do not:

$$P_N(n) = \frac{N!}{n!(N-n)!} p^n q^{N-n}. \tag{2.4a}$$

The number of different arrangements for which n elements satisfy the criterion and $N-n$ do not is given by the binomial coefficient in the above expression (Figure 2.2):

$$\text{number of such configurations} = \frac{N!}{n!(N-n)!}. \tag{2.4b}$$

Example 2.5 Consider five air molecules in an otherwise empty room. What is the probability that exactly two of them are in the front third of the room?

Figure 2.2 The number of different ways in which n of N elements can satisfy a criterion, illustrated here for: 1 of 3 (left); 2 of 4 (middle); and 2 of 5 (right). A plus sign indicates an element that satisfies the criterion and a blank indicates one that does not.

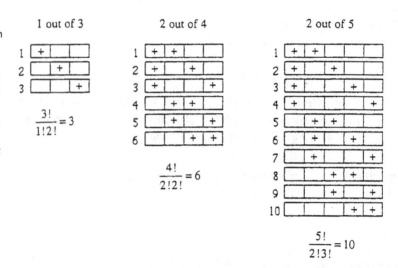

Also, how many different arrangements of these five molecules are there such that exactly two are in the front?

For each molecule, the probability of being in the front third is $1/3$, so $p = 1/3, q = 2/3$. Our system has $N = 5$ molecules. The probability for $n = 2$ of them to be in the front third is

$$P_5(2) = \frac{5!}{2!3!} \left(\frac{1}{3}\right)^2 \left(\frac{2}{3}\right)^3 = \frac{80}{243} = 0.33.$$

The number of different arrangements is given by the binomial coefficient

$$\frac{5!}{2!3!} = 10.$$

B.2   Handling factorials

Although these probability calculations are correct for systems of any size, they become cumbersome if there are more than a few elements. For example, if we wanted to know the probability that exactly 40 out of 100 flipped coins land heads, the answer would be

$$P_{100}(40) = \frac{100!}{40!60!} \left(\frac{1}{2}\right)^{40} \left(\frac{1}{2}\right)^{60}.$$

Although numbers like $(1/2)^{40}$ can be calculated using logarithms, the factorials (e.g. $100! = 100 \times 99 \times 98 \times 97 \times \cdots$) become overwhelming when the numbers are this large.

Fortunately, an approximation known as Stirling's formula allows us to calculate factorials accurately for larger numbers. Stirling's formula is

$$m! \approx \sqrt{2\pi m} \left(\frac{m}{e}\right)^m, \tag{2.5a}$$

or, using logarithms,

$$\ln m! \approx m \ln m - m + \frac{1}{2} \ln 2\pi m. \qquad (2.5b)$$

As you will show in a homework problem, this approximation is accurate to 0.8% for $m = 10$, and its accuracy increases as m increases.

B.3   Many criteria

How does our treatment apply to the probabilities for distributions involving *more* than two possibilities? For example, what if we are interested in the following distribution of air molecules between the following three parts of a room, from front to back,

- the front third ($p_1 = 1/3$),
- the next sixth ($p_2 = 1/6$),
- the back half ($p_3 = 1/2$)?

In the problems you will extend the treatment given here to show that for a system of N elements with a complete set of m mutually exclusive criteria[2] whose probabilities are respectively $p_1, p_2, \ldots p_m$, the probability that n_1 satisfy the first criterion, n_2 satisfy the second, etc. is given by

$$P_N(n_1, n_2, \ldots, n_m) = \frac{N!}{n_1! n_2! \ldots n_m!} p_1^{n_1} p_2^{n_2} \cdots p_m^{n_m}. \qquad (2.6)$$

Summary of Sections A and B

If f is a function that has the value f_s when the system is in state s and if P_s is the probability that the system is in state s, then the mean value of f is given by (equation 2.1)

$$\overline{f} = \sum_s f_s P_s,$$

where the sum is over all states s accessible to the system.

If f and g are functions of the state of a system and c is a constant, then (equations 2.2, 2.3)

$$\overline{f + g} = \overline{f} + \overline{g},$$
$$\overline{cf} = c\overline{f}.$$

Suppose that we are interested in some criterion for the behavior of a single element of a system, for which p is the probability that the criterion is satisfied and q is the probability that it is not satisfied ($q = 1 - p$). Then for a system of N

---

[2] This means that each element must satisfy one criterion, but only one, so $p_1 + p_2 + \cdots + p_m = 1$, $\quad n_1 + n_2 + \cdots + n_m = N$.

elements, the probability that this system is in a state for which n elements satisfy the criterion and the remaining $N - n$ elements do not is given by (equation 2.4a)

$$P_N(n) = \frac{N!}{n!(N-n)!} p^n q^{N-n}.$$

The binomial coefficient $N!/[n!(N-n)!]$ is the number of different configurations of the individual elements for which n satisfy the criterion and $N - n$ do not.

A useful tool for calculating the factorial of large numbers is Stirling's formula (equations 2.5a, b),

$$m! \approx \sqrt{2\pi m} \left(\frac{m}{e}\right)^m,$$

or equivalently

$$\ln m! \approx m \ln m - m + \frac{1}{2} \ln 2\pi m.$$

For a complete set of m mutually exclusive criteria, whose probabilities are respectively $p_1, p_2, \ldots, p_m$, the probability that, out of N particles or elements, n_1 satisfy the first criterion, n_2 the second, and so on is given by (equation 2.6)

$$P_N(n_1, n_2, \ldots, n_m) = \frac{N!}{n_1! n_2! \cdots n_m!} p_1^{n_1} p_2^{n_2} \cdots p_m^{n_m}.$$

C   Statistically independent behaviors

So far, we have assumed that the behaviors of the individual elements of a system are statistically independent, that is, that the behavior of each is independent of the others. For example, we assumed that the probability that coin 2 lands heads up does not depend on how coin 1 landed.

There are many systems, however, for which the behaviors of the individual elements are *not* independent. For example, suppose that you are drawing aces from a single deck of cards. The probability for the first draw to be an ace is $4/52$, because there are four aces in a deck of 52 cards. For the second card, however, the probability depends on the first draw. If it was an ace, then there are only three aces left among the 51 remaining cards. If not, then there are still four aces left. So the probabilities for the second card would be $3/51$ or $4/51$, depending on the first draw. The two behaviors are not statistically independent[3] (Figure 2.3).

In physical systems, interactions among particles often mean that any particle is influenced by the behaviors of its neighbors. Consequently, when we use the results of this chapter we must take care to ensure that the behaviors of the individual elements are indeed statistically independent. We may have to choose groups of particles as our elements perhaps an entire nucleus, or a molecule, or a group of molecules. But when the criteria are statistically independent, the total

[3] You can still use probabilities to handle these situations but not the preceding method, because there we assumed that p and q for any one element are independent of the behavior of the others.

Figure 2.3 What is the probability that the very next card dealt will be a queen? Does it depend on what has already been dealt? How?

probability with respect to all the criteria is simply the product of the individual probabilities.

Example 2.6 Consider a single air molecule in an empty room. What are the probabilities for the positions of that molecule with respect to the front third and the top half of the room?

- *Criterion 1: the molecule is in the front third, $p_1 = 1/3$, $q_1 = 2/3$.*
- *Criterion 2: the molecule is in the top half, $p_2 = 1/2$, $q_2 = 1/2$.*

(The subscripts on the probabilities here indicate the criterion to which they belong.) The various probabilities with respect to both these criteria are then:

- front third, top half, $p_1 p_2 = (1/3)(1/2) = 1/6$;
- front third, bottom half, $p_1 q_2 = (1/3)(1/2) = 1/6$;
- rear two thirds, top half, $q_1 p_2 = (2/3)(1/2) = 2/6$;
- rear two thirds, bottom half, $q_1 q_2 = (2/3)(1/2) = 2/6$.

Example 2.7 Suppose that you flip three coins and roll two dice. What is the probability that exactly two of the coins land heads up and that one of the dice shows a six?

The probability that two of three coins land heads up ($p = 1/2$, $q = 1/2$) is, using result 2.4a,

$$P_3(2) = \frac{3!}{2!1!} \left(\frac{1}{2}\right)^2 \left(\frac{1}{2}\right)^1 = \frac{3}{8}.$$

and the probability that one of the two dice shows a six ($p = 1/6, q = 5/6$) is

$$P_2(1) = \frac{2!}{1!1!} \left(\frac{1}{6}\right)^1 \left(\frac{5}{6}\right)^1 = \frac{5}{18}.$$

Because the behavior of the dice is independent of the behavior of the coins, we simply multiply the two together. The answer is

$$\left(\frac{3}{8}\right)\left(\frac{5}{18}\right) = \frac{5}{48}.$$

Using the definition 2.1 of mean values, we can prove that, for two functions f and g having two statistically independent behaviors, the mean value of the product of the two functions is simply the product of the mean values:

$$\overline{fg} = \overline{f}\,\overline{g}. \tag{2.7}$$

To show this, take P_i to be the probability that the system is in state i with respect to the first behavior and W_j to be the probability that the system is in state j with respect to the second behavior. The combined probability for the system to be in the state (i, j) with respect to the two behaviors is $P_i W_j$. The mean value of the product fg is then given by

$$\overline{fg} = \sum_{i,j}(P_i W_j)f_i g_j = \sum_i P_i f_i \sum_j W_j g_j = \overline{f}\,\overline{g}.$$

Example 2.8 What would be the mean value of the product of the numbers showing upwards for two rolled dice?

Let n_1 and n_2 be the numbers showing on the first and second die, respectively. The two are statistically independent, because how the second die lands is independent of the first. The mean value of the number showing on either die is

$$(1 + 2 + 3 + 4 + 5 + 6)/6 = 3.5.$$

Therefore,

$$\overline{n_1 n_2} = \overline{n}_1\,\overline{n}_2 = (3.5)(3.5) = 12.25.$$

In the homework problems this same thing is calculated the hard way – namely, by finding the mean value of $n_1 n_2$ for all 36 different configurations for the two dice.

Summary of Section C

When the behavior of one element of a system is unaffected by the behavior of another, or when an element's behavior with respect to one criterion is unrelated to its behavior with respect to another, then the two behaviors are statistically independent. For statistically independent behaviors, the probabilities are multiplicative. That is, if P and W are the probabilities for two statistically independent behaviors satisfying their respective criteria, then the probability that both behaviors satisfy the respective criteria is given by the product PW.

If f is a function of one behavior and g is a function of a statistically independent behavior then the mean value of the product is the product of the mean values (equation 2.7):

$$\overline{fg} = \overline{f}\,\overline{g}.$$

Problems

Section A

1. Suppose that P_s is the probability that a system is in state s, c is a constant; and f and g are two functions that have the values f_s and g_s, respectively, when the system is in state s. Using the definition of mean values 2.1 prove that:
   (a) $\overline{(f+g)} = \overline{f} + \overline{g}$;
   (b) $\overline{cf} = c\overline{f}$.

2. A coin is flipped many times. If $f_{\text{heads}} = 5$ and $f_{\text{tails}} = 27$, what is the mean value of f (i.e., the average value of f per flip)?

3. The number of dots showing on a die is n, and $f(n)$ is some function of n. If you were to roll many many dice, what would be the mean value of f for
   (a) $f = (n+2)^2$,
   (b) $f = (n-2)^2$,
   (c) $f = n^2 - 5n + 1$,
   (d) $f = n^3 - 10$?

4. A weighted die is rolled in such a way that the probability of getting a six is $1/2$ and the probability of getting each of the other five faces is $1/10$. What would be the average value per roll of.
   (a) the number of dots ($f_n = n$),
   (b) the square of the number of dots ($f_n = n^2$)?

5. You have a die that is weighted in such a way that the probability of a six is $3/8$ and the probability of each of the other five states is $1/8$. Consider a function $f(n) = (n-2)^2$, where n is the number showing. What would be the average value of this function per roll, if you were to roll the die many times?

6. Consider a spin-$1/2$ particle of magnetic moment μ in an external magnetic field B. Its energy is $E = -\mu B$ if it is spin up and $E = +\mu B$ if spin down. Suppose the probability that this particle is in the lower energy state is $3/4$ and that it is in the higher energy state is $1/4$. Find the average value of the energy of such a particle, expressed in terms of μB.

Section B

7. Consider a system of four flipped coins.
   (a) What is the probability that two land heads and the other two tails?
   (b) Label the four coins 1, 2, 3, and 4. Make a chart that lists the various possible configurations that have two heads and two tails. Is the number of configurations on your chart the same as that predicted by the binomial coefficient in equation 2.4?

8. Consider a system of five molecules. The probability that any one is in an excited state is 1/10. Find the probability that there are
   (a) none in an excited state,
   (b) one and only one in an excited state,
   (c) two in excited states.

9. If you roll two dice, what is the probability of throwing "snake eyes" (each die showing one dot up)?

10. If you were to roll four dice, find the probability that
    (a) none lands with six dots up,
    (b) one and only one lands with six dots up.

11. If you roll eight dice, find the probability that
    (a) five and only five have four dots up and the number of different configurations that give this outcome,
    (b) five or more have four dots up?

12. Consider five spin-1/2 elementary particles (distinguishable and with no external fields present). What is the probability that four have spin up and the other has spin down, and how many different configurations of the five could give this result?

13. What is the probability for exactly three of five flipped coins to land heads, and in how many different ways can they land to give this result?

14. Consider five air molecules in an otherwise empty room. What is the probability that at any instant exactly three of them are in the front third of the room and the other two are in the back two thirds?

15. For 16 flipped coins, how many different ways could they land with 12 heads and four tails?

16. Roughly what is the numerical value of 200! in powers of 10? ($e^x = 10^{0.4343x}$)

17. Using Stirling's formula, calculate the probability of getting exactly 500 heads and 500 tails when flipping 1000 coins.

18. If you flip 100 coins, what is the probability that exactly 42 land heads up?

19. Suppose you roll 500 dice. Using Stirling's formula, calculate the probability of rolling exactly: (a) 50 sixes, (b) 80 sixes, (c) 200 sixes.

20. Test the accuracy of Stirling's formula by comparing its results percentage-wise with explicit calculations of $n!$, for the following values of n: (a) 2, (b) 5, (c) 10, (d) 20.

21. Consider a system of 100 coins which you can tell apart. (The ability to tell them apart is important, as we'll see later in the book.) How many different configurations are there that give a total of 50 heads and 50 tails?

22. Suppose that you roll three dice and flip three coins. Find the probability of getting exactly:
    (a) one six and one head,
    (b) no sixes and no heads,
    (c) two sixes and two heads.

23. Calculate the probability of getting exactly two sixes and one five when rolling five dice. Do this in two different ways, as follows.
    (a) First calculate the probability that two of the five dice land sixes, and then multiply this by the probability that one of the remaining three lands a five. (note: The remaining three dice have only five ways in which they can land.)
    (b) Next calculate the probability that one of the five dice lands a five, and then multiply this by the probability that two of the remaining four dice land a six.
    (c) Are the two results the same? (If not, you have made a mistake.)

24. Consider two mutually exclusive criteria, such as the criteria in the previous problem. An element of a system cannot satisfy both simultaneously. Suppose that there are r equally probable outcomes, so that $p = 1/r$ is the probability of satisfying the first criterion. After those elements satisfying the first have been excluded, there are only $r - 1$ possibilities for the remaining elements, so the probability for the second criterion to be satisfied becomes $1/(r - 1)$. What is the probability $P_N(n, m)$ that n out of N elements satisfy the first criterion, and m of the remaining $N - n$ elements satisfy the second?

25. Start with the binomial probability distribution 2.4, and look at what might happen to those particles that did *not* satisfy the criterion (for example, those molecules that were *not* in the front third of the room). These failures with respect to the first criterion might themselves be split into two groups with respect to *another*, mutually exclusive, criterion, with probabilities p' and p'', respectively, of satisfying that criterion. For example, those molecules not in the front third of the room might be in the next one sixth ($p' = 1/6$) or the back half ($p'' = 1/2$), so that $q = 1 - p = p' + p''$. Now expand the q^{N-n}

term in the binomial distribution by using another binomial expansion for $q^{N-n} = (p' + p'')^{N-n}$ to find the probability that, out of these $N - n$ failures with respect to the first criterion, n' satisfy the probability-p' criterion and the remaining n'' ($= N - n - n'$) satisfy the last criterion (i.e., satisfy neither of the first two). You should wind up with equation 2.6 for the case of three criteria. You can keep splitting up each of the criteria into subgroups in the above manner to get the probabilities with respect to any number of mutually exclusive criteria.

Section C

26. You are dealing cards from a full 52-card, freshly shuffled, deck. You are interested in whether the first two cards dealt will be clubs. Criterion 1: the first card is a club. Criterion 2: the second card is a club.
    (a) Are these two criteria statistically independent?
    (b) If you return the first card to the deck and reshuffle before dealing the second card, would the two criteria be statistically independent?

27. Answer the questions in Figure 2.3.

28. Suppose that you have two freshly shuffled full decks of cards, and you deal one card from each.
    (a) What is the probability that the first card dealt is an ace?
    (b) What is the probability that the second card dealt is a club?
    (c) Are the two criteria statistically independent?
    (d) What is the probability that the first card dealt is an ace *and* the second card dealt is a club?

29. You are involved in a game where two cards are dealt in the manner of the previous problem. Suppose that the dealer pays you $3 if the second card dealt is a club, regardless of the first card and that you pay him $1 if the second card is not a club and the first card is not an ace. (Otherwise, no money changes hands.) Use equation 2.1 to compute the mean value of the money you win per game if you play it many times.

30. Consider 10 air molecules in an otherwise empty room. Find the probability that
    (a) exactly four molecules are in the front third and exactly six in the top half,
    (b) exactly three molecules are both in the front third and the top half (that is, the same molecules satisfy both criteria).

31. You roll two dice many times and are interested in the average value of the product of the two numbers showing, $n_1 n_2$. Calculate this product for all 36 possible different configurations of the two dice and take the average of these 36 values. How does your answer compare with that in Example 2.8?

32. You roll two dice at a time. Die 1 has 6 different possible states and die 2 also has 6, making a total of $6 \times 6 = 36$ different ways the two can land. Suppose that f is the sum of the number of dots showing on the two dice. Calculate the mean value of f per roll using two approaches.

(a) Sum over all 36 configurations the probability of occurrence of a configuration (1/36) times the sum of the dots showing in that configuration.

(b) Noting that the outcomes for the two dice are statistically independent (the probabilities for the second die are independent of the results for the first roll), use the result that $\overline{f + g} = \overline{f} + \overline{g}$. (The two answers should be the same.)

Chapter 3
Systems with many elements

The techniques developed in Chapter 2 for predicting the behaviors of small systems from the behaviors of their individual constituents are correct for systems of any size; they become cumbersome, though, when applied to systems with more than a few elements. Fortunately, there is an easy way of streamlining our calculations.

A Fluctuations

Suppose that we are interested in the outcomes of flipping 1000 coins. Equation 2.4a gives us the correct probabilities for all 1001 possible outcomes, ranging from 0 heads to 1000 heads; we get (Figure 3.1)

$$P_{1000}(0) = 9.3 \times 10^{-302}, \qquad P_{1000}(1) = 9.3 \times 10^{-299},$$
$$P_{1000}(2) = 4.6 \times 10^{-296}, \qquad \cdots$$

But these 1001 separate calculations are a great deal of work and give more information than would normally be useful. What if the system had a million elements, or a billion?

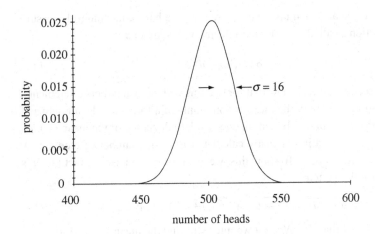

Figure 3.1 Probabilities for various numbers of heads when 1000 coins are flipped. The distribution peaks at 500 heads, for which the probability is 0.0252. As you can see, the chances of getting less than 450 or more than 550 heads are negligible.

It would be less work, less confusing, and nearly equally informative if we could just calculate the following two numbers:

- the average number of coins that would land heads if the coin-flip experiment were repeated many times;
- some measure of the fluctuations we could expect around this value.

In the above case of 1000 flipped coins, for example, it is extremely likely that the number of heads will fall between 450 and 550 (Figure 3.1). But the probability of getting *exactly* 500 heads is only 0.0252.

A.1 Mean value and standard deviation

We now investigate how to calculate mean values and characteristic fluctuations for any system. We will imagine that we have a large number of such systems, which have been prepared in the same way (an "ensemble"). For example, we might have many systems of 1000 flipped coins. Equivalently, we might flip the same set of 1000 coins many different times.

For large numbers of identically prepared systems having N elements each, the average number of elements per system that satisfy a criterion is given by

$$\bar{n} = pN, \tag{3.1}$$

where p is the probability for any given element to satisfy the criterion. We can think of this as the definition of the probability p: it is the fraction of the total number of elements that satisfy the criterion. Alternatively, this relationship can be derived from the definition of mean values 2.1.

The average fluctuation of n about its mean value must be zero, because the definition of the mean value guarantees that the positive fluctuations cancel the negative ones. But the squares of the fluctuations are all positive numbers. So if we average these and then take the square root, we have a meaningful measure of the deviations, called the "standard deviation" (symbol σ):

$$\sigma = \sqrt{\overline{(n - \overline{n})^2}}. \tag{3.2}$$

In the next section we will show that for systems with large numbers of elements, the distribution of n about the mean is commonly of a form called Gaussian and that the probability for n to be within one standard deviation of the mean is 0.68.

The standard deviation is easily calculated from the number of elements N and the probabilities p, q. To show this, we examine σ^2 and use the fact that $\overline{n}$ is a constant for the system:

$$\sigma^2 = \overline{(n - \overline{n})^2} = \overline{n^2 - 2\overline{n}n + \overline{n}^2} = \overline{n^2} - 2\overline{n}\,\overline{n} + \overline{n}^2 = \overline{n^2} - \overline{n}^2. \tag{3.3}$$

We already know that $\overline{n} = Np$, but we must still find the mean value of n^2:

$$\overline{n^2} = \sum_n n^2 P_N(n) = \sum_n n^2 \frac{N!}{n!(N-n)!} p^n q^{N-n}.$$

The easiest way to evaluate this sum is to use the binomial expansion

$$(p + q)^N = \sum_n \frac{N!}{n!(N-n)!} p^n q^{N-n}$$

and the trick that

$$n^2 p^n = \left(p \frac{\partial}{\partial p} \right)^2 p^n,$$

where we treat p and q as independent variables and evaluate the partial derivative at the point $p = 1 - q$. With these, the above expression for the mean value of n^2 becomes

$$\overline{n^2} = \sum_n n^2 \frac{N!}{n!(N-n)!} p^n q^{N-n}$$

$$= \left(p \frac{\partial}{\partial p} \right)^2 \sum_n \frac{N!}{n!(N-n)!} p^n q^{N-n} = \left(p \frac{\partial}{\partial p} \right)^2 (p + q)^N.$$

In this last form, we take the two derivatives and use $p + q = 1$ to get (homework)

$$\overline{n^2} = (Np)^2 + Npq = \overline{n}^2 + Npq.$$

Putting this into the last expression in equation 3.3 gives

$$\sigma^2 = \overline{n^2} - \overline{n}^2 = Npq \quad \text{or} \quad \sigma = \sqrt{Npq}. \tag{3.4}$$

According to equations 3.1 and 3.4, as the number of elements in a system increases, the mean value $\overline{n}$ increases linearly with N, whereas the standard deviation σ increases only as the square root of N:

$$\overline{n} \propto N, \quad \sigma \propto \sqrt{N}.$$

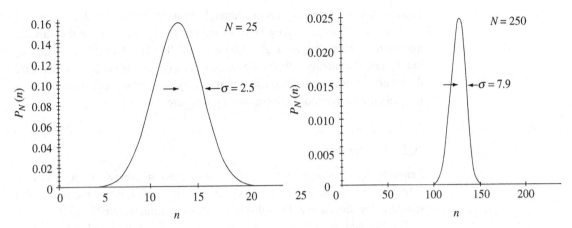

Therefore, the fluctuations do not increase as fast as the mean value. The "relative fluctuation" compares the fluctuations with the mean value and is given by

$$\frac{\sigma}{\bar{n}} = \frac{\sqrt{Npq}}{Np} = \sqrt{\frac{q}{Np}} \approx \frac{1}{\sqrt{N}}. \tag{3.5}$$

As systems get larger, the fluctuations become a smaller fraction of the mean values. Thus the larger the system, the more predictable its behavior. This is illustrated in Figure 3.2.

Figure 3.2 Plots of probabilities as a function of n for systems of 25 elements (on the left) and 250 elements (on the right), with $p = 1/2$. As N increases, the absolute width of the peak increases, but its relative width decreases.

Summary of Section A

Consider many identically prepared systems having N elements each. If p is the probability that any one element satisfies a criterion of interest and q is the probability that it does not, then the average number of elements that satisfy the criterion is given by (equation 3.1)

$$\bar{n} = pN;$$

the standard deviation for the fluctuations about the mean value is given by (equations 3.2, 3.4)

$$\sigma = \sqrt{\overline{(n - \bar{n})^2}} = \sqrt{Npq},$$

and the relative fluctuation is given by (equation 3.5)

$$\frac{\sigma}{\bar{n}} = \frac{\sqrt{Npq}}{Np} = \sqrt{\frac{q}{Np}} \approx \frac{1}{\sqrt{N}}.$$

We will soon show that for sufficiently large systems, the values of n are within one standard deviation of $\bar{n}$ 68% of the time.

The fluctuation of a variable is often more interesting than its mean value. For example, the average electrical current from an AC source is zero, because the current goes in each direction half the time. Similarly, the average velocity of an air molecule in your room is zero, although the individual molecules are moving very fast. In these and many other examples, the mean value of the variable may be misleading, suggesting no motion at all. The standard deviation may be much more illuminating. When the mean value of a variable is zero, its standard deviation is sometimes called its "root mean square" value, because in this case it equals the square root of the mean of the squares.

A.2 Examples

Example 3.1 Consider systems of 100 molecules in otherwise empty rooms. What is the average number of molecules in the front third of the rooms, the standard deviation about this value, and the relative fluctuation?

For this case, $N = 100$, $p = 1/3$, and $q = 2/3$. Therefore we have

$$\bar{n} = pN = \left(\frac{1}{3}\right)(100) = 33.3,$$

$$\sigma = \sqrt{Npq} = \sqrt{(100)\left(\frac{1}{3}\right)\left(\frac{2}{3}\right)} = 4.7, \qquad \frac{\sigma}{\bar{n}} = \frac{4.7}{33.3} = 0.14.$$

Example 3.2 Repeat the above for typical real systems of 10^{28} molecules in otherwise empty rooms. For this case $N = 10^{28}$, and p and q remain the same as above, so

$$\bar{n} = pN = \left(\frac{1}{3}\right)(10^{28}) = 3.3 \times 10^{27},$$

$$\sigma = \sqrt{Npq} = \sqrt{(10^{28})\left(\frac{1}{3}\right)\left(\frac{2}{3}\right)}$$

$$= 4.7 \times 10^{13},$$

$$\frac{\sigma}{\bar{n}} = 1.4 \times 10^{-14}.$$

Notice the tiny relative fluctuation for this system of 10^{28} particles; the larger the system, the smaller the relative fluctuations. Macroscopic systems are very predictable, even though their individual elements are not.

B The Gaussian distribution

We have seen that for systems of more than a few elements, calculating the probabilities $P_N(n)$ from the binomial formula 2.4a can be an extremely tedious task. Fortunately, there is an easier way. The entire distribution of probabilities over all possible configurations, or states, can be expressed in terms of the two parameters $\bar{n}$ and σ, which we can calculate from equations 3.1 and 3.4. This simplified Gaussian distribution involves approximations that become increasingly reliable as the number of elements in the system gets larger. Therefore, the Gaussian distribution is useful in those cases where the binomial formula is not. For small systems, only the binomial approach is correct. For larger systems, both approaches are accurate but the Gaussian approach is much simpler.

B.1 The Taylor series approach

Our derivation of this simplified formula will involve a Taylor series expansion. The derivation is given in Appendix B and goes as follows. Consider a smooth differentiable function $f(x)$. Suppose that we know the value of this function and all its derivatives at some point $x = a$. Then we can calculate the value of the function at any other point through the formula

$$f(x) = \sum_{m=0}^{\infty} \frac{1}{m!} f^{(m)}(a)(x-a)^m, \tag{3.6}$$

where

$$f^{(m)}(a) = \frac{d^m f}{dx^m}\bigg|_{x=a}.$$

Writing out the first few terms explicitly gives

$$f(x) = f(a) + f'(a)(x-a) + \tfrac{1}{2}f''(a)(x-a)^2 + \cdots. \tag{3.6'}$$

Notice that if the function is a constant, only the first term contributes. If the function is linear in x, only the first two terms contribute, and so on. Notice also that the higher-order terms become smaller as $x - a$ becomes smaller. These two observations tell us that the smoother the function and the closer x is to a, the more accurately the first few terms approximate the function at the point x. Therefore, in using the Taylor series expansion, it is advantageous to:

- apply it to functions which are as smooth as possible; and
- choose a to be close to the values of x in which we are interested.

In our case we consider $P_N(n)$ to be a continuous function of n. To satisfy the first criterion, we expand the logarithm of $P_N(n)$, because the logarithm of a function varies much more slowly and smoothly than does the function itself. To satisfy the second criterion, we expand around the point of highest probability. If $n_{\max}$ represents the state of the system for which $P_N(n)$ is a maximum then we are most likely to be interested in values of $P_N(n)$ for n near $n_{\max}$, because they occur more frequently.

We assume that the probability peaks at $\bar{n}$ ($n_{\max} = \bar{n}$). This may not be true for small systems with skewed distributions. But for larger systems, the

Figure 3.3 Relative frequency of occurrence vs. the fraction of the flipped coins that land heads, for systems of 10, 100, and 1000 coins. Larger systems have more peaked distributions and smaller relative fluctuations.

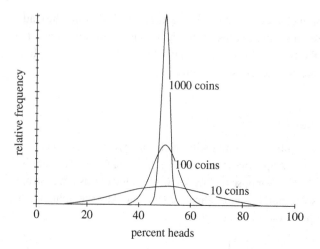

probability distribution is more sharply peaked (Figures 3.2 and 3.3), forcing n_{max} and $\bar{n}$ closer together. Consequently, this approximation is indeed justified in those larger systems for which we are seeking an alternative to the binomial expansion.

B.2   Derivation

We begin our derivation by expanding the logarithm of $P_N(n)$ about the point $n = \bar{n}$, according to the Taylor series formula (we drop the subscript N to avoid clutter):

$$\ln P(n) = \ln P(\bar{n}) + \left.\frac{d}{dn}\ln P(n)\right|_{n=\bar{n}} (n - \bar{n})$$
$$+ \frac{1}{2}\left.\frac{d^2}{dn^2}\ln P(n)\right|_{n=\bar{n}} (n - \bar{n})^2 + \cdots$$

To evaluate these terms, we write[1]

$$\ln P(n) = \ln \frac{N!}{n!(N-n)!}p^n q^{N-n}$$
$$= \ln N! - \ln n! - \ln(N-n)! + n \ln p + (N-n)\ln q$$

and then use Stirling's formula 2.5a to write out the logarithms of the factorials (equation 2.5):

$$\ln m! \approx m \ln m - m + \frac{1}{2}\ln 2\pi m.$$

---

[1] $\ln(ab) = \ln a + \ln b$, $\ln(a/b) = \ln a - \ln b$, and $\ln a^b = b \ln a$.

In this form, we can take the derivatives and evaluate the first few terms at the point $n = \bar{n}$ (with $\bar{n} = Np$ and $q = 1 - p$) to get (homework):

$$\ln P(\bar{n}) = \frac{1}{2} \ln \frac{1}{2\pi Npq} = \frac{1}{2} \ln \frac{1}{2\pi\sigma^2},$$

$$\frac{d}{dn} \ln P(n) \Big|_{n=\bar{n}} = 0,$$

$$\frac{d^2}{dn^2} \ln P(n) \Big|_{n=\bar{n}} = -\frac{1}{Npq} = -\frac{1}{\sigma^2}.$$

The first derivative is zero and the second derivative is negative, as must be true if the function indeed has its maximum at the point $n = \bar{n}$. We ignore third- and higher-order terms, because the expansion to second order already gives us amazingly accurate results for $P(n)$, even for values of n far away from $\bar{n}$. Our expansion is now

$$\ln P(n) = \frac{1}{2} \ln \frac{1}{2\pi\sigma^2} + 0 + \frac{1}{2} \left(-\frac{1}{\sigma^2}\right)(n - \bar{n})^2 + \cdots$$

and, taking the antilogarithm,

$$P(n) = \frac{1}{\sqrt{2\pi}\sigma} e^{-(n-\bar{n})^2/2\sigma^2} \qquad \text{with } \sigma^2 = Npq, \bar{n} = Np. \qquad (3.7)$$

We will encounter Gaussian distributions like this several times in this course, so we add a summary paragraph for future reference.

Any function of the form

$$F(z) = C e^{-Bz^2} \qquad \text{(Gaussian)} \qquad (3.8)$$

is called a Gaussian distribution and has a characteristic bell-curve form. As we saw above, the constant B is related to the standard deviation σ and to the second derivative of the logarithm of the function through

$$B = \frac{1}{2\sigma^2} = -\frac{1}{2} \frac{\partial^2}{\partial z^2} \ln F \Big|_{z=0}. \qquad (3.9)$$

If the total area under the curve is unity, as must be true for probability distributions,[2] then the height and width of the bell curve are related through $C = (B/\pi)^{1/2}$ (homework), so

$$F(z) = \sqrt{\frac{B}{\pi}} e^{-Bz^2}, \qquad \text{for } \int_{-\infty}^{+\infty} F(z)dz = 1. \qquad (3.10)$$

---

[2] That is, we are absolutely sure that the system must be in one of its possible configurations, so the sum of the probabilities over all possible configurations must equal unity.

Figure 3.4 Comparison of the Gaussian predictions 3.7 with the correct binomial result 2.4 for the probabilities $P_N(n)$ for systems with $N = 4$ and $N = 10$ elements, respectively ($p = 1/2$).

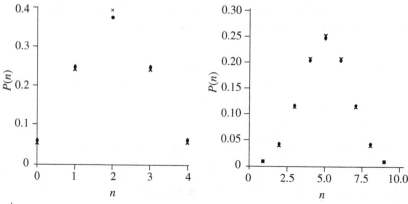

Figure 3.5 For a Gaussian distribution, 68.3% of all events are within one standard deviation of the mean ($\overline{n}$), 95.4% are within two standard deviations, and 99.7% are within 3 standard deviations.

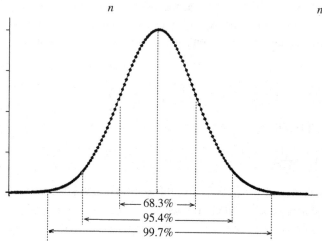

B.3   Accuracy and spread

We know that the Gaussian distribution 3.7 should be more accurate for larger systems. But how large must they be? In Figure 3.4 the predictions of the Gaussian approximation (crosses) and the correct predictions of the binomial formula (dots) are compared for systems of $N = 4$ and $N = 10$ elements, with $p = q = 1/2$. It is seen that even for $N = 4$ the Gaussian approximation is remarkably accurate, and for $N = 10$ the two are almost indistinguishable.

To find the probability that n is within one standard deviation of $\overline{n}$, we sum (numerically) the probabilities over all n between $\overline{n} - \sigma$ and $\overline{n} + \sigma$:

$$\sum_{n=\overline{n}-\sigma}^{\overline{n}+\sigma} P(n) = \sum_{n=\overline{n}-\sigma}^{\overline{n}+\sigma} P(n)\Delta n \approx \int_{\overline{n}-\sigma}^{\overline{n}+\sigma} P(n)\mathrm{d}n = 0.683. \tag{3.11}$$

That is, for a Gaussian distribution 68.3% of all events lie within one standard deviation of the mean. In a similar fashion we find that 95.4% of all events lie within two standard deviations of $\overline{n}$, 99.7 within three standard deviations, etc. (Figure 3.5).

Summary of Section B

For a system of N elements, each of which has probability p of satisfying the criterion of interest and probability q of not satisfying it, the probability that n elements satisfy the criterion and the remaining $N - n$ do not is given by (equation 2.4a)

$$P_N(n) = \frac{N!}{n!(N-n)!} p^n q^{N-n}$$

and, for systems with more than just a few elements, we can write this as (equation 3.7)

$$P(n) = \frac{1}{\sqrt{2\pi}\sigma} e^{-(n-\bar{n})^2/2\sigma^2},$$

where (equations 3.1, 3.4)

$$\bar{n} = Np \quad \text{and} \quad \sigma = \sqrt{Npq}.$$

Equation 2.4a is correct for all systems, but it is most useful for very small systems. For larger systems the second form, 3.7, is easier to use, and its accuracy increases as the size of the system increases.

Any function of the form (equation 3.8)

$$F(z) = Ce^{-Bz^2} \quad \text{(Gaussian)}$$

is called Gaussian. The constant B is related to the standard deviation σ and to the second derivative of the logarithm of the function through (equation 3.9)

$$B = \frac{1}{2\sigma^2} = -\frac{1}{2}\frac{\partial^2}{\partial z^2} \ln F \bigg|_{z=0}.$$

If the area under the curve is equal to unity, as must be true for probability distributions, then (equation 3.10)

$$C = \sqrt{\frac{B}{\pi}} \quad \text{and therefore} \quad F(z) = \sqrt{\frac{B}{\pi}} e^{-Bz^2}.$$

B.4 Examples

We illustrate the application of the Gaussian distribution to probability calculations with the following examples.

Example 3.3 Suppose that there are 3000 air molecules in an otherwise empty room. What is the probability that exactly 1000 of them are in the front third of the room at any instant?

For the Gaussian distribution, we need the values of $\bar{n}$ and σ. For this problem,

$$N = 3000, \quad n = 1000, \quad p = 1/3, \quad q = 2/3,$$

so

$$\bar{n} = pN = 1000, \quad \sigma = \sqrt{Npq} = 25.8.$$

With these, we have

$$\frac{1}{\sqrt{2\pi}\sigma} = 0.0155, \qquad \frac{(n-\bar{n})^2}{2\sigma^2} = 0,$$

and therefore

$$P_{3000}(1000) = \frac{1}{\sqrt{2\pi}\sigma} e^{-(n-\bar{n})^2/2\sigma^2} = 0.0155 e^{-0} = 0.0155.$$

Example 3.4 For the preceding case of 3000 air molecules in an otherwise empty room, what is the probability that exactly 1100 air molecules will be in the front third of the room at any instant?

Everything is the same as above except that $n = 1100$. So the exponent in the Gaussian formula 3.7 is

$$-\frac{(n-\bar{n})^2}{2\sigma^2} = -7.50.$$

Consequently, the answer to the question is

$$P_{3000}(1100) = \frac{1}{\sqrt{2\pi}\sigma} e^{-(n-\bar{n})^2/2\sigma^2} = 0.0155 e^{-7.5} = 8.55 \times 10^{-6}.$$

C The random walk

C.1 The problem

One important further application of probabilities is the study of motion that occurs in individual discrete steps. If each step is random, independent of the other steps, then the study of the net motion is referred to as the "random walk problem."

The problem relates to an ensemble of drunkards who begin their random strolls from a single light post. The lengths and directions of their steps might be influenced by such things as the wind, the slope of the ground, etc. But, given the probabilities of the various directions and lengths for a single step, we can use the tools of this chapter to answer the following two questions:

- after each person has taken N steps, what will be the average position relative to the starting point?
- How spread out will the drunks be? That is, what will be the standard deviation of their positions around their average position?

Motion in more than one dimension can be broken up into its individual components, so we develop the formalism for motion in one dimension.

Among the studies that fit into the random walk framework is molecular diffusion, for which a step would be the distance traveled between successive collisions with other molecules. A molecule may go in any direction and may go various distances between collisions. Similarly, the travel of electrons through a metal, of "holes" through a semiconductor, and of thermal vibrations through a solid are all random walk problems.

Sometimes the motion for any one step is not completely random. In molecular diffusion, for example, a molecule is more likely to scatter forward than backward, indicating that its motion after a collision is not completely independent of its motion before the collision. But after some number of collisions, any trace of its previous motion will be lost. So we could fit this type of problem into the random walk framework simply by letting a single step encompass the appropriate number of collisions.

C.2 One step

The most difficult part of this problem is to find the average distance traveled and the standard deviation for one single step, which we label $\bar{s}$ and σ, respectively. These two parameters answer the following two questions: "if a large number of drunks had all started out at the same spot and had all taken one step of variable size, where would they be, on average, and how spread out would they be? Once these questions are answered, it is relatively easy to answer these questions for the average position and spread after N steps, which we label $\bar{S}_N$ and σ_N, respectively.

Suppose that P_s is the probability that a step is of length s in the direction of interest. Alternatively, suppose that $P(s)ds$ is the probability that the length of the step falls within the range ds. Then by the definition of mean values, the average distance traveled and the average of the square of the distance are given by

$$\bar{s} = \sum_s sP_s \quad \text{or} \quad \bar{s} = \int sP(s)ds \tag{3.12}$$

and

$$\overline{s^2} = \sum_s s^2 P_s \quad \text{or} \quad \overline{s^2} = \int s^2 P(s)ds. \tag{3.13}$$

From these we also get the standard deviation (equation 3.3):

$$\sigma^2 = \overline{(s - \bar{s})^2} = \overline{s^2} - 2\bar{s}\,\bar{s} + \bar{s}^2 = \overline{s^2} - \bar{s}^2. \tag{3.14}$$

C.3 N steps

The equations (3.12)–(3.14) refer to a single step. We now find the average distance traveled and the standard deviation after each drunk has taken "N" steps.

The total distance gone by any particular drunkard is the sum of the distances gone during each step:

$$S_N = \sum_{i=1}^{N} s_i = s_1 + s_2 + s_3 + \cdots + s_N$$

It is easy to average this over all drunks. Each step is completely random and governed by the same probabilities as all other steps. So the average length of each is the same:

$$\overline{s_1} = \overline{s_2} = \overline{s_3} = \cdots = \overline{s}.$$

Therefore, the average distance traveled after N steps is simply the product of the number of steps times the average distance traveled in any step, $N\overline{s}$:

$$\overline{S_N} = \overline{s_1 + s_2 + s_3 + \cdots + s_N} = \overline{s} + \overline{s} + \overline{s} + \cdots + \overline{s} = N\overline{s}. \tag{3.15}$$

To calculate the standard deviation after N steps, we start from the definition. As above,

$$\sigma_N^2 = \overline{(S_N - \overline{S_N})^2} = \overline{S_N^2} - \overline{S_N}^2.$$

The term $\overline{S_N}^2$ is simply the square of the above result for $\overline{S_N}$, but to find $\overline{S_N^2}$, we first write

$$S_N^2 = (s_1 + s_2 + s_3 + \cdots)^2.$$

When we square the expression on the right, we get N^2 terms altogether. N of these terms are squared terms, such as s_1^2, s_2^2, etc., and the remaining $N(N-1)$ terms are cross terms, such as s_1s_2, s_1s_3, etc. Thus

$$S_N^2 = \left(s_1^2 + s_2^2 + s_3^2 + \cdots\right) + (s_1s_2 + s_1s_3 + \cdots + s_2s_1 + s_2s_3 + \cdots + s_3s_1 + \cdots)$$
$$= \text{squared terms} + \text{cross terms}.$$

In this form, the averaging is easy. Since the probability is the same for each step, we have for the N squared terms

$$\overline{s_1^2} = \overline{s_2^2} = \overline{s_3^2} = \cdots = \overline{s^2}.$$

For the $N(N-1)$ cross terms, we use the fact that the steps are independent of each other. Therefore we use result 2.7 that $\overline{fg} = \overline{f}\,\overline{g}$ for statistically independent behaviors and get

$$\overline{s_1s_2} = \overline{s_1s_3} = \overline{s_2s_3} = \cdots = \overline{s}\,\overline{s} = \overline{s}^2.$$

Combining these results for the N squared terms and the $N(N-1)$ cross terms, we have

$$\overline{S_N^2} = \overline{\left(s_1^2 + s_2^2 + s_3^2 + \cdots\right)} + \overline{(s_1s_2 + s_1s_3 + \cdots + s_2s_1 + s_2s_3 + \cdots + s_3s_1 + \cdots)}$$
$$= N\overline{s^2} + N(N-1)\overline{s}^2.$$

With this and the result 3.15 that $\overline{S_N} = N\overline{s}$, we have for the square of the standard deviation

$$\sigma_N^2 = \overline{S_N^2} - \overline{S_N}^2 = [N\overline{s^2} + N(N-1)\overline{s}^2] - (N\overline{s})^2$$
$$= N\left(\overline{s^2} - \overline{s}^2\right) = N\sigma^2.$$

Taking the square root of both sides of this equation we find that the standard deviation for N steps is equal to the product of $\sqrt{N}$ and the standard deviation for any one step:

$$\sigma_N = \sqrt{N}\sigma. \tag{3.16}$$

Notice that the average distance traveled, S_N, increases linearly with the number of steps N, whereas the standard deviation increases only as the square root $\sqrt{N}$. In comparison with the distance gone, the position of a random walker gets relatively more predictable as N increases (provided that $\bar{s} \neq 0$) but absolutely less predictable (Figure 3.6).

$$\frac{\sigma_N}{S_N} = \frac{\sqrt{N}\sigma}{N\bar{s}} \approx \frac{1}{\sqrt{N}} \tag{3.17}$$

Note the similarity to the binomial probability distribution (equations 3.1, 3.4, 3.5), where the mean value increases linearly with N, the standard deviation as $\sqrt{N}$, and the relative fluctuation as $1/\sqrt{N}$ (i.e., it decreases with N).

Summary of Sections B and C

For the random walk problem in any one dimension, if P_s is the probability that a step covers a distance s or $P(s)\mathrm{d}s$ is the probability that the length of the step falls within the range $(s, s + \mathrm{d}s)$, then the average distance traveled and the average squared distance traveled after each drunk takes one step are given by (equation 3.12)

$$\bar{s} = \sum_s sP_s \quad \text{or} \quad \bar{s} = \int sP(s)\mathrm{d}s$$

and (equation 3.13)

$$\overline{s^2} = \sum_s s^2 P_s \quad \text{or} \quad \overline{s^2} = \int s^2 P(s)\mathrm{d}s.$$

The square of the standard deviation is given by (equation 3.14)

$$\sigma^2 = \overline{(s - \bar{s})^2} = \overline{s^2} - \bar{s}^2.$$

After N steps the average distance traveled and the standard deviation are (equation 3.15)

$$\overline{S_N} = N\bar{s}$$

and (equation 3.16)

$$\sigma_N = \sqrt{N}\sigma.$$

Figure 3.6 Illustration of the average distance gone (shown by arrows) and the standard deviation (shown by the shaded disks) after 1, 5, and 20 steps, with $\bar{s} = 1$ and $\sigma = 2$. See equation 3.17. Note that for one step, the standard deviation is large compared with average distance traveled but that after 20 steps the reverse is true.

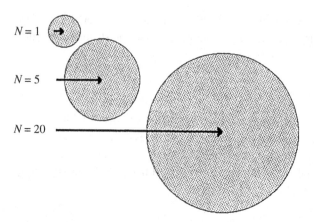

C.4   Conduction in metals

As an example of the random walk problem, we consider a very approximate picture for the motion of conduction electrons in a metal. We will assume that the random thermal motion of the electrons causes each to undergo about 10^{12} collisions per second, on average, and that typically they travel a root mean square distance σ in any direction of about 10^{-8} m between collisions. Normally, if all directions of travel are equally likely, the average length of a step in any single direction is zero.

In the presence of an electric field, however, one direction would be slightly favored. In a typical case of conduction, the average net distance $\bar{s}$ traveled between collisions is 10^{-15} m in the direction favored by the field. This is ten million times smaller than the root mean square length of a single step due to thermal motion, so you can see that the influence of the electric field is very small compared with the random thermal motion. To sum up,

$$\bar{s} = 10^{-15}\,\text{m}, \qquad \sigma = 10^{-8}\,\text{m}.$$

We can use the random walk method to calculate the motion of the conduction electrons over an extended time period. Let's see what happens after 10 minutes (600 seconds) have passed, for example. At 10^{12} collisions per second, the number of steps taken in 600 seconds is $N = 6 \times 10^{14}$, so the average distance gone and the standard deviation are

$$\overline{S_N} = N\bar{s} = 6 \times 10^{14}(10^{-15}\,\text{m}) = 0.6\,\text{m},$$
$$\sigma_N = \sqrt{N}\sigma = \sqrt{6 \times 10^{14}}(10^{-8}\,\text{m}) = 0.2\,\text{m}.$$

Notice that after 6×10^{14} steps the standard deviation is smaller than the average displacement, even though it was ten million times larger than the average displacement for one single step. The larger the number of steps, the more predictable is the behavior. Does this sound familiar?

Problems

Section A

1. Consider many systems, each having 100 rolled dice. Suppose that we are interested in the number of dice per system showing sixes. For these systems, calculate
   (a) the mean number of sixes,
   (b) the standard deviation about this value,
   (c) the relative fluctuation.

2. Repeat the above problem for systems of 10^8 rolled dice.

3. Using the theorem that the mean value of a constant times a function equals the constant times the mean value of the function $(\overline{cf} = c\overline{f})$, prove that the mean value of $(n - \overline{n})^2$ is $\overline{n^2} - \overline{n}^2$ ($\overline{n}$ is a constant).

4. For air at room temperature, the probability that any one molecule is in an excited electronic state is about 10^{-10} ($p = 10^{-10}$, $q \approx 1$). In a typical room there are about 10^{28} air molecules. For this case, calculate
   (a) the mean number of excited molecules,
   (b) the standard deviation,
   (c) the relative fluctuation.

5. There are just 30 air molecules in an otherwise empty room. Calculate
   (a) the average number that will be in the front third of the room at any time,
   (b) the standard deviation about this value,
   (c) the relative fluctuation.

6. Repeat the above problem for 3×10^{27} air molecules in an otherwise empty room.

7. Suppose there are 100 ammonia molecules in a room. Find
   (a) the average number that are in the front half of the room,
   (b) the standard deviation about this number,
   (c) the probability that exactly 50 are in the front half of the room at any instant,
   (d) the probability that exactly 53 are in the front half of the room at any instant.

8. In a certain semiconductor, the probability that an electron jumps from the filled "valence band" to the empty "conduction band" is 10^{-10} (i.e., 1 chance in 10^{10}). If there are 10^{24} electrons in the valence band, find

(a) the average number of electrons in the conduction band at any instant,

(b) the standard deviation for this number of electrons in the conduction band.

9. In deriving the expression for the standard deviation following equation 3.3 we showed that

$$\overline{n^2} = \sum_n n^2 \frac{N!}{n!(N-n)!} p^n q^{N-n} = \left(p\frac{\partial}{\partial p}\right)^2 \sum_n \frac{N!}{n!(N-n)!} p^n q^{N-n}$$

$$= \left(p\frac{\partial}{\partial p}\right)^2 (p+q)^N.$$

(a) Explain the justification for each of the three steps in this series of equations.

(b) Derive a corresponding expression for calculating $\overline{n}$.

(c) Evaluate both expressions to show that $\overline{n} = Np$ and $\sigma^2 = Npq$.

Section B

10. Consider identically prepared systems, each having 600 rolled dice. Suppose we are interested in the number of dice per system that are showing sixes. Find
(a) the average number of sixes per system,
(b) the standard deviation σ,
(c) the values of A and B in the probability distribution $P_{600}(n) = Ae^{-B(n-\overline{n})^2}$,
(d) the probability that exactly 100 of 600 will show sixes,
(e) the probability that exactly 93 of 600 will show sixes.

11. In the above problem, what is the number of different possible combinations of the dice such that 100 show sixes and 500 do not?

12. In the derivation of the Gaussian form of the probability distribution, $P_N(n)$, we showed, using $\ln ab = \ln a + \ln b$, $\ln a/b = \ln a - \ln b$, $\ln a^b = b \ln a$, that

$$\ln P(n) = \ln \frac{N!}{n!(N-n)!} p^n q^{N-n}$$

$$= \ln N! - \ln n! - \ln(N-n)! + n \ln p + (N-n)\ln q.$$

(a) Rewrite this expression, expanding all the factorials on the right-hand side using Stirling's formula,

$$\ln m! \approx m \ln m - m + \frac{1}{2}\ln 2\pi m.$$

(b) Show that

$$\ln P(n = \overline{n}) = \frac{1}{2}\ln \frac{1}{2\pi Npq} = \frac{1}{2}\ln \frac{1}{2\pi\sigma^2},$$

using $\overline{n} = Np$, $q = 1 - p$, $\ln ab = \ln a + \ln b$.

(c) Take the derivative of the expression for $\ln P(n)$ in part (a) and drop terms that go to zero as n gets very large. Then evaluate this at $n = \bar{n} = Np$ to show that $\frac{d}{dn} \ln P(n)\big|_{n=\bar{n}} = 0$.

(d) Following a procedure like that in part (c), take the second derivative and show that $\frac{d^2}{dn^2} \ln P(n)\big|_{n=\bar{n}} = -\frac{1}{Npq} = -\frac{1}{\sigma^2}$.

(e) Combine these results and the Taylor series expansion to show that $P(n) = \frac{1}{\sqrt{2\pi}\sigma} e^{-(n-\bar{n})^2/2\sigma^2}$.

13. For large systems, we can turn the sum over probabilities for all possible configurations into an integral, as indicated in equation 3.11. Because the integrand is nearly zero outside the range $(0, N)$, we can expand the range of integration to $(-\infty, +\infty)$ with negligible effect and use $\int_{-\infty}^{\infty} e^{-ax^2} dx = \sqrt{\pi/a}$. Do this for the Gaussian approximation and show that this sum is equal to unity. Why should it equal unity?

14. You are interested in the number of heads when flipping 100 coins. In the Gaussian approach, with $P_N(n) = Ae^{-B(n-\bar{n})^2}$, what are the values of the constants A and B? Find the probability of obtaining exactly the following numbers of heads: (a) 50, (b) 48, (c) 45, (d) 40, (e) 36.

15. What is the ratio of $P_N(n = \bar{n} \pm \sigma)$ and $P_N(n = \bar{n})$?

16. Suppose you flip 400 coins many times. Find
    (a) the average number of heads per time,
    (b) the standard deviation about this value,
    (c) the probability that exactly 200 would land heads,
    (d) the probability that exactly 231 would land heads.

17. Imagine that you were to roll 360 dice many times. Find
    (a) the average number of sixes showing each time,
    (b) the standard deviation around this value,
    (c) the probability of getting exactly 60 sixes,
    (d) the probability of getting exactly 74 sixes.

18. Suppose that we have 10 000 spin-1/2 particles, which are either spin up or spin down. Thermal agitation causes them to flip around, so that any one particle spends roughly half the time up and half down. On average, at any instant there will be 5000 up and 5000 down.
    (a) What is the standard deviation for fluctuations around this value?
    (b) What is the probability that at a given instant there are exactly 4900 up and 5100 down?

19. Consider 10 000 atoms, each of which has a probability 0.1 of being in an excited state. Assuming a Gaussian distribution, calculate the probability that the number of atoms in an excited state is (a) 1000, (b) 100.

20. Suppose you roll 180 dice, and you are interested in how many land with six dots up. Assume that the distribution can be approximated as being Gaussian.
    (a) What is the probability that exactly 30 land with six dots up?
    (b) According to the Gaussian formula, what is the probability that 181 of the 180 land with six dots up? (Note that the answer is impossible, which means that the Gaussian result is technically in error.)

21. Compare the Gaussian prediction for the probability with the correct (binomial) result for the following cases:
    (a) 1 of 6 dice lands a six,
    (b) 3 of 6 dice land sixes,
    (c) 0 of 6 dice lands a six.

22. Compare the Gaussian prediction for the probability with the correct (binomial) result, for the following cases:
    (a) 10 of 60 dice land sixes.
    (b) 8 of 60 dice land sixes.
    (c) 15 of 60 dice land sixes.
    (d) 0 of 60 dice lands a six.

23. Compare the Gaussian prediction for the probability with the correct (binomial) result for the following cases:
    (a) 10 of 20 coins land heads,
    (b) 12 of 20 coins land heads,
    (c) 15 of 20 coins land heads,
    (d) 2 of 20 coins land heads.

24. The Gaussian distribution that we derived is of the form $P_N(n) = Ae^{-B(n-\bar{n})^2}$, where A and B are constants. Suppose that we have for the first derivative in the Taylor series expansion

$$\frac{\mathrm{d}}{\mathrm{d}n} \ln P(n)\bigg|_{n-\bar{n}} = \varepsilon,$$

where ε is small but not zero. What would the corresponding form of $P_N(n)$ be in this case?

25. You are now going to show that, in the Gaussian distribution $P(x) = Ae^{-Bx^2}$, the constant A is equal to $\sqrt{B/\pi}$. (E.g., if $B = 1/2\sigma^2$ then $A = 1/\sqrt{2\pi}\sigma$.) Do this by insisting that the sum over probabilities must equal unity, $\int P(x)\mathrm{d}x = 1$. To make this difficult integral easier, first square it: $\int P(x)\mathrm{d}x \int P(y)\mathrm{d}y = 1^2 = 1$. Then combine the integrands and turn the area integral, over x and y into an area integral over polar coordinates. This integral is easy to do and should give you the desired result.

26. A certain crystal contains 400 defects, which migrate randomly throughout its volume. We are interested in how many of these are in the crystal's top layer,

which makes up one tenth of the crystal's total volume. If we approximate the probability that n of the 400 defects are in this top layer by $P_N(n) = Ae^{-B(n-\bar{n})^2}$, find

(a) the numerical value of B,

(b) the numerical value of A,

(c) the probability of there being exactly 48 defects in the top layer at any instant.

Section C

27. A bunch of "digital drunks" can only take steps of $(+1, 0, -1)$ meters in the x direction. A strong wind is blowing, so the probabilities are not symmetrical, being given by $P(-1) = 0.3$, $P(0) = 0.2$, $P(+1) = 0.5$. What are the average distance gone and the standard deviation for

(a) one step,

(b) 400 steps?

(c) What is the ratio $\sigma/\bar{s}$ for one step and for 400 steps?

28. Consider "digital drunks" as in the previous problem, except that now they can take steps in one dimension of lengths $(0, \pm1, \pm2)$ meters. Suppose that the probabilities for each of these step lengths are $P(-2) = 0.1$, $P(-1) = 0.1$, $P(0) = 0.3$, $P(1) = 0.3$, $P(2) = 0.2$. What are the average distance gone and the standard deviation for (a) one step, (b) 400 steps? (c) What is the ratio $\sigma/\bar{s}$ for one step and for 400 steps?

29. An ammonia bottle is opened very briefly in the center of a large room, releasing many ammonia molecules into the air. These ammonia molecules go on average 10^{-5} m between collisions with other molecules, and they collide on average 10^7 times per second. After each collision they are equally likely to go in any direction.

(a) What is the average displacement in one dimension (say the z-dimension) for a single step? (Hint: If one step is of length a, then the z-component of this step is $(a \cos\theta)$. Averaging any function f over all solid angles gives $\bar{f} = (1/4\pi) \int f \sin\theta \, d\theta d\phi$.)

(b) What is the square of the standard deviation for any one step? (Hint: $\sigma^2 = \bar{s^2} - \bar{s}^2$. The second of these terms is the square of that calculated in part (a). For the other term, see the hint in part (a) for averaging a function.)

(c) What is the average displacement in the z direction of the escaped ammonia molecules after 2 seconds?

(d) What is the standard deviation of the value obtained in part (c)?

(e) If you were on the z-axis and 6 m from the bottle, how long would it take before more than 32% of the ammonia molecules had positions that were farther from the bottle in the z direction than you?

30. A tiny drop of dye is put in a very still tub of water. The dye molecules travel about 10^{-11} m between collisions and undergo about 4×10^{13} collisions per second. The direction of any step is completely arbitrary; all directions are equally probable. For molecular motion in one direction, calculate the average distance traveled and the standard deviation after (a) one step, (b) one minute, (c) one year (3.17×10^7 seconds). (Refer to the hints in the previous problem if necessary.)

31. A large number of holes at a particular point start to migrate through a semiconductor that has no external field. After each collision with a lattice site, a hole is equally likely to go in any direction. Such collisions occur roughly 10^{13} times per second, and the hole goes an average of 3×10^{-10} m between collisions. Using the hints in problem 29, calculate for motion in any one dimension the average distance traveled and the standard deviation after (a) one step, (b) one second.

32. Consider the motion of some electrons in a semiconductor that has an electric field across it. Suppose that between collisions an average electron goes 10^{-16} m in the direction favored by the field, with a standard deviation of 10^{-9} m. It undergoes 10^{14} collisions per second. Find the average distance traveled and the standard deviation after (a) 1 second, (b) 5 minutes. (c) Suppose that the probability distribution for a single step is of the form $P(x)\mathrm{d}x = Ae^{-B(x-x_0)^2}\mathrm{d}x$. What are the values (with units) of A, B, and x_0?

33. You put a voltage across a metal wire and examine the progress of a group of electrons that begin at a certain point. You notice that after 4 seconds they have gone an average distance of 0.10 m in the $+x$ direction through the wire and have spread out to the point where their standard deviation about this location is 10^{-3} m. They undergo 10^{12} collisions per second.
    (a) What is the average distance gone in the $+x$ direction between any two successive collisions?
    (b) What is the standard deviation for the average distance traveled between any two successive collisions?
    (c) If the density of conduction electrons is 10^{27} m^3 and the wire has a radius of 1 mm, what is the electrical current through this wire?

34. Energy produced in the center of the Sun has a hard time finding its way out. We can estimate roughly how long it takes an average photon to get out by looking at the motion in one dimension only. On average, a photon goes about 1 cm between collisions with hydrogen nuclei or electrons and undergoes about 10^8 such collisions per second. (Use the hints in problem 29 if necessary.)
    (a) What is the average distance traveled in any dimension per step?
    (b) What is the standard deviation about this value?

(c) The radius of the Sun is about 7.0×10^8 m. About how many steps must a photon take before having a 32% chance of being outside the Sun in this dimension?

(d) To how many years does this number of steps correspond? (1 year = 3.17×10^7 seconds).

35. Imagine that there is a brief radiation leak at a nuclear power plant. The radioactive gas molecules have an average speed of 360 m/s and a mean free path (i.e., an average distance between collisions) of 3×10^{-3} m. The root mean square step length, projected in any one direction (e.g., in the x direction) is 1.73×10^{-3} m. Suppose that this leak happens to occur when the air is perfectly still, with no turbulence or other mixing at all. What is the characteristic radius of the radioactive cloud after 1 minute, 1 hour, and 1 day?

36. Consider the expression $(s_1 + s_2 + s_3 + \cdots + s_N)^2$. By writing out the terms explicitly, show that there are (i) N^2 terms altogether, (ii) N squared terms, and (iii) $N(N-1)$ cross terms for (a) $N = 2$, (b) $N = 3$.

37. Given that $\sigma_N^2 = \overline{S_N^2} - \overline{S_N}^2$, where $\overline{S_N^2} = N\overline{s^2} + N(N-1)\overline{s}^2$ and $\overline{S_N} = N\overline{s}$, show that $\sigma_N^2 = N\sigma^2$, if $\overline{s}$ and σ are the average distance traveled and the standard deviation for one step.

Part III
Energy and the first law

Chapter 4
Internal energy

A The general idea

Our investigations of larger systems begin with their internal energy, which involves the relative motion and interactions among the system's own particles. It does *not* include interactions with or motion relative to objects outside the system. The internal energy of a nail, for example, would include the energy of vibration of the atoms and the motions and interactions of the conduction electrons (Figure 4.1). But it would not include the nail's potential energy or motion relative to the Earth, for example. Of course, if you enlarge the system to include both the Earth and the nail, then these would be part of the internal energy of this larger system, but they are not part of the internal energy of the nail by itself.

B Potential energies

We now examine the energies of the individual particles in solids, liquids, and gases, by means of models that are useful in developing intuition for these systems.

B.1 General thoughts

Imagine a particle that is anchored in place by interactions with its neighbors. We can use a Taylor series expansion (Appendix B) to write its potential

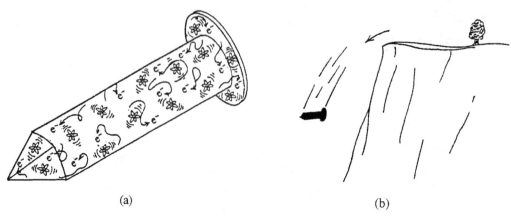

(a)   (b)

Figure 4.1 (a) The internal energy of a nail includes such things as the vibrations of the iron atoms and the potential and kinetic energies of the conduction electrons. (b) If the nail were thrown over a cliff, its motion relative to the Earth and its potential energy due to the Earth's gravity would *not* be part of its internal energy, because they involve more than just the nail itself.

energy $u(x)$ (Figure 4.2) as a function of the displacement x from its equilibrium point:

$$u(x) = u(0) + \left.\frac{du}{dx}\right|_{x=0} x + \frac{1}{2}\left.\frac{d^2u}{dx^2}\right|_{x=0} x^2 + \cdots.$$

The first derivative is zero, because the potential energy is a minimum at equilibrium. For sufficiently small values of x, terms of order x^3 and higher can be ignored, so we can write the particle's potential energy as

$$u(x) \approx u_0 + \tfrac{1}{2}\kappa x^2, \tag{4.1}$$

where $u_0 = u(0)$ and κ are the potential energy and its second derivative at the equilibrium point, respectively. We get similar expressions for displacements in the y and z dimensions, so each anchored particle is a tiny harmonic oscillator in all three dimensions.

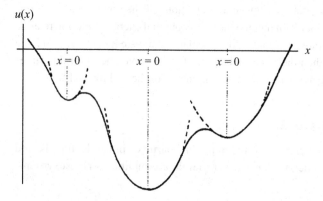

$u(x)$

$x=0$ $x=0$ $x=0$ x

Figure 4.2 Plot of potential energy versus x for an arbitrary potential energy function for a particle. Near any relative minimum, at which we can choose $x=0$, the potential is parabolic, i.e., $u(x) \approx u_0 + (1/2)\kappa x^2$ for small displacements x. This can be shown mathematically by using a Taylor series.

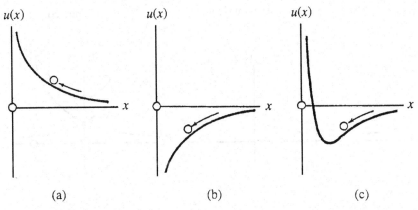

Figure 4.3 The potential energy of a particle due to just one neighbor. We can imagine that one particle is anchored at the origin and the other is coming from infinitely far away. It is convenient to think of the particle and its potential energy as a ball on a hill. (a) When the forces are repulsive, the particle must go "uphill" against these repulsive forces, and its potential energy is positive. (b) When forces are attractive, the incoming particle is going "downhill" and its potential energy is negative. Plot (c) is typical for the interactions between most atoms and molecules – weakly attractive at long ranges but strongly repulsive at very short ranges, where their electron clouds overlap. If the interactions were repulsive at all ranges, as in plot (a), the particles of the system would all fly apart unless they were confined under pressure in some container.

The depth of the potential well u_0[1] depends on the strength of the interactions (Figures 4.3, 4.4). Because these interactions depend on the motions and spacings of the particles (Figure 4.5), u_0 depends on temperature, pressure, and particle concentration:[2]

$$u_0 = u_0(T, p, N). \tag{4.2}$$

For gases u_0 is nearly zero, because on average the molecules are far apart and their mutual interactions are negligible. For most solids, u_0 is negative and nearly constant, because the atoms are bound to one another and the interatomic spacing changes only very slightly with large changes in temperature and pressure.

In liquids, however, the molecules are both mobile and close together. At low temperatures, they move more slowly and have time to seek the preferred orientations of lower potential energy. At higher temperatures, the increased molecular

---

[1] Potential energies can be measured relative to any arbitrary reference level, but the standard convention is that the potential energy is zero when a particle is all by itself and not interacting with anything else. Using this convention, u_0 is negative when particle interactions are attractive, and positive when they are repulsive.

[2] If there is more than one type of particle, we would have to specify the number of each: $N \rightarrow N_1, N_2, \ldots$

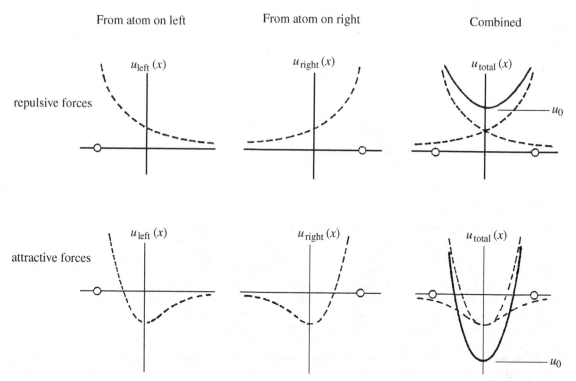

Figure 4.4 The potential energy of a particle due to two neighbors. In any one dimension, the combined potential energy of the particle (solid line) is the sum of that due to the particles on its left and on its right (broken lines). (Top) If the interactions are repulsive, u_0 is positive. (Bottom) If the interactions are attractive, u_0 is negative.

speeds and randomized motions tend to reduce their time in such preferred orientations, so the potential wells become shallower. As heat is added to liquids, then, not all goes into the motions of the particles. Some goes into raising u_0. This extra avenue for storing energy gives liquids correspondingly larger heat capacities.

The depth of the potential well u_0 may change abruptly at phase transitions if there are large changes in interparticle spacing and/or interactions.

B.2 Solids, liquids, and gases

In a solid, the individual atoms are held in place by electromagnetic interactions with neighboring atoms as if they were bound in place by tiny springs (Figure 4.5). Each atom vibrates in all three dimensions around its equilibrium position, and its energy is given by

$$\varepsilon = \varepsilon_{\text{potential}} + \varepsilon_{\text{kinetic}}$$
$$= u_0 + \frac{1}{2}\kappa x^2 + \frac{1}{2}\kappa y^2 + \frac{1}{2}\kappa z^2 + \frac{1}{2m}p_x^2 + \frac{1}{2m}p_y^2 + \frac{1}{2m}p_z^2. \tag{4.3}$$

It can be seen that $\varepsilon_{\text{potential}}$ is made up of two parts, the potential energy reference level u_0 and the potential energy of vibration. Some solids are not isotropic, and then the constants κ may be different in different directions.

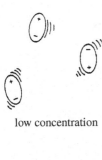

low concentration

high concentration

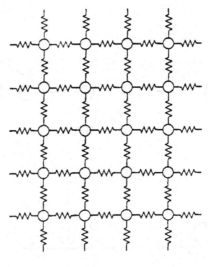

solid

Figure 4.5 (Left) The depth of the potential well u_0 (Figures 4.4, 4.7) depends on the motions and spacings of the molecules, which change with temperature and pressure. (Right) Interactions among neighboring atoms in solids make them behave as if they were connected together by tiny springs.

In liquids, the potential energies of the mobile molecules fluctuate rapidly as the configurations of the other molecules around them change. So the potential energy reference level u_0 is an average or "mean field" value, and the total energy of a molecule in a liquid can be written as

$$\varepsilon = \varepsilon_{\text{potential}} + \varepsilon_{\text{kinetic}} = u_0 + \frac{1}{2m}p_x^2 + \frac{1}{2m}p_y^2 + \frac{1}{2m}p_z^2. \tag{4.4}$$

In gases, neighboring particles are usually so far apart that interactions are negligible. So, for most cases, the potential energy is minuscule and we can treat a particle's energy as purely kinetic:

$$\varepsilon = \varepsilon_{\text{kinetic}} = \frac{1}{2m}p_x^2 + \frac{1}{2m}p_y^2 + \frac{1}{2m}p_z^2. \tag{4.5}$$

C Quantum effects

C.1 Rotations and vibrations

Collisions with other molecules might cause polyatomic molecules in a liquid or gas rotate and/or vibrate internally. This would provide additional modes of energy storage beyond the potential energies and translational kinetic energies discussed in the preceding section:

$$\varepsilon = \varepsilon_{\text{pot}} + \varepsilon_{\text{trans}} + \varepsilon_{\text{rot}} + \varepsilon_{vib}. \tag{4.6}$$

The fact that, according to quantum mechanics, particles behave as waves puts restrictions on the allowed energies, which is particularly evident in molecular

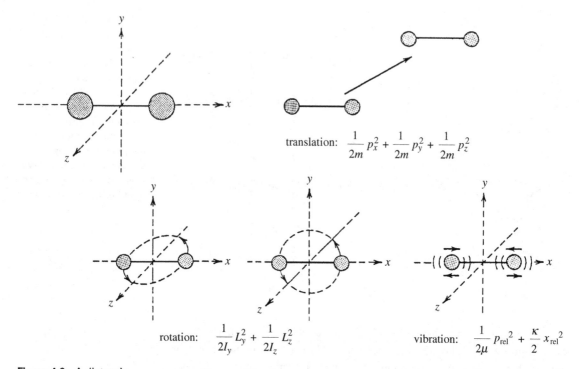

translation: $\dfrac{1}{2m} p_x^2 + \dfrac{1}{2m} p_y^2 + \dfrac{1}{2m} p_z^2$

rotation: $\dfrac{1}{2I_y} L_y^2 + \dfrac{1}{2I_z} L_z^2$ vibration: $\dfrac{1}{2\mu} p_{rel}^2 + \dfrac{\kappa}{2} x_{rel}^2$

Figure 4.6 A diatomic gas molecule is like a little dumbbell. It can store translational kinetic energy by moving in all three dimensions. But it can rotate around only two axes. The rotational inertia around the third (the x-axis in this figure) is so small, that the energy of even the first excited state is too high to reach. Normally, it can store no energy in vibrations, because the excitations require too much energy.

vibrations. Molecular binding puts each atom in a potential well that is a fraction of an angstrom wide. Therefore, the wavelengths of the standing waves are short (subsection 1B.8, Figure 1.9), corresponding to energies so high that vibrational excitations are usually not possible at normal temperatures.[3]

Similar quantum effects also appear in molecular rotations. As we saw in subsection 1B.6, angular momentum **L** is quantized in terms of $\hbar$. Kinetic energies for rotations around any axis are inversely related to the rotational inertia I. For rotations around the ith axis,

$$\varepsilon_{rot} = \frac{1}{2I_i} L_i^2, \qquad \text{where } L_i = (0, \pm 1, \pm 2\ldots)\hbar \qquad (4.7)$$

(Because we can only know the exact value of L_i for one axis at a time, it is sometimes more convenient to use the total angular momentum, $L^2 = l(l+1)\hbar^2$.)

The smaller the rotational inertia, the larger the energy of the first excited rotational state. Sometimes the rotational inertia around one or more axes is so small that excitation requires more energy than is available through molecular collisions. So these particular rotational motions do not occur.

---

[3] No energy can be extracted from the "zero point motion" (Section 1B.8), because there is no lower state for the particle to fall into. We often measure energies relative to this level, calling this "the state of zero energy." Energy can only be stored by excitations into higher levels.

C.2   Example – the diatomic gas molecule

As a specific example, consider diatomic gas molecules, such as nitrogen or oxygen. The translational kinetic energy is as given in equation 4.5 and, as discussed above, the vibrational levels are too high to be accessible. The rotational kinetic energy is particularly interesting, however, because the rotational inertia of a diatomic molecule about the axis going through both atomic nuclei is very small, and so the corresponding rotational excitations would require too much energy. Around the other two molecular axes, however, the rotational inertia is much larger, making these rotational states more accessible at ordinary temperatures. With these considerations for the rotational modes, we can write the total energy of a typical diatomic molecule as (Figure 4.6)

$$\varepsilon = \varepsilon_{\text{trans}} + \varepsilon_{\text{rot}} = \frac{1}{2m}p_x^2 + \frac{1}{2m}p_y^2 + \frac{1}{2m}p_z^2 + \frac{1}{2I_1}L_1^2 + \frac{1}{2I_2}L_2^2 \qquad (4.8)$$

D   Degrees of freedom

In all the examples above, the energies of individual particles are of the form

$$\varepsilon = u_0 + \sum_i b_i \xi_i^2 \qquad (4.9)$$

where the b_i are constants (e.g., $\kappa/2, 1/2m, 1/2I$, etc.) and the ξ_i are position or momentum coordinates (x, p_x, L_1, etc.) Each of these $b_i\xi_i^2$ terms represents a distinct way in which a particle can store energy, called a "degree of freedom." For example, each atom in a solid has six degrees of freedom, because there are six $b_i\xi_i^2$ terms in equation 4.3. For the diatomic gas molecule of equation 4.8 there are five such terms, so each diatomic gas molecule has five degrees of freedom.

In this book, we use the standard notation, whereby ν represents the number of degrees of freedom per particle and N represents the number of particles in the system:

$$\text{degrees of freedom per particle} = \nu$$
$$\text{degrees of freedom for a system of } N \text{ particles} = N\nu. \qquad (4.10)$$

E   Equipartition

Consider the distribution of energy between the various degrees of freedom. In collisions there is a tendency for energy to be transferred from the faster particle to the slower one, so the energies even out. (Think of collisions on a pool table.) From our studies of harmonic oscillators in introductory physics courses, we know that the average potential and kinetic energies are the same. So terms like $\kappa x^2/2$, and $p_x^2/2m$ carry equal energies when a average is taken. Finally, motion in all three directions is equally likely, so the average kinetic or potential energies in the $x, y,$ and z directions must all be equal.

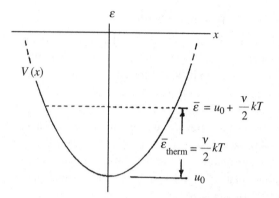

For these reasons, it should not be surprising that the average energy stored in
each degree of freedom is the same. This is known as the "equipartition theorem."
We will prove it in a later chapter, where we will also show that if the energy
expression for a particular degree of freedom is of the form $b_i \xi_i^2$ (equation 4.9),
then the average energy stored in this degree of freedom is given by

$$\bar{\varepsilon} = \frac{1}{2}kT \tag{4.11}$$

where T is the temperature and k is Boltzmann's constant:

$$k = 1.381 \times 10^{-23} \text{ J/K} = 8.63 \times 10^{-5} \text{ eV/K}.$$

Although we will soon give temperature a more formal definition, we can see
from 4.11 that it measures the average energy stored in each degree of freedom.

F Thermal energy

The "thermal energy" of a system is taken to be the energy stored in the wiggles
and jiggles of its particles and does not include the potential energy reference
level u_0 (Figure 4.7).[4] The energy of a single atom in a solid can be written as
(equation 4.3)

$$\varepsilon = u_0 + \underbrace{\frac{1}{2}\kappa x^2 + \frac{1}{2}\kappa y^2 + \frac{1}{2}\kappa z^2 + \frac{1}{2m}p_x^2 + \frac{1}{2m}p_y^2 + \frac{1}{2m}p_z^2}_{\varepsilon_{thermal}}.$$

Since each degree of freedom carries an average energy of $kT/2$ (equation 4.11),
the average thermal energy of an atom in a solid is $\bar{\varepsilon}_{thermal} = 6kT/2$.

---

[4] In liquids the distinction between the reference level and thermal energy is more difficult. As we have
seen before, continually changing motions and orientations between neighboring molecules cause
rapidly changing potential energies. So in liquids the u_0 reference level represents a time-averaged
mean feild value for a molecule's potential energy and is sensitive to temperature.

Generalizing this example, the average energy of a particle in any system can be written as

$$\bar{\varepsilon} = u_0 + \frac{v}{2}kT, \tag{4.12}$$

where v is the number of degrees of freedom, and the internal energy of a system of N such particles is

$$E = N\bar{\varepsilon} = Nu_0 + \frac{Nv}{2}kT. \tag{4.13}$$

The "thermal energy" E_{therm} of the system is the second term in 4.13:

$$E_{\text{therm}} = N\varepsilon_{\text{therm}} = \frac{Nv}{2}kT.$$

you can see that the thermal energy is stored in the various degrees of freedom, and is proportional to the temperature.

Notice that if we add energy ΔE to a system of N molecules then the rise in temperature ΔT depends on the number of degrees of freedom per molecule, v:

$$\Delta E = N\Delta u_0 + \frac{Nv}{2}k\Delta T \tag{4.14}$$

As we have seen, $\Delta u_0 \approx 0$ for most solids and gases, so by measuring the increase in temperature ΔT when energy ΔE is added, we can usually determine the number of degrees of freedom per molecule. For liquids, however, u_0 increases with temperature and acts like additional degrees of freedom for energy storage.

At phase transitions we may add large amounts of heat to a system, without any change in temperature. Where does this added energy go? According to equation 4.13, we can see two possibilities: since N and T are constant, only u_0 and v can change. There is always a change in u_0 during a phase transition, because molecular arrangements in the new phase are different, resulting in different potential energies. The number of degrees of freedom per molecule, v might also change, because the change of phase could change the constraints on the motions of the individual molecules.

Summary of Chapter 4

Internal energy is the energy stored in the motions and interactions of the particles entirely within a system. The interactions between neighboring particles gives rise to potential energies. The reference level from which potential energies are measured is given the symbol u_0. In solids, the atoms are anchored in place by electromagnetic interactions with their neighbors. They oscillate around their equilibrium positions as tiny harmonic oscillators. The molecules in liquids are fairly free to roam through the liquid. In most gases the potential energy is nearly zero, owing to minimal interactions with the distant neighbors. If the interactions in a system are predominantly attractive then u_0 is negative, and if they are predominantly repulsive then u_0 is positive.

In the detailed energy equation, each term that involves a position or momentum coordinate represents a "degree of freedom." Such terms normally have the form $b\xi^2$, where b is a constant and ξ is a coordinate. The fact that energies are quantized may limit the number of degrees of freedom available to a particle because, for some cases, even the first excited state might require too much energy.

On average, the internal energy of a system is distributed equally between all degrees of freedom. The average energy per degree of freedom is proportional to the temperature (equation 4.11):

$$\bar{\varepsilon} = \frac{1}{2}kT,$$

where

$$k = 1.381 \times 10^{-23} \text{ J/K} \qquad \text{(Boltzmann's constant)}.$$

The thermal energy of a particle is carried in its various degrees of freedom and is measured relative to the potential energy reference level u_0.[5] If there are ν degrees of freedom per particle then the average energy per particle and the total internal energy of a system of N such particles are given by (equations 4.12, 4.13)

$$\bar{\varepsilon} = u_0 + \frac{\nu}{2}kT$$

and

$$E = N\bar{\varepsilon} = Nu_0 + \frac{N\nu}{2}kT.$$

Problems

Sections A and B

1. Give examples of types of energy that would be part of your body's internal energy, and of types of energy that would not, unless the system were enlarged to include your environment.

2. Consider the average potential energy of a water molecule in an ice crystal and of one in the liquid state. Which is lower? How do you know?

3. (a) Show that the function $f(x) = x^3 + x^2 - 2$ has a local minimum at $x = 0$. (Hint: Show that the first derivative is zero and that the second derivative is positive at $x = 0$.)
   (b) Expand this function in a Taylor series around the point $x = 0$, up to the fourth-order term (the term in x^4).

---

[5] The situation in liquids is a little different, as discussed earlier. See also footnote 4.

(c) If we keep terms only to order x^2, what is the range in x for which our error is less than 10%?

4. Repeat the above problem for the function $f(x) = -e^{-x^2}$.

5. Expand the functions $\sin x$, $\cos x$, $\ln(1 + x)$, and e^x to order x^4 in Taylor series expansions around the origin. Do you see any pattern in these expansions that would allow you to continue the expansion to any order? Write out each of these infinite series in closed form.

$$\left(\text{E.g., } \sin x = \sum_{n=0}^{\infty} \frac{(-1)^n}{(2n + 1)!} x^{2n+1} \right)$$

6. We are going to make a very rough estimate of how much pressure must be applied to a typical solid to compress it to the point where the potential energy reference level u_0 of the individual atoms becomes positive. Ordinarily, for a typical solid, u_0 is around -0.2 eV.
   (a) If the interatomic spacings are typically 0.2 nm, how many atoms are there per cubic meter?
   (b) Roughly, how much work (in joules) must be done on one cubic meter of this solid to raise u_0 to zero? (Hint: You have to raise the potential energy per atom by about 0.2 eV.)
   (c) Work is equal to force times distance parallel to the force (Fdx) but, by multiplying and dividing by the perpendicular surface area, this can be changed into pressure times volume ($-pdV$). Because solids are elastic, the change in volume is proportional to the change in applied pressure, $dV = -Cdp$, and the constant C is typically 10^{-17} m^5/N. With this background, calculate the work done on a solid as the external pressure is increased from 0 to some final value p_f.
   (d) With your answer to part (c) above, estimate the pressure that must be exerted on a typical solid to compress it to the point where u_0 becomes positive (Figure 4.4, top right).
   (e) What is a typical value for the variation of u_0 with pressure, $\partial u_0 / \partial p$, at constant temperature and atmospheric pressure in a solid?

7. Consider the following three systems: (A) the water molecules in a cold soft drink, (B) the copper atoms in a brass doorknob, and (C) the helium atoms in a blimp (a small cigar-shaped airship). Below are listed expressions of the energy for an atom in each system. In each case, fill in the blank with the letter of the most appropriate system.

$$\underline{\quad} \quad \varepsilon = u_0 + \frac{1}{2}\kappa x^2 + \frac{1}{2}\kappa y^2 + \frac{1}{2}\kappa z^2 + \frac{1}{2m}p_x^2 + \frac{1}{2m}p_y^2 + \frac{1}{2m}p_z^2$$

$$\underline{\quad} \quad \varepsilon = \frac{1}{2m}p_x^2 + \frac{1}{2m}p_y^2 + \frac{1}{2m}p_z^2$$

$$\underline{\quad} \quad \varepsilon = u_0 + \frac{1}{2m}p_x^2 + \frac{1}{2m}p_y^2 + \frac{1}{2m}p_z^2$$

Section C

8. We are going to estimate the vibrational energy of the first excited state for a nitrogen molecule (N_2). Each atom finds itself in a potential well due to its interactions with the other atom. This potential well is roughly 0.02 nm across.

   (a) What are the wavelengths of the longest two standing waves that would fit in this well?

   (b) What momenta do these correspond to?

   (c) Considering kinetic energy only, how much energy in eV would be required to excite an atom from the ground state to the first excited state? (The mass of a nitrogen atom is 2.34×10^{-26} kg.)

   (d) To estimate the minimum temperature needed for excitations to occur, we compare kT (where $k = 1.38 \times 10^{-23}$ J/K $= 8.63 \times 10^{-5}$ eV/K) with the energy required to reach the first excited state. Roughly what is the minimum temperature needed for vibrational excitations in nitrogen gas?

9. Consider the rotation of diatomic molecules around an axis that runs perpendicular through the midpoint of the line that joins the two atoms (see Figure 4.6). The mass of a nitrogen atom is 2.34×10^{-26} kg, and the interatomic separation in an N_2 molecule is 1.10×10^{-10} m.

   (a) What is the rotational inertia of an N_2 molecule around this axis? ($I = \sum m_i r_i^2$.)

   (b) Find the energy in eV required to excite this molecule from the non-rotating state to the first excited rotational state (i.e., from $l = 0$ to $l = 1$, where $L^2 = l(l+1)\hbar^2$).

   (c) What is the minimum temperature for rotational excitations in nitrogen? See problem 8(d).

10. Repeat problem 9 for an oxygen molecule, O_2, given that the mass of an oxygen atom is 2.67×10^{-26} kg and the interatomic spacing is 1.21×10^{-10} m.

11. Repeat problem 9 for a hydrogen molecule, given that the mass of a hydrogen atom is 1.67×10^{-27} kg and the interatomic spacing is 7.41×10^{-11} m.

Section D

12. How many degrees of freedom has a sodium atom in a salt crystal?

13. Why do you suppose that, at high temperatures, a molecule of water vapor (H_2O) has three rotational degrees of freedom and a molecule of nitrogen gas (N_2) has only two?

14. Assuming that a conduction electron in a metal is free to roam anywhere within the metal (not being constrained to any small region by a particular well), how many degrees of freedom does it have?

15. Consider the phase change for iron from solid to liquid forms.
    (a) How many degrees of freedom does each iron atom have in the solid state?
    (b) After it has melted?
    (c) Did the number of degrees of freedom of the conduction electrons change?
    (d) Did the number of degrees of freedom of the whole system increase or decrease?
    (e) On a microscopic scale, what happens to the energy put into the iron to melt it?

16. The heat capacities of some diatomic gas molecules show that they have three degrees of freedom at very low temperatures, five degrees of freedom at intermediate temperatures, and seven degrees of freedom at very high temperatures. How would you explain this?

Sections E and F

17. Estimate the molar heat capacity of a diatomic gas with five degrees of freedom per molecule, by calculating how many joules of energy must be added to raise the temperature by 1 °C. (Assume that the volume is constant, so that only heat is added and no work is done.)

18. (a) Make an estimate of u_0 (in eV) for a water molecule in the liquid state at 100 °C. Assume that there are six degrees of freedom per molecule in both the liquid and the vapor states and that 2260 kJ of energy per kg are released when it condenses. Ignore any work done on the molecule due to the change in volume.
    (b) What is the average thermal energy per molecule?
    (c) What is the average total energy per molecule for liquid water at 100 °C?

19. At 0 °C, a water molecule in both ice and liquid water has six degrees of freedom. One mole of water has mass 18 grams and a latent heat of fusion equal to 6025 joules per mole. Given this information, calculate the following in units of eV:
    (a) the average thermal energy per molecule in liquid water at 0 °C and in ice at 0 °C,
    (b) the amount of energy per molecule added in making the phase change,
    (c) the change in the potential energy reference level u_0 in going from the solid to the liquid state.
    Does the water's thermal energy increase, decrease, or remain the same as ice melts?

20. Using equipartition, calculate the root mean square value of the following quantities in a gas at room temperature (295 K).
    (a) The speed of a nitrogen molecule ($m = 4.68 \times 10^{-26}$ kg).

(b) The speed of a hydrogen molecule ($m = 3.34 \times 10^{-27}$ kg).

(c) The angular momentum of a diatomic oxygen molecule around one of the two rotational axes, for which its moment of inertia is 1.95×10^{-46} kg m^2.

(d) If the axis in part (c) is the z-axis, what would be the root mean square value of the quantum number l_z?

21. What is the total thermal energy at room temperature (293 K) in a gram of (a) lead, (b) dry air (78% N_2, 21% O_2, 1% Ar)?

22. You are climbing a mountain and you and your equipment weigh 700 N. Suppose that of the food energy you use, one quarter goes into work (getting you up the mountain) and three quarters into waste heat. Half the waste heat goes into evaporating sweat. For every kilometer of elevation that you gain, how many kilograms of food do you burn, and how many kilograms of water do you lose? (Very roughly, food provides 4×10^6 J/kg, and the latent heat of evaporation at the ambient temperature is about 2.5×10^6 J/kg.)

Chapter 5
Interactions between systems

Energy can be transferred between systems by the following three mechanisms

- the transfer of heat ΔQ;
- the transfer of work ΔW (i.e., one system does work on another);
- the transfer of particles ΔN.

These are called thermal, mechanical, and diffusive interactions, respectively (see Figure 5.1). The first three sections of this chapter introduce these interactions in a manner that is intuitive and qualitatively correct, although lacking in the mathematical rigor of the chapters that follow.

A   Heat transfer – the thermal interaction

In the preceding chapter we learned that thermal energy gets distributed equally among all available degrees of freedom, on average. So the energy of interacting systems tends to flow from hot to cold until it is equipartitioned among all degrees of freedom. The energy that is transferred due to such temperature differences is called heat, and it travels via three distinct mechanisms: conduction, radiation, and convection.

Conduction involves particle collisions (Figure 5.2a). On average, collisions transfer energy from more energetic particles to less energetic ones. Energy flows from hot to cold.

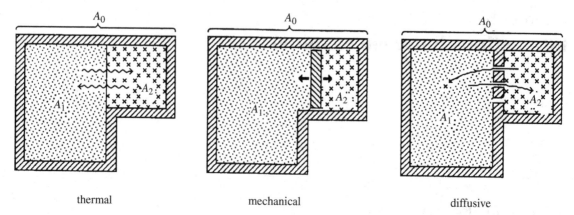

Figure 5.1 Pictorial representations of thermal, mechanical, and diffusive interactions between systems. The combined system, $A_0 = A_1 + A_2$, is completely isolated from the rest of the Universe.

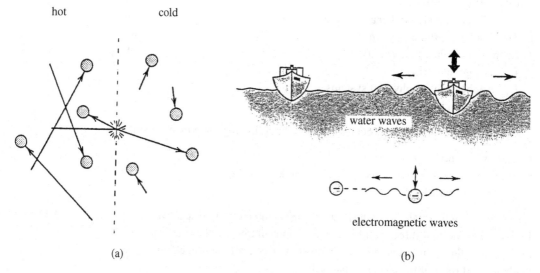

Figure 5.2 (a) Conduction: collisions transfer energy from particles with higher kinetic energies to those with lower kinetic energies, on average. (b) Radiation, illustrated by toy boats in a bathtub. As one is jiggled up and down, it sends out waves which cause other toy boats to bob up and down as they pass by. Similarly, accelerating electrical charges generate electromagnetic waves, which radiate outward and transfer energy to other electrical charges that they encounter.

Energy transfer via radiation can be illustrated by toy boats in a tub. (Figure 5.2b). If one is jiggled up and down, it sends out waves. Other toy boats will oscillate up and down as these waves pass by. In a similar fashion (but at much higher speeds), electromagnetic waves are generated by accelerating electrical charges, and this energy is absorbed by other electrical charges that these waves encounter.

Owing to the thermal motions of particles, all objects both radiate energy into their environment and absorb energy from their environment. Hotter objects radiate energy more intensely due to the more energetic motion of their particles. Therefore, there is a net transfer of radiated energy away from hotter objects towards cooler ones.

Heat transfer by convection involves the movement of particles from one point to another. A system gains the energy of particles that enter and loses the energy of particles that leave. Examples of convection include all fluid motions. Within solids, mobile conduction electrons also engage in heat transfer via convection. (How does this differ from the diffusive interaction? We'll soon find out.)

B   Work – the mechanical interaction

Another way to increase a system's internal energy is to do work on it, $dW = \mathbf{F} \cdot d\mathbf{s}$; for example, if you compress a gas then its temperature rises. Some sort of external force must cause a displacement of the system's particles.[1] It is customary to use the symbol ΔW for work done *by* the system. Since forces come in equal and opposite pairs,

$$\text{work done } by \text{ the system} = -\text{work done } on \text{ the system.} \qquad (5.1)$$

Whereas ΔQ represents heat added *to* the system, therefore increasing its internal energy, ΔW represents work done *by* the system (i.e., against the external force $\mathbf{F}$), therefore decreasing the internal energy.[2] Combining both types of interactions we have, for the change in internal energy,

$$\Delta E = \Delta Q - \Delta W \qquad \text{(thermal and mechanical interactions)} \qquad (5.2)$$

It is sometimes convenient to write the product of force times displacement in other ways. Examples include:

$$dW = F dx \qquad \text{(an external force } F \text{ pushing over a distance } dx)$$
$$dW = p dV \qquad \text{(an external pressure } p \text{ forcing a change in volume } dV)$$
$$dW = -\mathbf{B} \cdot d\mathbf{\mu} \qquad \text{(a magnetic field } \mathbf{B} \text{ causing a change in magnetic moment } \mathbf{\mu})$$
$$dW = -\mathbf{E} \cdot d\mathbf{p} \qquad \text{(an electric field } \mathbf{E} \text{ causing a change in electric dipole moment } \mathbf{p})$$
$$(5.3)$$

All have the same general form. We normally use $p dV$ as the prototype. Any change dV, $d\mathbf{\mu}$, $d\mathbf{p}$, etc., means that particles within the system have moved because of the external force. Therefore work has been done and kinetic energies have changed.

---

[1] Why couldn't a displacement due to an *internal* force cause a change in the internal energy?

[2] This sign convention reflects early interest in the conversion of heat (input) to work (output) by engines. The name "thermodynamics" and much of the early progress in the field can be traced to these studies.

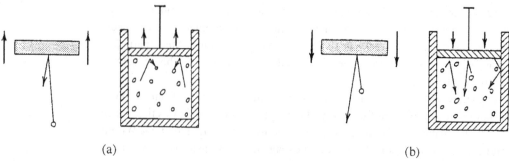

(a)   (b)

Figure 5.3 (a) When a particle collides elastically with receding particles or boundaries, it loses kinetic energy. So, as a system under pressure expands the molecules lose energy. (b) Conversely, elastic collisions with approaching particles or boundaries cause an increase in kinetic energy. So, as a system under pressure is compressed its molecules gain energy.

How can the internal energy of a system increase when the displacement of particles due to external forces causes a decrease in their potential energies? The answer is that potential energies relative to *external* forces are not part of the *internal* energy of a system, but the increased kinetic energy *is* part of it.

If the system is expanding, there is a tendency for each particle to collide with things moving outward, be those other particles or receding boundaries. Collisions with receding objects cause a loss in kinetic energy (Figure 5.3a). Conversely, when a system is being compressed, collisions with things moving inward cause the kinetic energy to grow (Figure 5.3b). Therefore the expansion of systems under pressure causes a decrease in internal energy, and compression causes an increase. You may wish to speculate (homework) about systems under *tension* rather than pressure.

Summary of Sections A and B

The transfer of energy between systems is accomplished in any or all of three ways: the exchange of heat, work, and/or particles. These are referred to as thermal, mechanical, and diffusive interactions, respectively. Heat tends to flow from hotter to colder until temperatures are equalized.

The exchange of heat is accomplished through conduction, radiation, or convection. Conduction involves collisions between particles. Radiation involves the emission of electromagnetic waves by accelerating charges and the absorption of this energy by charged particles that these waves encounter. Convection involves energy transfer of particles as they enter or leave a system.

Work is achieved by the action of a force over a distance. Many different kinds of force may act on a system, but the work done has the same general form – the product of an external force and the change in the conjugate internal coordinate. It is customary to use pdV as the prototype for mechanical interactions.

ΔQ represents heat added *to* the system, and ΔW represents work done *by* the system. Therefore (equation 5.2)

$$\Delta E = \Delta Q - \Delta W \qquad \text{(thermal and mechanical interactions).}$$

C Particle transfer – the diffusive interaction

C.1 The chemical potential

We now examine the diffusive interaction. When particles enter a system they may carry energy in different ways, two of which we have already encountered: heat transfer and work. Any energy transfer that is *not* due to either of these mechanisms is described by the "chemical potential" μ as follows. When ΔN particles enter a system, the energy delivered via this third mechanism is given by

$$\Delta E = \mu \Delta N \qquad \text{(diffusive interaction only; no work or heat transfer).} \qquad (5.4)$$

To understand this term, we first review the two types of energy transfer that are excluded. As you well know, work is the product of force times distance. Perhaps, though, you have not yet encountered a formal definition of heat. Two important aspects of heat that will be introduced and quantified in future chapters are the following.

(1) A heat input increases the number of states accessible to a system.[3] (You might think of this as allowing the particles more ways to wiggle and jiggle.)
(2) As we will learn in subsection 9B.1 (equation 9.6), there are *three* different ways in which heat (ΔQ) may enter or leave a system. Only one of these is the familiar thermal interaction. For the other two mechanisms, it is *not* necessarily true that the heat lost by one system is equal to that gained by the other (that is, $dQ_1 \neq -dQ_2$).

When a particle goes from one system to another, it experiences a new environment and new interactions (e.g., it might fall into a deeper potential well that releases kinetic energy to the new system). This will change the number of states accessible to the system, so heat (ΔQ) will be gained or lost. Note that this is *not* due to temperature differences between the two system, so it is *not* part of the thermal interaction. Rather it is due to the new environment that the transferred particle experiences. Also note that if the environments of the two systems differ, then the heat lost by one system will *not* be the same as that gained by the other.

To quantify this idea, consider the transfer of ΔN particles from system A_2 to system A_1 (Figure 5.1, on the right). For simplicity, we exclude thermal and mechanical interactions by assuming either that both systems are insulated and rigid or both are at the same temperature and pressure. Combining equations 5.2 and 5.4 for the change in internal energy of the two, we have

$$\Delta E_1 = \Delta Q_1 + \mu_1 \Delta N \qquad (\Delta N \text{ particles enter region 1}),$$
$$\Delta E_2 = \Delta Q_2 - \mu_2 \Delta N \qquad (\Delta N \text{ particles leave region 2}).$$

---

[3] We will find that the heat entering a system is directly proportional to the increase in entropy, which is a measure of the number of states that are accessible to the system.

Adding the two together and using energy conservation ($\Delta E_1 + \Delta E_2 = 0$), we have

$$0 = \underbrace{(\Delta Q_1 + \Delta Q_2)}_{\Delta Q_0} + \underbrace{(\mu_1 - \mu_2)\Delta N.}_{\substack{\text{energy transferred by} \\ \text{the diffusive interaction}}}$$

This shows how the energy transferred by the diffusive interaction relates to the net amount of heat ΔQ_0 that is released or absorbed.

Suppose, for example, that $\mu_1 < \mu_2$, so that the particles' chemical potential decreases as they enter their new environment. The above equation tells us that heat will be released ($\Delta Q_0 > 0$). That is, a *decrease* in chemical potential corresponds to the *release* of heat. As we saw above, the release of heat increases the number of accessible states, and this increase may happen in either or both of two distinct ways (see equation 1.4):

(1) an increase in the accessible volume in momentum space, V_p;
(2) an increase in the accessible volume in coordinate space, V_r.

The first happens when particles entering a new system fall into deeper potential wells, owing to their interactions in the new environment. The loss in potential energy produces a corresponding gain in kinetic energy, hence a larger accessible volume V_p in momentum space. For example, think of the heat released when concentrated sulfuric acid is mixed with water, or think of what happens to the individual atoms when hydrocarbons are burned.

The second happens when particles move into regions where they have more room and hence a larger accessible volume V_r in coordinate space. For example, they might move to regions of lower concentration or into an evacuated chamber ("free expansion").

C.2 Particle distributions

In our macroscopic world, systems seek configurations of lower potential energy. Rocks fall down (Figure 5.4). A boulder that has fallen to the valley floor would never jump back out. Things are different in the microscopic world, however, due to the thermal motions of the particles. The smaller the particle, the more violent

Figure 5.4 When forces between particles are attractive, a new particle entering the system falls into a potential well, like a boulder falling off a cliff. And just as the boulder's kinetic energy is transferred to the dust and debris on the valley floor, so is the kinetic energy of the incoming particle transferred to others.

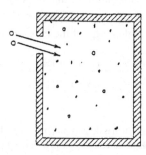
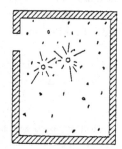

the motion. Particles are continually jumping back out of wells into which they have fallen.

These thermal motions cause particles to flow towards regions of lower concentrations, simply because there are more particles in the region of higher concentration ready to move out than there are particles out in the region of lower concentration ready to move back in.

Consequently, when a system is in equilibrium, the two considerations that govern the particle distribution are that they tend to move towards regions of

- lower potential energy
- lower concentration

As you might infer from the preceding section, the chemical potential μ of equation 5.4 is the appropriate measure of these two tendencies. Although the details of this measure will be developed in Chapter 14, for now we can think of it as follows:

μ depends on: 1. depth of the potential well and 2. particle concentration.   (5.5)

Deeper potential wells and smaller concentrations mean smaller (i.e., more negative) values for the chemical potential.

Because particles seek configurations that minimize these two factors, the chemical potential governs diffusive interactions in the same way that temperature and pressure govern thermal and mechanical interactions, respectively. Particles flow towards regions of *lower* chemical potential, just as heat flows toward regions of *lower* temperature, and movable boundaries move toward regions of *lower* pressure. The underlying principle for all these interactions is the "second law of thermodynamics" (Chapter 7), whose consequences are so familiar that we call them "common sense."

The configuration of lowest chemical potential usually involves a compromise in trying to minimize both potential energy and particle concentration; a gain in one area may be offset by a reduction in another. So, in the microscopic world, we may find some particles in regions of higher potential energy, albeit in correspondingly lower concentrations. An example is the water vapor in our atmosphere. The potential energy of a water molecule in the liquid phase is much lower because of the strong interactions between closely neighboring molecules:

$$u_{0,\,\text{liquid}} \approx -0.4 \text{ eV},$$

$$u_{0,\text{vapor}} \approx 0.$$

If water sought the configuration of lowest potential energy, all water molecules would be in the ocean. None would be in the atmosphere.

Now consider the evaporation of water into dry air. Initially $\mu_{\text{vapor}} < \mu_{\text{liquid}}$, owing to the very small concentration of water molecules in the vapor phase (see equation 5.5). The molecules diffuse toward the lower chemical potential – that is, the water evaporates. But as the concentration in the vapor phase increases, its

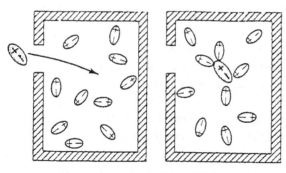 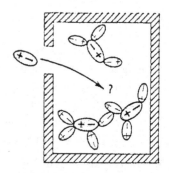

Figure 5.5 When a salt ion or molecule enters a solution, it falls into a potential well owing to the electrostatic attraction between it and the oppositely charged parts of neighboring water molecules. As more salt is added, the potential well's depth decreases, because there are fewer remaining free water molecules with which it can interact. Furthermore, the concentration of the dissolved salt increases. Both the raising of the potential well and the rising concentration cause the chemical potential of the dissolved salt to rise, eventually reaching that in the crystalline salt. At this point the solution is saturated and there will be no further net transfer of salts between the two.

chemical potential rises accordingly. When μ_{vapor} rises to the point where it equals μ_{liquid}, diffusion in both directions is the same and so the net evaporation stops. For molecules leaving the liquid phase, the decrease in particle concentration is no longer enough to offset the increase in potential energy. The air has become "saturated," and the two phases are in "diffusive equilibrium."

A similar thing happens as a crystalline salt dissolves in water, as illustrated in Figure 5.5. Other examples are the electronic devices that rely on the diffusion of electrons across p-n junctions into regions of lower concentration and higher potential energy. If there were no diffusion the devices would not work.

We do not see this same behavior in the macroscopic world, where thermal motions are minuscule. For large objects, potential energies rule and thermal motions are irrelevant. But things are different in the microscopic world, owing to the random thermal motion of the atoms and molecules. This motion is the reason for diffusion – why gases expand to fill their containers, why not all water molecules are in the ocean, etc. It has a firm statistical basis, which we will quantify in later chapters.

C.3 Particle transfer and changes in temperature

When particles fall into deeper potential wells, the potential energy lost is converted into increased thermal energy and the temperature rises (Figure 5.6). This happens when water condenses or a fire burns. Conversely, when particles diffuse into regions of higher potential energy, the thermal energy decreases and the temperature falls.

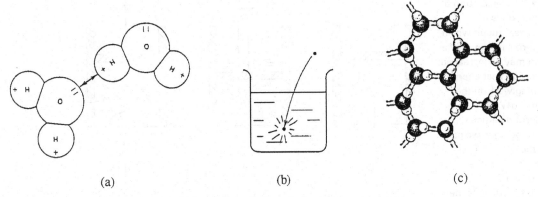

Figure 5.6 (a) Water molecules, which are electrically polarized, attract each other strongly. (b) So a water molecule falls into a potential well and releases thermal energy as it joins the liquid state. (c) In ice, reduced thermal motion allows the water molecules to maintain arrangements that reduce their potential energy still further by keeping like charges close, so additional thermal energy is released as water freezes.

Summary of Section C

A system's chemical potential μ measures the average change in internal energy per entering particle that is *not* due to the transfer of heat or work. For ΔN entering particles (equation 5.4),

$$\Delta E = \mu \Delta N \qquad \text{(diffusive interaction only — no work or heat transfer)}.$$

There are two sources for this energy: changes in potential energy due to particle interactions, and changes in particle concentrations (equation 5.5):

μ depends on: 1. depth of potential well and 2. particle concentration.

We will quantify these two aspects in a later chapter. Particles seek regions of lower chemical potential since interparticle forces favor configurations of lower potential energy and thermal motion tends to carry particles towards lower concentrations. Consequently, the chemical potential governs diffusive interactions in the same way that temperature and pressure govern thermal and mechanical interactions:

- thermal interaction – heat flows towards lower temperature;
- mechanical interaction – boundaries move toward lower pressure;
- diffusive interaction – particles move toward lower chemical potential.

The configuration of lowest chemical potential may involve a compromise between potential energy and particle concentration. A drop in one may be sufficient to offset a gain in the other. When particles move into regions of different potential energy, thermal energy is absorbed or released and temperatures change accordingly.

Figure 5.7 Illustration of the three kinds of processes through which the internal energy E of a system may be increased: transferring heat energy dQ, doing work on the system $-dW$, or transferring particles $\mu\,dN$. Altogether, we can write the change in internal energy of a system as $dE = dQ - dW + \mu\,dN$, which is the first law of thermodynamics.

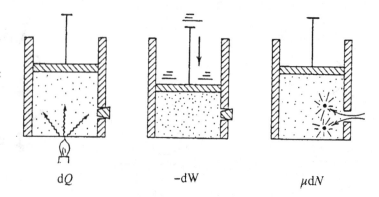

dQ $-dW$ μdN

D   The first law of thermodynamics

In the preceding sections we examined each of the three ways by which the internal energy of a system may be changed (Figure 5.7):

- by transferring heat in or out of the system;
- by having work done on or by the system;
- by adding or removing particles from the system.

These are expressed in results 5.2 and 5.4, which together constitute the first law of thermodynamics:

First law of thermodynamics

The change in internal energy of a system is given by

$$dE = dQ - dW + \mu\,dN. \tag{5.6}$$

The dW term is preceded by a negative sign, because dW is the work done *by* the system. If more than one kind of work is being done (e.g., equation 5.3) then dW must be replaced by a sum over different kinds of work dW_j. Similarly, if there are several kinds of particles in the system then the last term becomes $\sum_i \mu_i dN_i$, where the sum is over the different types of particle:

$$dE = dQ - \sum_j dW_j + \sum_i \mu_i dN_i. \tag{5.7}$$

E   Exact and inexact differentials

The differentials appearing in the first law, (5.6), are of two types, "exact" and "inexact." An exact differential is the differential of a well-defined function, but an inexact differential is not. This difference has several implications, which we now explore.

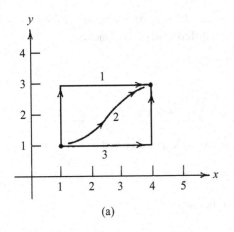

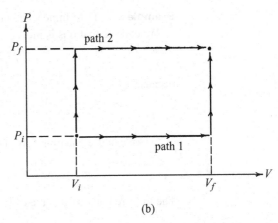

(a)   (b)

Figure 5.8 (a) When integrating a differential between initial and final points, the result is independent of the path (paths 1, 2, and 3 are shown) if the differential is exact, but not if it isn't. (b) Two of the infinite number of possible paths for a system to follow between given initial and final values of the pressure and volume.

E.1 The general case

Consider the change in value of a given function, $F = F(x, y)$, as we move along a path from an initial point (x_i, y_i) to a final point (x_f, y_f). The total change ΔF is the integral of the exact differential dF:

$$\Delta F = \int_i^f dF = F(x_f, y_f) - F(x_i, y_i).$$

It is uniquely determined by the two endpoints, (x_f, y_f) and (x_i, y_i), and does not depend on the route taken (Figure 5.8). Examples would include the exact differentials $\Delta E, \Delta N, \Delta V, \Delta p$. By contrast, the integral of an *inexact* differential does depend on the path taken.

The differential of a function is given by

$$dF = \frac{\partial F}{\partial x}\, dx + \frac{\partial F}{\partial y}\, dy \tag{5.8}$$

Therefore, one way to determine whether a differential

$$d\Phi = g(x, y)dx + h(x, y)dy \tag{5.9}$$

is exact is to see whether we can find some function $F(x, y)$ such that

$$\frac{\partial F}{\partial x} = g(x, y) \quad \text{and} \quad \frac{\partial F}{\partial y} = h(x, y).$$

If we can, the differential is exact, and if we can't, it is inexact.

Alternatively, we can use the identity

$$\frac{\partial^2 F}{\partial y \partial x} = \frac{\partial^2 F}{\partial x \partial y}$$

Combining this with equations 5.8 and 5.9, we can see that for exact differentials

$$\frac{\partial g}{\partial y} = \frac{\partial h}{\partial x}. \tag{5.10}$$

Example 5.1   Determine whether $d\Phi = 2xy\,dx + x^2\,dy$ is an exact differential. We can see that this is indeed an exact differential of the function

$$F = x^2 y + \text{constant},$$

because

$$\frac{\partial F}{\partial x} = 2xy \qquad \text{and} \qquad \frac{\partial F}{\partial y} = x^2.$$

Or we can use equation 5.10. For this example $g = 2xy$ and $h = x^2$, so

$$\frac{\partial g}{\partial y} = 2x \qquad \text{and} \qquad \frac{\partial h}{\partial x} = 2x.$$

The two are the same, so the differential is exact.

E.2   Applications to physical systems

In real physical systems, examples of exact differentials would include changes in particle number, in internal, energy, or in volume (ΔN, ΔE, ΔV):

$$\Delta N = N_f - N_i, \qquad \Delta E = E_f - E_i, \qquad \Delta V = V_f - V_i.$$

These are all properties of the system that can be measured at any time. Changes can be determined from initial and final values alone without knowing what happened in between. This is not true, however, for heat or work. Although we can determine changes in internal energy during a process from its initial and final values alone, we would not know how much of this change was caused by heat entering the system and how much by work done on the system, unless we knew the particular path followed by the process.

For example, suppose that we measure the internal energy of an iron bar and then leave the room while a friend altered its energy. When we return, it might be hotter, and from the increased temperature we can determine the increase in internal energy, ΔE. But we will not be able to tell whether that change was made by adding heat to the bar, or by doing work on it, such as by hitting it with a hammer. That is, we would not know ΔQ or ΔW.

If we find the bar squeezed in a clamp, we might suspect that work had been done on it (i.e., the clamp squeezed it). However, it could be that the bar was slipped into the clamp when cooled and contracted, then was reheated, expanding and becoming stuck against the clamp, without the clamp having moved at all. So our guess would be wrong.

In summary, changes in internal energy ΔE, volume ΔV, and number of particles ΔN during a process can be determined from initial and final values alone, independently of the particular details of the process (i.e., the path taken). This makes dE, dV, dN exact differentials. By contrast, the heat added, ΔQ, or work done, ΔW, do depend on the details of how the process is carried out. So dQ and dW are inexact differentials.

Notice that we can write the inexact differential dW in terms of an exact differential dV: $dW = pdV$. In Chapter 8 we will find a way of doing the same for dQ.

F Dependent and independent variables

In thermodynamics we often have many different system variables, such as internal energy, temperature, pressure, volume, chemical potential, number of particles, entropy, and others, of which only two or three are independent. This could cause much confusion, particularly when the partial derivatives are involved.

For example, consider some function F that is a function of many different variables $q, r, s, t, u, v, w, x, y, \ldots$, only two of which are independent. To find the partial derivative with respect to one independent variable, for example $\partial F / \partial u$, we must hold the other independent variable constant. But how will we indicate which of the various possible "other" variables is the one held constant?

There are two customary ways of doing this. The notations

$$\frac{\partial F(u, x)}{\partial u} \quad \text{or} \quad \left(\frac{\partial F}{\partial u} \right)_x$$

are the normal ways of indicating that there are only two independent variables involved and that x is the one held constant while the partial derivative with respect to u is taken. By extension, the notations

$$\frac{\partial F(w, x, y)}{\partial w} \quad \text{or} \quad \left(\frac{\partial F}{\partial w} \right)_{x,y}$$

indicate there are three independent variables and that x and y are held constant while the partial derivative with respect to w is taken. To take the desired partial derivative, the function must first be written entirely in terms of the chosen set of independent variables.

Example 5.2 Consider a function of three variables, $F(x, y, z)$, only two of which are independent. The three are interrelated by $y = x^2 z$. Given that

$$F(x, y, z) = x + xyz$$

find $(\partial F / \partial x)_y$.

First, we must write the function in terms of the chosen independent variables, x and y. Writing z in terms of x and y gives

$$z = \frac{y}{x^2}, \quad \text{so} \quad F(x, y) = x + xy \left(\frac{y}{x^2} \right) = x + \frac{y^2}{x}.$$

Now that F is written in terms of x and y we can take the appropriate partial derivative:

$$\left(\frac{\partial F}{\partial x} \right)_y = 1 - \frac{y^2}{x^2}.$$

Summary of Sections D–F

The change in internal energy for a system undergoing thermal, mechanical, and/or diffusive interactions is given by the first law of thermodynamics (equation 5.6):

$$dE = dQ - dW + \mu dN \quad \text{(first law)}.$$

A differential $d\Phi = g(x, y)dx + h(x, y)dy$ is exact if it meets any of these criteria:

- the integral from (x_i, y_i) to (x_f, y_f) is independent of the path taken for all values of initial and final points;
- there exists some function $F(x, y)$ such that $\partial F/\partial x = g$ and $\partial F/\partial y = h$;
- $\partial g/\partial y = \partial h/\partial x$.

In the first law, the change in internal energy dE is an exact differential, as is the change in volume dV and the change in the number of particles dN. However, the heat added dQ and the work done by the system dW are *not* exact differentials. One consequence of this is that when the internal energy of a system is changed by a finite measured amount ΔE there is no way of knowing how much of that energy entered as heat ΔQ and how much as work done on the system, $-\Delta W$, without knowing the particular thermodynamic path followed in going from the initial to the final state.

For partial derivatives, we must indicate which variables are being held constant, and we must write all other variables in terms of the chosen independent variables before the derivative can be taken.

Problems

Sections A and B

1. Consider a small hot rock at 390 K inside a building with cold air and cold walls at 273 K. Air is a very poor conductor of heat, so the bulk of the energy transfer is radiative.

   (a) Are the molecules of the walls and air sending out electromagnetic waves?

   (b) Since the cooler system is much larger, doesn't it radiate much more energy altogether than the rock?

   (c) Considering your answer to part (b), why is there a net flow of energy from the rock to the air, rather than vice versa?

2. Consider a system that is *not* under pressure but whose volume decreases as a container wall moves inward.

   (a) Is any work done on the system? Why or why not?

   (b) Explain from a microscopic point of view why the internal energy is not increased in this case. (Hint: If it is not under pressure, are there any molecules colliding with the container walls?)

3. Consider a system under tension, such as a stretched rubber band, or a stretched steel bar. Now suppose that you let it contract somewhat.
   (a) Does its internal energy increase or decrease?
   (b) If its temperature increases, how might you explain this? (Hint: potential wells, such as those in Figure 4.4, might be useful.)

4. Calculate the work done by a gram of water when it vaporizes at atmospheric pressure. Use the fact that a mole of water vapor at 100 °C and atmospheric pressure occupies a volume of 30.6 liters. How does this work compare with the latent heat of vaporization, which is 2260 joules per gram?

5. When the gasoline explodes in an automobile cylinder, the temperature is about 2000 K, the pressure is about 8×10^5 Pa, and the volume is about $100 \, cm^3$. The piston has cross sectional area $80 \, cm^2$. The gas then expands adiabatically (i.e., no heat leaves or enters the cylinder during the process) as the piston is pushed downward, until its volume increases by a factor of 10. For adiabatic expansion of the gas, $pV^\gamma = $ constant, where $\gamma = 1.4$.
   (a) How much work is done by this gas as it pushes the piston downward?
   (b) Assuming it behaves as an ideal gas ($pV/T = $ constant), what is the final temperature of the gas?

6. A certain insulating material has 5×10^{22} atoms, each having six degrees of freedom. It initially occupies a volume of $10^{-6} \, m^3$ at a pressure of 10^5 Pa. The pressure and volume are related by $p(V - V_0) = $ constant, where $V_0 = 0.94 \times 10^{-6} \, m^3$.
   (a) If the pressure on the system is increased ten fold, how much work is done by the system?
   (b) Suppose that the potential energy per particle u_0 remains constant, and that the pressure increases sufficiently quickly that no heat enters or leaves the system during the process, i.e., the process is adiabatic. By how much does the temperature of the insulator rise?

Section C

7. Is the potential energy reference level u_0 for an H_2SO_4 (sulfuric acid) molecule entering fresh water positive or negative? (The temperature rises.) For the molecules entering the solution, is the change in chemical potential positive or negative?

8. The boiling point of water is considerably higher than the boiling point for other liquids composed of light molecules such as NH_3 or CH_4. It is even much higher than the boiling point of molecules nearly two or three times as massive, such as N_2, O_2, or CO_2. Why do you suppose this is?

9. We are going to examine mechanisms for the cold packs that are used for athletic injuries, where two chemicals are mixed and the resulting temperature of the pack drops remarkably.

(a) Suppose that when you mix two chemicals, some of the molecules dissociate, giving the system more degrees of freedom without changing the total energy or the system's overall potential energy reference level, Nu_0. What would happen to the temperature, and why?

(b) What would happen to the temperature if the number of degrees of freedom didn't change, but the overall potential energy reference level, Nu_0, rose as a result of the mixing?

10. Initially, system B has 2×10^{25} particles, each having $u_{0,B} = -0.35\,\text{eV}$ and five degrees of freedom. System A has 10^{26} particles each having $u_{0,A} = -0.40\,\text{eV}$ and three degrees of freedom. After these two systems are combined, the situation for particles of system A doesn't change but the particles of system B have $u_{0,B'} = -0.25\,\text{eV}$ and three degrees of freedom each. If the two systems are both at temperature 290 K before they are mixed, what will be the temperature of the combined system after mixing?

11. A certain material vaporizes from the liquid phase at 700 K. In both phases, the molecules have three degrees of freedom. If u_0 in the liquid phase is $-0.12\,\text{eV}$, what is the latent heat of vaporization in joules per mole?

12. Repeat the above problem for the case where the molecules have three degrees of freedom in the liquid phase but five in the gas phase.

13. Consider a solid that sublimes (goes from solid to gas) at 300 K. In the solid phase, the molecules have six degrees of freedom and $u_0 = -0.15\,\text{eV}$, and in the gas phase, they have three degrees of freedom (and $u_0 = 0$, of course). What is the latent heat of sublimation in joules per mole?

14. Consider a system of three-dimensional harmonic oscillators, for which the energy of each is given by

$$\varepsilon = u_0 + (1/2)\kappa(x^2 + y^2 + z^2) + (1/2m)\left(p_x^2 + p_y^2 + p_z^2\right).$$

Suppose that their mutual interactions change in such a way that u_0 drops by 0.012 eV, without any energy entering or leaving the system as a whole. For any one oscillator, find the average change in (a) thermal energy, (b) kinetic energy, (c) Potential energy.

15. Estimate the temperature inside a thermonuclear explosion in which deuterium nuclei fuse in pairs to form helium nuclei. Assume that each nucleus has three degrees of freedom and that 10% of the deuterium fuses into helium. Assume that the average drop in u_0 upon fusing is 1 MeV per nucleon.

16. Consider the burning of carbon in oxygen. Estimate the value of the change in potential energy reference levels for the molecular ingredients, $u_0(CO_2) - [u_0(C) + u_0(O_2)]$, assuming that the carbon and oxygen start out at room temperature (295 K) and that the temperature of the flame is 3000 K. Assume that the carbon (graphite) atoms initially have six degrees of freedom, and

that both oxygen and carbon dioxide molecules have five degrees of freedom apiece.

17. The electrical polarization of the water molecule makes water an exceptional material in many ways. The interactions between neighboring molecules are extremely strong, yielding very deep potential wells. Furthermore, these interactions (and hence u_0) change noticeably with temperature. To demonstrate this, calculate the number of degrees of freedom per molecule that would be needed to give the observed specific heat, 4186 J/(kg K) if u_0 does not change.

18. Suppose that 0.15 moles of some acid are added to a liter of water; both are initially at room temperature. The addition raises the temperature by 0.1 K. How much deeper is the potential well of the acid molecules when in water than when in the concentrated acid? (The heat capacity of a liter of water is $4186 \, J/K$.)

19. From a molecular point of view, why is u_0 for:
    (a) liquid water less (more negative) than that of water vapor?
    (b) ice less than that of liquid water?
    (c) salt in water less than that of salt in oil?

20. The latent heat of vaporization for a certain acid at room temperature is 19260 J/mole. The molecules have six degrees of freedom in both phases. The work done on it due to its change in volume during condensation is $p\Delta V \approx nRT$.
    (a) Assuming that self-interactions (and hence u_0) are negligible in the vapor state, use the above information to calculate u_0 for this acid (in eV) in the liquid state at room temperature.
    (b) When 10^{-2} mole of this acid in the liquid form is added to one liter of water, both at room temperature, the temperature of the water rises by 0.1 °C. What is the depth of the potential well for this acid in water?

Section D

21. One mole of air at 0 °C and atmospheric pressure (1.013×10^5 Pa) occupies 22.4 liters of space. Suppose we compress it by 0.2 liter (a small enough amount that we can assume the pressure remains constant.)
    (a) How much work have we done?
    (b) How many joules of thermal energy would we have to remove for the internal energy to remain unchanged?
    (c) If we did both of these, but found that the temperature decreased slightly, what would we conclude about the behavior of u_0 as molecules get closer together?

22. A magnetic moment is induced in most materials when they are placed in a magnetic field. For aluminum, this induced magnetic moment is directly

proportional to the applied field and aligned parallel to it, $\mu = c\mathbf{B}$, where the constant of proportionality c is equal to 1830 $A^2 m^3/N$ for a 1 m^3 sample. This sample of aluminum is placed in an external magnetic field, whose strength is increased from 0 T to 1.3 T; 1 T $= 1\,N/(A\,m)$.

(a) How much work is done on the sample?

(b) Since this work is done *on* the sample, the sample's internal energy increases. But the potential energies of the atomic magnets decrease as they align with the imposed external field. How can the system's internal energy increase if the sample's potential energy decreases? (Hint: Think of what the internal energy does and does not include.)

23. A gas with 10^{27} degrees of freedom is under a pressure of 150 atm. It is allowed to expand by 0.01 m^3 without any heat or particles being added or removed in the process. Calculate (a) the work done by the gas, (b) the change in temperature. (Hint: Find the change in internal energy first.)

24. As liquid water is compressed at constant temperature, its pressure increases according to the formula: $p = [1 + 2.5 \times 10^4(1 - V/V_0)]$ atm, where V_0 is its volume under atmospheric pressure, 1.013×10^5 Pa.

(a) If some water has a volume of 1 liter at atmospheric pressure, what will be its volume at the bottom of the ocean, where the pressure is 500 atm?

(b) How much work is done by a liter of water that is brought to the surface from the ocean bottom? (Hint: The pressure is not constant, so you will have to integrate $p\,dV$.)

(c) Knowing that to change the temperature of water by 1 °C requires a change in internal energy of 4186 J per kg, calculate the change in temperature of the water sample brought up from the bottom of the ocean. Assume that it is closed and insulated, so that no heat or particles enter or leave the sample as it is raised.

Section E

25. Test each of the following differentials to see whether they are exact, using two methods for each:

(a) $-y \sin x\,dx + \cos x\,dy$,

(b) $y\,dx + x\,dy$,

(c) $yx^3 e^x\,dx + x^3 e^x\,dy$,

(d) $(1 + x)ye^x\,dx + xe^x\,dy$,

(e) $4x^3 y^{-2}\,dx - 2x^4 y^{-3}\,dy$.

26. State which of the following differentials are exact:

(a) $3x^2 y^2\,dx + 2x^3 y\,dy$,

(b) $3x^2 e^y\,dx + 2x^3 e^y\,dy$,

(c) $[1 + \ln(x)] \sin y\,dx + x \ln x \cos y\,dy$,

(d) $e^{2x^2}[4xy \ln(y)]\,dx + e^{2x^2}[\ln(y) + 1]\,dy$.

27. Consider the path integral of the exact differential $dF = 2xy\,dx + x^2\,dy$. Integrate this from $(1, 1)$ to $(4, 3)$ along both the paths 1 and 3 in Figure 5.8a. Are the two results the same?

28. Consider the path integral of the differential $dG = 3xy\,dx + x^2\,dy$. Integrate this from $(1, 1)$ to $(4, 3)$ along paths 1 and 3 of Figure 5.8a. Are the two results the same? Is this differential exact?

29. The work done by a system can be written as $dW = p\,dV$, where p is the pressure and dV the change in volume. Compute the work done by a system as its pressure and volume change from the initial values p_i, V_i to the final values p_f, V_f, by evaluating the integral $\Delta W = \int p\,dV$:
    (a) along path 1 in Figure 5.8b.
    (b) Along path 2 in Figure 5.8b.
    (c) Does the amount of work done depend on the path taken?

30. Find the integral of the differential $dF = y^2\,dx + x\,dy$ from point $(2, 2)$ to point $(6, 5)$:
    (a) along a path that first goes straight from $(2, 2)$ to $(6, 2)$ and then straight from $(6, 2)$ to $(6,5)$,
    (b) along a path that first goes straight from $(2, 2)$ to $(2, 5)$ and then straight from $(2, 5)$ to $(6, 5)$,
    (c) along a path that goes along the diagonal line $y = (3/4)x + (1/2)$ from $(2, 2)$ to $(6, 5)$.

31. Repeat the above problem for the function $dG = y^2\,dx + 2xy\,dy$.

Section F

32. Suppose that $w = xy$ and $z = x^2/y$. Express:
    (a) z as a function $z(w, y)$ of w and y,
    (b) z as a function $z(w, x)$ of w and x,
    (c) w as a function $w(z, y)$ of z and y,
    (d) w as a function $w(z, x)$ of z and x,
    (e) x as a function $x(y, z)$ of y and z.

33. For the problem above, evaluate the following partial derivatives:
    (a) $(\partial z/\partial w)_y$, (b) $(\partial w/\partial y)_z$, (c) $(\partial x/\partial y)_z$, (d) $(\partial y/\partial w)_z$.

34. Consider the variables x, y, z, u, v, where

$$x(u, v) = u^2 v, \qquad y(u, v) = u^2 + 2v^2, \qquad z(x, y) = xy.$$

Any of these five variables can be expressed as a function of any other two. Find the correct expression for:
    (a) y as a function $y(x, v)$ of x and v,
    (b) z as a function $z(u, v)$ of u and v,
    (c) v as a function $v(y, u)$ of y and u,
    (d) $(\partial y/\partial x)_v$, (e) $(\partial x/\partial u)_v$, (f) $(\partial x/\partial u)_y$.

35. Given that $x = y^2z - w$ and $w = x + y^2$, what is $(\partial w/\partial y)_z$?

36. Given that $w = xe^y$ and $x = y^2z$, what is $(\partial w/\partial z)_y$?

37. Consider the function $f(q, r, s, t) = qst - e^r$, where $r = st$ and $s = q^2r$. What is $(\partial f/\partial t)_s$?

38. For a certain ideal gas, the variables E, p, V, N, T are interrelated by the following equations: $E = (3/2)NkT$, $pV = NkT$, $N = $ constant. What is $(\partial E/\partial p)_V$?

Part IV
States and the second law

Chapter 6
Internal energy and the number of accessible states

We now begin our study of the possible configurations or states for macroscopic systems. Because the volume of a quantum state in six-dimensional phase space is extremely small, a very large number of such states are available to the particles of most systems.

One mole of material contains roughly 10^{24} atoms, each of which could be in a large number of different quantum states. Consequently, the total number of possible states for any macroscopic system is huge. Numbers like $10^{10^{24}}$ are typical. They are probably much larger than any numbers you have encountered before (Table 6.1).

When energy is added to a system it gives the particles access to additional states of higher energy. Even a small increase in the number of states per particle results in a very large increase in the number of states for the system as a whole. As an example, consider a system of coins, each of which has two possible states: heads or tails. A system of N such distinguishable coins has $2 \times 2 \times 2 \times 2 \times \cdots = 2^N$ different possible heads–tails configurations. If the number of states per coin were increased from 2 to 3 (three-sided coins?), the number of possible states for the system would increase from 2^N to 3^N, an increase by a factor $(3/2)^N$. If $N = 10^{24}$, this would be a factor of $(3/2)^{10^{24}} \approx 10^{10^{23}}$. For macroscopic systems, any small increase in the number of states per particle results in a huge increase in the number of states for the system.

Table 6.1. *Some large numbers*

grains of sand in Waikiki beach	10^{18}
age of Universe in microseconds	10^{24}
water molecules in Atlantic Ocean	10^{46}
atoms in Earth	10^{50}
volume of Universe in cubic microns	10^{97}
states for molecules in a glass of water	$10^{1,000,000,000,000,000,000,000,000}$

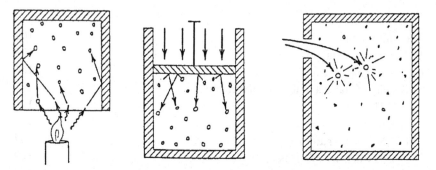

Figure 6.1 The addition of energy or particles usually occurs in a small localized region of a system, such as along a boundary. The system is not in equilibrium until interactions among particles have provided the opportunity for the added energy and particles to be anywhere, so that all possible distributions are equally likely. The time required for this to happen is the relaxation time. From that point on, the probabilities for the various possible configurations remain the same.

A Equilibrium

An isolated system is said to be in equilibrium when the probabilities for the various possible configurations of its elements do not vary with time. For example, suppose that we release ammonia molecules at the front of a room. Initially, the states near the point of release have relatively high probabilities of containing these molecules, whereas those at the rear of the room have no chance at all. But the molecules migrate and the probabilities change until the molecules are equally likely to be found anywhere in the room. At this point the probabilities stop changing and the system is in equilibrium.

Similarly, energy transfer through heating or compression (Figure 6.1) may initially affect only particles in one local region. In time, however, interactions among the particles distribute this energy throughout the system until all possible distributions are equally probable. From this point on, the system is in equilibrium.

The characteristic time needed for a perturbed system to regain equilibrium is called the "relaxation time." When applying statistical tools to processes involving macroscopic systems, it is helpful if the systems are near equilibrium. This requires that the transfer of heat, work, or particles must proceed at a rate that is slow compared with the relaxation time for that particular process. When this happens, we say the process is "quasistatic."

Summary of Section A

The number of quantum states that are accessible to physical systems is usually extremely large, but finite. When the number of states accessible to the individual elements is increased even very slightly, the number of states accessible to the system as a whole increases immensely.

An isolated system is said to be in equilibrium when the probabilities for it to be in the various accessible states do not vary with time. If a system is perturbed, the characteristic time required for it to come into equilibrium is called the relaxation time for that process. When interactions between systems proceed slowly enough that the systems are always near equilibrium, the process is called quasistatic.

B The fundamental postulate

The tools for the statistical analysis of the equilibrium behaviors of large systems are based on one single, very important fundamental postulate.

Fundamental postulate

An isolated system in equilibrium is equally likely to be in any of its accessible states, each of which is defined by a particular configuration of the system's elements.

This postulate seems quite reasonable, but this in itself does not justify its adoption. Rather, we must validate it by comparing the results of experiments with predictions based on the postulate. This has been done for a huge number of systems and processes, and we find that the predictions are correct every time.

If the number of states accessible to the entire system is given by Ω, and all are equally probable, then the probability for the system to be in any one of them must be (Figure 6.2)

$$P_{\text{any one state}} = 1/\Omega, \tag{6.1}$$

and if a subset has Ω_i states then the probability for the system to be in this subset is

$$P_{\text{subset } i} = \frac{\Omega_i}{\Omega}. \tag{6.2}$$

Example 6.1 Consider the orientations of three unconstrained and distinguishable spin-1/2 particles. What is the probability that two are spin up and one spin down at any instant?

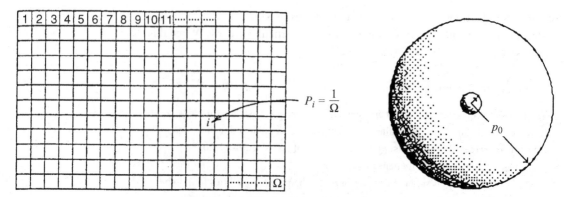

Figure 6.2 (Left) A representation of the Ω states that are available to a system. If all are equally probable then the probability that the system is in any one state is $P = 1/\Omega$, and the probability that it is in a subset containing Ω_i of these states is $P_i = \Omega_i/\Omega$. (Right) If the magnitude of a particle's momentum is constrained to be less than p_0, what is the probability that its momentum is less than $p_0/10$?

Of the eight possible spin configurations for the system,

$$\uparrow\uparrow\uparrow \quad \uparrow\uparrow\downarrow \quad \uparrow\downarrow\uparrow \quad \downarrow\uparrow\uparrow \quad \uparrow\downarrow\downarrow \quad \downarrow\uparrow\downarrow \quad \downarrow\downarrow\uparrow \quad \downarrow\downarrow\downarrow$$

the second, third, and fourth comprise the subset "two up and one down". Therefore, the probability for this particular configuration is

$$P_{2\text{ up and 1 down}} = \tfrac{3}{8}.$$

Example 6.2 Consider electrons in a plasma whose momenta are constrained to have magnitudes less than some maximum value p_0. What is the probability that the momentum of any particular electron is less than one tenth this maximum value (Figure 6.2)?

The number of accessible states is proportional to p^3 (equation 1.4):

$$\Omega = \frac{V_r V_p}{h^3} = \frac{V_r(4\pi p^3/3)}{h^3} \propto p^3.$$

Therefore, the probability that the electron has momentum less than $p_0/10$ is

$$P_i = \frac{\Omega_i}{\Omega} = \frac{p^3}{p_0^3} = \left(\frac{p_0/10}{p_0}\right)^3 = 10^{-3}.$$

C   The spacing of states

It is sometimes convenient to identify the state of a small system by its energy. If several different states have the same energy, we say that the energy level is "degenerate."

For large systems, the degeneracy of each energy level is huge. In the homework you will be able to show that there are about $10^{10^{27}}$ different ways of arranging the air molecules (if you could distinguish them) in your room so that half of them are in the front and half in the back. All these different arrangements have exactly the same energy (i.e., they are degenerate), and clearly their number is very small in comparison to the number of degenerate arrangements among quantum states (as opposed to arrangements between the front and back of the room).

Not only is the degeneracy of any one level huge, but neighboring energy levels are very tightly packed as well. The spacing between the levels for the air in a room is about 10^{-24} eV, or about 10^{-50} of the total internal energy of the air. This is far too small to be measurable.

When dealing with large systems, then, we cannot control the particular quantum energy level for the system as a whole. The best we can do is to hold the system to within some small energy range ΔE, which is very large in comparison with the spacing of the levels. For the number of states in the range between E and $E + \Delta E$, we write

$$\Omega(E, \Delta E) = g(E)\Delta E \qquad (6.3)$$

where the "density of states," $g(E)$, is the number of states per unit energy (subsection 1B.5).

D   Density of states and the internal energy

We shall now show that for macroscopic systems, the number of accessible states is extremely sensitive to the system's thermal energy, volume, and number of particles.

D.1   The model

The success of the statistical approach depends only upon the feature that the number of accessible states is extremely sensitive to the internal energy. This overall feature can be derived from any reasonable model, so we illustrate it with a familiar one, in which the energy in each degree of freedom has the form $b\xi^2$, where b is a constant and ξ is a position or momentum coordinate (e.g., $p_x^2/2m$, $kx^2/2$, etc.)

D.2   Identical particles

As we saw in Chapter 2, the number of states for a system is the product of the number of states for the individual particles. If the system has N distinguishable particles and each particle has ω states available to it, then the number of states for the total system is given by

$$\Omega = \omega \times \omega \times \omega \times \cdots = \omega^N \qquad (6.4)$$

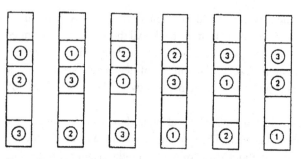

Figure 6.3 Shown here are six of the 60 different ways in which three distinguishable particles can be arranged among five states, one per state. These six arrangements are the same if the particles are identical. So for three *identical* particles the number of different arrangements is reduced by a factor of $3! = 6$ compared with the number of arrangements for distinguishable particles.

Figure 6.4 There are $N!$ ways of permuting N distinguishable particles among N states so that there is one per state, as shown here for $N = 1, 2, 3,$ and 4. In each case, all $N!$ permutations are identical if the particles are identical.

$1! = 1$ $\boxed{1}$

$2! = 2$ $\boxed{1\,2} + \boxed{2\,1}$

$3! = 6$ $\boxed{1\,2\,3} + \boxed{1\,3\,2} + \boxed{2\,1\,3} + \boxed{2\,3\,1} + \boxed{3\,1\,2} + \boxed{3\,2\,1}$

$4! = 24$ $\boxed{1\,2\,3\,4} + \boxed{1\,2\,4\,3} + \boxed{1\,3\,2\,4} + \boxed{1\,3\,4\,2} + \boxed{1\,4\,2\,3}$

$+ \boxed{1\,4\,3\,2} + \boxed{2\,1\,3\,4} + \boxed{2\,1\,4\,3} + \boxed{2\,3\,1\,4} + \boxed{2\,3\,4\,1}$

$+ \boxed{2\,4\,1\,3} + \boxed{2\,4\,3\,1} + \boxed{3\,1\,2\,4} + \boxed{3\,1\,4\,2} + \boxed{3\,2\,1\,4}$

$+ \boxed{3\,2\,4\,1} + \boxed{3\,4\,1\,2} + \boxed{3\,4\,2\,1} + \boxed{4\,1\,2\,3} + \boxed{4\,1\,3\,2}$

$+ \boxed{4\,2\,1\,3} + \boxed{4\,2\,3\,1} + \boxed{4\,3\,1\,2} + \boxed{4\,3\,2\,1}$

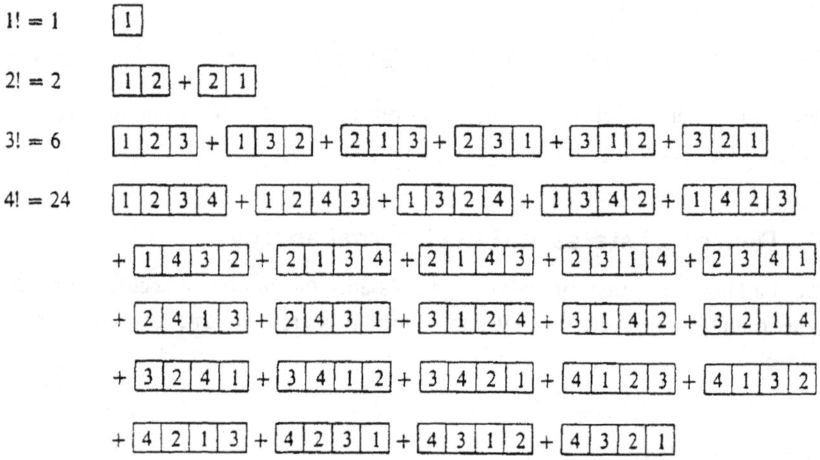

If the particles are indistinguishable, however, the number of states is reduced by $1/N!$

$$\Omega = \frac{\omega^N}{N!} \tag{6.5}$$

The reason for this is illustrated in Figures 6.3 and 6.4. There are $N!$ ways of arranging N particles among N boxes. If the particles are identical, these arrangements are all the same. So the number of different quantum states for identical particles is reduced by a factor of $N!$ relative to that for distinguishable particles.[1]

---

[1] As we will see in Chapter 19, the $1/N!$ factor is only correct if there are many more states than particles, so that there is little likelihood that two or more particles will occupy the same state simultaneously.

If we use Stirling's approximation for $N!$ (equation 2.5),[2]

$$N! \approx \sqrt{2\pi N} \left(\frac{N}{e}\right)^N \approx \left(\frac{N}{e}\right)^N, \qquad (6.6)$$

we can write result 6.5 as

$$\Omega = \frac{\omega^N}{N!} \approx \left(\frac{e\omega}{N}\right)^N. \qquad (6.5')$$

With this, equations 6.4 and 6.5' can be written jointly as

$$\Omega = \omega_c^N \qquad (6.7)$$

where ω_c is the number of states per particle, "corrected" for the case of identical particles:

$$\omega_c = \omega \qquad \text{(distinguishable particles)}$$

$$\omega_c = \frac{e\omega}{N} \qquad \text{(corrected for identical particles)} \qquad (6.8)$$

Equations 6.7 and 6.8 reduce the problem of finding the number of states Ω for a large system of N particles to that of finding the number of states ω for one single particle.

D.3 States per particle

The number of states accessible to a system is calculated with full mathematical rigor in Appendix C. But we can obtain the same answer using simple physical reasoning, as we now show.

For any one particle, the number of accessible states equals the total accessible phase-space volume $V_r V_p$ divided by the volume of one state, h^3 (Equations 1.4 and 1.5). For a particle that moves in three dimensions,

$$\omega = \int \frac{dx\,dy\,dz\,dp_x\,dp_y\,dp_z}{h^3}. \qquad (6.9)$$

The evaluation of this integral is constrained by the fact that the total thermal energy of the entire system is fixed, which means that the permitted values of the various coordinates are interdependent. Although it is theoretically possible that the energy of the entire system is held in just one degree of freedom (i.e., one position or momentum coordinate), this is very unlikely as it would force all the other degrees of freedom into their one zero-energy state. A more even distribution of energies is much more likely (just as it is much more likely for large systems of flipped coins to have nearly fifty–fifty heads–tails distributions). That is, most accessible states correspond to a more even distribution of energies

---

[2] The last step here requires us to show that $\sqrt[2N]{2\pi N} = 1$ in the limit of large N or, equivalently, that $\sqrt{2\pi N}$ is negligible when compared to $(N/e)^N$. This can be done by applying L'Hospital's rule to the logarithm of $\sqrt[2N]{2\pi N}$ to show that the limit of the logarithm is zero.

among the various degrees of freedom. Since the average thermal energy for each of the $N\nu$ degrees of freedom is given by

$$\overline{b\xi^2} = \left(\frac{E_{\text{therm}}}{N\nu}\right) \qquad \Rightarrow \qquad \xi_{\text{rms}} = \left(\frac{E_{\text{therm}}}{bN\nu}\right)^{1/2}$$

we would guess that the integral over each coordinate variable ξ would be proportional to this weighted average:

$$\int d\xi \propto \left(\frac{E_{\text{therm}}}{N\nu}\right)^{1/2}.$$

(We include the constant b in the constant of proportionality.)

For atoms in a solid, the potential energy degrees of freedom involve the position coordinate terms

$$\frac{\kappa}{2}x^2 + \frac{\kappa}{2}y^2 + \frac{\kappa}{2}z^2,$$

so the integral over each of the variables x, y, z gives the above result. But for a gas the molecular positions are unconstrained, so the integral over x, y, z simply gives the volume of the container:

$$\int dx\,dy\,dz = V \qquad \text{(for a gas)}.$$

For any rotational degrees of freedom, the rotational kinetic energy about any particular axis also has the above form ($b\xi^2 \to L^2/2I$), so we would expect the same result

$$\int d\theta\,dL \propto 2\pi \left(\frac{E_{\text{therm}}}{N\nu}\right)^{1/2}$$

(The angular orientation is unconstrained, so the integral over $d\theta$ gives 2π, which we can include in the constant of proportionality.)

All this should tell us to expect that equation 6.9 for the number of states accessible to an individual particle will be given by

$$\omega = \int \frac{dx\,dy\,dz\,dp_x\,dp_y\,dp_z\cdots}{h^3}$$

$$= \begin{cases} CV\left(\dfrac{E_{\text{therm}}}{N\nu}\right)^{\nu/2} & \text{for a gas,} \\[2ex] C\left(\dfrac{E_{\text{therm}}}{N\nu}\right)^{\nu/2} & \text{for a solid,} \end{cases}$$

where C is a constant of proportionality and where the ν degrees of freedom per particle might sometimes include rotations and vibrations. If we correct in the case of a gas for identical particles (equation 6.8), noting that the atoms in a solid

Table 6.2. *Number of accessible states for ideal gases corrected for identical particles, and for solids*

monatomic ideal gas ($v = 3$)	$\omega_c = C\left(\dfrac{V}{N}\right)\left(\dfrac{E_{\text{therm}}}{N}\right)^{3/2}$	$C = \dfrac{em^{3/2}}{h^3}\left(\dfrac{4\pi e}{3}\right)^{3/2}$
diatomic ideal gas ($v = 5$)	$\omega_c = C\left(\dfrac{V}{N}\right)\left(\dfrac{E_{\text{therm}}}{N}\right)^{5/2}$	$C = \dfrac{em^{3/2}4\pi^2 I}{h^5}\left(\dfrac{4\pi e}{5}\right)^{5/2}$
solid ($v = 6$)	$\omega_c = C\left(\dfrac{E_{\text{therm}}}{N}\right)^{3}$	$C = \left(\dfrac{m}{\kappa}\right)^{3/2}\left(\dfrac{2\pi e}{3h}\right)^{3}$

For the entire system of N particles, $\Omega = \omega_c^N$.

To find the dependence on temperature, make the replacement $E_{\text{therm}} = \dfrac{v}{2}kT$.

are distinguishable by their positions in the solid,[3] we get the following result for the corrected number of states per particle:

$$\omega_c = \begin{cases} C\left(\dfrac{V}{N}\right)\left(\dfrac{E_{\text{therm}}}{Nv}\right)^{v/2} & \text{for a gas,} \\[4mm] C\left(\dfrac{E_{\text{therm}}}{Nv}\right)^{v/2} & \text{for a solid,} \end{cases} \tag{6.10}$$

where we have included the factor e in the constant of proportionality C.

It is comforting to know that these intuitive results are the same as those of the mathematically rigorous treatment in Appendix C (See equation C.5). The only advantage of the rigorous treatment is that it gives us explicit expressions for the constants of proportionality C. We summarize these results for the particular cases of monatomic gases, diatomic gases, and the atoms in solids in Table 6.2.

The important result here is that the number of states accessible to a system is extremely sensitive to the system's energy and number of particles (and also to the volume for a gas). Each of these variables is raised to an extremely large exponent, which means that small changes in these variables cause huge changes in the number of accessible states.

D.4 The density of states

We now use the results of Table 6.2 to calculate the density of states, which is the number of states per unit energy interval (cf. equation 6.3):

$$g(E) = \frac{d\Omega}{dE}, \qquad \text{where} \quad \Omega = \omega_c^N \propto \left(\frac{E_{\text{therm}}}{N}\right)^{Nv/2}.$$

---

[3] If the atoms in a solid could migrate to other sites or exchange places, then our integral over the position coordinates, $\int dx\,dy\,dz$, would increase by $N!$ times, because the first atom could be at any of the N lattice sites, the second at any of the $N - 1$ remaining sites, etc. But this factor $N!$ would just cancel the identical-particle correction factor of $1/N!$, and so the answer would be the same.

In taking the derivative, we use E, V, N as the independent variables. We assume that the potential energy reference level u_0 remains unchanged (so that $\partial/\partial E = \partial/\partial E_{\text{therm}}$) and, upon differentiating, we write the exponent as $Nv/2 - 1 \approx Nv/2$. Combining all the constants together, the results are

$$g(E)_{\text{gas}} \approx C\left(\frac{V}{N}\right)^N \left(\frac{E_{\text{therm}}}{N}\right)^{Nv/2}, \qquad g(E)_{\text{solid}} \approx C\left(\frac{E_{\text{therm}}}{N}\right)^{Nv/2}. \qquad (6.11)$$

Example 6.3 Consider a room full of air which has typically 10^{28} molecules and 5×10^{28} degrees of freedom. (Each molecule has three translational and two rotational degrees of freedom.) If room temperature is increased by 1 K, by what factor does the number of states accessible to the system increase?

Room temperature is around 295 K, so an increase from 295 K to 296 K is an increase of about 0.34% in temperature – and hence in thermal energy. Consequently, according to Table 6.2 the number of accessible states would increase by a factor[4]

$$\frac{\Omega_2}{\Omega_1} = \left(\frac{E_2}{E_1}\right)^{Nv/2} = (1.0034)^{2.5 \times 10^{28}} = 10^{3.7 \times 10^{25}}.$$

So an increase of just 1 K in room temperature causes a phenomenal increase in the number of accessible states for the system.

D.5 Liquids

We now look at liquids. In contrast with gases, the volume available to one particle in a liquid is limited by the volume occupied by others. Also, the variation of u_0 with temperature make liquids act as if they have additional degrees of freedom, which makes the number of states even more sensitive to the thermal energy or temperature. Although these things make it difficult to model liquids accurately, they have no effect on the most powerful tools of thermodynamics, which are independent of the models and depend only on the fact that the number of accessible states is extremely sensitive to the system's internal energy.

Summary of Sections B–D

The fundamental postulate asserts that for a system in equilibrium, all accessible states are equally probable. Therefore, the probabilities that an isolated system is in any one state, or in any subset Ω_i of all accessible states, are (equations 6.1, 6.2)

$$P_{\text{any one state}} = \frac{1}{\Omega}, \qquad P_{\text{subset } i} = \frac{\Omega_i}{\Omega}.$$

---

[4] To handle large numbers, use logarithms. It is also convenient to note that $e^x = 10^{0.4343x}$ and $\ln(1 + \varepsilon) \approx \varepsilon$ for small ε.

For small systems, either energy levels are nondegenerate or the degeneracy is small. For large systems, however, energy levels have huge degeneracies, and the spacings between neighboring levels are nearly infinitesimal in comparison with the total internal energy. For large systems, we write the number of accessible states in the range between E and $E + \Delta E$ in terms of the density of states $g(E)$ and the size of the energy interval ΔE (equation 6.3):

$$\Omega(E, \Delta E) = g(E)\Delta E.$$

The number of states for the entire system is the product of the numbers of states ω for the N individual particles ($\Omega = \omega^N$). But if the particles are identical then the number of distinct states available to the system is reduced by a factor $1/N!$, which gives (equation 6.7)

$$\Omega = \omega_c^N,$$

with (equation 6.8)

$$\omega_c = \omega \qquad \text{(distinguishable particles)},$$
$$\omega_c = \frac{e\omega}{N} \qquad \text{(identical particles)}.$$

The number of states accessible to any one particle is the total accessible volume in phase space divided by the volume of a single quantum state. If the particles are moving in three dimensions then (equation 6.9)

$$\omega = \int \frac{dx\,dy\,dz\,dp_x\,dp_y\,dp_z}{h^3}.$$

Upon evaluating these integrals, we find that the number of states accessible to each particle ω_c and the number of states accessible to the entire system Ω are as listed in Table 6.2, and the density of states for these systems is (equation 6.11)

$$g(E)_{\text{gas}} \approx C \left(\frac{V}{N}\right)^N \left(\frac{E_{\text{therm}}}{N}\right)^{N\nu/2}, \qquad g(E)_{\text{solid}} \approx C \left(\frac{E_{\text{therm}}}{N}\right)^{N\nu/2},$$

where C is a constant and ν is the number of degrees of freedom per particle.

We end with a comment about relativistic gases, such as might be found in the interiors of stars or other plasmas. At very high temperatures, the molecules are torn apart (as are perhaps the atoms as well), so the particles behave as a monatomic gas with no rotational or vibrational degrees of freedom. This leaves each with only $\nu = 3$ translational kinetic degrees of freedom. Because the energy of each particle is linear in its momentum rather than quadratic ($\varepsilon = cp$), the integral over each momentum coordinate is directly proportional to E, rather than $\sqrt{E}$. So $\int dp_x\,dp_y\,dp_z \propto E^3$. Consequently, the corrected number of states per particle for a relativistic gas of identical particles is

$$\omega_{\text{c, relativistic gas}} \approx C \left(\frac{V}{N}\right) \left(\frac{E}{N}\right)^3. \qquad (6.12)$$

Problems

For handling the large numbers in some of these problems, it may be helpful to use logarithms. Also, it may be helpful to note that $\log_{10} e = 0.4343$ so that $e^x = 10^{0.4343x}$ and that $\ln(1 + \varepsilon) \approx \varepsilon$ for small ε.

Section A

1. A system has 10^{24} identical particles.
   (a) If the number of states available to each particle increases from two to three, by what factor does the number of states available to the entire system increase? (Give the answer in powers of ten.)
   (b) If the number of states available to each particle increases from 100 to 101, by what factor does the number of states available to the entire system increase?
   (c) If the number of states available to each of these particles increases from 1.000×10^{30} to 1.001×10^{30}, by what factor does the number of states available to the entire system increase?

2. Consider a room containing 10^{28} distinguishable air molecules.
   (a) If you are interested in only their distributions between the front and back halves of the room, then each molecule has two possible states: either it is in the front or the back. How many different configurations are possible for the entire system? (Answer in powers of ten.)
   (b) What would be the answer to part (a) if you had divided the room into quarters?
   (c) What would be the answer to part (a) if you had divided the room into 156 compartments?

3. How many times larger is (a) 10^4 than 10^3, (b) 10^{75} than 10^{63}, (c) 10^{1000} than 10^{100}, (d) 10^{10^3} than 10^{10^2}, (e) $10^{10^{24}}$ than $10^{10^{22}}$?

4. The definition of equilibrium states that the *probabilities* of all possible distributions are the same, but that is different from stating that the particles will be uniformly distributed between the states. Explain the difference, using the 16 different ways in which four different molecules can be distributed between the front and back halves of an otherwise empty room.

Section B

5. Consider a system of three spin-1/2 particles, each having z-component of magnetic moment equal to $\pm\mu$. If there is no external magnetic field ($B = 0$), all spin states are of the same energy and are equally accessible.
   (a) Write down all possible spin configurations of the three. (See Example 6.1.)
   (b) What is the probability that the z-component of the magnetic moment of the system is $-\mu$?

(c) What is the probability that the spin of particle 1 is up (+) *and* the spin of particle 2 is down (−)?

6. Consider a system of four flipped pennies.
   (a) Find the probability that two are heads and two are tails by writing down all 16 possible heads–tails configurations and counting. Compare with the binomial prediction $P_N(n)$ of equation 2.4.
   (b) What is the probability that all four are tails?
   (c) Suppose someone told you that two of them were heads and two were tails, but not which were which. In this case, what is the probability that penny 1 is heads and penny 3 is tails?

7. A particle is confined to be within a rectangular box of dimensions 1 cm by 1 cm by 2 cm. Furthermore, the magnitude of its momentum is constrained to be less than 3×10^{-24} kg m/s.
   (a) What is the probability that it is in any one particular quantum state?
   (b) What is the probability that the magnitude of its momentum is less than 2×10^{-24} kg m/s?
   (c) What is the probability that the magnitude of its momentum is greater than 10^{-24} kg m/s and less than 3×10^{-24} kg m/s *and* that it is in the right-hand half of the box?

8. An otherwise empty room contains 6×10^{26} air molecules. How many times more probable is it that they are split exactly 50–50 between front and back halves of the room than that there is a 49–51 split? (Hint: Use $P_N(n)$ and Stirling's formula.)

Section C

9. We are going to estimate the spacing between states for a system of 10^{28} air molecules in a room that is 5 meters across. We will use the fact that a particle confined to a certain region of space is represented by a standing wave, and only certain wavelengths fit. (See Figure 1.9.) The momentum of an air molecule is given by h/λ and its energy by $(1/2m)p^2$, where $m = 4.8 \times 10^{-26}$ kg.
   (a) What is the smallest possible energy of an air molecule in this room? (Hint: What is the longest possible wavelength?)
   (b) What is the next smallest possible energy?
   (c) Suppose we were to add a tiny amount of energy to the air and that this added energy just happened to be taken by one of the slowest moving molecules in the room. What is the smallest amount of energy we could add?

10. Above, we calculated the minimum energy that could be added to air molecules in a room if that energy happened to go to the slowest moving molecule. Now calculate the minimum added energy if that energy goes

to a molecule which is moving at a more typical speed, 400 m/s. (Hint: $KE = p^2/2m$, $\Delta KE = 2p\Delta p/2m$. What is the smallest Δp?)

Section D

11. Make a table showing the number of different possible heads–tails configurations for (a) two, (b) three, (c) four coins.
    (d) Is the number of different configurations equal to the product of the numbers of states available to the individual coins in each case?

12. If you roll two dice, how many different configurations are possible if the dice are:
    (a) distinguishable?
    (b) indistinguishable? (Notice that 2! is *not* the right correction factor. We will see later that $N!$ is *not* correct if there is an appreciable likelihood that two particles may occupy the same state, e.g., two dice can show the same number of dots.)

13. How many different heads–tails configurations are possible for a system of six coins if the coins are:
    (a) distinguishable?
    (b) indistinguishable? (See the previous problem.)

14. Assuming that the elements are distinguishable from each other, find the number of different states available to
    (a) a system of five dice,
    (b) a system of three coins and two dice,
    (c) a system of 10^{24} molecules, each of which could be in any one of five different states? (Give the answer in powers of ten.)

15. System A has 30 particles, each having two possible states. System B has 20 particles, each having three possible states. All the particles are distinguishable from each other. Calculate the number of states available to (a) system A, (b) system B, (c) the combined system, $A + B$?

16. How many different ways are there of putting two particles in three boxes (no more than one per box), if the particles are (a) distinguishable, (b) identical?

17. Repeat the above problem for
    (a) two particles in 5 boxes,
    (b) two particles in 100 boxes,
    (c) three particles in 100 boxes,
    (d) three particles in 1000 boxes.

18. In a certain system, the density of states doubles when the internal energy is doubled.
    (a) Is this a macroscopic or a microscopic system?
    (b) How many degrees of freedom does it have?

19. In a certain system, the density of states increases by a factor of $10^{10^{22}}$ when the system's thermal energy is increased by 0.01%.
    (a) Is this a macroscopic or a microscopic system?
    (b) How many degrees of freedom does it have?

20. A certain system has 4×10^{24} degrees of freedom. By what factor does the number of available states increase if the internal energy is (a) doubled, (b) increased by 0.1%? (Assume that the energy range, ΔE, is the same for both cases. Give the answer in powers of ten.)

21. Consider a cup (1/4 liter) of water. Each molecule has six degrees of freedom (three translational and three rotational).
    (a) Roughly how many water molecules are in this system?
    (b) If you raise the temperature of the cup of water from room temperature (295 K) to boiling, the system's thermal energy increases by what factor? (Assume that u_0 remains constant.)
    (c) Estimate the factor by which the density of states increases when the temperature of a cup of water is raised from room temperature to the boiling point.

22. Consider the air in an oven at 500 K. The oven has a volume of $0.15 \, \text{m}^3$ and contains 2.2×10^{24} identical nitrogen molecules, each having five degrees of freedom and a mass of 4.8×10^{-26} kg.
    (a) What is the thermal energy of this system?
    (b) The magnitude of the momentum of any molecule can range from 0 to p_0. Estimating p_0 to be roughly twice the root mean square momentum, what is the volume in momentum space that is available to any particle?
    (c) We are going to calculate the number of accessible quantum states for the molecules' translational motions, ignoring the rotational states because the latter turn out to be relatively few in comparison. Considering the translational motion only, how many different quantum states would be accessible to any particle, if it were all by itself? ($V_r V_p / h^3$, where $V_p = (4/3)\pi p_0^3$.)
    (d) What is the number of states per particle ω_c corrected for the case of identical particles?
    (e) How many different quantum states are accessible to the entire system?

23. In a classroom at 290 K, there are typically 3×10^{27} air molecules each having five degrees of freedom and mass 4.8×10^{-26} kg. The room has dimensions 3 m by 6 m by 7 m. You are going to calculate how many quantum states are accessible to this system.
    (a) What is the thermal energy of this system?
    (b) The magnitude of the momentum of any molecule can range from 0 to p_0. Estimating p_0 to be roughly twice the root mean square momentum, what is the volume in momentum space that is available to any particle?

(c) For reasons cited in the previous problem, we are going to calculate the number of accessible quantum states for translational motion and ignore the relatively fewer states for rotational motion. How many different quantum states would be accessible to any particle if it were all by itself?

(d) If all the particles were identical, what would be the corrected number of states per particle, ω_c?

(e) How many different quantum states are accessible to the entire system?

(f) Answer parts (d) and (e) but now assuming that 79% are identical nitrogen molecules and 21% are identical oxygen molecules (of about the same mass).

24. Repeat parts (a)–(e) of the above problem for the system of iron atoms in a nail at room temperature (295 K). Each iron atom has six degrees of freedom and mass 8.7×10^{-26} kg, and its center of mass movement is confined to a volume of 10^{-30} m^3. There are about 10^{23} such iron atoms in a typical nail.

25. Suppose that as a certain material melts, the number of degrees of freedom per molecule is reduced from six to three. Therefore, the system's thermal energy, E_{therm}, is reduced by a factor $1/2$ and the exponent $v/2$ in Equation 6.9 is also reduced by a factor $1/2$. That would seem to indicate that as you add heat to this system the number of states gets reduced immensely. Later we will show, however, that as you add heat to any system, the number of accessible states must increase. How can you explain this apparent contradiction for the melting of this material? (Hint: Although a particle in the two phases has the same accessible volume in momentum space, since it has the same temperature and the same kinetic energy, this cannot be said for the accessible volume in coordinate space.)

26. (a) Show that in our result 6.8 for ω_c, for systems with identical particles we have left out a factor $\sqrt[2N]{2\pi N}$.

(b) Show that this factor goes to 1 in the limit of large N. (Hint: Show that the logarithm of this factor goes to zero by applying L'Hospital's rule to it.)

(c) Stirling's approximation is often simplified for large N to the form $N! = (N/e)^N$ or equivalently $\ln N! = N \ln N - N$. This simplification ignores the factor involving $\sqrt{2\pi N}$. How is this justified?

Chapter 7
Entropy and the second law

A   Interacting systems

We now examine interacting systems. We will find that the number of states for the combined system is extremely sensitive to the distribution of energy among the interacting subsystems, having a very sharp narrow peak at some "optimum" value (Figure 7.1). Configurations corresponding to a greater number of accessible states are correspondingly more probable, so the distribution of energy is most probably at the peaked optimum value. Even the slightest deviation would cause a dramatic reduction in the number of accessible states and would therefore be very improbable.

This chapter is devoted to developing this statement of probabilities, which underlies the most powerful tools of thermodynamics. We elevate it to the stature of a "law." Even though there is some small probability that the law may be broken, it is so minuscule that we can rest assured that we will never see it violated by any macroscopic system. Rivers will flow uphill and things will freeze in a fire if the law is broken. No one has ever seen it happen, and you can bet that you won't either.

B   Microscopic examples

We now investigate some examples of how the number of states is affected by the distribution of energy between interacting systems. Consider the situation

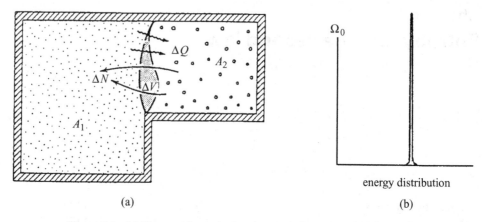

Figure 7.1 (a) We consider an isolated system A_0 composed of two subsystems, A_1 and A_2, which are interacting with each other thermally, mechanically, and/or diffusively. (b) As a result of these interactions, the number of states Ω_0 accessible to the combined system is a very sharply peaked function of the distribution of energy between the two interacting subsystems.

of Figure 7.1, where an isolated system A_0 is composed of two subsystems, A_1 and A_2, which may be interacting in any manner. The internal energies of the two subsystems may change as a result of their interaction, but the energy of the combined system is constant:

$$E_1 + E_2 = E_0 = \text{constant.}$$

In the last chapter (Table 6.2) we found that the number of states that are accessible to a system with $N\nu$ degrees of freedom is extremely sensitive to its thermal energy:[1]

$$\Omega \propto E^{N\nu/2}. \tag{7.1}$$

Furthermore, as we learned in Chapter 2, the number of states for the combined system is the product of the numbers of states for the subsystems:

$$\Omega_0 = \Omega_1 \Omega_2 \propto E_1^{N_1 \nu_1/2} E_2^{N_2 \nu_2/2}. \tag{7.1'}$$

We now apply this result to two examples. For simplicity, we take the constant of proportionality to be unity, and we assume that the energy comes in units of 1, with the energy of the combined system being 5:

$$E_1 + E_2 = E_0 = 5.$$

---

[1] For simplicity we set the reference level for potential energies, u_0, equal to zero, so that $E = E_{\text{therm}}$.

Table 7.1. *The number of accessible states for two small interacting systems with six and 10 degrees of freedom, respectively* $(E_0 = 5)$

E_1	E_2	$\Omega_1 = E_1^3$	$\Omega_2 = E_2^5$	$\Omega_0 = \Omega_1\Omega_2$
0	5	0	3125	0
1	4	1	1024	1024
2	3	8	243	1944
3	2	27	32	864
4	1	64	1	64
5	0	125	0	0
			Total	3896

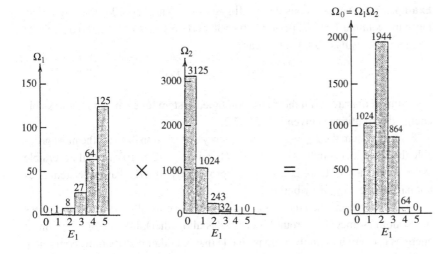

Figure 7.2 Plots of Ω_1, Ω_2, and Ω_0 vs. E_1 for interacting systems A_1 and A_2, which have six and 10 degrees of freedom, respectively. For simplicity, we assume that the energy comes in units of 1, and that the total energy of the combined system is $E_1 + E_2 = E_0 = 5$.

Example 7.1 For our first example we choose very small systems of perhaps two or three particles each: A_1 has 6 degrees of freedom and A_2 has 10. According to the above,

$$\Omega_1 = (E_1)^3, \qquad \Omega_2 = (E_2)^5. \qquad (7.2)$$

Equations 7.1 and 7.2 give us the number of accessible states for each of the six possible ways of distributing this energy between the two subsystems. The results are displayed in Table 7.1 and Figure 7.2.

Since the combined system is equally likely to be in any of these states, the most probable energy distribution is $E_1 = 2$, $E_2 = 3$. Half of the available states $(1944/3896 = 0.50)$ have this energy distribution, so the system will be in this configuration half the time.

Table 7.2. *The number of accessible states for two small interacting systems with 60 and 100 degrees of freedom, respectively* $(E_0 = 5)$

E_1	E_2	$\Omega_1 = E_1^{30}$	$\Omega_2 = E_2^{50}$	$\Omega_0 = \Omega_1 \Omega_2$
0	5	0	8.88×10^{34}	0
1	4	1	1.27×10^{30}	1.27×10^{30}
2	3	1.07×10^9	7.18×10^{23}	7.71×10^{32}
3	2	2.06×10^{14}	1.13×10^{15}	2.32×10^{29}
4	1	1.15×10^{18}	1	1.15×10^{18}
5	0	9.31×10^{20}	0	0
			Total	7.72×10^{32}

Example 7.2   Next we consider small systems that are just 10 times larger than those in Example 7.1 (i.e., perhaps 20 or 30 particles each). A_1 has 60 degrees of freedom and A_2 has 100. For this case,

$$\Omega_1 = (E_1)^{30}, \qquad \Omega_2 = (E_2)^{50}.$$

The number of states available to the combined system for each of the six possible energy distributions is given in Table 7.2.

For this system the $E_1 = 2$, $E_2 = 3$ energy distribution is by far the most probable, because it accounts for 99.9% ($7.71 \times 10^{32}/7.72 \times 10^{32}$) of all accessible states. Nonetheless, there is still a small probability (0.001) that the system does *not* have this energy distribution.

Notice that even when the combined system has only $60 + 100 = 160$ degrees of freedom (somewhere around 50 particles altogether), the one distribution of energies is overwhelmingly more probable than all other distributions combined. In the next section we will learn that for macroscopic systems, the probability for anything other than the "optimal" energy distribution is impossibly small.

In these examples, we assumed that energy comes in units of 1 for simplicity. Had we allowed a continuum of energy distributions, we would have arrived at the smooth distribution illustrated in Figure 7.3. But the important point is that as the number of degrees of freedom increases, the distribution of states becomes increasingly peaked.

C   Macroscopic systems

C.1   A typical system

We now consider a *macroscopic* example, using the same simplifying assumptions as before (energy increments of 1, the energy of the combined system is 5)

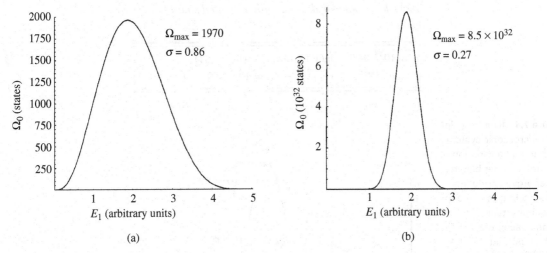

Figure 7.3 Plots of the number of accessible states vs. energy distribution for two very small interacting systems. The systems are the same as those in Tables 7.1 and 7.2, respectively, except that the energy can have a continuum of values rather than coming in units of 1. (a) The two systems have six and 10 degrees of freedom, respectively. (b) The two systems have 60 and 100 degrees of freedom, respectively.

Table 7.3. *The number of accessible states for two macroscopic interacting systems with 6×10^{24} and 10×10^{24} degrees of freedom, respectively ($E_0 = 5$)*

E_1	E_2	$\Omega_1 = E_1^{3 \times 10^{24}}$	$\Omega_2 = E_2^{5 \times 10^{24}}$	$\Omega_0 = \Omega_1 \Omega_2$
0	5	0	$10^{3.49 \times 10^{24}}$	0
1	4	1	$10^{3.01 \times 10^{24}}$	$10^{3.01 \times 10^{24}}$
2	3	$10^{0.90 \times 10^{24}}$	$10^{2.39 \times 10^{24}}$	$10^{3.29 \times 10^{24}}$
3	2	$10^{1.43 \times 10^{24}}$	$10^{1.51 \times 10^{24}}$	$10^{2.94 \times 10^{24}}$
4	1	$10^{1.81 \times 10^{24}}$	1	$10^{1.81 \times 10^{24}}$
5	0	$10^{2.09 \times 10^{24}}$	0	0
			Total	$10^{3.29 \times 10^{24}}$

but for systems that are 10^{24} times larger. That is, the number of degrees of freedom for A_1 and A_2 are now 6×10^{24} and 10×10^{24}, respectively, so that

$$\Omega_1 \approx (E_1)^{3 \times 10^{24}}, \qquad \Omega_2 \approx (E_2)^{5 \times 10^{24}}.$$

In this case, the numbers of states for each of the six possible distributions of energy between the two subsystems is as given in Table 7.3. The apparent similarity of these numbers is deceiving. For example, the two numbers $10^{3.01 \times 10^{24}}$

Table 7.4. *Tools for dealing with large and small numbers*

$a^x = 10^{x \log_{10}(a)}$	$\ln(1 + \varepsilon) \approx \varepsilon \quad$ (for $\varepsilon \ll 1$)
$e^x = 10^{0.4343x}$	$(1 + \varepsilon)^x \approx e^{\varepsilon x} = 10^{0.4343\varepsilon x} \quad$ (for $\varepsilon \ll 1$)

Figure 7.4 (Not to scale) For macroscopic systems we find that a variation of just one part per billion in the energy distribution causes a huge change in probability, typically being a factor of $10^{220\,000}$ at the peak and $10^{360\,000\,000\,000\,000}$ at each steeply sloping side.

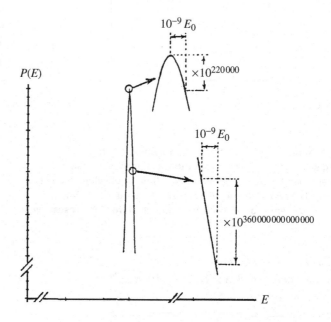

and $10^{3.29 \times 10^{24}}$ differ by $10^{0.28 \times 10^{24}} = 10^{280\,000\,000\,000\,000\,000\,000\,000\,000}$. The distribution $(E_1 = 2, E_2 = 3)$ is this many times more probable than all the other distributions combined! Clearly, the odds are overwhelming.

C.2 Smaller energy increments

In the preceding examples, the energy increments were one fifth of the total. Suppose that we have extremely accurate instruments that can measure energies to *parts per billion*. Even then the difference between two neighboring distributions would be astronomical. Using the tools of Table 7.4, we can see that the difference between the two distributions differing in energy by one part per billion ($\delta = 10^{-9} E_0$) is given by

$$\frac{\Omega_0(E_1 = 2)}{\Omega_0(E_1 = 2 + \delta)} = 10^{3.6 \times 10^{14}} = 10^{360\,000\,000\,000\,000}. \tag{7.3}$$

The energies in this example lie on the steeply sloping side of the energy distribution (Figure 7.4). Had we started at the peak, the numbers would have been different but the conclusions the same, as we demonstrate next.

C.3 The general case

We now calculate the probabilities for *all* possible distributions of energy between two interacting systems, making E_1 a *continuous* variable. To eliminate clutter, we use the symbol n_i for the total number of degrees of freedom of system i:

$$n_i = N_i \nu_i, \tag{7.4}$$

and n_0 for that of the combined system ($n_0 = n_1 + n_2$). The probability that system 1 has energy E_1 is proportional to the number of accessible states. From Table 6.2 and Equation 7.1',

$$\Omega_0(E_1) = E_1^{n_1/2} E_2^{n_2/2}, \quad \text{where } E_2 = E_0 - E_1. \tag{7.5}$$

We expand the logarithm of the number of states[2] in a Taylor series around its maximum, just as we did for probabilities in Section 3B. The details of this calculation are also in Appendix D and give the result[3]

$$P(E_1) = \frac{1}{\sqrt{2\pi}\sigma} e^{-(E_1 - \overline{E}_1)^2/2\sigma^2} \tag{7.6}$$

where

$$\overline{E}_1 = \frac{n_1}{2} kT \quad \text{and} \quad \sigma = \sqrt{\frac{n_1 n_2}{2 n_0}} kT.$$

As we have seen before, the standard deviation σ increases and the relative fluctuation $\sigma/\overline{E}$ decreases with the square root of the system's size. For example, if A_2 is a relatively large system, i.e., a "reservoir," so that $n_1 \ll n_2 \approx n_0$, then the standard deviation and relative fluctuation become

$$\sigma \approx \sqrt{\frac{n_1}{2}} kT, \quad \frac{\sigma}{\overline{E}_1} \approx \sqrt{\frac{2}{n_1}} \quad (A_2 \gg A_1). \tag{7.7}$$

So, for a small macroscopic system, having 10^{24} degrees of freedom, the relative fluctuation would be given by

$$\frac{\sigma}{\overline{E}_1} \approx 10^{-12} \quad \text{(for } 10^{24} \text{ degrees of freedom)}.$$

Even if we could measure energies to parts per billion, this would still be 1000 times larger than the standard deviation. The probability for a fluctuation this large would be

$$\frac{P(\overline{E}_1 \pm 1000\sigma)}{P(\overline{E}_1)} = e^{-(\pm 1000\sigma)^2/2\sigma^2} \approx 10^{-217\,000}. \tag{7.8}$$

---

[2] Do you remember why we expand $\ln \Omega$ rather than Ω itself? $\ln \Omega$ is a smoother function than is Ω, so the first few terms of the expansion approximate it better.

[3] Not surprisingly, this result is identical to the binomial probability results 3.7 and 3.4, except that the energy distribution is in units of kT. For comparison, simply make the substitution $N \to n_0/2$ and note that the probabilities for any one quantum of energy to be in systems 1 or 2, respectively, are $p = n_1/n_0$ and $q = n_2/n_0$.

Compare this with the numbers in Table 6.1. It is clear that, even for a continuous distribution of energies, the probability for the smallest measurable fluctuation from the distribution's peak is impossibly small. A quick calculation shows that if we had an extremely fast and accurate instrument that could measure energies in increments of one part per billion, and if this instrument could give us readings at microsecond intervals, we still would have to wait nearly $10^{217\,000}$ (more precisely, $10^{216\,976}$) times longer than the age of the Universe to see the distribution of energies fluctuate just once!

The odds are overwhelming. Interacting macroscopic systems in equilibrium will always be in a state with the one optimal distribution of energies between them. The chance of our seeing a different distribution of energies is infinitesimal.

Summary of Sections B and C

Consider an isolated system A_0 that is composed of two interacting subsystems A_1 and A_2. When in equilibrium, the number of states for the combined system is given by (equation 7.1′)

$$\Omega_0 = \Omega_1 \Omega_2 \propto E_1^{N_1 \nu_1/2} E_2^{N_2 \nu_2/2}$$

This function of the energy distribution is so sharply peaked for macroscopic systems that the probability of ever seeing a fluctuation is impossibly small.

We can calculate the probabilities for all possible energy distributions by expanding the logarithm of Ω_0 in a Taylor series about the peak. Writing Ω_0 as a function of the energy in system A_1 and using the simplifying notation (equation 7.4)

$$n_i = N_i \nu_i$$

we get the following distribution in probabilities (7.6),

$$P(E_1) = \frac{1}{\sqrt{2\pi}\sigma} e^{-(E_1 - \overline{E}_1)^2 / 2\sigma^2}$$

where

$$\overline{E}_1 = \frac{n_1}{2} kT \qquad \text{and} \qquad \sigma = \sqrt{\frac{n_1 n_2}{2n_0}} kT.$$

If A_1 is interacting with a large reservoir, for example, the standard deviation and relative width are (equation 7.7)

$$\sigma \approx \sqrt{\frac{n_1}{2}} kT, \qquad \frac{\sigma}{\overline{E}_1} \approx \sqrt{\frac{2}{n_1}} \qquad (A_2 \gg A_1).$$

The relative fluctuation is inversely proportional to the square root of the number of degrees of freedom. Therefore, for a typical small but macroscopic system having 10^{24} degrees of freedom, the fluctuations are about 10^{-12} of the system's internal energy – a long way below our ability to detect them.

D   The second law of thermodynamics

We have learned that, when two interacting macroscopic systems are in equilibrium, the distribution of energy must be such that the number of states available to the combined system is a maximum. Their energies can be expressed as functions of other parameters, such as temperature, pressure, chemical potential, the numbers of particles of the various types, volumes, magnetic fields, electric fields, etc. Since the energies are functions of these "system variables," we can reword our previous result as follows.

When two interacting macroscopic systems are in equilibrium, the values of the various system variables will be such that the number of states Ω_0 available to the combined system is a maximum.

There is a very important corollary to this result, which is called the "second law of thermodynamics." There are many equivalent ways of stating it, but they all rely on the following: if the number of states Ω_0 is a maximum when two systems are in equilibrium, then Ω_0 must be increasing as they approach equilibrium.

> **Second law of thermodynamics**
>
> As two interacting macroscopic systems approach equilibrium, the changes in the system variables will be such that the number of states Ω_0 available to the combined system increases. More simply, in the approach to equilibrium,
>
> $$\Delta\Omega_0 > 0. \tag{7.9}$$

Notice that the second law is based on probabilities, whereas the first law reflects inviolable fact (we think). The second law does not apply to small systems, whereas the first law does. For large macroscopic systems there is some small but finite probability that the second law could be violated. But we will never see it happen.

For example, there is some small probability that all the air in your room might rush over to one corner, leaving you to suffocate. Or water might flow uphill, or heat might flow from cold to hot and water might boil as ice cubes are added. There is some small probability that your blood might transport carbon dioxide to your cells and oxygen away rather than vice versa. The likelihood of any of these happening would be similar to that of flipping 10^{24} coins and having them all land heads. As we saw in the previous section, even the inception of such anomalous behavior is extremely improbable – even fluctuations at the parts per billion level. Clearly, we can base our studies on a law whose chance of violation is so minuscule that we can rest assured that we will never witness a violation. In fact, in every moment of our existence we bet our very lives on these odds and, needless to say, we always win.

We should mention that there are increasingly important fields where the systems studied are microscopic, having far fewer than Avogadro's number of

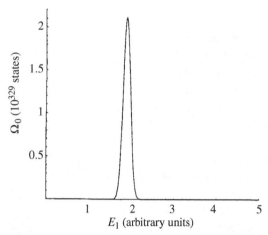

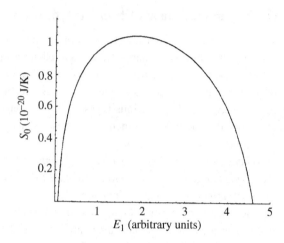

Figure 7.5 Plot of the number of accessible states Ω_0 and the entropy S_0 versus energy distribution for two small interacting systems with 600 and 1000 degrees of freedom, respectively ($\Omega \approx E^{Nv/2}$). The entropy is clearly a smaller and smoother function, even for this very small system. The difference between the two functions becomes even more pronounced as the size of the system increases.

elements. Examples include microelectronics, digital information storage and signal processing, cellular biology, low-temperature studies, thin films and surface physics, nanotechnology, and many others. For these smaller systems, the relative fluctuations are larger (equation 7.7), and there is a correspondingly larger chance for violation of the second law. But you should be able to handle these smaller systems using either the results 7.6 or the techniques for small systems that were introduced at the beginning of this book.

E Entropy

E.1 Definition and properties

The number of states for a macroscopic system is extremely large and unwieldy. The logarithm of a large number is smaller and more manageable, so we find it more convenient to work with the logarithm of Ω rather than with Ω itself. A particular multiple of this logarithm is called the "entropy." It is given the symbol S and is defined as follows (Figure 7.5):

$$\text{entropy } S \equiv k \ln \Omega. \tag{7.10}$$

Boltzmann's constant k (equation 4.11) gives entropy the magnitude and units that are convenient for typical macroscopic systems.

Let's take a moment to appreciate how nice this is to work with. If Ω is a number like $10^{10^{24}}$, then $\ln \Omega$ is a number like 10^{24}. If we multiply this by k, which is a number like 10^{-23}, the result is a number like 10. That is,

$$\Omega \approx 10^{10^{24}},$$
$$\ln \Omega \approx 10^{24},$$
$$k \ln \Omega \approx 10 \text{ J/K}.$$

Isn't that a big improvement? Notice that no information is lost. Given this number, we can easily work backwards to find Ω, if we wish:

$$S = k \ln \Omega \quad \Rightarrow \quad \Omega = e^{S/k}. \tag{7.11}$$

Entropy also has another convenient property. Whereas the number of states accessible to the combined system is multiplicative,

$$\Omega_0 = \Omega_1 \Omega_2,$$

the entropy of the combined system is additive (since $\ln ab = \ln a + \ln b$):

$$k \ln \Omega_0 = k \ln \Omega_1 + k \ln \Omega_2 \quad \Rightarrow \quad S_0 = S_1 + S_2.$$

That is, the whole is the sum of the parts. This makes it algebraically similar to some other system variables, such as internal energy, volume, or number of particles:

$$S_0 = S_1 + S_2, \qquad V_0 = V_1 + V_2,$$
$$E_0 = E_1 + E_2, \qquad N_0 = N_1 + N_2.$$

Also like the energy, volume, and number of particles, the entropy of a system can be determined at any time unambiguously, independently of what it was in the past or what it will be in the future. It is a measure of the number of accessible states, which in principle can be counted at any time, just like you could count number of chairs in a room at any time. That makes its differential dS exact, just like the differentials dE, dV, dT, dp and dN.

Example 7.3 Consider a system A_0 consisting of interacting subsystems A_1 and A_2 for which $\Omega_1 = 10^{20}$ and $\Omega_2 = 2 \times 10^{20}$. What is the number of states available to the combined system A_0? Also, what are the entropies S_1, S_2, and S_0 in terms of Boltzmann's constant?

The number of states accessible to the combined system is

$$\Omega_0 = \Omega_1 \Omega_2 = 2 \times 10^{40}.$$

The entropies S_1 and S_2 are given by

$$S_1 = k \ln \Omega_1 = k \ln 10^{20} = 46.1k,$$
$$S_2 = k \ln \Omega_2 = k \ln 2 \times 10^{20} = 46.7k.$$

The entropy of the combined system is

$$S_0 = k \ln \Omega_0 = k \ln 2 \times 10^{40} = 92.8k.$$

which you could also get by adding the results for S_1 and S_2 above.

So far we have considered just two interacting subsystems. But we would get the same results for three, four, five, or any number of interacting subsystems.[4] The number of states accessible to the combined system would be given by

$$\Omega_0 = \Omega_1 \Omega_2 \Omega_3 \cdots$$

and its entropy would be given by

$$S_0 = S_1 + S_2 + S_3 + \cdots. \tag{7.12}$$

E.2 Entropy and the second law

The word "entropy" is not a part of our everyday vocabulary, as is the word "energy," but there is nothing inherently magical or mystical about it. The entropy is simply a convenient measure of the number of states accessible to a system. As explained above, it is more convenient than using the number of states directly for two reasons:

- it is smaller and more manageable;
- entropies are additive, whereas numbers of states are multiplicative.

The second law, equation 7.9, tells us that the number of accessible states must increase as systems approach equilibrium and that it is a maximum when they are in equilibrium. Because the entropy is the logarithm of the number of accessible states, when Ω increases so does S and when Ω is a maximum so is S (Figure 7.5). Consequently, an *alternative and equivalent* statement of the second law is as follows.

> **Second law of thermodynamics**
>
> For systems interacting in any way (whether or not they are yet in equilibrium) the entropy of the combined system cannot decrease:
>
> $$\Delta S_0 \geq 0. \tag{7.13}$$

Notice that the second law applies to the combined entropy of all the interacting systems, not to the entropy of just one of them. For example, our Sun loses entropy as it radiates energy out into space. But the entropy of the Universe as a whole increases as a result.

This chapter on entropy and the second law is the most important chapter in the whole book, and it is worthwhile to review what has been covered.

---

[4] In fact, you can prove that if the result is correct for two systems then it has to be true for any number of systems, by calling the combined system "system 1," and then adding one more as "system 2." In this way you can keep adding one more system for as long as you like, and the results must be true each time.

Summary of Chapter 7

Earlier, it was promised that the fundamental postulate would form the basis of all statistical tools used in the study of large systems. This postulate states that a system in equilibrium is equally likely to be found in any of its accessible states. Hence configurations corresponding to a greater number of states are correspondingly more probable.

For two or more interacting macroscopic systems, the number of states for the combined system is extremely sensitive to the distribution of energy between them, being very sharply peaked. Hence, as energy is exchanged between interacting systems, they will tend toward more probable configurations (i.e., those with a greater number of accessible states) until they reach the peak in probabilities. At that point, they are in equilibrium. The peak is so narrow that any deviation from this optimal energy distribution will never be seen.

The distribution of energies is a function of system variables such as pressures, temperatures, volumes, numbers of particles, magnetic moments, etc., depending on the types of interactions between the subsystems. The above statement that the combined system in equilibrium must have some "optimum" distribution of energies among the subsystems, is also a statement that these system variables must have "optimal" values.

We define a quantity called "entropy," which is proportional to the logarithm of the number of accessible states, making it much smaller and easier to work with (equation 7.10):

$$\text{entropy } S \equiv k \ln \Omega.$$

The fact that Ω must be a maximum when the system is in equilibrium means that S must also be a maximum.

We will rarely refer directly to the fundamental postulate itself again, but the needed information is carried in the law that S is a maximum when interacting systems are in equilibrium. This is what we will use henceforth.

E.3 Examples

In the last chapter we found the number of states Ω accessible to some common systems in terms of the temperature, volume, and number of particles (Table 6.2). According to equation 7.10, the dependence of the corresponding entropies on (V,T) would be:[5]

for a monatomic ideal gas, $\quad S = \ln C + Nk \ln V + \frac{3}{2} Nk \ln T;$

for a diatomic ideal gas, $\quad S = \ln C + Nk \ln V + \frac{5}{2} Nk \ln T;$

for a solid, $\quad S = \ln C + 3Nk \ln T.$

$$(7.14)$$

---

[5] You might wonder how we can take the logarithm of units, such as K for the temperature or m^3 for the volume. We don't have to: for every "unit" that appears in the temperature or volume, there must be a "1/unit" in the constant C, because the total number of states Ω has no units at all. And, adding the two, you get $\ln(\text{unit}) + \ln(1/\text{unit}) = 0$, so they cancel, no matter what they are.

where we have lumped all constants together in C. You can see that entropy increases with volume and temperature, owing to the increased volumes in coordinate and momentum space, respectively. For solids, the increased volume in coordinate space is accessed through larger amplitude vibrations and is therefore contained in the temperature term.

But there is trouble on the horizon. The system must have at least one state available to it (at least the one that it is in). So $k \ln \Omega$ cannot be negative. Yet, in the above expressions, the entropy could be negative for sufficiently small volumes or temperatures. So we must conclude that the way in which we counted states in Chapter 6 is incorrect if either the system is extremely dense or the temperature is extremely low. The physics of these highly condensed systems (condensed in either coordinate or momentum space) is very interesting, and we will return to it after we have developed more appropriate machinery later in this book.

Problems

Section A

1. Consider the following continuous functions of the energy of a system:
$$f_1(E) = E^2(5 - E)^3, \qquad f_2(E) = E^{20}(5 - E)^{30},$$
$$f_3(E) = E^{2 \times 10^{23}}(5 - E)^{3 \times 10^{23}}.$$

   (a) Show that for each of these, the maximum is at the same place, $E = 2$.
   (b) Show that the peak widths differ, by calculating the ratios $f(2)/f(1)$ and $f(2)/f(3)$ for each.

2. For what value of E_1 does the function $E_1^{n_1}(E_0 - E_1)^{n_2}$ peak? Answer in terms of n_1, n_2, and E_0.

Section B

3. Consider a system A_0 composed of subsystems A_1 and A_2, which have three and four degrees of freedom, respectively. The energy comes in units of 1, and the total energy of the combined system is 4.

   (a) Construct a table similar to Tables 7.1 or 7.2, illustrating Ω_1, Ω_2, and Ω_0 for the various possible energy distributions between the two subsystems.
   (b) What is the probability that the system will be in a state with $E_1 = 3$, $E_2 = 1$?
   (c) Which distribution of energies is most probable, and what is the probability for this distribution?

4. Repeat the above problem for interacting subsystems with five and six degrees of freedom, respectively.

5. Consider a system A_0 composed of subsystems A_1 and A_2 for which the number of degrees of freedom are 4 and 6, respectively. Energy comes in units of 1 and the energy of the combined system is 6.

   (a) Construct a table like Table 7.1 or 7.2 illustrating the seven possible energy distributions for the system and Ω_1, Ω_2, and Ω_0 for each.

(b) Which distribution of energies is most probable, and what is the probability for this distribution?

6. Repeat the above problem for systems 10 times as large.

7. Consider a system A_0 composed of three interacting subsystems, A_1, A_2, and A_3. Suppose that these subsystems have four, five, and six degrees of freedom, respectively, and that the energy comes in units of 1, with total energy $E_0 = E_1 + E_2 + E_3 = 4$. Using $\Omega = E^{Nv/2}$,
   (a) make a table showing Ω_1, Ω_2, Ω_3, and Ω_0 for each possible distribution of the energy among these three systems (15 different distributions in all).
   (b) find which energy distribution is the most probable and its probability.

8. Consider two interacting systems, isolated from the rest of the Universe. System 1 has four degrees of freedom and system 2 has ten. Their combined energy is 4, and comes in units of 1.
   (a) Make a table similar to that of Tables 7.1 and 7.2 for the energies and states available to each system and the states available to the combined system.
   (b) What is the probability that this system will at any instant be in the $(E_1, E_2) = (1, 3)$ state?

9. Review the definition of equilibrium. Explain why it is necessary for a combined system to be in equilibrium for the probability calculations relating to Tables 7.1–7.3 to be valid. (Hint: You will need to use the fundamental postulate.)

10. In Chapter 6 we saw that the number of quantum states available to a gas particle is proportional to the system's volume, V. Therefore, we found we could write the volume dependence of the number of states accessible to a system of N such particles as $\Omega \approx V^N$ (equation 6.11). We now consider two interacting systems, having identical particles and identical temperatures but separated by a partition. If the combined volume V_0 is fixed, $V_1 + V_2 = V_0$, then we can write $\Omega_0 = \Omega_1 \Omega_2 = \text{constant} \times V_1^{N_1}(V_0 - V_1)^{N_2}$. Find for what value of V_1 this is a maximum. Answer in terms of N_1, N_2, and V_0.

Section C
11. How many times larger is
   (a) 10^{12} than 10^{11},
   (b) $10^{6.02 \times 10^{25}}$ than $10^{6.01 \times 10^{25}}$,
   (c) $10^{10^{25}}$ than $10^{10^{23}}$?

12. How many times larger is:
   (a) 10^3 than 10^2,
   (b) 10^{45} than 10^{35},
   (c) 10^{1000} than 10^{998},
   (d) 10^{1000} than 10^{100},

(e) $10^{1.56 \times 10^{24}}$ than $10^{1.41 \times 10^{24}}$,

(f) $10^{3.56 \times 10^{25}}$ than $10^{8.67 \times 10^{24}}$?

13. What number is a billion times larger than $10^{2 \times 10^{20}}$?

14. In constructing Table 7.3 we assumed that the numbers of degrees of freedom for each of the two subsystems were 6×10^{24} and 10×10^{24}, respectively. Construct a similar table for two interacting systems that are twice as large.

15. Consider a system consisting of two interacting subsystems with 24×10^{24} and 20×10^{24} degrees of freedom, respectively. Construct a table similar to Table 7.3 listing Ω_1, Ω_2, and Ω_0.

16. Suppose that $\Omega_1 = E_1^{10^{24}}$ and $\Omega_2 = E_2^{2 \times 10^{24}}$. Find how much a change of 0.01 in the energy distribution affects the probabilities, by calculating the ratio $\Omega_0(E_1 = 2, E_2 = 3)/\Omega_0(E_1 = 2.01, E_2 = 2.99)$. (Hint: Use logarithms and the tools in Table 7.4.)

17. For the case illustrated in Table 7.3, where the subsystems have 6×10^{24} and 10×10^{24} degrees of freedom, respectively, compute the ratio $\Omega_0(E_1 = 2 + \delta, E_2 = 3 - \delta)/\Omega_0(E_1 = 2, E_2 = 3)$, for $\delta = 5 \times 10^{-9}$. (Hint: Use logarithms and the tools in Table 7.4.) Did you get the answer given in equation 7.3?

18. Given that $\Omega_0 = \Omega_1 \Omega_2 = E_1^{m_1/2}(E_0 - E_1)^{m_2/2}$, where $m = N\nu$ is the number of degrees of freedom for a system, use the fact that $d\Omega_0/dE_1 = 0$ when Ω_0 is a maximum to show that, in equilibrium, $E_1 = [m_1/(m_1 + m_2)]E_0$. How is this consistent with equipartition?

19. Consider two interacting systems, isolated from the rest of the Universe. System 1 has 4×10^{24} degrees of freedom and system 2 has 1.2×10^{25}. Their combined energy is 4, and comes in units of 1.

    (a) Making the same assumptions as in the text, construct a table similar to that of Table 7.3 for the energies and states available to each system and the states available to the combined system.

    (b) What is the probability that this system at any instant is *not* in the $(E_1, E_2) = (1, 3)$ state?

20. Prove the relationships given in Table 7.4. (Hint: Use a Taylor series expansion for $\ln(1 + \varepsilon)$.)

21. In the chapter we stated that $10^{217\,000}$ is $10^{217\,000}$ times longer than the age of the universe in microseconds (10^{24}). That is just an approximation. What is really the ratio of these two numbers?

22. Suppose we have a detector that is capable of measuring fluctuations in the thermal energy of a system to parts per billion. We wish to find the maximum number of particles in a system for which we are capable of detecting such fluctuations.

(a) Roughly how many degrees of freedom would a system have if we were just barely able to detect fluctuations of one standard deviation in its internal energy? (Hint: See equations 3.5 or 7.7.)

(b) If this is a solid whose atoms have six degrees of freedom each and are separated by 10^{-10} m (which is typical), what would be the volume of this solid?

(c) If it were a cube, what would be the length of a side? How does this compare with the width of a hair, which is about 50 μm?

(d) If the temperature is 295 K and each degree of freedom carries an average energy of $(1/2)kT$, what is the total thermal energy of this system, and what is the minimum fluctuation in energy that our measuring device can detect? (Give both answers in joules.)

23. Consider a Taylor series expansion (Appendix D, equation D.2) of the logarithm of the number of states accessible to a combined system.

(a) Why should the first derivative be zero?

(b) Show that by setting the first derivative equal to zero, you get the result D.3.

(c) Evaluate the second derivative at $E_1 = \overline{E}_1$ and show that the answer is $-n_0/(2E_1E_2)$ as claimed in equation D.4. (You will need to use $n_1/E_1 = n_2/E_2 = n_0/E_0$, which follows from equation D.3.)

(d) Show that the Gaussian form D.6 follows from the results for the first and second derivatives.

24. Show that the coefficients of the exponentials in equations 3.7 and 7.6 must have the form $1/(\sqrt{2\pi}\sigma)$ if the sum over all distributions gives a total probability of unity.

25. Starting with equation 7.6, show that the probability calculation in equation 7.8 is correct.

Section D

26. If water were to flow uphill, what would have to happen to its temperature if the first law is not to be violated? (Hint: Total energy is conserved.)

27. List some processes that would violate the second law without violating the first law.

28. Write out the number $10^{100} = 10\,000\,000\,000\,000 \cdots$ longhand (i.e., don't use exponential notation) and time how long it takes you to do it. How long would it take you to write out $10^{10^{24}} = 10^{1\,000\,000\,000\,000\,000\,000\,000\,000}$ longhand? (1 year $= 3.17 \times 10^7$ s.)

Section E

29. Consider two small interacting systems, A_1 and A_2, for which $\Omega_1 = 2$ and $\Omega_2 = 4$. What are (a) Ω_0, (b) S_1 and S_2, (c) S_0?

30. Repeat the above problem for $\Omega_1 = 200$ and $\Omega_2 = 400$.

31. Repeat the above problem for $\Omega_1 = 10^{2 \times 10^{24}}$ and $\Omega_2 = 10^{3 \times 10^{24}}$.

32. Consider five small interacting systems for which $\Omega_1 = 1$, $\Omega_2 = 2$, $\Omega_3 = 3$, $\Omega_4 = 4$, and $\Omega_5 = 5$.
    (a) What is the number of states Ω_0 accessible to the combined system?
    (b) What are S_1, S_2, S_3, S_4, and S_5 in terms of Boltzmann's constant k?
    (c) Compute the entropy of the combined system S_0 in two ways (in units of Boltzmann's constant),
       (1) by using the answer to part (a),
       (2) by adding the entropies in part (b).

33. A certain system has 6×10^{24} degrees of freedom. Its internal energy increases by 1%.
    (a) By what factor does the number of accessible states increase?
    (b) Given that the change of entropy is $\Delta S = S_f - S_i$, where $S = k \ln \Omega$, what is the increase in the system's entropy?

34. How many quantum states are accessible to a system if its entropy is
    (a) 1 J/K, (b) 42 J/K?

35. Consider two interacting systems, whose combined energy $E_0 = E_1 + E_2 = 64$ J. System 1 has 2×10^{24} degrees of freedom, and system 2 has 6×10^{24}. The states accessible to each system are given by $\Omega = (E/C)^{n/2}$, where $C = 10^{-35}$ J and $n = N\nu$ is the number of degrees of freedom for the system.
    (a) When the two are in equilibrium, what is the value of E_1?
    (b) At equilibrium, what are the entropies of the individual and combined systems?

36. Find the number of degrees of freedom of a system if
    (a) the entropy changes by 0.1 J/K when the thermal energy is tripled,
    (b) the entropy changes by 100 J/K when the thermal energy is doubled.

37. Consider an ideal monatomic gas of 10^{26} atoms.
    (a) What is the internal energy of this system at a temperature of 27 °C?
    (b) If the temperature is raised by 1 °C, by what factor does the number of accessible states increase?
    (c) What is the increase in entropy in this case?

38. What is the entropy of a freshly shuffled deck of cards?

39. In the results for ideal gases given in equation 7.14, the constant C contains a factor $m^{3/2}$, where m is the molecular mass (see equation C.5). This means that for two gases having the same number of particles and the same volume and temperature, the one with the more massive molecules has the greater entropy. Why should this be?

Chapter 8
Entropy and thermal interactions

In this chapter we study the changes in entropy during thermal interactions. We will assume all interactions are quasistatic, so that the systems can be considered to be in equilibrium at all times.

A Temperature

A.1 Definition and consequences

Temperature measures the dependence of entropy on internal energy for purely thermal interactions (Figure 8.1). It is defined by

$$\frac{1}{T} \equiv \left(\frac{\partial S}{\partial E} \right)_{V,N}. \tag{8.1}$$

According to the first law ($dE = dQ - pdV + \mu dN$, cf. equation 5.7), the variation in E for purely thermal interactions (i.e., V, N constant) is precisely what we define as the heat transfer, dQ.[1] Therefore, for quasistatic processes[2] definition 8.1

---

[1] It is customary to refer to the heat gained or lost by a system as "heat transfer", which is slightly misleading. Only in purely thermal interactions (for which $dE = dQ$, as we consider here) is the heat gained by one system necessarily equal to that lost by another. It is energy that is conserved, not heat.

[2] Of course, we must assume that the interactions are quasistatic. We cannot even define the temperature (or pressure or chemical potential) for a system that is not in equilibrium.

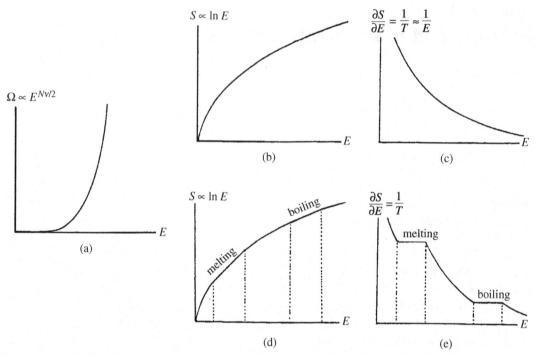

Figure 8.1 (a) The number of accessible states increases extremely rapidly with E. (b) The entropy increases more slowly, as it is proportional to the logarithm of E. (c) $\partial S/\partial E$ decreases with increasing E and equals $1/T$. (d), (e) The slope of S vs. E is constant during the phase transitions indicated.

is equivalent to

$$\frac{1}{T} = \frac{dS}{dQ}, \tag{8.2}$$

because for $dV = 0$ and $dN = 0$ we have $dE = dQ$. Equation 8.2 is more often written as

$$dS = \frac{dQ}{T}, \qquad \text{or} \qquad dQ = T dS. \tag{8.3}$$

This definition of temperature makes the change in entropy directly proportional to the heat transfer. We don't have to indicate what parameters are held constant because, no matter what else is going on and which other parameters are changing, the definition of temperature ensures that the changes in dS and dQ are proportional to each other.

Because we defined entropy (equation 7.10) as a measure of the number of quantum states, you might think that to determine changes in a system's entropy you would have to work microscopically and be infinitely patient in order to count this enormous number of tiny states. But equation 8.3 allows us to do it with simple tools like thermometers and calorimeters. What a relief!

We can now replace dQ by TdS, so that

$$dE = T dS - p dV + \mu dN \qquad \text{(first law)} \tag{8.4}$$

involves three pairs of conjugate variables, (T, S), (p, V), and (μ, N), each term having the same mathematical form. Each conjugate pair includes one "intrinsic" and one "extrinsic" variable, defined as follows. If you divide a system in equilibrium into two or more pieces, the intrinsic variable for the entire system is *equal* to that of the individual parts, whereas the extrinsic variable for the entire system is the *sum* of the individual parts. We have for two systems

$$
\begin{aligned}
\text{intrinsic,} \quad & T = T_1 = T_2, \quad & p = p_1 = p_2, \quad & \mu = \mu_1 = \mu_2; \\
\text{extrinsic,} \quad & S = S_1 + S_2, \quad & V = V_1 + V_2, \quad & N = N_1 + N_2.
\end{aligned}
\tag{8.5}
$$

In the form 8.4, the first law involves only exact differentials, dE, dS, dV, and dN. The inexact differentials, dQ and dW, have been replaced by TdS and pdV, respectively.

Because the second law deals with changes in entropy, the consequences of the second law are often more transparent if we rearrange the first law 8.4 in the following form:

$$
dS = \frac{1}{T}dE + \frac{p}{T}dV - \frac{\mu}{T}dN \quad \left(= \frac{1}{T}dQ \right).
\tag{8.6}
$$

This equation also provides a practical way of determining changes in a system's entropy by measuring changes in energy, volume, or particles. At a more fundamental level, of course, this equation is telling us how these measured changes are affecting the volume in phase space to which the particles have access.

Looking at the coefficients of the three terms in equation 8.6, we see that

$$
\frac{1}{T} = \left(\frac{\partial S}{\partial E} \right)_{V,N}, \quad \frac{p}{T} = \left(\frac{\partial S}{\partial V} \right)_{E,N}, \quad \frac{\mu}{T} = -\left(\frac{\partial S}{\partial N} \right)_{E,V}.
\tag{8.7}
$$

The first of these is the definition of temperature 8.1. You can think of the second and third as corresponding definitions of pressure and chemical potential. They tell us that p/T measures how the entropy varies with volume and that μ/T measures how the entropy varies with the number of particles.

A.2 Thermal equilibrium

When two thermally interacting systems are in equilibrium with each other, we say that they are in "thermal equilibrium." And, according to the second law, their combined entropy must be a maximum (Figure 8.2).

$$
S_0 = S_1 + S_2 \text{ is a maximum} \quad \text{(at equilibrium).}
$$

Now suppose that while in equilibrium a small amount of energy is transferred from system A_2 to system A_1 ($dE_2 = -dE_1$). When a function is at a maximum its first derivatives are zero, so to first order in the energy transfer, there will be no change in the entropy. Using equation 8.6 for the changes dS_1 and dS_2 we find that, for purely thermal ($dV = dN = 0$) interactions between the systems,

$$
0 = dS_0 = \frac{dE_1}{T_1} + \frac{dE_2}{T_2} = \left(\frac{1}{T_1} - \frac{1}{T_2} \right) dE_1 \quad \text{(thermal interactions)}
$$

$$
\Rightarrow \quad T_1 = T_2 \quad \text{(thermal equilibrium).}
$$

$$
\tag{8.8}
$$

That is, in thermal equilibrium, the temperatures are equal.

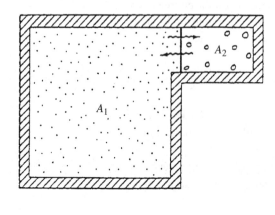

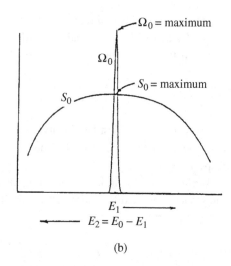

(a)   (b)

Figure 8.2 (a) Consider two systems, A_1 and A_2, that are interacting thermally but not mechanically or diffusively. (b) As we learned in the preceding chapter, when the two are in equilibrium, the distribution of energy between them is such that the total entropy $S_0 = k\ln\Omega_0$ is a maximum.

A.3 Equipartition

We are now in a position to show that an average energy of $(1/2)kT$ is associated with each degree of freedom, as was asserted in Section 4E. In Section 6D we found that the number of states accessible to a system with $N\nu$ degrees of freedom is given by[3]

$$\Omega = f(V, N)(E_{\text{therm}})^{N\nu/2},$$

(see Table 6.2) where the function $f(V, N)$ depends on the particular system. If we insert this into the definition of entropy given in equation 7.10 and use $E_{\text{therm}} = E - Nu_0$, we get

$$S = k \ln \Omega = k \ln f(V, N) + \frac{N\nu}{2}k \ln(E - Nu_0).$$

Now we apply the definition of temperature given in equation 8.1 and get (for u_0 constant)

$$\frac{1}{T} = \left(\frac{\partial S}{\partial E}\right)_{V,N} \quad \Rightarrow \quad \frac{1}{T} = \frac{N\nu k}{2E_{\text{therm}}} \quad \Rightarrow \quad E_{\text{therm}} = \frac{N\nu}{2}kT. \quad (8.9)$$

This is equipartition: in each degree of freedom there is an average energy $(1/2)kT$.

As we learned in Chapter 4, the assumption that u_0 is constant is usually valid for solids and gases. But in liquids u_0 increases as energy is added, and this extra avenue for storing energy makes the system act as if it had additional degrees of freedom. It is customary, however, to include in E_{therm} only those terms *not* associated with increases in the potential energy reference level u_0, so that the result 8.9 becomes appropriate for all systems.

According to equation 8.9, our choice of temperature scale uniquely determines Boltzmann's constant k. Choosing a system for which u_0 is constant, we

---

[3] To be precise, this result is valid only if the energy in each degree of freedom is of the form $\varepsilon = b\xi^2$. We saw that, for example, for relativistic particles $\varepsilon = b\xi$ and the power $N\nu/2$ becomes $N\nu$.

simply add energy ΔE and measure the resulting change in temperature ΔT. Then we determine k from

$$\Delta E = \frac{N\nu}{2} k \Delta T. \tag{8.10}$$

Summary of Section A

We define the temperature of a system by (equation 8.1)

$$\frac{1}{T} \equiv \left(\frac{\partial S}{\partial E} \right)_{V,N}.$$

This definition makes the change in entropy directly proportional to the heat transfer (equation 8.3),

$$dS = \frac{dQ}{T} \quad \text{or} \quad dQ = T dS,$$

and it enables us to write the first law entirely in terms of exact differentials (equation 8.4):

$$dE = T dS - p dV + \mu dN.$$

Because the second law deals with changes in entropy, it is often convenient to display these changes explicitly, by writing the first law in the form (equation 8.6)

$$dS = \frac{1}{T} dE + \frac{p}{T} dV - \frac{\mu}{T} dN.$$

This shows how the entropy (and so the number of accessible states) varies with changes in energy, volume, and particle number. In analogy with the definition of temperature for thermal interactions, pressure and chemical potential can be defined in terms of how the entropy varies during mechanical and diffusive interactions, respectively (equation 8.7):

$$\frac{1}{T} = \left(\frac{\partial S}{\partial E} \right)_{V,N}, \quad \frac{p}{T} = \left(\frac{\partial S}{\partial V} \right)_{E,N}, \quad \frac{\mu}{T} = -\left(\frac{\partial S}{\partial N} \right)_{E,V}.$$

The second law tells us that when two systems are in equilibrium, their combined entropy is a maximum. Using equation 8.6 for the changes in their entropies when energy dE is transferred between them, we can show that when two systems are in thermal equilibrium, their temperatures are equal (equation 8.8):

$$T_1 = T_2 \quad \text{(thermal equilibrium)}.$$

If the energy per degree of freedom has the usual form $b\xi^2$ then the definitions of entropy and temperature ensure that an average energy of $(1/2)kT$ is associated with each degree of freedom (equation 8.9):

$$E_{\text{therm}} = \frac{N\nu}{2} kT.$$

B Heat transfer and accessible states

The change in the entropy of a system during any process is given by

$$\Delta S = S_f - S_i = k \ln \Omega_f - k \ln \Omega_i = k \ln \left(\frac{\Omega_f}{\Omega_i} \right). \tag{8.11}$$

Taking the antilogarithm, we find that a change in entropy $\Delta S = \Delta Q / T$ implies that the number of accessible states has changed by a factor

$$\frac{\Omega_f}{\Omega_i} = e^{\Delta S/k} = e^{\Delta Q/kT}. \tag{8.12}$$

A rearrangement of the first law, $\Delta Q = \Delta E + p\Delta V - \mu \Delta N$, allows the following equivalent form of the above expression

$$\frac{\Omega_f}{\Omega_i} = e^{(\Delta E + p\Delta V - \mu \Delta N)/kT}. \tag{8.12'}$$

This form has the advantage that it includes only the exact differentials ΔE, ΔV, ΔN, meaning that we would only have to measure initial and final values rather than monitoring the values along the way as we would for ΔQ.[4] Equation 8.12' will also be the central feature later in the book when we examine small systems interacting with reservoirs.

Example 8.1 By what factor does the number of states increase if 1 joule of heat is added to a system at room temperature (295 K)?

According to equation 8.12, the required factor is

$$\frac{\Omega_f}{\Omega_f} = e^{\Delta Q/kT}.$$

In this case $\Delta Q = 1$ J and $T = 295$ K, so $\Delta Q/kT = 2.5 \times 10^{20}$ and therefore

$$\frac{\Omega_f}{\Omega_i} = e^{2.5 \times 10^{20}} = 10^{1.1 \times 10^{20}}.$$

What a phenomenal change that is for just one joule of energy!

C Heat capacities

If you put a pot of water on one hot burner and a slice of bread on another, you will find that the bread is charred to a crisp before the pot of water is even lukewarm. Likewise, hot toast cools off much more quickly than a pot of hot water. Some objects have a greater capacity for holding heat than others. We quantify this concept by defining the "heat capacity" of an object to be a measure of how much heat energy must be added or removed in order to change its temperature by one degree.

In general, a system may be interacting in many ways with its environment. Adding heat energy may also stimulate interactions of other types. For example,

---

[4] We assume changes sufficiently small that T, p, μ do not change appreciably. Otherwise we would need to integrate.

Table 8.1. *Specific heats of various common substances at constant pressure. Except for the last two, they are measured at room temperature*

Substance	Specific heat		Substance	Specific heat	
	kcal/(kg K)	kJ/(kg K)		kcal/(kg K)	kJ/(kg K)
aluminum	0.21	0.90	sugar	0.27	1.13
copper	0.093	0.39	table salt	0.21	0.88
gold	0.030	0.126	wood	0.42	1.8
acetone (liquid)	0.53	2.22	air	0.25	1.05
ethyl alcohol (liquid)	0.58	2.43	helium (gas)	1.25	5.23
marble	0.21	0.88	water	1.00	4.19
dry leather	0.36	1.5	ice (below 0 °C)	0.50	2.09
synthetic rubber	0.45	1.9	liquid nitrogen	0.47	1.97

the volume may increase, or some particles may leave the system. These other interactions may add or remove energy from the system, so they will influence the change in temperature. Consequently, the heat capacity of a system depends on what else is happening.

We identify what else is *not* happening by subscripts. If y is a parameter that remains constant as we add the heat, then we define the heat capacity at constant y as

$$C_y = \left(\frac{\partial Q}{\partial T}\right)_y. \tag{8.13}$$

For example, the symbol C_p indicates that the pressure is held constant (although the volume might be changing) and the symbol C_V indicates that the volume is held constant (although the pressure might be changing). We seldom measure the heat capacities of systems interacting diffusively with their environment, so the number of particles is assumed to be constant unless otherwise stated. Heat capacities are not defined at ordinary phase transitions, because the latent heat added or subtracted causes no change in temperature.

You can see that the heat capacity of a system depends on its size. In order to raise its temperature by one degree, the ocean requires more heat than does a teaspoon of water. There are two common measures of heat capacity which depend only on the materials of a system and not its size. One is the heat capacity per mole, or molar heat capacity, and the other is the heat capacity per kilogram or specific heat (capacity) (Table 8.1). We define them as follows. For a system consisting of n moles of a substance and having total mass m,

$$\text{molar heat capacity } C_y = \frac{1}{n} C_y,$$
$$\text{specific heat (capacity) } c_y = \frac{1}{m} C_y. \tag{8.14}$$

D   Entropy and the third law

As we found in the last chapter and as is stated explicitly in the second law, entropy plays the central role in controlling the behaviors of systems. The entropy of any system at temperature T can be determined by integrating equation 8.3 $(dS = dQ/T')$ from $T' = 0$ to $T' = T$:

$$S(T) = S(0) + \int_0^T \frac{dQ}{T'} \qquad (8.15)$$

But to evaluate this, we must know the value of $S(0)$ and how to do the integral. The first is a theoretical matter, answered by the third law of thermodynamics. The second is an experimental matter. We examine each of these questions in the following two subsections.

D.1   The entropy at $T = 0$

As we remove energy from a system, we force it into states of ever decreasing total energy. As long as there are still some lower-energy states for the system to fall into, we can continue to remove energy. But eventually, it will reach that one state of lowest energy[5], and that is as far as it can go. At this point $\Omega = 1$ and so its entropy is zero. This defines $T = 0$. Thus

$$S = k \ln \Omega = k \ln 1 = 0 \qquad \text{(at absolute zero, } T = 0\text{)}.$$

Notice that this result does not depend on the nature of the system, nor the size of the box that it is in, nor what pressure it is under, nor the strength of the magnetic field, nor the number or nature of its particles, etc. No matter what conditions or constraints the system is held under, there will be one state[5] of lowest energy. And when the system is in that one state, its entropy is zero. This observation provides us with the "third law of thermodynamics."

> **Third law of thermodynamics**
>
> The entropy of a system goes to zero as the temperature goes to zero,
>
> $$S(T = 0) = 0, \qquad (8.16)$$
>
> no matter what the values of the external parameters are.

You may think that the lowest lying state could be degenerate. However, there seems to be always some form of weak interaction that breaks the degeneracy at very low energies, leaving one state lower than the others. For this reason, we will use the above form in this book. You can see that the third law automatically

---

[5] If there were n states all having the very lowest energy, the third law would read $S(T = 0) = k \ln n =$ constant. But the constant would be extremely tiny – on the order of 10^{-23} J/K.

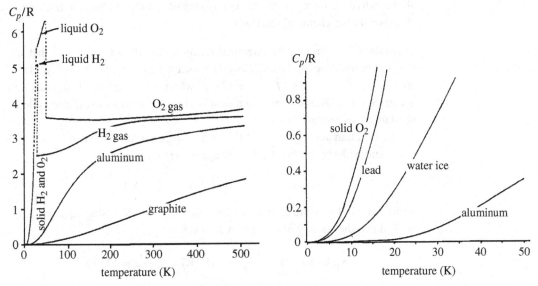

Figure 8.3 Plot of molar heat capacities at constant pressure vs. temperature for various materials, with some very-low-temperature behaviors enlarged on the right. Why are those for liquids so high?

takes care of the first of our two concerns regarding determining the entropy of a system, since $S(0) = 0$.

D.2 The entropy at finite temperatures

Having solved the first problem, $S(0)$, we now face the second – how to evaluate

$$S(T) = 0 + \int_0^T \frac{dQ}{T'}. \tag{8.15'}$$

From the definition of heat capacity (equation 8.13), we can write the heat added as

$$dQ = C_y dT,$$

so the value of the entropy for temperature T and other parameter values y is

$$S(T, y) = \int_0^T \frac{C_y dT'}{T'}. \tag{8.17}$$

For example, suppose we wish to know the entropy $S(T, p, N)$ for a glass of water ($N = 10^{25}$ molecules, which is $n = 16.6$ moles or $m = 0.30$ kg) at room temperature ($T = 295$ K) and atmospheric pressure ($p = 1$ atm). So, to evaluate the integral we would need to know the heat capacity of water at atmospheric pressure for temperatures from 0 to 295 K. We could either look this up in tables or measure it ourselves. Examples of measured molar heat capacities (C_p) for various materials are plotted in Figure 8.3.

You can see that determining a system's entropy requires some calculation, as does determining other extrinsic properties, such as energy, volume, or the number of particles. But it is not difficult, and it certainly beats trying to count all the individual quantum states. As an example, we look at the heat released or absorbed during chemical reactions.

Example 8.2 Consider the chemical reaction $A + B \rightarrow C$. Suppose that the molar heat capacities of the reactants (at constant pressure) in units of J/(mole K) are $C_A = 5\sqrt{T}$, $C_B = 8\sqrt{T}$, $C_C = 12\sqrt{T}$, where T is in kelvins. If this reaction is carried out at 300 K, how much heat is absorbed or released if one mole of substance C is produced?

The heat transfer is given by $\Delta Q = T\Delta S$. Since we start with reactants A and B and end with the product C, this change in entropy is given by

$$\Delta S = S_{\text{final}} - S_{\text{initial}} = S_C - (S_A + S_B)$$

So we need to know the entropies of S_A, S_B, S_C at 300 K. Using equation 8.17, we find that the entropy of reactant A at 300 K is

$$S_A(300\text{K}) = \int_0^{300} \frac{C_A dT}{T} = (5\text{ J/K}) \int_0^{300} \frac{dT}{\sqrt{T}} = 173\text{ J/K}.$$

Similar calculations for the entropies of reactant B and the product C give

$$S_B(300\text{ K}) = 277\text{ J/K}, \qquad S_C(300\text{ K}) = 416\text{ J/K}.$$

Therefore

$$\Delta S = S_C - (S_A + S_B) = 416 - (173 + 277)\text{ J/K} = -34\text{ J/K},$$

and the amount of heat transferred is

$$\Delta Q = T\Delta S = (300\text{ K})(-34\text{ J/K}) = -10\,200\text{ J}.$$

The negative sign indicates that the entropy decreases as heat leaves the system, (i.e., the reaction is *exothermic*).

D.3 Heat capacities at low temperatures

Because the integrand in Equation 8.17 has temperature in the denominator, the integral diverges at zero temperature unless the heat capacity is zero at that point. Consequently, the heat capacities of all systems must go to zero as the temperature goes to zero, no matter what (Figure 8.3):

$$C_y \rightarrow 0 \quad \text{as} \quad T \rightarrow 0. \tag{8.18}$$

This result has interesting consequences at low temperatures. A small heat capacity means that a relatively small amount of added energy will cause a large increase in temperature, and hours of patient effort to obtain a very low temperature may be wasted. Low-temperature equipment must be carefully isolated from very small and seemingly innocent energy sources, such as voices, vibrations from cars outside the building, stray electromagnetic fields in the room, or even an insect landing on the apparatus.

Summary of Sections B–D

Because the entropy and the number of accessible states are related through $\Omega = e^{S/k}$, the factor by which the number of accessible states changes during a process is (equation 8.12)

$$\frac{\Omega_f}{\Omega_i} = e^{\Delta S/k} = e^{\Delta Q/kT} = e^{(\Delta E + p\Delta V - \mu\Delta N)/kT},$$

where the first of these expressions comes from the definition of entropy, the second from the definition of temperature, and the third from a rearrangement of the first law.

The heat capacity of a system measures how much heat is required to raise its temperature by one degree (equation 8.13).

$$C_y = \left(\frac{\partial Q}{\partial T}\right)_y.$$

The subscript y indicates the parameters held constant as the measurement is being made. Two common measures that depend on the nature of the substance but not its size are (equation 8.14)

$$\text{molar heat capacity } C_y = \frac{1}{n}C_y$$

$$\text{specific heat capacity } c_y = \frac{1}{m}C_y$$

where n is the number of moles, and m is the mass.

Since the change in entropy dS is equal to dQ/T, the entropy of a system at any temperature T can be calculated through (equation 8.15)

$$S(T) = S(0) + \int_0^T \frac{dQ}{T'}.$$

The third law states that the entropy must go to zero at absolute zero (equation 8.16), because the system must be in that one state of lowest possible energy; i.e.,

$$S(T = 0) = 0$$

no matter what the conditions are. Given this and the fact that the heat capacity relates the heat added, dQ, to the change in temperature dT, we can write the entropy of a system as a function of temperature T and other parameters y (equation 8.17):

$$S(T, y) = \int_0^T \frac{C_y dT'}{T'}.$$

Because entropy is finite, the integral cannot diverge at the zero temperature limit. Therefore, we know that heat capacities must go to zero at absolute zero (equation 8.18):

$$C_y \to 0 \quad \text{as} \quad T \to 0.$$

Problems

Section A

1. (a) For a monatomic ideal gas of N molecules, $\Omega(E)$ is proportional to E raised to what power?

   (b) Using equation 8.1 show that, for a monatomic ideal gas of N molecules, $E = (3/2)NkT$.

   (c) Starting with $\Omega(E) = \text{constant} \times (E_{\text{therm}})^{\alpha N\nu}$ (equations 6.7 and 6.10), where α is any number, use the definitions of entropy and temperature to derive the relationship between E_{therm} and T. (Assume that $E = E_{\text{therm}}$.)

2. Suppose that there are 10^{28} diatomic air molecules in a room.

   (a) How many degrees of freedom does this system have?

   (b) What is the internal energy of this system at room temperature (295 K)?

3. Two small systems, A_1 and A_2, are in thermal equilibrium. The number of states accessible to each increases with its energy according to $\Omega_1 = (E_1/C)^{10}$ and $\Omega_2 = (E_2/C)^8$, where $C = 10^{-23}$ J. The total energy of the combined system is fixed at $E_0 = E_1 + E_2 = 10^{-18}$ J.

   (a) How many degrees of freedom have systems A_1 and A_2, respectively?

   (b) Use the fact that $\partial \Omega_0 / \partial E_1 = 0$ when Ω_0 is a maximum to find E_1 and E_2 when the combined system is in equilibrium.

   (c) What is the entropy of the combined system in equilibrium?

   (d) Using the definition of temperature, and the fact that in equilibrium the temperature of either system is the same, find the temperature of the system.

4. Suppose that you don't like the way in which the temperature scale is defined, and you wish to define a scale that gives a nice round number for the value of Boltzmann's constant. You measure your temperatures on this scale in °R (for "degrees round"). In units of °R, what would be the boiling point of water if:

   (a) $k = 1.0 \times 10^{-16}$ erg/ °R?

   (b) $k = 10^{-4}$ eV/ °R?

   (c) $k = 1.0$ J/ °R?

   (d) $k = 1.0$ eV/ °R?

5. What would be the value of Boltzmann's constant if the temperature of the triple point (273.16 K) were defined as 100 K? What would be the boiling point of water on this scale?

6. For a monatomic ideal gas, each molecule has three translational degrees of freedom only. Suppose you calibrate your temperature scale by saying that

300 K is defined to be the temperature of a mole of this ideal gas when it has a thermal energy of 3740 J. What would be the value of Boltzmann's constant, k?

7. (a) If you add 20 J of heat to a system at $-20\,°C$, what is the change in its entropy?
   (b) By what factor does the number of states accessible to the system increase?

8. (a) How many joules of heat energy would you have to add to the Pacific Ocean (average temperature $T = 4\,°C$, volume $V = 0.70 \times 10^9\ \mathrm{km}^3$) to double the number of states accessible to it?
   (b) Would your answer be the same if you were dealing with a cup of water at $4\,°C$ instead?

9. Consider some ice at $-1\,°C$ in a glass of water at $+10\,°C$. For each joule of heat energy that flows from the water to the ice, find the change in entropy of (a) the ice, (b) the water, (c) the total system.

10. A 10^{-2} kg ice cube, initially at $0\,°C$, melts in the Atlantic Ocean, where the water temperature is $10\,°C$. After melting, the ice melt heats up to match its $10\,°C$ environment. (The latent heat of fusion $= 333$ J/g and the specific heat $= 4.18$ J/(g K). Find the change in entropy of
    (a) the water that was originally in the ice cube,
    (b) the Atlantic Ocean.
    (c) By what factor does the number of states available to the combined system change?

11. The number of states accessible to an ideal gas having energy in the range between E and $E + \delta E$ is given by $\Omega_0 = \text{constant} \times V^N E^{3N/2}\delta E$, where V is the volume of the gas and N the number of molecules.
    (a) Using equation 8.7, show that $pV = NkT$.
    (b) This is sometimes written as $pV = nRT$, where n is the number of moles of the gas and R is called the "gas constant." What is R in terms of Boltzmann's constant and Avogadro's number?

12. The number of states is expressed as a function of various parameters for three systems below. For each, find an "equation of state," which gives the relationship between p, V, N, and T. (C and b are constants.)
    (a) $\Omega = Ce^{bNV^2}(EV)^N$,
    (b) $\Omega = \frac{\pi}{2}\left(\frac{2}{h}\right)^{3N} V^{N/2}e^{bNV} E^{2N}$,
    (c) $\Omega = Ce^{-b/V^{10}} E^{3N}$.

13. For each system in problem 12, find the dependence of the internal energy E on V, N, and T.

14. For each system in problem 12, find the dependence of the chemical potential μ on E, V, N, and T.

15. Consider two gaseous systems interacting mechanically and thermally but not diffusively. They are isolated from the rest of the Universe, and their total volume is fixed at V_0.
    (a) Show that the total number of accessible states, Ω_0, is a very sensitive function of the distribution of volume between them. (See equation 6.11 and the argument preceding it for the dependence of Ω on V.)
    (b) Since $dV_1 = -dV_2$, show that $\partial S_1/\partial V_1 = \partial S_2/\partial V_2$ when the systems are in equilibrium.
    (c) From this, what can you conclude about how p_1, T_1, p_2, and T_2 are related in equilibrium?

16. A hundred grams of water are heated from $10\,°C$ to $95\,°C$. If the specific heat of water is $4.19\ J/(g\,°C)$, what is the increase in entropy of the water during this process?

17. Three hundred grams of aluminum are heated from $-50\,°C$ to $300\,°C$. If the specific heat of aluminum is $0.88\ J/(g\,°C)$, what is the change in entropy of the aluminum?

18. Consider the differentials listed below. For each, state whether it is exact. For those that are not exact, find a multiplicative factor $f(x, y)$ such that $f(x, y)dF$ is exact:
    (a) $dF = 2x\,dx + (x^2/y)\,dy$,
    (b) $dF = 2xy\,dx + x^2\,dy$,
    (c) $dF = 2xy^2\,dx + x^2y\,dy$,
    (d) $dF = 3x^3y^2\,dx + x^4y\,dy$,
    (e) $dF = dx/x + dy/y$,
    (f) $dF = p\,dV + V\,dp$,
    (g) $dF = p^2\,dV + pV\,dp$.

19. Estimate the total thermal energy of the following systems at 290 K:
    (a) the air in your bedroom (one mole occupies 22.4 liters at $0\,°C$);
    (b) the iron atoms in a 5 gram nail (the atomic mass number for iron is 56);
    (c) A diamond of mass 0.1 gram (the atomic mass number of carbon is 12);
    (d) the Pacific Ocean ($0.7 \times 10^9\ km^3$ of water, each molecule having six degrees of freedom).

20. A system has 10^{25} degrees of freedom and initial volume $1\ m^3$ and is under a pressure of $10^5\ N/m^2$. While held at a constant temperature (assume constant internal energy) of $17\,°C$, it expands by $1\ mm^3$.
    (a) Does the number of accessible states increase or decrease?
    (b) By how many times?

21. In a certain system, the number of accessible states increases by a factor $10^{10^{20}}$ when 1 joule of energy is added at constant V and N.
    (a) What is the increase in entropy, ΔS?
    (b) What is the temperature of the system?

22. Which of the following differentials, dE, dV, dQ, dp, dT, $d\mu$, dN, dW, dS, are exact? For each that is inexact, find a multiplicative factor involving E, V, T, p, N, etc. that would make it exact.

23. Suppose that the heat entering a system can be expressed in terms of the temperature and volume as $dQ = b(T^2 dV + TV dT)$ and that the work done by the system can be expressed in terms of its pressure and volume as $dW = c(4p^4 V^2 dp + 2p^5 V dV)$, where b and c are constants. Find the multiplicative factors that turn these inexact differentials into exact ones.

24. Consider a system of 10^{25} particles at temperature 295 K, pressure 10^5 Pa, and chemical potential -0.3 eV. It experiences the following very small changes: 2.2×10^{-2} J of heat are added, it expands by 10^{-7} m^3 and it gains 10^{17} particles. What is the change in its (a) internal energy, (b) entropy?

Section B

25. Consider a system at 300 K to which 1 joule of heat is added.
    (a) What is its change in entropy?
    (b) By what factor does the number of accessible states increase?

26. For any system, the following quantities are all interrelated: the number of degrees of freedom, $N\nu$, the change in entropy, ΔS, the ratio of final to initial thermal energies, E_f/E_i, and the ratio of final to initial states, Ω_f/Ω_i. Below is a table, each horizontal row representing some process for some system. Can you fill in the missing numbers?

$N\nu$	ΔS (J/K)	E_f/E_i	Ω_f/Ω_i
10^{24}	1		
	2	1.01	
		1.03	$10^{10^{23}}$
10^{23}		1.02	

27. (a) By how much does the entropy of the Atlantic Ocean ($T = 280$ K, $V = 0.36 \times 10^9$ km^3) change when 0.1 joule of heat is added?
    (b) What about a cup of water at 280 K?
    (c) By what factor does the number of states available to each system increase?

28. A system of 10^{24} particles with $\mu = -0.2$ eV is at room temperature (295 K). Find the factor by which the number of accessible states increases in the following cases.
   (a) The number of particles is increased by 0.01%, without adding energy to, or doing work on, the system. (That is, the incoming particles have zero total energy $u_0 + \varepsilon_{therm} = 0$, and the system's volume is unchanged.)
   (b) A single energyless particle is added.

29. A rubber ball is in contact with a heat reservoir that keeps its temperature constant at 300 K. (So assume that the internal energy is constant.) At a pressure of 1.001 atm its volume decreases by 10^{-3} cm^3.
   (a) What is the change in entropy?
   (b) By what factor does the number of accessible states change?
   (c) Repeat for a temperature of 20 °C, a pressure of 1.02×10^5 Pa, and a volume reduction of 10^{-10} m^3.

30. A magnet is in contact with a reservoir that keeps it at 300 K. The magnet has a magnetic moment $\mu_z = 10^{-3}$ J/T and is sitting in an external field oriented along the z-axis of strength $B_z = 0.1$ T. The external field is increased by 1%, and the induced magnetic moment also increases by 1%.
   (a) What is the change in entropy?
   (b) By what factor does the number of accessible states change?

31. By how much does the entropy of a system increase if the number of accessible states doubles?

32. Consider 1 m^3 of steam held at a temperature of 600 °C and a pressure of 50 atm. By how much must its volume expand if the number of accessible states increases a billionfold?

33. Consider an insulated ideal gas, such that no heat energy can enter or leave. It is slowly compressed to 96% of its original volume, and its temperature rises correspondingly.
   (a) What is the change in its entropy?
   (b) An increase in temperature indicates more thermal motion and larger momenta. That would imply that the particles have more accessible room in momentum space with a corresponding increase in number of accessible quantum states. How, then, can you justify your answer to part (a)?

34. When one joule of heat energy is added to a system, the number of accessible states increases by a factor of $10^{10^{19}}$. What is the temperature of the system?

35. You have a system of 10^{24} particles at 300 K. If you hold the volume constant and add 10^{19} more perfectly energyless particles, you notice the temperature

of the system rises slightly, as if you had added 0.4 J of heat. Estimate the chemical potential for the particles of this system (in eV).

Section C

36. Why can't heat capacities be defined at ordinary phase transitions?

37. Calculate the change in entropy for 0.24 kg of each of the following as it is heated from 295 K to 296 K (use Table 8.1): (a) Water, (b) gold, (c) marble, (d) wood.

38. Which of the above materials would make the most efficient reservoir for the storage of solar heat? Why do you suppose daily and seasonal temperature changes are greater inland than on the coast?

39. For a material that expands as its temperature rises, do you think C_p or C_V would be larger? Why?

40. (a) One helium atom has mass 6.7×10^{-27} kg. Estimate the specific heat at constant volume c_V for helium in kJ/(kg K). (Note that if the volume is constant, no work is done as the heat is added.)
    (b) Why does your answer differ from that in Table 8.1?

41. For an ideal gas having a fixed number N of particles, the internal energy depends only on the temperature and not at all on the volume: $E = (v/2)NkT$. Furthermore, the pressure, volume, and temperature are related by the ideal gas law, $pV = NkT$. The first law for a system of a fixed number of particles reads $dE = dQ - pdV$. Using this information and the definition of molar heat capacities, show that (a) $C_V = (v/2)R$, (b) $C_p = C_V + R$, where $R = N_A k$ is the molar gas constant.

Section D

42. (a) If 2000 J of heat are added to 100 g of gold, initially at 0 °C, what is the change in entropy?
    (b) By what factor does the number of accessible states increase?

43. Repeat the above problem for 500 g of aluminum.

44. Compute the change in entropy of 1 gram of water as it goes from solid ice at 0 °C to water vapor at 100 °C. The latent heat of fusion is 330 J/g and of vaporization is 2260 J/g.

45. Consider the chemical reaction $2A + B \rightarrow C$. You produce 1 mole of C at 500 K and atmospheric pressure. The molar heat capacities of these substances at atmospheric pressure are all zero between 0 K and 10 K and constant above 10 K, being given for $T > 10$ K by

$$C_A = 19.4 \text{ J/(mole K)}, \qquad C_B = 35.9 \text{ J/(mole K)},$$
$$C_C = 67.7 \text{ J/(mole K)}.$$

(a) How much heat will be released in this reaction?

(b) How can the system lose entropy and not violate the second law?

46. Consider the chemical reaction $2A + 2B \rightarrow C + D$. You produce 1 mole of C at 300 K and atmospheric pressure. How much heat will be released in this reaction if the molar heat capacities of these substances, in units of J/(mole K), at atmospheric pressure are as follows?:

$$C_A = 18.6(T/200 \text{ K})^{1/3}, \qquad C_B = 15.7(T/200 \text{ K})^{1/3},$$
$$C_C = 23.4(T/200 \text{ K})^{1/3}, \qquad C_D = 39.7(T/200 \text{ K})^{1/3}.$$

47. The latent heat of fusion for iron at 1809 K is 246 J/g, or 13 790 J/mole. Its atomic mass number is 56. In the solid state, each iron atom has six degrees of freedom, and in the liquid state each has only three.

(a) What is the change in the potential energy reference level, u_0, when iron goes from solid to liquid?

(b) What is the change in entropy per gram?

48. Repeat the above problem, but for the sublimation of dry ice. CO_2 has a mass number of 44 and a latent heat of sublimation of 25 200 J/mole at 194.6 K. In the solid state there are six degrees of freedom per molecule, but there are only five in the vapor state. The average work done per molecule in subliming is $pv = kT$, where v is the volume per molecule.

49. Suppose that the specific heat at constant pressure for ice in units of J/(g K) were given by $0.2613\sqrt{T}$ from 0 K to 64 K, 2.09 from 64 K to 273 K, and 4.18 from 273 K to 293 K. The latent heat of fusion of ice at 273 K is 333 J/g. Use this information to calculate the entropy of a liter of water at 20 °C.

Part V
Constraints

Chapter 9
Natural constraints

A Overview

The preceding chapters introduced the fundamental ideas that connect the microscopic and macroscopic behavior of systems. They also gave an overview of the three types of interactions between systems and how the second law controls them. These concepts form the statistical basis of thermodynamics, and the tools are so general that they can be applied to almost any system imaginable. This is the single most impressive feature of the subject. Unfortunately, it is also the single most *confusing* feature of the subject. There are so many different kinds of systems and such a variety of parameters – internal energy, temperature, pressure, entropy, volume, chemical potential, number of particles, and many more. Furthermore, the interdependence among these parameters varies from one system to the next and in ways that are usually not specified. Consequently, we often deal with general and abstract expressions, each involving many parameters whose interrelationships are either vague or unknown.

But the large number of parameters can be turned to our advantage. We don't need them all, so we can choose to use whichever we wish and ignore the rest. Furthermore, their behaviors and interrelationships are heavily constrained. In this and the following chapters we learn how to make order out of chaos through a judicious choice of parameters and the application of constraints.

A.1   Parameters and constraints

There is one independent variable for each type of interaction, as we see in the first law. Because they appear as differentials, we ordinarily choose the independent variables as follows:

- thermal interaction, the entropy S;
- mechanical interaction, the volume V;
- diffusive interaction, the number of particles N.

If a system were simultaneously undergoing several different mechanical interactions

$$-p\mathrm{d}V + \mathbf{B} \cdot \mathrm{d}\boldsymbol{\mu} + \mathbf{E} \cdot \mathrm{d}\mathbf{P} + \cdots,$$

or diffusive interactions

$$\mu_1 \mathrm{d}N_1 + \mu_2 \mathrm{d}N_2 + \cdots,$$

there would be one independent variable for each interaction (V, $\boldsymbol{\mu}$, $\mathbf{P}$, ... or N_1, N_2, ..., respectively). For clarity, we use the three standard variables S, V, N, where V and N may each represent more than one independent variable if the system is engaging in more than one type of mechanical or diffusive interaction.

Of the interrelationships between the many parameters, some are universal and inviolable, applying equally to all systems. We investigate these "natural constraints" in this chapter. Many other interrelationships depend on the particular nature of the system. For example, gases behave differently from liquids and solids, and within each of these groups there are further differences. In Chapter 10 we investigate the use of models to express relationships that vary from one system to the next.

The first law expresses how a system's internal energy varies with the entropy, volume, and number of particles: $E = E(S, V, N)$. But we may wish to study other properties and other variables. In Chapter 11 we will learn how to express any property in terms of whichever variables we wish.

Table 9.1. *Common constraints on systems*

Type	constraint
adiabatic[a]	$dQ = 0$
isobaric	$dp = 0$
isochoric	$dV = 0$
isothermal	$dT = 0$
nondiffusive or closed	$dN = 0$

[a] This is equivalent to $dS = 0$ *if* the process is quasistatic.

Individual systems may incur constraints imposed by such things as their natural environment or our experimental setup. For example, we may do an experiment at atmospheric pressure ($p = $ constant), or we might study a system that is neither gaining nor losing particles ($N = $ constant). Constraints that are imposed upon certain systems in certain environments are called "imposed constraints." Table 9.1 lists some of the more common ones. We will learn how to deal with them in Chapter 12.

Each imposed constraint reduces the number of independent variables by one: if one variable is held constant, all thermodynamic properties depend on only two parameters rather than three; if two are held constant then only one independent variable is left. And if any three parameters are fixed, no property of the system can change at all. For example, the first law, equation 8.4,

$$dE = TdS - pdV + \mu dN,$$

expresses the interrelationship between the seven parameters E, T, S, p, V, μ, N, of which no more than three are independent. If you hold any two constant then all seven can be expressed as a function of just one. You can pick whichever you wish, using the techniques to be described in Chapter 11.

In Chapter 13 we study engines and refrigerators. Engines revolutionized our society and gave a strong incentive to the study and development of thermodynamics. They offer useful illustrations of the application of constraints, both natural and imposed. In Chapter 14 we investigate the effects of constraints that are common in diffusive interactions. The chapter includes an overview of diffusive equilibrium followed by important applications such as osmosis, chemical equilibrium, and phase equilibrium.

A.2   The focus of this chapter

We now turn our attention to the subject of the present chapter. With such a great deal of interdependence among the many thermodynamic variables, any constraint placed on one of them must affect the others. For example, the fact that

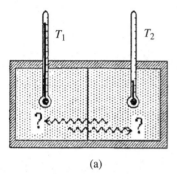
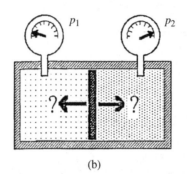
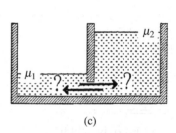

(a)   (b)   (c)

Figure 9.1 When two interacting system are not yet in equilibrium, which way will (a) the heat flow, (b) the piston move, (c) the particles flow? The answer to all these lies in the second law requirement that the total entropy must increase.

entropy cannot decrease demands that heat must flow from hot to cold and not vice versa (Figure 9.1). This is just one of a large number of interrelationships resulting from the second law's constraint on the entropy. Likewise, the zeroth, first, and third laws also impose constraints that result in additional interdependencies. The laws are inviolate and therefore the resulting interrelationships must apply to all macroscopic systems and all processes, no matter what. Here we group these "natural and universal" constraints into five categories according to their origins:

- those arising from Nature's desire to maximize the entropy (the second law);
- identities (zeroth law);
- those arising from energy conservation (the first law);
- those that must exist in any mathematical expression when there are more parameters than independent variables (Maxwell's relations);
- those arising because the entropy goes to zero as $T \to 0$ (the third law).

B Second law constraints

The second law requires that when interacting systems are at equilibrium, their total entropy is a maximum. In this section we will demonstrate that this demands the following universal and inviolable behaviors of all systems, regarding their (a) thermal, (b) mechanical, and (c) diffusive interactions.

1. As two interacting systems approach equilibrium,
   - (a) heat flows toward the system with the lower temperature,   (9.1a)
   - (b) boundaries move toward the system with the lower pressure,   (9.1b)
   - (c) particles flow toward the system with lower chemical potential.   (9.1c)
2. After two interacting systems have reached equilibrium,
   - (a) their temperatures are equal,   (9.2a)
   - (b) their pressures are equal,   (9.2b)
   - (c) their chemical potentials are equal.   (9.2c)
3. When heat, volume, or particles (ΔQ, ΔV, ΔN) are transferred, one of these by itself with the other two held constant, the following must be true:
   - (a) $\Delta T \Delta Q > 0$ (if heat is added, the temperature must rise);   (9.3a)
   - (b) $\Delta p \Delta V < 0$ (if the volume is increased, the pressure must fall); (9.3b)
   - (c) $\Delta \mu \Delta N > 0$ (if particles are added, the chemical potential must rise). (9.3c)

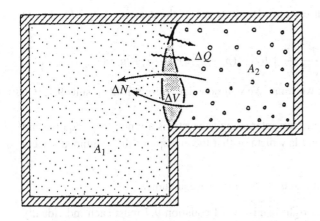

Figure 9.2 When systems A_1 and A_2 are interacting thermally, mechanically, and diffusively, any energy, volume, or particles gained by one of them must come from the other. The volume change ΔV during the interaction is shaded. According to the second law, the entropy of the combined system must be a maximum when it reaches equilibrium.

All these results are simply "common sense" to us, because we live in a world in which the second law rules. The reason for deriving these results here is to point out how much of our "common sense" world is governed by just one simple idea – the second law.

B.1 Approaching equilibrium

To derive the above results, we consider the general interaction shown in Figure 9.2. An isolated system, A_0, is composed of two interacting subsystems, A_1 and A_2. We are interested in the change in entropy of the combined system,

$$dS_0 = dS_1 + dS_2 = \frac{dQ_1}{T_1} + \frac{dQ_2}{T_2}, \tag{9.4}$$

and we express dQ_2 in terms of the changes dQ_1, dV_1, dN_1 in system 1.[1]

Any energy, volume, and/or particles gained by one subsystem come from the other:

$$dE_2 = -dE_1, \qquad dV_2 = -dV_1, \qquad dN_2 = -dN_1. \tag{9.5}$$

The first of these relationships ($dE_2 = -dE_1$) can be rewritten using the first law:

$$dQ_2 - p_2 dV_2 + \mu_2 dN_2 = -(dQ_1 - p_1 dV_1 + \mu_1 dN_1).$$

We solve for dQ_2, replacing dV_2 and dN_2 by $-dV_1$ and $-dN_1$, respectively:

$$dQ_2 = -dQ_1 + (p_1 - p_2)dV_1 - (\mu_1 - \mu_2)dN_1. \tag{9.6}$$

We now substitute this expression for dQ_2 into equation 9.4 to get an expression

---

[1] In Chapter 5 we learned that the differential dQ is *not* exact because Q is not a function of state, making it quite different from E, V, and N. So we have to be careful, employing the first law to ensure that we don't leave anything out.

for the change in total entropy, dS_0, in terms of the changes dQ_1, dV_1, dN_1 in system 1:

$$dS_0 = \frac{1}{T_2}\left[\left(\frac{T_2 - T_1}{T_1}\right)dQ_1 + (p_1 - p_2)\,dV_1 - (\mu_1 - \mu_2)\,dN_1\right]. \qquad (9.7)$$

This is the expression we want. We will now derive from it the results 9.1, 9.2, and 9.3.

First, imagine that the two interacting subsystems are not yet in equilibrium (Figure 9.1). The second law dictates that the entropy of the combined system must increase,

$$dS_0 > 0 \qquad \text{(approaching equilibrium)}.$$

The three terms on the right-hand side of equation 9.7 must each individually obey the inequality, because the changes dQ, dV, dN are independent; for example, the systems could be interacting only thermally, or only mechanically, or only diffusively. The result is the first set (9.1a–c) of promised results.

(a) If the two are interacting thermally, then $[(T_2 - T_1)/T_1]\,dQ_1 > 0$. That is, heat must flow toward the lower temperature (e.g., $dQ_1 > 0$ if $T_2 > T_1$).
(b) If the two are interacting mechanically, then $(p_1 - p_2)\,dV_1 > 0$. That is, boundaries must move toward the lower pressure (e.g., $dV_1 > 0$ if $p_1 > p_2$).
(c) If the two are interacting diffusively, then $-(\mu_1 - \mu_2)\,dN_1 > 0$. That is, particles must flow towards the lower chemical potential (e.g., $dN_1 > 0$ if $\mu_1 < \mu_2$).[2]

B.2   At equilibrium

An important corollary applies to two systems that have reached equilibrium. Because the entropy is a maximum, it remains unchanged with small changes in the variables:

$$dS_0 = 0 \qquad \text{(at equilibrium)}.$$

Again, the three terms in equation 9.7 are independent. Setting each term individually equal to zero gives the second set (9.2a–c) of the promised results: for two systems in thermal, mechanical, and diffusive equilibrium, respectively,

$$T_2 = T_1, \qquad p_2 = p_1, \qquad \mu_2 = \mu_1.$$

B.3   Second order constraints

Now we examine second-law constraints on second derivatives. At a function's maximum, the first derivative in any variable is zero but the second derivative is

---

[2] Often there are many different particles and their transfer from one system to another is not one to one (e.g., chemical reactions). Following through the above development with each type of particle in each system labeled separately, the condition becomes $\mu_1 dN_1 + \mu_2 dN_2 + \mu_3 dN_3 + \cdots < 0$.

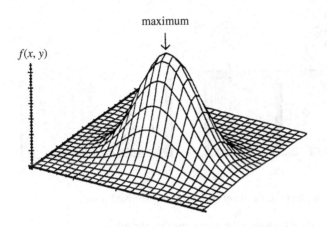

maximum

$f(x, y)$

Figure 9.3 When a differentiable function is at its maximum, the first derivative with respect to any variable is zero, and the second derivative is negative ($f' = 0$ and $f'' < 0$).

negative (Figure 9.3). Hence when two interacting systems are displaced slightly from equilibrium, the resulting change in entropy must be negative:

$$\Delta S_0 = \Delta S_1 + \Delta S_2 < 0 \qquad \text{(if displaced away from equilibrium)}.$$

To reveal the implications, we begin with two systems in equilibrium at temperature T, pressure p, and chemical potential μ. We then transfer either a small amount of heat, volume, or particles (ΔQ, ΔV, or ΔN) from one to the other. To keep the mathematics simple, we consider system 2 to be a huge reservoir, so that its temperature, pressure, and chemical potential remain unchanged. We take system 1 to be much smaller, so that the transfer may cause a noticeable change ΔT, Δp, or $\Delta \mu$.[3] During the transfer, then, the *average* temperature, pressure, or chemical potential for the two systems would be: for the small system A_1,

$$\overline{T_1} = T + \frac{\Delta T}{2}, \qquad \overline{p_1} = p + \frac{\Delta p}{2}, \qquad \overline{\mu_1} = \mu + \frac{\Delta \mu}{2}, \qquad (9.8)$$

and for the huge reservoir, A_2,

$$\overline{T_2} = T, \qquad \overline{p_2} = p, \qquad \overline{\mu_2} = \mu.$$

If we put these values into equation 9.7 for the change in total entropy and use $(1 + \varepsilon)^{-1} \approx 1 - \varepsilon$ (with $\varepsilon = \Delta T / 2T$ in the heat transfer term), we get

$$\Delta S_0 = \frac{1}{2T} \left(-\frac{\Delta T \Delta Q}{T} + \Delta p \Delta V - \Delta \mu \Delta N \right) < 0. \qquad (9.9)$$

This tells us immediately that for the small system, if we allow only one type of

[3] If we allowed only *one* type of interaction at a time, there would indeed be change ΔT, Δp, or $\Delta \mu$ even at phase transitions. You might think of freezing water, where you remove heat and the temperature does not change. But it expands, so it is undergoing both thermal and mechanical interactions. If you did *not* allow it to expand, its temperature would indeed fall as heat is removed.

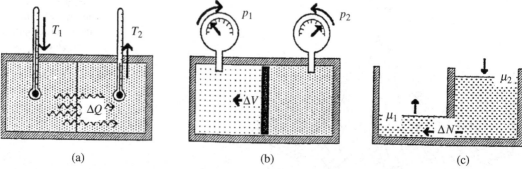

(a)   (b)   (c)

Figure 9.4 For thermal, mechanical, or diffusive interactions, the second law guarantees the following. (a) Heat flows from hot to cold; as a result, the higher temperature falls and the lower temperature rises. (b) The piston moves from high pressure toward low. As a result, the higher pressure falls and the lower pressure rises. (c) Particles flow from high chemical potential to low. As a result, the higher chemical potential falls and the lower one rises. That is, all changes are in the direction that approaches equilibrium.

interaction at a time, we get the third set (9.3a–c) of promised results:

(a) $\Delta T \Delta Q > 0$ (e.g., if heat is added, the temperature must rise);
(b) $\Delta p \Delta V < 0$ (e.g., if the volume is increased, the pressure must fall);
(c) $\Delta \mu \Delta N > 0$ (e.g., if particles are added, the chemical potential must rise).

Although we have used a reservoir to simplify our calculations, these interrelationships must in fact hold for any interacting systems, because the amount of heat, volume, or particle transfer (ΔQ, ΔV, or ΔN) does not depend on where it comes from.

Results 9.1 and 9.3 tell us that when systems are interacting, the transfer of heat, volume, and or particles must bring their temperatures, pressures, and/or chemical potentials closer to equilibrium (Figure 9.4). In thermal interactions, for example, heat must flow from hot to cold. As it does, the hotter system cools off and the cooler one warms up until their temperatures are equal. The same happens to pressures in mechanical interactions and chemical potentials in diffusive interactions.

B.4   Fluctuations

We now examine fluctuations in any one property χ (e.g., T, p, V, N, E, etc.) for a system in equilibrium with a large reservoir. We are free to choose whichever three independent variables we wish, but we must specify the two that are being held constant while fluctuations in the third are being studied. We begin by noting that the probability for any particular configuration is proportional to the number of accessible states (equations 6.2, 7.11):

$$P \propto \Omega = e^{S_0/k},$$

where S_0 is the total entropy of the combined system ($S_0 = S + S_R$, where S_R is the entropy of the reservoir). We expand S_0 in a Taylor series expansion for small displacements $\Delta \chi$ from equilibrium:

$$S_0 = S_{0,\max} + \left. \frac{\partial S_0}{\partial \chi} \right|_{\max} \Delta \chi + \frac{1}{2} \left. \frac{\partial^2 S_0}{\partial \chi^2} \right|_{\max} (\Delta \chi)^2 + \cdots.$$

If we ignore the higher-order terms, note that the first derivative of S_0 is zero when in equilibrium (i.e., S_0 is a maximum), and incorporate a factor $\exp(S_{0,\text{max}}/k)$ into the constant of proportionality, the probability for a fluctuation $\Delta\chi$ is then given by

$$P \approx C \exp\left(\frac{1}{2k} \frac{\partial^2 S_0}{\partial \chi^2}\bigg|_{\text{max}} (\Delta\chi)^2 \right).$$

We recognize this as the familiar Gaussian form (3.7)–(3.10), which we can write as

$$P(\Delta\chi) = \frac{1}{\sqrt{2\pi}\,\sigma} e^{-(\Delta\chi)^2/2\sigma^2}, \qquad \text{where} \qquad \frac{1}{\sigma^2} = -\frac{1}{k} \frac{\partial^2 S_0}{\partial \chi^2}\bigg|_{\text{max}}. \qquad (9.10)$$

To illustrate this result, we look at the fluctuations in energy, volume, or number of particles (ΔE, ΔV, or ΔN) for a system in equilibrium with a reservoir. We can write the entropy in terms of these variables using the first law in the form 8.6. Because energy, volume, and particles are conserved (i.e., $dE_R = -dE$, $dV_R = -dV$, $dN_R = -dN$), the change in entropy of the combined system is

$$dS_0 = dS + dS_R = \left(\frac{1}{T} - \frac{1}{T_R} \right) dE + \left(\frac{p}{T} - \frac{p_R}{T_R} \right) dV - \left(\frac{\mu}{T} - \frac{\mu_R}{T_R} \right) dN.$$

From this, we see that the first derivatives with respect to our chosen variables are

$$\frac{\partial S_0}{\partial E} = \left(\frac{1}{T} - \frac{1}{T_R} \right), \qquad \frac{\partial S_0}{\partial V} = \left(\frac{p}{T} - \frac{p_R}{T_R} \right), \qquad \frac{\partial S_0}{\partial N} = -\left(\frac{\mu}{T} - \frac{\mu_R}{T_R} \right).$$

The second derivatives are found by noting that the reservoir is very large, so that T_R, p_R, μ_R are constants, unaffected by small transfers of energy, volume, or particles. Hence

$$\frac{\partial^2 S_0}{\partial E^2} = \left(\frac{\partial}{\partial E}\left(\frac{1}{T} \right) \right)_{V,N}, \qquad \frac{\partial^2 S_0}{\partial V^2} = \left(\frac{\partial}{\partial V}\left(\frac{p}{T} \right) \right)_{E,N}, \qquad \frac{\partial^2 S_0}{\partial N^2} = -\left(\frac{\partial}{\partial N}\left(\frac{\mu}{T} \right) \right)_{E,V}.$$

Consequently, for any system in thermal, mechanical, or diffusive equilibrium with a large reservoir, the probabilities for fluctuations in its internal energy, volume, or number of particles are given by equation 9.10, with standard deviations

$$\frac{1}{\sigma_E^2} = -\frac{1}{k}\left(\frac{\partial}{\partial E}\left(\frac{1}{T} \right) \right)_{V,N}, \qquad \frac{1}{\sigma_V^2} = -\frac{1}{k}\left(\frac{\partial}{\partial V}\left(\frac{p}{T} \right) \right)_{E,N},$$

$$\frac{1}{\sigma_N^2} = \frac{1}{k}\left(\frac{\partial}{\partial N}\left(\frac{\mu}{T} \right) \right)_{E,V}. \qquad (9.11a)$$

To evaluate these, we could use experimental data (e.g., on how $1/T$ varies as small amounts of energy ΔE are added), or we could use a model that interrelates the appropriate parameters. For example, in a standard model (Section 4F), E and T are related through

$$E = \text{constant} + \frac{N\nu}{2} kT \qquad \Rightarrow \qquad \frac{1}{T} = \frac{N\nu}{2\,(E - \text{constant})}$$

and, from this, equation 9.11a yields

$$\sigma_E = \sqrt{\frac{N\nu}{2}}\, kT, \tag{9.11b}$$

a result we have seen before (equation 7.7). For an ideal gas, with $pV = NkT$, evaluation of equation 9.11a for σ_V and σ_N gives (homework)

$$\sigma_V = \frac{V}{\sqrt{N}}, \qquad \sigma_N = \sqrt{N} \qquad \text{(ideal gas).} \tag{9.11c}$$

For all three, we notice the familiar result that the *relative* fluctuations decrease as the square root of the system's size:

$$\frac{\sigma_E}{E} = \frac{1}{\sqrt{N\nu/2}}, \qquad \frac{\sigma_V}{V} = \frac{1}{\sqrt{N}}, \qquad \frac{\sigma_N}{N} = \frac{1}{\sqrt{N}}. \tag{9.11d}$$

Summary of Section B

For two mutually interacting systems, we can write the change in entropy of the combined system in terms of the changes in system 1 (equation 9.7):

$$dS_0 = \frac{1}{T_2}\left[\left(\frac{T_2 - T_1}{T_1}\right) dQ_1 + (p_1 - p_2)\, dV_1 - (\mu_1 - \mu_2)\, dN_1\right].$$

The second law has the following requirements for systems interacting (a) thermally, (b) mechanically, or (c) diffusively:

1. As two interacting systems approach equilibrium (equations 9.1):
   (a) heat flows toward the lower temperature;
   (b) boundaries move toward the lower pressure;
   (c) particles flow toward the lower chemical potential.
2. After two interacting systems have reached equilibrium (equations 9.2)
   (a) their temperatures are equal;
   (b) their pressures are equal;
   (c) their chemical potentials are equal.
3. When heat, volume, or particles (ΔQ, ΔV, ΔN) are transferred, one of these by itself with the other two held constant, it must have the following effect (equations 9.3)
   (a) $\Delta T \Delta Q > 0$ (if heat is added, the temperature must rise);
   (b) $\Delta p \Delta V < 0$ (if volume is increased, the pressure must fall);
   (c) $\Delta \mu \Delta N > 0$ (if particles are added, the chemical potential must rise).

The results 9.1 and 9.3 guarantee that interacting systems will converge toward equilibrium rather than diverging away from it.

When a system is interacting purely thermally, mechanically, or diffusively with a large reservoir, fluctuations in its internal energy, volume, or number of particles (or any other parameter) may be calculated from the following probability distributions (9.10):

$$P(\Delta\chi) = \frac{1}{\sqrt{2\pi}\sigma} e^{-\Delta\chi^2/2\sigma^2}, \qquad \text{where} \qquad \frac{1}{\sigma^2} = -\frac{1}{k}\frac{\partial^2 S_0}{\partial\chi^2}\bigg|_{max}.$$

The respective standard deviations for a system interacting with a reservoir are given by (9.11a)

$$\frac{1}{\sigma_E^2} = -\frac{1}{k}\left(\frac{\partial}{\partial E}\left(\frac{1}{T}\right)\right)_{V,N}, \qquad \frac{1}{\sigma_V^2} = -\frac{1}{k}\left(\frac{\partial}{\partial V}\left(\frac{p}{T}\right)\right)_{E,N},$$

$$\frac{1}{\sigma_N^2} = \frac{1}{k}\left(\frac{\partial}{\partial N}\left(\frac{\mu}{T}\right)\right)_{E,V}.$$

The value of these depends on the system. For example, we find that (9.11b–d)

$$\text{in the general case,} \quad \sigma_E = \sqrt{\frac{Nv}{2}}kT;$$

$$\text{for an ideal gas,} \quad \sigma_V = \frac{V}{\sqrt{N}}, \qquad \sigma_N = \sqrt{N}.$$

C   Thermometers and gauges

The results 9.1, 9.2, and 9.3 make gauges possible. When a system interacts with a gauge, the two exchange heat, volume, or particles until they reach equilibrium. At that point, the temperature, pressure, or chemical potential of the gauge is the same as that of the system. Therefore, we only need to calibrate one system – the gauge – and then we can use it to measure the temperature, pressure, or chemical potential of everything else.

The choice of an appropriate gauge generally involves a compromise over concerns of convenience, price, size, accuracy, and range of utility. We prefer a gauge that is small, because we want to minimize its impact on the system whose properties we are measuring.

Gauges to measure pressure and chemical potential are calibrated in direct comparison with metric standards of mass, length, and time, because their units are newtons per square meter and joules, respectively. Temperature is different, however. Equations 8.1 ($1/T = (\partial S/\partial E)_{V,N}$) and 7.10 ($S = k\ln\Omega$) define the combination $kT : (kT = (\partial E/\partial\ln\Omega)_{V,N})$ in terms of the measurable quantities Ω and E and give it units of energy. For either k or T to have individual significance requires a further definition. This is accomplished by defining the triple point for water to be at temperature 273.16 K (Figure 9.5).[4] These two definitions of kT and T also determine the value of the constant k. Figure 9.5 also shows a comparison of the three most common temperature scales.

---

[4] The triple point is produced by reducing the pressure on liquid water until it begins to freeze and boil simultaneously. This happens at a pressure of about 0.006 atm, or 600 N/m^2.

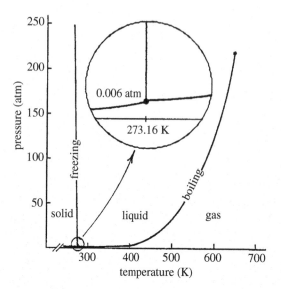

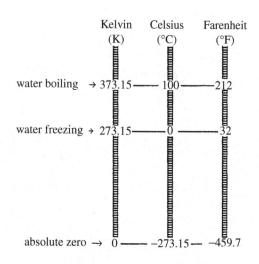

Figure 9.5 (Left) A plot of the boiling and freezing points of water (horizontal axis) as a function of the pressure (vertical axis). The area around the triple point ($T = 273.16$ K, $p = 0.006$ atm) is enlarged. The lines trace the locus of points (T, p) at which the two respective phases are in equilibrium. (Right) A comparison of Kelvin, Celsius, and Fahrenheit temperature scales.

D Zeroth law constraints

An algebraic theorem states that if $T_1 = T_2$ and $T_2 = T_3$ then $T_1 = T_3$. This is the original "zeroth law of thermodynamics." Just as equal temperatures indicate thermal equilibrium, equal pressures indicate mechanical equilibrium, and equal chemical potentials indicate diffusive equilibrium. The same algebraic theorem can also be applied to these. Consequently, a broader statement of the zeroth law is:

Zeroth law of thermodynamics

If two systems are each in thermal, mechanical, and/or diffusive equilibrium with a third system, then they are in thermal, mechanical, and/or diffusive equilibrium, respectively, with each other.

One consequence is that we can use gauges to tell whether two systems are in equilibrium, without having to bring the two together.

Summary of Sections C and D

The second law requires that heat, volume, and/or particles be transferred between appropriately interacting systems until their temperatures, pressures, and/or chemical potentials are equal. This makes gauges possible. In each case, all we need is one small calibrated system, and then we can use it to measure the temperature, pressure, or chemical potential of any other system.

Standards for pressure and chemical potential already exist in the metric standards for mass, length, and time. Temperature is different, however, because only the combination kT has physical significance. So the value of Boltzmann's constant depends on the temperature scale. The "absolute" scale of temperature is established by defining the triple point of water to be 273.16 K exactly.

The zeroth law of thermodynamics states that if two systems are each in thermal, mechanical, and/or diffusive equilibrium with a third system, then they are in thermal, mechanical, and/or diffusive equilibrium, respectively, with each other.

E First law constraints

The first law ($dE = TdS - pdV + \mu dN$) interrelates changes in E, S, V, and N. We can express this interrelationship in alternative ways that are sometimes quite useful.

E.1 The integrated internal energy

All the differentials in the first law involve extrinsic variables. If we divide a system up into a myriad of tiny pieces, as in Figure 9.6, the total energy, entropy, volume, and number of particles for the system are each the sum of the parts. For a system in equilibrium, the temperature, pressure, and chemical potential (T, p, μ) are the same throughout. Therefore we can construct a system by adding together all its parts, using the first law to calculate the change in internal energy as the ith part is added on:

$$E = \sum_i \Delta E_i = \sum_i (T\Delta S_i - p\Delta V_i + \mu\Delta N_i) = TS - pV + \mu N. \qquad (9.12)$$

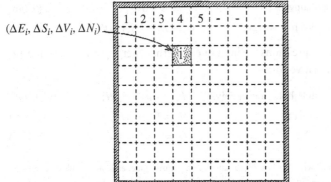

$(\Delta E_i, \Delta S_i, \Delta V_i, \Delta N_i)$

$E_{\text{total}} = \sum_i \Delta E_i$

$S_{\text{total}} = \sum_i \Delta S_i$

$V_{\text{total}} = \sum_i \Delta V_i$

$N_{\text{total}} = \sum_i \Delta N_i$

Figure 9.6 If we divide a system into a myriad of little pieces, the total energy, entropy, volume, and number of particles are sums of those from the individual pieces. We can use the first law to write the system's total internal energy as the sum of the contributions from the pieces.

E.2 Changes in chemical potential

If systems are interacting, changes in the internal energy of one of them can be calculated by writing down the differential of result 9.12:

$$dE = TdS + SdT - pdV - Vdp + \mu dN + Nd\mu. \tag{9.12'}$$

Comparing this with the first law ($dE = TdS - pdV + \mu dN$) yields the following relationship between changes in the three intrinsic variables:

$$SdT - Vdp + Nd\mu = 0$$

or

$$Nd\mu = -SdT + Vdp \qquad \text{(Gibbs–Duhem equation)} \tag{9.13}$$

Notice that none of the intrinsic variables T, p, or μ can change all by itself. If two remain constant, so does the third. And the changes in any one can be determined from the measured changes in the other two.[5] If there were two or more kinds of particle, this would become

$$\sum_i N_i d\mu_i = -SdT + Vdp. \tag{9.13'}$$

F Thermodynamic potentials

We now introduce three more functions, all of which have units of energy and are defined in terms of parameters that we have considered already. In this sense, they are not new; in fact, they are superfluous. However, they are often convenient and useful for the study of common isothermal and/or isobaric (constant-pressure) processes, into which they bring the authority of the second law. Together with the internal energy, they are called the thermodynamic potentials. Their definitions and mathematical properties are summarized in Table 9.2.

F.1 Definitions and differential forms

We name and define these special energy functions as follows:

$$\text{Helmholtz free energy,} \qquad F \equiv E - TS; \tag{9.14a}$$

$$\text{enthalpy,} \qquad H \equiv E + pV; \tag{9.14b}$$

$$\text{Gibbs free energy,} \qquad G \equiv E - TS + pV. \tag{9.14c}$$

If we put into these expressions the integrated internal energy from equation 9.12, we have these alternative forms:

$$\text{Helmholtz free energy,} \qquad F = -pV + \mu N; \tag{9.14a'}$$

$$\text{enthalpy,} \qquad H = TS + \mu N; \tag{9.14b'}$$

$$\text{Gibbs free energy,} \qquad G = \mu N. \tag{9.14c'}$$

---

[5] How does this square with our assertion that there are *three* independent variables? There are only two independent *intrinsic* variables, because they cannot completely describe the system. For example, they cannot tell us the system's *size*. Hence, at least one *extrinsic* variable is needed.

Table 9.2. *Thermodynamic potentials, defined in terms of the internal energy*
$E = TS - pV + \mu N$, *with* $dE = T dS - p dV + \mu dN$

Helmholtz free energy F	Enthalpy H	Gibbs free energy G
"work function"	"heat function"	"Gibbs function"
$F \equiv E - TS$	$H \equiv E + pV$	$G \equiv E - TS + pV$
$\quad = -pV + \mu N$	$\quad = TS + \mu N$	$\quad = \mu N$
$dF = -S dT - p dV + \mu dN$	$dH = T dS + V dp + \mu dN$	$dG = -S dT + V dp + \mu dN$
if T constant, $F = \min$	if p constant, $H = \max$	if T, p constant, $G = \min$
if T, N constant, $dF = -p dV$	if p, N constant, $dH = T dS$	if T, p constant, $dG = \mu dN$

If we take the differential of the Helmholtz free energy 9.14a, we get

$$dF = dE - T dS - S dT.$$

Then, using the first law for dE, we get the differential form

$$dF = -S dT - p dV + \mu dN. \tag{9.15a}$$

Doing the same for the enthalpy and the Gibbs free energy gives

$$dH = T dS + V dp + \mu dN, \tag{9.15b}$$
$$dG = -S dT + V dp + \mu dN. \tag{9.15c}$$

F.2 Helmholtz free energy

The Helmholtz free energy can be useful in the analysis of *isothermal* processes, as we will now see. Consider two systems ($A_0 = A_1 + A_2$, see Figure 9.2) interacting mechanically and diffusively but held at constant temperature. For each, the change in Helmholtz free energy is given from equation 9.15a by

$$dF = -p dV + \mu dN \qquad (\text{isothermal, } dT = 0). \tag{9.16a}$$

Since the volume and particles gained by one are lost by the other,

$$dV_2 = -dV_1, \qquad dN_2 = -dN_1,$$

we can write the total change in F as

$$dF_0 = dF_1 + dF_2 = -(p_1 - p_2)dV_1 + (\mu_1 - \mu_2)dN_1.$$

The second law requires volume to be gained by the system with greater pressure, and particles to flow towards smaller chemical potentials. Therefore

$$-(p_1 - p_2)dV_1 \le 0, \quad (\mu_1 - \mu_2)dN_1 \le 0 \qquad \Rightarrow \qquad dF_0 \le 0.$$

That is, for systems interacting isothermally the second law demands that changes in their Helmholtz free energy must be negative, reaching a minimum at equilibrium.

From the differential form 9.15a, you can see that if the process is both isothermal and nondiffusive ($dT = dN = 0$), the Helmholtz free energy measures the work done:

$$dF = -pdV \qquad \text{(isothermal, nondiffusive)}.$$

For this reason, it is sometimes called the work function.

F.3 Enthalpy

The enthalpy can be useful in the analysis of *isobaric* processes, as we will now see. Consider two systems interacting thermally and diffusively but held at constant pressure. For each, the change in enthalpy is given by equation 9.14b:

$$dH = dE + pdV \qquad \text{(isobaric, } dp = 0). \tag{9.16b}$$

Since the energy or volume gained by one comes from the other,

$$dE_2 = -dE_1, \qquad dV_2 = -dV_1,$$

the change in enthalpy for the combined system can be written as

$$dH_0 = 0 + (p_1 - p_2)dV_1.$$

The second-law constraint 9.1b requires the term on the right to be positive when approaching equilibrium (i.e., volume is gained by the system under higher pressure):

$$dH_0 \geq 0.$$

Consequently, as systems approach equilibrium under isobaric conditions, changes in their enthalpy must be positive, reaching a maximum at equilibrium.

According to equation 9.15b, for processes that are nondiffusive as well as isobaric ($dp = dN = 0$), the enthalpy measures the heat transfer:

$$dH = TdS \qquad \text{(isobaric and nondiffusive, } dp = dN = 0).$$

For this reason, it is sometimes called the heat function. In Chapter 13 we will learn that enthalpy measurements are particularly useful for analyzing engine performance, revealing the amount of heat exchanged or the work done in various parts of a cycle.

F.4 Gibbs free energy

Finally, the Gibbs free energy is relevant in a wide variety of diffusive processes that reach equilibrium under *isothermal and isobaric* constraints ($dT = dp = 0$).

Consider two systems interacting diffusively. According to equation 9.15c, the change in Gibbs free energy for either system is[6]

$$dG = \mu dN \qquad \text{(isothermal and isobaric, } dT = dp = 0\text{)}. \qquad (9.16c)$$

Since particles gained by one are lost by the other ($dN_2 = -dN_1$), the change in Gibbs free energy for the combined system is

$$dG_0 = dG_1 + dG_2 = (\mu_1 - \mu_2)dN_1 \qquad \text{(isothermal, isobaric)}$$

The second law requires that particles are gained by the system with the lower chemical potential, so the term on the right must be negative for systems approaching equilibrium; thus

$$dG_0 \leq 0 \qquad \text{(isothermal and isobaric)}.$$

That is, when a system is in diffusive equilibrium its Gibbs free energy is a minimum for that particular temperature and pressure.

We will often be interested in diffusive equilibrium between systems involving different kinds of particles with different chemical potentials. In these cases, the Gibbs free energy is the sum over all the different types of particles,

$$G = \sum_i \mu_i N_i, \qquad (9.14c'')$$

and the condition that G must be a minimum at equilibrium is

$$\Delta G = \sum_i \mu_i \Delta N_i = 0 \qquad \text{at equilibrium } (T, p \text{ constant}). \qquad (9.17)$$

Summary of Sections E and F

We can integrate the first law to find (equation 9.12)

$$E = TS - pV + \mu N.$$

Comparing the differential of this with the first law, we find that (equation 9.13)

$$Nd\mu = -SdT + Vdp,$$

which tells how changes in the three intrinsic parameters μ, T, p are interrelated.

Thermodynamic potentials can be useful for certain isothermal and/or isobaric processes. Their definitions, differential forms, properties, and alternative names are given in Table 9.2.

---

[6] According to equation 9.14c', the differential of $G = \mu N$ is $dG = \mu dN + Nd\mu$, which seems to have an extra term $Nd\mu$ compared with the expression 9.16c. However, $d\mu = 0$ if temperature and pressure are fixed (equation 9.13), so in this case the two expressions are the same.

G   Maxwell's relations

G.1   Derivation

We are now going to examine the interrelationships between the many variables that must exist because no more than three of them are independent. We expose these interrelationships by exploiting the fact that the differentials dE, dF, dH, dG are exact. You remember that a differential is exact if you can find a function whose differential it is. We have already obtained these functions in the previous section (9.12, 9.14a, b, c):

$$E = TS - pV + N\mu,$$
$$F = E - TS,$$
$$H = E + pV,$$
$$G = E - TS + pV.$$

The property that we shall now use for each of these functions is illustrated as follows. Suppose that w is a function of (x, y, z): $w = w(x, y, z)$. Then its differential is given by

$$dw = f dx + g dy + h dz, \tag{9.18}$$

where

$$f = \left(\frac{\partial w}{\partial x}\right)_{y,z}, \qquad g = \left(\frac{\partial w}{\partial y}\right)_{x,z}, \qquad h = \left(\frac{\partial w}{\partial z}\right)_{x,y}.$$

Using the property that for exact differentials

$$\frac{\partial^2 w}{\partial y \partial x} = \frac{\partial^2 w}{\partial x \partial y},$$

we get

$$\left(\frac{\partial f}{\partial y}\right)_{x,z} = \left(\frac{\partial g}{\partial x}\right)_{y,z}. \tag{9.19a}$$

The corresponding relationships for the pairs of variables (x, z) and (y, z) are

$$\left(\frac{\partial f}{\partial z}\right)_{x,y} = \left(\frac{\partial h}{\partial x}\right)_{y,z} \qquad \text{and} \qquad \left(\frac{\partial g}{\partial z}\right)_{x,y} = \left(\frac{\partial h}{\partial y}\right)_{x,z}. \tag{9.19b, c}$$

We now apply this to our thermodynamic potentials, whose differential forms are (equations 8.4, 9.15a, b, c):

$$dE = T dS - p dV + \mu dN \qquad \text{(independent variables } S, V, N\text{)};$$
$$dF = -S dT - p dV + \mu dN \qquad \text{(independent variables } T, V, N\text{)};$$
$$dH = T dS + V dp + \mu dN \qquad \text{(independent variables } S, p, N\text{)};$$
$$dG = -S dT + V dp + \mu dN \qquad \text{(independent variables } T, p, N\text{)}.$$

Each of these is of the form 9.18, and so we can apply equations 9.19a, b, c to it. Altogether, this gives the 12 interrelationships listed in Table 9.3. The four that apply to the common case of nondiffusive interactions ($N=$ constant) are given in the left-hand column.

G.2 Meaning

Each of the 12 Maxwell's relations relates a variation in one property to a variation in another. Thus they offer us several different ways to measure the change in a parameter, so we can choose the option that is most convenient for us. (More on this in Chapter 11.) All must be true for all systems, and they are a consequence of there being many more parameters than independent variables. Each partial derivative is simply related to some easily measured property of the system. Many are tabulated in handbooks in the form of coefficients for volume expansion, compressibilities, heat capacities, etc., and others can be measured directly. For example, temperature and pressure are given by equation 8.7:

$$T = \left(\frac{\partial E}{\partial S}\right)_{V,N}, \qquad p = T\left(\frac{\partial S}{\partial V}\right)_{E,N}.$$

Thermometers and pressure gauges are common tools, and their measure gives us the corresponding partial derivatives. In principle, you could devise gauges or equipment to measure the quantities in all the other differential relationships appearing in Maxwell's relations.

It is interesting to think in more detail about how the various partial differentials are obtained (Figure 9.7). For example, in equation M1 in Table 9.3 we would

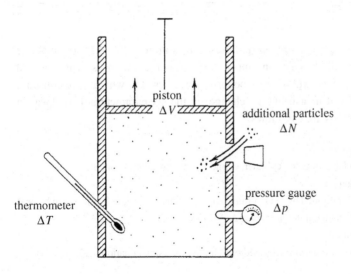

Figure 9.7 An apparatus for measuring the ratios of variations in Maxwell's relations, illustrating that we can easily measure ΔQ, ΔT, Δp, ΔV, and ΔN. How would we then determine ΔS or $\Delta \mu$?

Table 9.3. *Maxwell's relations, labeled "M1" through "M12." In the main text, the three relations in the first row are labeled M1–M3, those in the second row M4–M6, those in the third row M7–M9, and those in the fourth row M10–M12. The three relations on each line are derived from the given thermodynamic potential with the three independent variables indicated. In the left-hand column, all four apply to nondiffusive interactions (N = constant). In the center column, two apply to adiabatic processes (S = constant) and two to isothermal processes (T = constant). In the right-hand column, two apply to isochoric processes (V = constant) and two to isobaric processes (p = constant)*

Derived from internal energy, $E(S, V, N)$:

$$\left(\frac{\partial T}{\partial V}\right)_{S,N} = -\left(\frac{\partial p}{\partial S}\right)_{V,N}, \quad -\left(\frac{\partial p}{\partial N}\right)_{S,V} = \left(\frac{\partial \mu}{\partial V}\right)_{S,N}, \quad \left(\frac{\partial T}{\partial N}\right)_{S,V} = \left(\frac{\partial \mu}{\partial S}\right)_{V,N}$$

Derived from Helmholtz free energy, $F(T, V, N)$:

$$\left(\frac{\partial S}{\partial V}\right)_{T,N} = \left(\frac{\partial p}{\partial T}\right)_{V,N}, \quad -\left(\frac{\partial p}{\partial N}\right)_{T,V} = \left(\frac{\partial \mu}{\partial V}\right)_{T,N}, \quad -\left(\frac{\partial S}{\partial N}\right)_{T,V} = \left(\frac{\partial \mu}{\partial T}\right)_{V,N}$$

Derived from enthalpy, $H(S, p, N)$:

$$\left(\frac{\partial T}{\partial p}\right)_{S,N} = \left(\frac{\partial V}{\partial S}\right)_{p,N}, \quad \left(\frac{\partial V}{\partial N}\right)_{S,p} = \left(\frac{\partial \mu}{\partial p}\right)_{S,N}, \quad \left(\frac{\partial T}{\partial N}\right)_{S,p} = \left(\frac{\partial \mu}{\partial S}\right)_{p,N}$$

Derived from Gibb's free energy, $G(T, p, N)$:

$$-\left(\frac{\partial S}{\partial p}\right)_{T,N} = \left(\frac{\partial V}{\partial T}\right)_{p,N}, \quad \left(\frac{\partial V}{\partial N}\right)_{T,p} = \left(\frac{\partial \mu}{\partial p}\right)_{T,N}, \quad -\left(\frac{\partial S}{\partial N}\right)_{T,p} = \left(\frac{\partial \mu}{\partial T}\right)_{p,N}$$

need to measure how temperature varies with volume ($\Delta T/\Delta V$) in adiabatic expansions for which S is constant, or how pressure varies as heat is added ($\Delta p/\Delta S$ with $\Delta S = \Delta Q/T$) in isochoric processes, for which V is constant. Common means of measuring the changes in the six parameters that appear in Maxwell's relations are as follows:

- ΔT, thermometer
- ΔV, meter sticks, graduated cylinder, cylinder with movable piston, etc.
- Δp, pressure gauge
- ΔS, heat transfer divided by temperature, $\Delta Q/T$, where ΔQ might be measured by heat capacities and change in temperature, $\Delta Q = C\Delta T$
- ΔN, mass balance.
- $\Delta \mu$, voltmeter, or a thermometer and pressure gauge using equation 9.13.

Summary of Section G

One way to express the interdependence of the many different thermodynamic variables is to use the fact that the changes in internal energy, Helmholtz free energy, enthalpy, and Gibbs free energy of a system are exact differentials (equations 8.4, 9.15a–c):

$$dE = T dS - p dV + \mu dN \qquad \text{(independent variables } S, V, N\text{)}$$
$$dF = -S dT - p dV + \mu dN \qquad \text{(independent variables } T, V, N\text{)}$$
$$dH = T dS + V dp + \mu dN \qquad \text{(independent variables } S, p, N\text{)}$$
$$dG = -S dT + V dp + \mu dN \qquad \text{(independent variables } T, p, N\text{)}$$

Each of these are of the form (equation 9.18):

$$dw = f dx + g dy + h dz,$$

where

$$f = \left(\frac{\partial w}{\partial x} \right)_{y,z}, \quad g = \left(\frac{\partial w}{\partial y} \right)_{x,z}, \quad h = \left(\frac{\partial w}{\partial z} \right)_{x,y}.$$

Using the property that for exact differentials

$$\frac{\partial^2 w}{\partial x \partial y} = \frac{\partial^2 w}{\partial y \partial x},$$

we get (equations 9.19a–c)

$$\left(\frac{\partial f}{\partial y} \right)_{x,z} = \left(\frac{\partial g}{\partial x} \right)_{y,z}, \quad \left(\frac{\partial f}{\partial z} \right)_{x,y} = \left(\frac{\partial h}{\partial x} \right)_{y,z}, \quad \left(\frac{\partial g}{\partial z} \right)_{x,y} = \left(\frac{\partial h}{\partial y} \right)_{x,z}.$$

Applying these to each of the four exact differentials above, we get the 12 Maxwell's relations listed in Table 9.3. Each tells us how a variation in one parameter is related to a variation in another. Most involve easily measured properties of the system. As we will learn in Chapter 11, Maxwell's relations give us flexibility in choosing parameters to measure in studying the properties of a system.

H The third law and degenerate systems

H.1 Behaviors at absolute zero

We now examine behaviors near absolute zero. At $T = 0$ a system is in the very lowest state possible, and this has two important implications:

- its entropy is zero (the third law, see Section 8D and footnote 5) and
- no more energy can possibly be extracted from it.

The vanishing entropy is independent of all other factors, such as pressure, volume, or number of particles. The fact that the entropy is fixed means that it does not change with V, N, or p:

$$\left(\frac{\partial S}{\partial V}\right)_{T=0} = \left(\frac{\partial S}{\partial N}\right)_{T=0} = \left(\frac{\partial S}{\partial p}\right)_{T=0} = 0. \tag{9.20}$$

These might seem surprising. The entropy measures the number of accessible states. Shouldn't this number increase if you increase the accessible volume dV in coordinate space or decrease if you further constrain it by an increase dp in pressure? Wouldn't more particles (dN) give it more ways of rearranging itself? Apparently not. We will see why this is so.

Through the Maxwell relations M4, M6, M10, and M12, equations 9.20 imply that the following are also true for a closed system ($N = $ constant) at $T = 0$:

$$\left(\frac{\partial V}{\partial T}\right)_p = \left(\frac{\partial \mu}{\partial T}\right)_p = \left(\frac{\partial p}{\partial T}\right)_V = \left(\frac{\partial \mu}{\partial T}\right)_V = 0. \tag{9.20$'$}$$

When heated at absolute zero (giving rise to an increase dT), the system neither expands ($dV = 0$) nor undergoes an increase in pressure ($dp = 0$). And heating causes no change in the particles' propensity to enter or leave the system ($d\mu = 0$).

H.2   Quantum effects and degenerate systems

Some interesting and important quantum effects are observed near absolute zero. In subsection 1B.6 we learned that a particle's spin angular momentum about any axis can be described in terms of integer multiples of $\hbar$ for bosons, and odd-half-integer multiples of $\hbar$ for fermions. An important consequence of this behavior under rotation is that no two identical fermions ("identical includes identical spin orientations) may occupy the same quantum state, whereas any number of identical bosons can do so.

Any system of particles near absolute zero is said to be "degenerate" if all the particles are occupying the quantum states of lowest possible energy. Since all bosons can be in the ground state (Figure 9.8a), both the total kinetic energy of a degenerate boson system and the pressure it exerts are very small. (Not zero, because the uncertainty principle demands that there must be some motion even at absolute zero.)

For identical fermions, however, only one particle may occupy the quantum state of lowest energy, only one the state of next lowest energy, and so on (Figure 9.8b). Successive fermions must occupy higher states even at absolute zero. As a result, both the kinetic energy and the pressure of a "degenerate" fermion system can be substantial.

Degenerate fermions are like a glass of water. They fill the lowest energy states up to a certain energy, called the "Fermi level" or "Fermi surface" (i.e., the surface of the "Fermi sea"), and the states above that are empty. If the

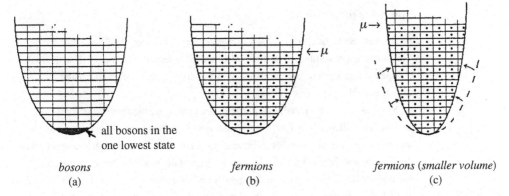

all bosons in the
one lowest state

bosons
(a)

fermions
(b)

fermions (smaller volume)
(c)

Figure 9.8 At low temperatures, systems become "degenerate," with all particles occupying the states of lowest energy. (a) All bosons fall into the one state of lowest energy. (b) Unlike the bosons, no two identical fermions may occupy the same state, so they fill the lowest N states, forming a "Fermi sea." (c) If identical fermions must fill a smaller spatial volume, the Fermi level is correspondingly higher.

fermions are compressed, the Fermi surface must rise (Figure 9.8c), just as squeezing an elastic water container makes the water level rise: the number of quantum states is fixed at $V_r V_p / h^3$, so a smaller volume in coordinate space means that a larger volume in momentum space is required to accommodate the fermions.

The conduction electrons in metals are sufficiently close together that the Fermi level is typically several electron volts – more than 100 times greater than thermal energy of an isolated electron at room temperature. So thermal energies are almost negligible by comparison, and the conduction electrons are nearly degenerate. The electrons in a white dwarf star are nearly degenerate even at millions of degrees. Therefore, when we discuss degenerate systems near absolute zero, the word "near" is relative.

To see how the third law bears on this subject, consider a particle entering a degenerate system of fixed volume. The first law gives the increase in energy of the system:

$$\Delta E = T\Delta S - p\Delta V + \mu \Delta N.$$

Because the volume of the system is fixed, $\Delta V = 0$, and because the system still remains in that one configuration or state of lowest possible energy, the entropy remains unchanged at zero ($\Delta S = 0$). Therefore, the change in internal energy due to a single incoming particle is

$$\varepsilon_{\text{added particle}} = 0 + 0 + \mu \quad \text{(degenerate system)}. \tag{9.21}$$

Because all the lower-lying states are full, the particle must enter at the Fermi surface. Therefore, the Fermi energy equals the chemical potential for a degenerate fermion system:

$$\mu = \text{Fermi energy} \quad \text{(degenerate fermion system)}. \tag{9.22}$$

Summary of Section H

At absolute zero, the entropy of any system is zero, irrespective of other factors. This implies that volume, pressure, and chemical potential do not change with small changes in temperature. As we also saw in an earlier chapter, all heat capacities vanish as well.

Systems in which virtually all particles are occupying the lowest energy states possible are called degenerate. No two identical fermions may occupy the same quantum state, but any number of bosons may. Therefore, as temperatures approach zero, all bosons fall into the one lowest quantum state, but this cannot happen for fermions. All fermion states are occupied, one fermion apiece, up to the Fermi level, and all are empty above that. At absolute zero, the Fermi level is equal to the chemical potential.

Problems

Sections A and B

1. How many independent variables are there if a process is:
   (a) nondiffusive and carried out at atmospheric pressure?
   (b) carried out at atmospheric pressure and held at constant temperature?

2. Consider the function $f(w, x, y, z) = xw + yz + xyz$, with the constraint $x = w + y$.
   (a) How many independent variables are there?
   (b) If in addition $y = 10$, how many independent variables are there?
   (c) Express f as a function of w, z, using the above two constraints.
   (d) Express f as a function of x, z, using the above two constraints.
   (e) Express z as a function of f, w, using the above two constraints.

3. Give two imaginary processes that would violate the second law (entropy increase) without violating the first (energy conservation).

4. Consider the function $f(x) = -x^2 + 4x + 7$.
   (a) For what value of x is this a maximum?
   (b) What is the value of f' at this point?
   (c) Is f'' positive or negative at this point?

5. Consider the function $e^{-\alpha(x^2+y^2)}$, where α is a positive constant. Like entropy, this function is positive everywhere but has a maximum for some value of the two variables.
   (a) For what value of (x, y) is there a maximum?
   (b) What is the value of $\partial f/\partial x$ at this point?
   (c) What is the value of $\partial f/\partial y$ at this point?
   (d) Is $\partial^2 f/\partial x^2$ at this point positive or negative?
   (e) Is $\partial^2 f/\partial y^2$ at this point positive or negative?

6. For two systems interacting thermally, mechanically, and diffusively with each other, show that

$$dQ_2 = -dQ_1 + (p_1 - p_2)dV_1 - (\mu_1 - \mu_2)dN_1.$$

Then use this to prove equation 9.7.

7. Consider two systems, held at the same temperature and pressure, that are interacting diffusively with each other. Show that if they are not yet in equilibrium then particles flow from the system of higher chemical potential toward the one of lower chemical potential. (That is, derive the result 9.1c.)

8. Can you think of any process (involving two or more systems) in which the entropy increases without the addition of any heat? (It would have to be a nonequilibrium process.)

9. In Chapter 14 there are some examples of systems for which the entropy increases (hence $\Delta Q > 0$) while the temperature drops ($\Delta T < 0$) as energyless particles are added. Explain why this doesn't violate condition 9.3a.

10. Give an example in which one of the conditions 9.3a, b, c is violated if a system is engaging in more than one type of interaction at a time.

11. Glass crystallizes as it ages. What does that say about the change in chemical potential between the amorphous and crystalline phases?

12. A salt (sodium chloride) crystal dissolves in fresh water but not in air. What does that say about the relative value of the chemical potential of a sodium or chloride ion in air, in a salt crystal, and in water?

13. A certain sugar has a chemical potential of -1.1 eV in water and -0.8 eV in oil. Suppose that some of this sugar is dissolved in oil and you wish to remove it. Can you suggest a way to do it, taking advantage of the difference in the two chemical potentials?

14. Consider two systems, 1 and 2, interacting thermally but not mechanically or diffusively. Their initial temperatures are 0 °C and 50 °C, respectively. The heat capacities of system 1 are $C_{V1} = 2R$, $C_{p1} = 2.8R$, and those of system 2 are $C_{V2} = 3R$, $C_{p2} = 4R$; R, the gas constant $= 8.31$ J/(K mole). Find
    (a) their temperature after reaching equilibrium,
    (b) the change in entropy of system 1,
    (c) The change in entropy of the total combined system.
    (d) Why is one of the heat capacities (C_V or C_p) irrelevant here?

15. Show that equation 9.9 follows from equations 9.7 and 9.8 and from the condition that $\Delta S_0 < 0$ for displacements away from equilibrium.

16. When water freezes in a closed jar both its volume and its pressure increase, eventually bursting the jar. Does this violate the second-order condition $\Delta p \Delta V < 0$? Explain.

17. When ice melts, temperature remains constant as heat is added. Does this violate the second-order condition $\Delta Q \Delta T > 0$? Explain.

18. Consider particles moving from system A_1 to system A_2, with $\mu_1 > \mu_2$. According to the second-order constraints, will the two chemical potentials both increase, both decrease, move closer together, or move further apart, as a result of this transfer?

19. Briefly explain how you could verify experimentally any two of the three second-order constraints for any system you wish.

20. Consider the burning of hydrogen to make water, $2H_2 + O_2 \rightarrow 2H_2O$, the ratios of the changes in the number of the three types of molecules being given by $\Delta N(H_2) : \Delta N(O_2) : \Delta N(H_2O) = -2 : -1 : 2$.
    (a) What does the fact that this reaction occurs tell us about the relative sizes of the following combinations of the chemical potentials indicated, $2\mu(H_2) + \mu(O_2)$ and $2\mu(H_2O)$?
    (b) Physically, what causes this difference in chemical potential? Why is one combination lower than the other?

21. Consider a small system having 100 degrees of freedom in thermal equilibrium with a large reservoir at room temperature (295 K). Consider the probability distribution $P(E)$ for the small system as a function of its thermal energy. (Use $E = E_{\text{therm}} = (N\nu/2)kT = C_V T$ to transform temperatures into energies.)
    (a) At what value of E does the distribution peak?
    (b) What are the standard deviations for fluctuations σ_E in the energy and σ_T in the temperature?
    (c) What are the relative widths of the peaks, σ_E/E and σ_T/T?

22. Repeat the preceding problem for a rather small macroscopic system, having 10^{20} degrees of freedom, that is in thermal equilibrium with a large reservoir at 20 °C.

23. Consider a 10 gram ice cube in equilibrium with some ice water in a glass. Each water molecule in ice has six degrees of freedom. For this ice cube, using $E = E_{\text{therm}} = (N\nu/2)kT = nC_V T$, find
    (a) the mean thermal energy,
    (b) the standard deviation,
    (c) the relative width of the peak, σ/E.

24. Consider a mole of helium atoms (helium is an ideal monatomic gas) in thermal equilibrium with a very large reservoir at 300 K. For this helium system, using $E = E_{\text{therm}} = (N\nu/2)kT = nC_V T$, find
    (a) the average internal energy,
    (b) the standard deviation.
    (c) What fraction of the total energy is the standard deviation?

25. A thin film of some solid material (with six degrees of freedom per atom) is deposited on a substrate of area 0.5 cm^2. The atomic spacing in the deposited material is 0.15 nm. Roughly what is the minimum thickness for this film if we want the relative fluctuations in its internal energy (σ_E / E) to be less than a tenth of a part per billion?

26. A canister containing N_A nitrogen (N_2) molecules (where N_A is Avogadro's number) is in thermal equilibrium with a large heat reservoir. Each molecule has three translational and two rotational degrees of freedom. Using $nC_V = (N\nu/2)k$, find
    (a) the number of degrees of freedom of the whole gas,
    (b) the temperature of the reservoir, given that the mean energy of the gas is 3200 J.
    (c) You want to bet that at any instant the internal energy of the gas is within $x\%$ of its mean value. What is the smallest percentage x that you could use here and still have at least a 68% chance of winning?

27. A tiny bacterium has dimensions of about 10^{-6} m in each direction. Assume that it is mostly water.
    (a) How many molecules are in it?
    (b) If each molecule has six degrees of freedom, what are the relative fluctuations in its temperature, σ_T / T, given that $nC_V = (N\nu/2)k$?

28. Later we will show that, for the adiabatic expansion ($\Delta Q = \Delta N = 0$) of an ideal gas, $pV^\gamma = \text{constant}$, where $\gamma = (\nu + 2)/\nu$. Using this result, show that $-V(\partial p/\partial V)_{Q,N}$ is equal to γp. Show that for the isothermal expansion of an ideal gas (for which $pV = NkT$), $-V(\partial p/\partial V)_{T,N}$ is equal to p. Use these two results to make a rough estimate of the ratio $\sigma_V(\text{adiabatic})/\sigma_V(\text{isothermal})$ for the fluctuation in volume of a gas bubble.

29. Using the result 9.11c, calculate the relative fluctuation σ_V / V in the volume of a bubble of an ideal diatomic gas in the ocean at $0\,°C$ and atmospheric pressure if the bubble's volume is (a) $10^{-3}\ \mu\text{m}^3$, (b) 1 liter.

30. A typical value of $\left(\frac{\partial}{\partial N} \left(\frac{\mu}{T} \right) \right)_{E,V}$ is Nk. The ocean's salinity is 3.5% by weight or about 2.0% by number of particles. About 45% of these particles are sodium ions. A tiny planktonic organism has a volume of $40\ \mu\text{m}^3$ and the same salinity as its oceanic environment and lives where the water's temperature is $6\,°C$.
    (a) Make a rough estimate of how many sodium ions are in its body at any one time, on average.
    (b) What is the standard deviation about this value?
    (c) What is the relative fluctuation?

31. For an ideal gas, $pV = NkT$ and $\mu = kT \left(\frac{v+2}{2} + \ln \omega_c \right)$, where ω_c is given in equation 6.10. Use these to derive the results 9.11c for σ_V and σ_N.

32. Show that for an ideal gas, $\frac{\sigma_E}{E} = \sqrt{\frac{2}{Nv}}$. Use $E = (Nv/2)kT$.

Sections C and D

33. You want to determine the temperature of the air in an oven without using a thermometer.
    (a) How will you do it? What would be the thermometric (measured) parameter that you would use?
    (b) Would this prevent you from using the oven to cook food?

34. When a gas is at constant pressure, the volume varies with temperature. When a gas is at constant volume, the pressure varies with temperature. A thermometer with which of these features would be useful over a wider range of temperatures? Why? (Hint: Think about intermolecular forces at low temperatures.)

35. Suppose you have a constant volume gas thermometer containing an ideal gas, for which $pV = NkT$. When the gas bulb is placed in ice water at $0\,°C$ the pressure gauge reads 730 mm of mercury, and when it is placed in boiling water at $100\,°C$ it reads 990 mm of mercury.
    (a) If it reads 850 mm of mercury when placed in your bath water, what is the temperature of the water?
    (b) According to this thermometer, what is the temperature of absolute zero?

36. Think about the densities of liquid water and of ice.
    (a) Which is denser?
    (b) Which is favored by higher pressures?
    (c) Considering the answer to part (b), why do you suppose ice is slippery when you step on it but a sheet of glass is not?
    (d) Why does the freezing point rise as pressure is reduced?
    (e) Why does the boiling point fall as pressure is reduced?

37. Write down the zeroth, first, second, and third laws of thermodynamics, putting each in any form you wish as long as it is correct.

Section E

38. Consider a system A_1 having N_1 particles at chemical potential μ_1, and a system A_2 having N_2 particles at chemical potential μ_2, with $\mu_1 > \mu_2$. If these two systems are briefly brought into diffusive contact, a small number of particles, ΔN, goes from one to the other.
    (a) Which system acquires these particles and why?

(b) In terms of N_1, N_2, μ_1, and μ_2, what is the Gibbs free energy of the two systems before the interaction?

(c) For small ΔN the change in Gibbs free energy is roughly $(\mu_2 - \mu_1)\Delta N$. But for larger ΔN, some second-order terms might become important. What are these second-order terms?

39. Heat transfer is given by $T\mathrm{d}S$. Work done is given by $p\mathrm{d}V$. We integrated these over an entire system, getting TS and pV, respectively. Why, then, would it not be meaningful to call TS the "heat content" and pV the "work content" of the system? (Hint: Starting near absolute zero, can you think of a way for the system to end up with TS without adding any heat, or pV without doing any work?)

40. The entropy of water at 25 °C and 1 atm is 188.8 joules/(mole K). The molecular weight of water is 18 and the specific heat is 4.186 J/(g K). Suppose that you raise the temperature of this water to 27 °C.
(a) What is the new molar entropy of the water?
(b) By how much would you have to change the pressure in order to keep the chemical potential unchanged?

41. Starting with the integrated form of the first law (9.12), prove that $N\mathrm{d}\mu = -S\mathrm{d}T + V\mathrm{d}p$.

Section F
42. Show that the differential expressions for the Helmholtz free energy, enthalpy, and Gibbs free energy all follow from the definitions of these functions. Use the differential of their definitions, and the first law for $\mathrm{d}E$.

43. Although the differential forms 9.15 of the special energy functions follow from the definitions 9.14 and the first law, show that they can be derived also from equations 9.14′ and the condition 9.13.

44. Prove that Equations 9.14′ follow from the definitions 9.14.

45. Using the differential forms of the thermodynamic potentials 9.15, and the second law constraints 9.1, prove that for two interacting systems in equilibrium:
(a) the Helmholtz free energy is a minimum if the interaction is isothermal;
(b) the enthalpy is a maximum if the interaction is isobaric.
(c) The Gibbs free energy is a minimum if the interaction is both isothermal and isobaric.

46. Explain why Helmholtz free energy is really only the "work function" for certain types of processes.

47. Explain why the enthalpy is really only the "heat function" for certain types of processes.

48. Consider the Helmholtz free energy of two systems interacting mechanically and diffusively but held at constant temperature. Suppose that the two systems are displaced slightly away from equilibrium by the transfer of a small amount of volume or small number of particles from one to the other. By examining both the first-order and second-order terms in this small displacement, show that the Helmholtz free energy is a minimum when the two are in equilibrium. (For simplicity, you might consider one of them to be a reservoir for which $\Delta p = \Delta \mu = 0$.)

49. Consider the Gibbs free energy of two systems interacting diffusively, but held at constant temperature and pressure. Suppose the two systems are displaced slightly away from equilibrium by the transfer of a small number of particles from one to the other. By examining both the first- and second-order terms in this small displacement, show that the Gibbs free energy is a minimum when the two are in equilibrium.

Section G

50. Derive a Maxwell relation from the exact differential $dx = f\,dy + g\,dz$.

51. Derive the three Maxwell relations that are obtained from (a) dE, (b) dF, (c) dH, (d) dG.

52. You wish to confirm Maxwell's relations experimentally for a system of fixed number of particles ($dN = 0$), using an apparatus like that of Figure 9.7. How would you measure the following: (a) $\left(\frac{\partial T}{\partial V}\right)_S \approx \left(\frac{\Delta T}{\Delta V}\right)_S$, (b) $\left(\frac{\partial p}{\partial S}\right)_V$, (c) $\left(\frac{\partial S}{\partial V}\right)_T$, (d) $\left(\frac{\partial p}{\partial T}\right)_V$?

53. You wish to confirm Maxwell's relations experimentally for a system that can receive heat, work, or particles from its environment, using an apparatus similar to that of Figure 9.7. How would you measure the following: (a) $\left(\frac{\partial V}{\partial N}\right)_{T,p}$, (b) $\left(\frac{\partial V}{\partial S}\right)_{p,N}$, (c) $\left(\frac{\partial p}{\partial T}\right)_{V,N}$?

Section H

54. From the third law, we have $(\partial S/\partial N)_{T,p} = 0$ at absolute zero. We were able to turn this into $(\partial\mu/\partial T)_{p,N} = 0$, using one of Maxwell's relations.
    (a) Which one?
    (b) We could have obtained this same result another way, simply by looking at another relationship we have that involves variations in μ, T, p. Explain.

55. For a single particle free to move about in a volume V_r, the number of accessible quantum states is given by (equation 1.4) $\omega = V_r V_p/h^3$. For a system of N identical particles, the corrected number of accessible states per particle is $\omega_c = e\omega/N = e(V_r/N)V_p/h^3$. Use this information to find the typical kinetic energy of electrons for the following degenerate systems

(spin-up and spin-down electrons are *not* identical, so work this problem for one spin orientation only. $m_e = 9.1 \times 10^{-31}$ kg):

(a) a typical metal, in which there are 10^{29} conduction electrons per cubic meter;

(b) a white dwarf, in which there are about 10^{36} free electrons per cubic meter.

Chapter 10
Models

A Equations of state

Any thermodynamic property of a system depends on as many parameters as there are kinds of interaction (Figure 10.1). If a system interacts in three ways – thermally, mechanically, and diffusively – then each property depends on three variables. For example, of the seven variables (E, T, S, p, V, μ, N) appearing in the first law (equation 8.4),

$$dE = T dS - p dV + \mu dN,$$

only three are independent. If we choose S, V, N as our independent variables, then we should be able to express all properties of the system in terms of these three:

$$T = T(S, V, N), \qquad p = p(S, V, N), \qquad G = G(S, V, N),$$
$$\mu = \mu(S, V, N), \qquad E = E(S, V, N), \qquad \text{etc.}$$

The independent variables need not be S, V, N. We could choose nearly any three that we wish. But whichever set we choose, all other parameters are functions of them. If we are clever, we can figure out these functional relationships. We employ a variety of techniques to help us discover them. The "universal constraints" of the previous chapter give interrelationships for all systems. In succeeding chapters, we study how to use "imposed constraints" to simplify the analysis further by reducing the number of independent variables.

But in this chapter, we introduce simplified conceptual pictures, called "models," to help us visualize the interrelationships between parameters for specific types of system. When expressed as equations, these interrelationships are called "equations of state." Equations of state can relate any set of variables, but often they involve easily measured quantities such as temperature, pressure, and volume. Owing to the excessive number of variables, many such interrelationships must exist, whether or not we are clever enough to discover them. We now examine various types of systems, using material that is already familiar to us.

B Ideal gases and solids

In Chapter 6 we found that we could write the number of states accessible to a system as (equation 6.7)

$$\Omega = \omega_c^N,$$

where ω_c is the number of states per particle corrected for identical particles. We used rather general considerations to find that (equation 6.10)

$$\omega_{c,\,solid} \approx C\left(\frac{E_{therm}}{N\nu}\right)^{\nu/2}, \qquad \omega_{c,\,gas} \approx \left(C\frac{V}{N}\right)\left(\frac{E_{therm}}{N\nu}\right)^{\nu/2},$$

where each factor C is some constant, and where E_{therm} is related to the internal energy E through (equation 4.13)

$$E_{therm} = \frac{N\nu}{2}kT = E - Nu_0.$$

In the ideal gas model, the particles are tiny and perfectly elastic billiard balls, whose average spacing is large compared with their size. They have no long-range interactions and so no potential energy. In solids, however, interparticle interactions are strong, and the potential wells are deep and sensitive to the particles' spacing. So we have

$$u_{0,\,ideal\,gas} = 0, \qquad u_{0,\,solid} = u_0(V). \tag{10.1}$$

We now insert these expressions into the definition of entropy (equation 7.10),

$$S = k\ln\Omega = k\ln\left(\omega_c^N\right) = Nk\ln\omega_c,$$

to find that the respective entropies are the following functions of E, V, N:

$$S = Nk\ln C\left(\frac{V}{N}\right)\left(\frac{E}{N}\right)^{\nu/2} \qquad \text{for ideal gases,} \tag{10.2a}$$

$$S = Nk\ln C\left(\frac{E - Nu_0}{N}\right)^{\nu/2} \qquad \text{for solids} \tag{10.2b}$$

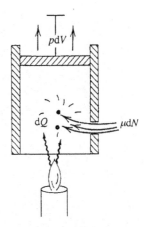

Figure 10.1 There is one independent variable for each type of interaction. In this illustration, the system is engaging in three types of interaction (thermal, mechanical, and diffusive), so every property of the system depends on three variables. What happens if it exchanges more than one kind of particle or does more than one kind of work?

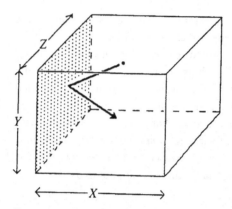

Figure 10.2 The change in momentum when a molecule collides elastically with a wall in the yz-plane is $\Delta p = 2mv_x$. The time it takes to go back and forth across the container is $\Delta t = 2X/v_x$. Therefore, the rate of change in momentum per molecule rebounding off this wall is $\Delta p/\Delta t = mv_x^2/X$. If we sum over all N particles for the total force and then divide this total force by the wall's area to get the pressure, we end up with the ideal gas law.

These equations of state relate the extrinsic variables S, E, V, N. We can now use equations 8.7 (from the first law),

$$\frac{1}{T} = \left(\frac{\partial S}{\partial E}\right)_{V,N}, \qquad \frac{p}{T} = \left(\frac{\partial S}{\partial V}\right)_{E,N}, \qquad \frac{\mu}{T} = -\left(\frac{\partial S}{\partial N}\right)_{E,V},$$

to expose further interrelationships. When applied to the entropy of an ideal gas (10.2a), these three equations give us for an ideal gas

$$E = \frac{N\nu}{2}kT, \tag{10.3a}$$

$$pV = NkT, \tag{10.3b}$$

$$N\mu = NkT\left(\frac{\nu+2}{2} - \ln\omega_c\right)$$

$$= E + pV - NkT\ln\omega_c, \tag{10.3c}$$

where the last expression uses the first two to express E and pV in terms of kT.

We have encountered all three results (10.3a, b, c) before. The first is equipartition; with each of the $N\nu$ degrees of freedom is associated an average energy $(1/2)kT$.[1] The second (10.3b) is the ideal gas law, which relates the four parameters p, V, N, T. (An alternate derivation is found in the problems and is outlined in Figure 10.2.) The third result (10.3c) uses equation 7.10 for $(S = k \ln \Omega = Nk \ln \omega_c)$ in the integrated form of the first law (9.12):

$$E = TS - pV + \mu N \qquad \Rightarrow \qquad N\mu = E + pV - TS.$$

---

[1] Notice that the internal energy depends on only two parameters N, T, and not three. The particles of an ideal gas do not interact at a distance, so their spacing (i.e., the volume) does not matter.

We now do the same for solids. When applied to the entropy of an ideal solid 10.2b, the three equations 8.7 on the preceding page give the following (homework):

$$E = Nu_0 + \frac{Nv}{2}kT, \tag{10.4a}$$

$$p = -N\left(\frac{\partial u_0}{\partial V}\right)_{E,N}, \tag{10.4b}$$

$$N\mu = Nu_0 + NkT\left(\frac{v}{2} - \ln\omega_c\right)$$
$$= E - NkT\ln\omega_c \quad \text{(solids)} \tag{10.4c}$$

Again, in the last equation we have used the first to express the internal energy in terms of kT.

As before, the first of the above equations displays equipartition for the thermal energy. And the last equation, for chemical potential, is the same as that for a gas (10.3c) except that the volume per particle is negligible for solids. The second equation is new, however. It tells us that in order to obtain for solids the counterpart of the ideal gas law (how pressure changes with volume), we would need to know how the depth of the potential wells depends on the interparticle spacing.

C Real gases

The ideal gas model works well for rarefied gases. But it must be modified to describe gases at higher densities where molecular sizes are *not* small compared with their spacing, and mutual interactions cannot be ignored. We begin this modification by rewriting the ideal gas law 10.3b in terms of the molar volume v and the gas constant R:

$$pv = RT, \tag{10.5}$$

$$\text{where} \quad R = N_A k = 8.31 \text{ J/(K mole)} = 0.0821 \text{ atm liter/(K mole)}. \tag{10.6}$$

This relation wrongly implies that we can make the volume arbitrarily small by applying a sufficiently large pressure. In fact, however, once the electron clouds of neighboring molecules begin to overlap it is almost impossible to compress the gas further. We can incorporate this limit through the replacement of v by $v - b$, where b is the minimum molar volume:

$$v \rightarrow v - b.$$

Alternatively, you can think of the volume available to any molecule as the entire volume, v, less that occupied by all the other molecules, b. This approach is used in the homework problems, where you examine the entropy of a real gas.

Another deviation from the ideal gas model is the weak long-range forces among molecules, which cause them to condense to liquids at sufficiently low temperatures. This attraction helps hold the molecules of the gas together, as does

Table 10.1. *Van der Waals constants for common gases*

Gas	$a \left(\frac{\text{liter}^2 \text{ atm}}{\text{mole}^2} \right)$	$b \left(\frac{\text{liter}}{\text{mole}} \right)$	Gas	$a \left(\frac{\text{liter}^2 \text{ atm}}{\text{mole}^2} \right)$	$b \left(\frac{\text{liter}}{\text{mole}} \right)$
acetone	13.91	0.0994	methane	2.253	0.04278
ammonia	4.170	0.03707	nitrogen	1.390	0.03913
carbon dioxide	3.592	0.04267	oxygen	1.360	0.03183
carbon monoxide	1.485	0.03985	propane	8.664	0.08445
chlorine	6.493	0.05622	sulfur dioxide	6.714	0.05636
ethyl alcohol	12.02	0.08407	water	5.464	0.03049
hydrogen	0.2444	0.02611			

Figure 10.3 There is long-range attraction between gas molecules. Electrostatic forces enhance the attraction between unlike charges by: (a) causing small distortions of their electron clouds, or (b) changing the orientations of polarized molecules.

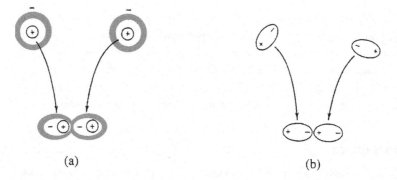

(a)   (b)

the external pressure exerted by the walls of the container. So the factor of p in equation 10.5 must be modified:

$$p \rightarrow p + \text{mutual attraction}$$

This mutual "van der Waals" attraction is caused by charge polarizations (Figure 10.3). It is inversely proportional to the sixth power of the average molecular separation[2] or, equivalently, to the square of the molar volume ($1/r^6 \approx 1/v^2$). Hence

$$p \rightarrow p + \frac{a}{v^2}$$

These two effects are incorporated into a modification of the ideal gas law $pv = RT$ called the "van der Waals equation of state" for a real gas:

$$\left(p + \frac{a}{v^2} \right) (v - b) = RT, \tag{10.7}$$

where the parameters a and b depend on the gas (Table 10.1). We see that the two modifications are most significant when the molar volume is small. For more

---

[2] Dipole–dipole interaction energy goes as $-1/r^3 \approx -1/v$. Since change in potential energy is given by work $= \int p dv \approx -1/v$, we expect that the appropriate pressure term would be $p \approx 1/v^2$.

rarefied real gases, we have $v \gg b$ and $p \gg a/v^2$, so the modifications can be ignored and the ideal gas law is good enough.

Because of their mutual attraction, the molecules are in a very shallow potential well. We can calculate the depth of this potential well per mole, U_0, by integrating $dU_0 = -\mathbf{F} \cdot d\mathbf{x} = -p_{vdw}dv$, where this "pressure" is due to the van der Waals attraction:

$$U_0 = N_A u_0 = -\int_{\infty}^{v} \left(-\frac{a}{v'^2}\right) dv' = -\frac{a}{v}. \tag{10.8}$$

D Liquids

We do not yet have a good model for liquids. Their molecules are mobile, like those of a gas, but their compressibilities are many orders of magnitude smaller. For these reasons, it is sometimes convenient to use a van der Waals model with molar volumes very near b, so that very large increases in the pressure p are required for very small changes in v. Molar volumes for liquids are typically three orders of magnitude smaller than for gases, and such large extrapolation requires modification of the parameters a, b.

Furthermore, the form of the pressure modification a/v^2 cannot be correct for the liquid phase, because other types of interaction dominate the van der Waals forces at small intermolecular distances (homework). But, because of the $v - b$ term, the overall feature that large changes in the applied pressure p cause only very small changes in the molar volume v is still correctly represented. Consequently, the van der Waals equation can be useful in acquiring qualitative insight into the behavior of liquids, particularly in studying gas–liquid phase transitions since it can be applied to both phases. However, it is not adequate for detailed or quantitative investigations of liquids.

E Further modeling of solids

One distinguishing feature of solids is that each individual atom is confined to a very small region of space, being stuck in a potential well that is caused by its electromagnetic interactions with neighbors. Each atom is located near the bottom of its potential well, and thermal agitation causes it to oscillate. Thus, each atom is a three-dimensional harmonic oscillator and has six degrees of freedom – three kinetic and three potential (see equation 4.3).

E.1 Phonons

It is sometimes convenient to think of a solid as being a lattice of atomic masses coupled by springs (Figure 4.5). These are not like ordinary springs, however, because an active atom can't simply send vibrations in all directions throughout the solid. Atoms confined in potential wells can have only certain energies (subsection 1B.8). Because the excitations come in only certain discrete amounts,

Figure 10.4 The atoms in solids can vibrate only with certain discrete energies. Vibrational excitations are either on or off, and cannot be shared among neighboring atoms. Whenever one atom starts shaking, another must stop. In this way, these vibrational excitations ("phonons") travel from one atom to the next. They act like little energy-carrying particles that wander throughout the lattice.

one atom must stop vibrating when its neighbor starts. The vibrations are either on or off. This means that vibrational excitations are passed from atom to atom throughout the solid, one at a time (Figure 10.4). Higher temperatures produce more excitations. If two happen to pass through the same place at the same time, then the atom at that spot occupies its second excited level for that moment. Like the particles of a gas, these vibrational excitations travel freely throughout the solid, passing from atom to atom and being constrained only by the solid's boundaries. Each traveling quantum of energy is called a "phonon," and together they can be treated as a gas – a "phonon gas."

E.2 Conduction electrons and holes

In addition to the lattice of harmonic oscillators, or phonon gas, conductors and semiconductors also have conduction electrons. The electrons of isolated atoms are in discrete energy levels. But when atoms are close together, perturbations from the neighbors cause these discrete levels to spread out into "bands" (Figure 10.5).

In conducting metals, the outermost electrons are mobile. They are either in a band of states that is only partially filled or in a filled band that overlaps with the next higher empty band. Either of these mechanisms gives these electrons freedom to move into empty neighboring states. Consequently, they are not tied to individual atoms but can gain energy and move from one atom to another. Because of their mobility, they are responsible for the high electrical and thermal conductivities of metals. In a typical metal, each atom contributes one or two electrons to this "conduction band," so the number of conduction electrons is typically one or two times the number of atoms. In the lower "valence bands," all states are filled, and so electrons have no neighboring states into which they can move. These valence electrons can be thought of as remaining bound to their parent atoms.

Conduction electrons, however, can be thought of as an "electron gas," confined only by the boundaries of the metal. In a later chapter, we will see that this is a very peculiar kind of gas. Not only is it nearly incompressible, but also, each electron's freedom of movement is heavily constrained by the other conduction

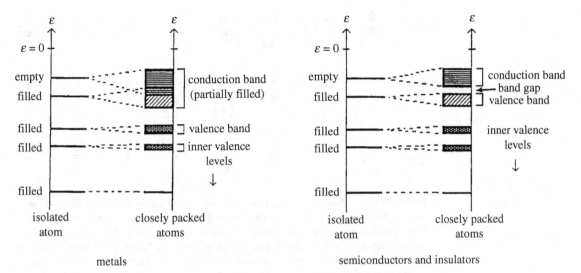

Figure 10.5 Illustration of the energies allowed the electrons in an atom. When isolated, the atom's electron energy levels are discrete. However, when atoms are closely packed, each atom's outer electrons are perturbed by its neighbors, causing these perturbed states to spread out into bands. (Left) In conductors, the outermost band that contains electrons is either partially empty or overlaps with the next higher band of empty states, giving the electrons in this "conduction band" considerable freedom to move into neighboring unfilled states. The mobility of these electrons is what gives these materials their high thermal and electrical conductivities. (Right) In other solids, the outermost band of completely filled electron states does not overlap with the next higher band of empty states. Either few (semiconductors) or no (insulators) electrons can jump the gap between these two bands. Any that do make it to the higher "conduction band" are surrounded by vacant states into which they can move, and so they have high mobilities.

electrons around it. Such an electron gas, therefore, is substantially different from the ideal gases we studied earlier.

In semiconductors the density of conduction electrons is much smaller than in metals. The reason is that the outermost filled band (the valence band) does not quite overlap with the next higher band (the conduction band) of empty electron states (Figure 10.5). To reach the empty states, electrons must jump this gap between bands. Relatively few electrons can do it, but those that do are then surrounded by empty states that give them considerable mobility.

Conduction in semiconductors may also be accomplished by vacancies or "holes" in the valence band (Figure 10.6). A hole is created when a valence electron jumps into the conduction band or onto an impurity atom, leaving its former state empty. The atom that lost the electron then carries a net positive charge. If another electron moves over to fill this hole, it creates another hole in the atom that it left. So the positive charge shifts from atom to atom in a direction backwards to the direction in which the electrons move. That is, the sequential migration of valence electrons in one direction causes the backwards migration

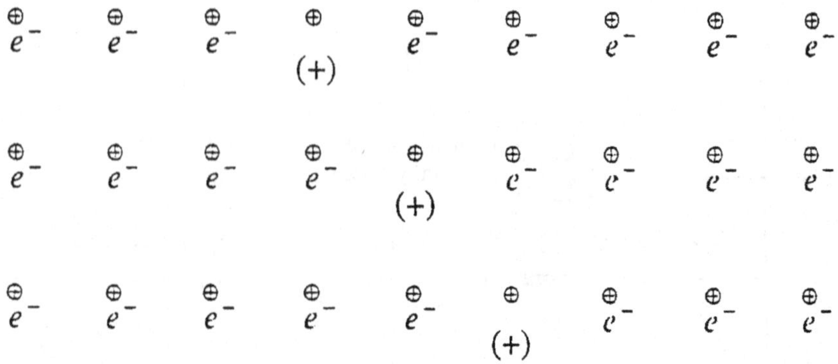

$\oplus$ $\oplus$ $\oplus$ $\oplus$ $\oplus$ $\oplus$ $\oplus$ $\oplus$ $\oplus$
e^- e^- e^- e^- e^- e^- e^- e^-
                       (+)

$\oplus$ $\oplus$ $\oplus$ $\oplus$ $\oplus$ $\oplus$ $\oplus$ $\oplus$ $\oplus$
e^- e^- e^- e^- e^- e^- e^- e^-
                            (+)

$\oplus$ $\oplus$ $\oplus$ $\oplus$ $\oplus$ $\oplus$ $\oplus$ $\oplus$ $\oplus$
e^- e^- e^- e^- e^- e^- e^- e^-
                                  (+)

Figure 10.6 Illustration of conduction by holes in the valence band, with the three sequences progressing from top to bottom. The encircled plus signs represent the fixed positive ions. When an electron moves in one direction to fill up a vacancy, it leaves another vacancy behind. Each vacancy is called a "hole" and carries a net positive charge, because it is an atom that is missing an electron. In the figure the electrons are moving to the left and the hole is moving to the right, this being the conventional direction of the electric current.

of these positively charged holes. But opposite charge moving in the opposite direction means that the direction of the electrical current is the same.

Since models for solids frequently involve more than one component, it is often helpful to simplify our studies by concentrating on one property of one component at a time. For example, we may wish to study the heat capacity of the lattice alone, or the magnetic properties of the electron gas alone, etc.

Summary of Sections A–E

Of all the parameters we use to describe a system, only a few are independent. Interdependencies can be revealed through universal constraints applying to all systems and by imposed constraints that further restrict the number of independent variables. In addition, there are interrelationships that depend on the particular nature of the system and vary from one system to the next. These system-dependent interrelationships are called "equations of state," and we use conceptual models to help us identify some of them.

In our model for ideal gases, each molecule is very tiny, noninteracting, and can access the entire volume of the gas. In our model for solids, each atom is a harmonic oscillator. We used these models in Chapter 6 to calculate the respective entropies (equations 10.2):

$$S = Nk \ln C \left(\frac{V}{N}\right)\left(\frac{E}{N}\right)^{v/2} \qquad \text{(ideal gases)},$$

$$S = Nk \ln C \left(\frac{E - Nu_0}{Nv}\right)^{v/2} \qquad \text{(solids)}.$$

To these we apply equations 8.7 from the first law

$$\frac{1}{T} = \left(\frac{\partial S}{\partial E}\right)_{V,N}, \qquad \frac{p}{T} = \left(\frac{\partial S}{\partial V}\right)_{E,N}, \qquad \frac{\mu}{T} = -\left(\frac{\partial S}{\partial N}\right)_{E,V},$$

to find the following equations of state for ideal gases (equations 10.3):

$$E = \frac{Nv}{2}kT,$$
$$pV = NkT,$$
$$N\mu = NkT\left(\frac{v+2}{2} - \ln\omega_c\right)$$
$$= E + pV - NkT\ln\omega_c,$$

and the following equations of state for solids (equations 10.4),

$$E = Nu_0 + \frac{Nv}{2}kT,$$
$$p = -N\left(\frac{\partial u_0}{\partial V}\right)_{E,N}$$
$$N\mu = Nu_0 + NkT\left(\frac{v}{2} - \ln\omega_c\right)$$
$$= E - NkT\ln\omega_c.$$

For real (as opposed to ideal) gases, intermolecular interactions cannot be neglected. If v represents the molar volume, b being its minimum value, and a/v^2 represents the van der Waals attraction among molecules, then we get the following modification (equation 10.7) of the ideal gas law:

$$\left(p + \frac{a}{v^2}\right)(v - b) = RT \qquad \text{(van der Waals equation)}.$$

Good models for liquids do not exist, although the van der Waals equation has some appropriate qualitative features and can give us insight into gas–liquid phase transitions.

In solids the atoms are anchored in place, each oscillating around some specific equilibrium position. Quantized vibrations called phonons travel throughout the solid.

Free-moving conduction electrons are an important constituent of metals and semiconductors. Perturbations from neighboring atoms cause outer electron states to spread out into bands. In metals, the outer electrons do not completely fill all the states in the band available to them, so some can easily move into neighboring empty states. This gives them a great deal of mobility and makes metals good thermal and electrical conductors. In semiconductors, the outer band of filled electron states is separated from the next higher band of vacant states by a small band gap. Relatively few electrons can jump this gap into the conduction band, where the vacant states allow them great mobility. Conduction in semiconductors is also achieved by positively charged holes in the valence band.

F Tests and applications

F.1 Easily measured properties

Many easily measured properties of systems provide insight for improving our models. Important examples include the molar heat capacities;

$$C_p = \frac{1}{n}\left(\frac{\partial Q}{\partial T}\right)_p, \qquad C_V = \frac{1}{n}\left(\frac{\partial Q}{\partial T}\right)_V, \qquad (10.9, 10.10)$$

where n is the number of moles, the isothermal compressibility,[3]

$$\kappa = -\frac{1}{V}\left(\frac{\partial V}{\partial p}\right)_T, \qquad (10.11)$$

and the coefficient of volume expansion,

$$\beta = \frac{1}{V}\left(\frac{\partial V}{\partial T}\right)_p. \qquad (10.12)$$

In these measures we use *molar* heat capacities and *fractional* changes in volume, $\Delta V/V$, because we want our tests to depend only on the nature of the material and not on its size.

We now examine molar heat capacities more closely. The internal energy of one mole ($n = 1$) of a substance is equal to Avogadro's number times the average energy per particle:

$$E_{\text{molar}} = N_A\left(u_0 + \frac{\nu}{2}kT\right) = N_A u_0 + \frac{\nu}{2}RT.$$

With this expression and the first law ($dE = dQ - pdV$, with $N = N_A$), we get for one mole

$$dQ = dE + pdv = N_A du_0 + \frac{\nu}{2}RdT + pdv,$$

so that the molar heat capacities at constant y become

$$C_y = \frac{1}{n}\left(\frac{\partial Q}{\partial T}\right)_y = N_A\left(\frac{\partial u_0}{\partial T}\right)_y + \frac{\nu}{2}R + p\left(\frac{\partial v}{\partial T}\right)_y. \qquad (10.13)$$

In solids and gases, the potential energy reference level is nearly constant ($\partial u_0/\partial T \approx 0$), so the molar heat capacities at constant volume and constant pressure are

$$C_V \approx \frac{\nu}{2}R, \qquad (10.14)$$

$$C_p \approx \frac{\nu}{2}R + p\left(\frac{\partial v}{\partial T}\right)_p$$

$$= \frac{\nu}{2}R + pv\beta \quad \text{(solids and gases)}. \qquad (10.15)$$

---

[3] We use standard symbols in this text, which sometimes leads to overlap. So you will have to distinguish them by context. For example, the symbol "κ" for isothermal compressibility is also the elastic constant in harmonic oscillators. Other examples include the μ used for both chemical potential and magnetic moments of particles, M used both for mass and molar magnetic moment, p used for pressure, momentum and probability, etc.

Table 10.2. *The ratio* $(C_p - C_v)/R$ *for various materials at* $0\,^{\circ}C$ *and for water at* $20\,^{\circ}C$

Material	$(C_p - C_V)/R$	Material	$(C_p - C_V)/R$
gases		solids	
air	1.00	aluminum	0.0068
ammonia	1.05	brass	0.0050
carbon dioxide	1.02	diamond	0.00015
water vapor	1.14	ice	0.027
liquids		gold	0.0052
ethyl alcohol	0.70		
water (at 20 °C)	0.045		
water (at 0 °C)	−0.014		

Therefore, the difference in the two measures of molar heat capacity C_V and C_p is

$$C_p - C_V \approx pv\beta \quad \text{(solids and gases).} \tag{10.16}$$

As illustrated in Table 10.2, the difference $C_p - C_V$ is large for gases. But it is so small for solids that often we don't bother to distinguish between the two. It is also small for liquids, although this is more difficult to demonstrate, as can be seen in the homework problems.

F.2 Heat capacities and equipartition

Because a molar heat capacity depends on the number of degrees of freedom per molecule, v, the study of this macroscopic property helps reveal microscopic structure. We find that the result 10.14 is correct for the molar heat capacities of most solids and gases,[4] but not for liquids.

For liquids, u_0 varies with temperature. This contributes to the molar heat capacity, as is seen from the $\partial u_0/\partial T$ term in equation 10.13:

$$C_V = N_A \left(\frac{\partial u_0}{\partial T} \right)_v + \frac{v}{2} R \quad \text{(molar volume v is nearly constant).}$$

The molecules of liquids tend to move and rotate to accommodate favored configurations of lower potential energy (Figure 10.7a). At higher temperatures, their motions are more violent and less ordered, reducing the time spent in favored orientations. Consequently, the time-averaged depth of the potential well u_0 gets shallower with increasing temperature, thus absorbing energy like additional degrees of freedom. If we use v' to indicate the effective number of degrees of freedom, then the molar heat capacity becomes

$$C_V = \frac{v'}{2} R.$$

---

[4] Provided that we can ignore quantum effects. More about this in later chapters.

Table 10.3. *Coefficients of volume expansion* $\beta = (1/V)(\partial V/\partial T)_p$ *at room temperature and* $p = 1$ atm

Solids	β(1/K)	Liquids	β(1/K)	Gases	β(1/K)
aluminum	5.5×10^{-5}	alcohol	9.0×10^{-4}	air	3.7×10^{-3}
brass	5.6×10^{-5}	olive oil	6.8×10^{-4}	CO_2	3.8×10^{-3}
marble	3.4×10^{-5}	water	2.1×10^{-4}	N_2	3.7×10^{-3}
steel	4.0×10^{-5}	mercury	1.8×10^{-4}	H_2	3.7×10^{-3}
quartz	3.4×10^{-5}			H_2O	4.2×10^{-3}
wood (oak)	11.2×10^{-5}				

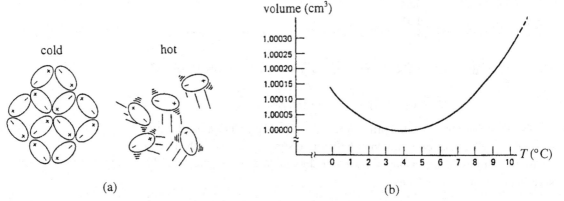

Figure 10.7 (a) In liquids the depth of the potential well varies with temperature. When cold, orientations with unlike charges close together are favored, and the attractive interactions make the average potential well deeper. (b) In water, the low temperature aggregates are porous, making the water less dense. As is seen on this plot of volume versus temperature for water, the volume per gram decreases with increased temperature between 0 and 4 °C, where the breakup of the tiny crystalline aggregates dominates. But above 4 °C, expansion due to the molecular thermal motion dominates.

For liquid water, v' is about 18, for liquid oxygen 12, and for liquid hydrogen 10.

The tendency of molecules in liquids to find favorable orientations at lower temperatures has an interesting consequence for the density of water (Figure 10.7b): the highly polarized water molecules tend to form porous hexagonal structures (Figure 5.6), which makes the water expand when it freezes. Fragments of these crystalline structures survive even after the ice melts. In heating up the ice melt, the continued breakup of these structures makes the water become denser up to a temperature of about 4 °C, after which the normal tendency to expand with increased thermal motion dominates.

F.3 Gases

We now examine the properties of gases in particular. The molar heat capacity C_V of any normal gas is nearly the same as that of an ideal gas, given in

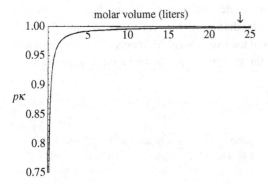

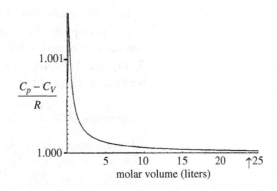

Figure 10.8 Plots of $p\kappa$ and $(C_p - C_V)/R$ vs. molar volume for nitrogen gas at 293 K (20 °C), for the van der Waals model. (The arrows indicate the molar volume at atmospheric pressure.) The ideal gas model gives a value of 1.000 for both. Notice that the predictions of the two models are nearly the same except at very small molar volumes.

equation 10.14.[5] Other properties depend more on the particular model chosen for the gas and therefore provide better tests of that model. Particularly interesting are the isothermal compressibility, the coefficient of volume expansion, and the difference in molar heat capacities (equations 10.11, 10.12, 10.16).

We have learned that the ideal gas model is appropriate for normal rarefied gases ($v \gg b$), and that when the molecules are closer together and mutual interactions become important the van der Waals model is better. We now list these two models in their differential forms. We solve each for dv, so that we can easily pick out its dependence on p and T as is required for obtaining the above properties. For the ideal gas model,

$$pv = RT.$$

The differential form of this is

$$pdv + vdp = RdT$$
$$\Rightarrow \quad dv = -\frac{v}{p}dp + \frac{R}{p}dT. \tag{10.17}$$

For the van der Waals model,

$$\left(p + \frac{a}{v^2}\right)(v - b) = RT.$$

The differential form of this is

$$Avdp + Bpdv = RdT$$
$$\Rightarrow \quad dv = -\frac{A}{B}\frac{v}{p}dp + \frac{1}{B}\frac{R}{p}dT \tag{10.18}$$

with

$$A = \left(1 - \frac{b}{v}\right), \qquad B = \left(1 - \frac{a}{pv^2} + \frac{2ab}{pv^3}\right). \tag{10.19}$$

The two expressions 10.17 and 10.18 for dv become the same in the limit of rarefied gases, because the van der Waals factors A and B become

$$A \approx 1, \qquad B \approx 1 \quad \text{for } v \gg b \text{ and } p \gg \frac{a}{v^2} \quad \text{(rarefied gas).} \tag{10.20}$$

---

[5] Constant volume means that average particle separations (hence interaction strength) are unchanged. So not only is u_0 small for real gases, but it hardly changes with temperature.

In the above differential forms, the dependence of v on p and T ($\mathrm{d}v = (\cdots)\mathrm{d}p + (\cdots)\mathrm{d}T$) is explicit.[6] The results for the various properties are listed in Table 10.4. You can see that the predictions of the two models differ only by the factors A and B. So when the molecules are far apart ($v \gg b$), the two models have the same predictions (Figure 10.8).

Summary of Section F

The relationships between certain parameters are easily measured experimentally and serve as good tests of our models. These interrelationships include (equations 10.9–10.12):

$$\text{molar heat capacities,} \quad C_p = \frac{1}{n}\left(\frac{\partial Q}{\partial T}\right)_p, \quad C_V = \frac{1}{n}\left(\frac{\partial Q}{\partial T}\right)_V;$$

$$\text{isothermal compressibility,} \quad \kappa = -\frac{1}{V}\left(\frac{\partial V}{\partial p}\right)_T;$$

$$\text{coefficient of volume expansion,} \quad \beta = \frac{1}{V}\left(\frac{\partial V}{\partial T}\right)_p.$$

For solids and gases, we can usually assume that $u_0 \approx$ constant. For these cases, the molar heat capacity at constant volume is given by (equation 10.14)

$$C_V \approx \frac{v}{2}R,$$

and the difference between C_p and C_V is (equation 10.16)

$$C_p - C_V \left(\approx p\frac{\partial v}{\partial T}\right)_p = pv\beta.$$

This difference is large for gases but very small for solids. Because the molar heat capacity of a system depends on the number of degrees of freedom per molecule, v, the study of this macroscopic property 10.16 helps to reveal the microscopic structure. For liquids, changes in the level of u_0 act as additional degrees of freedom for storing added energy.

The differential forms of the ideal and van der Waals gas models are (equations 10.17–10.19):

$$\text{ideal gas,} \quad \mathrm{d}v = -\frac{v}{p}\mathrm{d}p + \frac{R}{p}\mathrm{d}T;$$

$$\text{van der Waals gas,} \quad \mathrm{d}v = -\frac{A}{B}\frac{v}{p}\mathrm{d}p + \frac{1}{B}\frac{R}{p}\mathrm{d}T,$$

$$\text{with} \quad A = \left(1 - \frac{b}{v}\right), \quad B = \left(1 - \frac{a}{pv^2} + \frac{2ab}{pv^3}\right)$$

In these formulas the isothermal compressibility, coefficient of volume expansion, and difference in molar heat capacities are easily identified. They are listed in Table 10.4. In the limit of rarefied gases $A \approx B \approx 1$ so the two models give the same predictions in this region.

---

[6] The coefficients $(\ldots)$ are $(\partial v/\partial p)_T$ and $((\partial v/\partial T)_p$. Also note that $\partial v/v = \partial V/V$.

Table 10.4. *Isothermal compressibility, coefficient of volume expansion, and the difference in molar heat capacities according to ideal gas and van der Waals models*

Gas	Definition	Ideal gas	van der Waals
isothermal compressibility	$\kappa = -\dfrac{1}{V}\left(\dfrac{\partial V}{\partial p}\right)_T$	$\dfrac{1}{p}$	$\dfrac{A}{B}\dfrac{1}{p}$
coefficient of volume expansion	$\beta = \dfrac{1}{V}\left(\dfrac{\partial V}{\partial T}\right)_p$	$\dfrac{R}{pv} = \dfrac{1}{T}$	$\dfrac{1}{B}\dfrac{R}{pv}$
difference in heat capacities	$C_p - C_V \approx pv\beta$	R	$\dfrac{1}{B}R$

Problems

Section A

1. Given the system variables E, S, T, p, V, μ, N, if a system is interacting in three different ways, how many independent equations must exist (whether we know them or not) to interrelate these variables?

2. The two equations of state $E = (3/2)NkT$ and $pV = NkT$ for an ideal gas involve five parameters (E, N, T, p, V), so that three parameters are independent. Find
   (a) $p(T, V, N)$ and $(\partial p/\partial V)_{T,N}$,
   (b) $E(T, p, N)$ and $(\partial E/\partial p)_{T,N}$.

3. Suppose that there exists between the five parameters E, T, S, p, V the following interrelationship: $S = CV^2 \ln E$, where C is a constant. Evaluate the following: (a) $1/T = (\partial S/\partial E)_V$, (b) $p/T = (\partial S/\partial V)_E$, (c) $(\partial E/\partial V)_T$, (d) $(\partial p/\partial V)_S$.

4. Chemical potential is normally negative. According to equations 9.13 and 9.3c, it decreases with increasing temperature, increases with increasing pressure, and increases with increasing N. Invent any equation of state relating the parameters μ, T, p, N for which these requirements would be satisfied.

Section B

5. What is $(\partial E/\partial p)_{V,N}$ for an ideal gas? (Hint: Express E as a function of p and V.)

6. Show that equations 10.2a, b for the entropies of an ideal gas and a solid follow from equations 6.7 and 6.10 and the definition of entropy.

7. Using equation 10.2a for the entropy of an ideal gas, find the partial derivatives of equation 8.7 and so derive the three equations of state 10.3a, b, c.

8. Using equation 10.2b for the entropy of a solid, find the partial derivatives of equation 8.7 to derive the three equations of state 10.4a, b, c.

9. Consider a two-dimensional ideal gas of energy E and confined to area A, with identical molecules that are able to translate but not rotate or vibrate.
   (a) How many degrees of freedom does each molecule have?
   (b) How many degrees of freedom does a system of N molecules have?
   (c) What is the total internal energy of a system of 10^{22} such molecules at a temperature of 300 K?
   (d) Write down the number of quantum states Ω available to a system of N such particles as a function of A, N, and E. (Combine all the constants into one.)
   (e) What is the entropy as a function of A, N, and E?

10. Consider a one-dimensional ideal gas having total energy E. It is confined to a length L along the x-axis, with identical molecules that are able to translate but not rotate or vibrate.
   (a) How many degrees of freedom does a system of N molecules have?
   (b) Write down the number of quantum states Ω available to a system of N such particles as a function of L, N, and E. (Combine all the constants into one.)
   (c) What is the entropy of the system as a function of L, N, and E?
   (d) For a one-dimensional system, the work done by it in changing its length by dL is $dW = F dL$. Write dE as a function of T, F, dS, and dL, and write dS as a function of T, F, dE, and dL.
   (e) Derive an "ideal gas law" for this one-dimensional system from $F/T = (\partial S/\partial L)_{E,N}$.

11. The number of quantum states accessible to a certain system is given by $\Omega = Ce^{bV^{4/5}}E^{2N}$, where C and b are constants.
   (a) Write down an expression for the entropy in terms of V and E.
   (b) Using $1/T = (\partial S/\partial E)_{V,N}$, find how the internal energy depends on the temperature.
   (c) How many degrees of freedom does this system have?
   (d) Using $p/T = (\partial S/\partial V)_{E,N}$, find the interrelationship between p, V, and T.

12. Derive the ideal gas law by the method outlined in the caption to Figure 10.2.

13. The ideal gas law can be derived by a third method. Suppose that a container of volume V has N gas molecules, each of mass m, half of which are moving in the negative x direction with speed v_x. Consider the elastic collisions of these molecules with the wall of the container, which lies in the yz-plane and has area A, as in Figure 10.2.

(a) Show that the density of particles moving in the negative x direction is $N/2V$.

(b) Show that the number of molecules colliding with the wall per second is $(N/2V)Av_x$.

(c) Now show that the impulse given to the wall per collision is $2mv_x$.

(d) What is the average force exerted on the wall altogether?

(e) What is the average pressure exerted on the wall?

(f) Suppose that there is a distribution of molecular speeds, so that we have to average over v_x^2. Rewrite the answer to (e) using the equipartition theorem to express the average value of v_x^2 in terms of T. (If your answer isn't the ideal gas law, you have made a mistake somewhere.)

14. The entropy of an ideal gas is given by $S = Nk \ln \omega_c$, with $\omega_c = C(V/N)(E/Nv)^{v/2}$. This means that we can write out the entropy as $S = Nk \ln C + Nk \ln V - Nk \ln N + (Nv/2)k \ln E - (Nv/2)k \ln Nv$. How can you justify taking the logarithm of C, V or E? Doesn't it depend on the units you use? How can you take the logarithm of units?

15. Calculate the chemical potential of an ideal gas from $\mu/T = -(\partial S/\partial N)_{E,V}$. (See the above problem for the entropy of an ideal gas.) Show that the answer you get is the same as is obtained from the integrated form of the first law, $E = TS - pV + \mu N$.

16. For a monatomic ideal gas of N particles, we found that the internal energy depends on the temperature only, $E = (3/2)NkT$, not on the volume. How can you square this with the first law, which says that changes in the volume cause changes in internal energy, $dE = dQ - pdV + \mu dN$?

17. The number of states accessible to a certain system is given by $\Omega = Ce^{aV^{1/2}}E^{bV}$, where C, a, and b are constants. In terms of E, V, and these constants, find (a) the temperature, (b) the pressure (i.e., equations of state).

18. The number of states accessible to a certain system is given by $\Omega = Ce^{aNV^2}E^{bNV}$, where C, a, and b are constants. In terms of E, V, and these constants, find (a) the temperature, (b) the pressure, (c) the chemical potential.

19. Assuming that water vapor behaves as an ideal gas with six degrees of freedom per molecule, what is the enthalpy of one mole of water vapor at 500 °C and 3 atm? (See Chapter 9 for the definition of enthalpy.)

20. At room temperature (295 K) and atmospheric pressure (1.013×10^5 Pa), estimate the number of air molecules in a room 8 m by 10 m by 3 m and the average separation between the molecules.

21. Dry air is 78% N_2, 21% O_2 and 1% Ar.

(a) What is the molar mass of dry air?

(b) What is the molar mass on a humid day when the water vapor content is 3%?

22. For a certain hot air balloon, the mass of the balloon, harness, basket and payload is 600 kg. The air temperature inside the balloon is 70 °C and that outside it is 18 °C. Estimate the minimum volume of hot air needed if the balloon is to lift off the ground. (The molar mass of air is about 0.029 kg and atmospheric pressure is 1.013×10^5 Pa.)

23. We are interested in the molar entropies of some typical systems at room temperature and atmospheric pressure (295 K, 1.013×10^5 Pa). From Table 6.2, we can calculate the values of the constants C in equations 10.2:

$$C_{\text{monatomic gas}} = [3.59 \times 10^{101}(\text{J s})^{-3}] \, m^{3/2},$$
$$C_{\text{diatomic gas}} = [1.025 \times 10^{170}(\text{J s})^{-5}] \, m^{3/2} I,$$
$$C_{\text{solid}} = [6.34 \times 10^{101}(\text{J s})^{-3}](m/\kappa)^{3/2},$$

where m is the molecular (or atomic) mass, I the rotational inertia, and κ the elastic constant. Assume that the ideal gas law is valid for gases (and hence $V/N = kT/p$) and that the thermal energy per particle, E_{therm}/N, is $(\nu/2)kT$. Calculate the corrected number of states per particle ω_c and the molar entropy for:
(a) helium, a monatomic gas with $m = 6.7 \times 10^{-27}$ kg;
(b) nitrogen gas (N_2) with $m = 47 \times 10^{-27}$ kg, assuming that $I \approx 3.0 \times 10^{-46}$ kg m^2;
(c) a typical solid with $\sqrt{m/\kappa} \approx 6 \times 10^{-14}$ s.

24. In problem 23 above we calculated the entropy of nitrogen gas at room temperature and atmospheric pressure (295 K, 1.013×10^5 Pa) to be 213 J/(K mole). Using this and assuming nitrogen to be an ideal diatomic gas, calculate the internal energy, the Helmholtz free energy, the enthalpy, and the Gibbs free energy for one mole of nitrogen at room temperature and atmospheric pressure.

25. Calculate the change in the molar enthalpy, entropy, and internal energy for water at atmospheric pressure (1.013×10^5 Pa) in the following processes: At the liquid–gas transition, assume that the volume of the liquid is negligible compared with that of the gas and that the ideal gas law applies to the gas. The molar specific heat C_p is for water 75.4 J/(K mole) and for steam 35.7 J/(K mole), and the latent heat of vaporization for water is 4.07×10^4 J/mole.)
(a) going from a liquid at 0 °C to a liquid at 100 °C;
(b) going from a liquid at 100 °C to a gas at 100 °C;
(c) going from a gas at 100 °C to a gas at 200 °C.

Section C

26. Suppose that the accessible volume for a particle in a mole of a gas is given by $v - b$. Using this for the volume in equation 10.2a, and the relationship $p/T = (\partial S/\partial V)_{E,N}$, find the equation of state for this gas.

27. At standard temperature and pressure (0 °C, 1 atm), a mole of liquid water occupies a volume of 18 cm³ and a mole of water vapor occupies 22.4 liters.
    (a) Find the characteristic dimension of a single water molecule by considering the volume of a mole of the liquid.
    (b) Roughly what is the characteristic separation of molecules in the water vapor?
    (c) In the gas phase, how many times larger is the characteristic separation than the molecular size?

28. Consider the gases listed in Table 10.1. For each gas, the coefficient a is related to a molecule's polarizability and the parameter b is related to its size.
    (a) Which material has the most highly polarized or polarizable molecules?
    (b) For which would the a/v^2 term be largest when the molecules are touching ($v \approx b$)?

29. Consider a plasma of charged particles, such as ionized oxygen atoms. They all have positive charges, so they all repel each other with a force that falls off with separation according to $1/r^2$. How would you modify the van der Waals model for this system? (Identify constants by letters such as $a, b, c, \ldots$, without giving them values.)

30. Electrons have negative charges, so they experience a mutual repulsion that falls off with increasing separation according to $1/r^2$. They seem to be true point particles. That is, they seem to occupy no volume at all. How would you modify the van der Waals model to fit a pure electron gas? (Identify constants by letters such as $a, b, c, \ldots$, without giving them values.)

31. The density of liquid water is 10^3 kg/m³ and its molecular mass is 18. The density of liquid ethyl alcohol is 0.79×10^3 kg/m³ and its molecular mass is 46. Find the molar volumes for water and ethyl alcohol, and compare them with the van der Waals parameter b for the gas phase of these two materials.

Section D

32. The intermolecular interactions in liquids are such that they are repulsive if you try to compress the liquid and attractive if you try to expand it. Invent a self-interaction term for liquids, to replace the van der Waals term, that still has the approximately a/v^2 behavior at large molecular separation (i.e., large molar volume) but becomes repulsive for intermolecular separations less than $(2b)^{1/3}$ (i.e., $v < 2b$). Your invented term might have some constants to be determined by experiment.

33. In an earlier chapter we found that, for liquid water, large pressure increases give only very small changes in volume. In particular, we found that the pressure and volume were related by an equation of the form $p = A[1 + B(1 - V/V_0)]$, where A, B, and V_0 are constants.
    (a) Write down an expression that shows how the water's entropy depends on the volume. (Assume that E, V, and N are the independent variables, with E proportional to T.)
    (b) Write down any expression for the number of accessible states that has the correct volume dependence.

34. The water molecule in the liquid state has six degrees of freedom (three translational and three rotational). The variation in the potential energy reference level u_0 makes water behave in a way that is equivalent to 12 additional degrees of freedom. At 0 °C, $u_0 = -0.441$ eV. Write down an expression for u_0 for liquid water as a function of temperature in the range 0−100 °C.

Section E

35. An atom in a particular solid is anchored in place by electrostatic interactions with its neighbors. For small displacements from its equilibrium position, the restoring force is directly proportional to the displacement: $F_x = -\kappa_1 x$, $F_y = -\kappa_2 y$, $F_z = -\kappa_3 z$, where the force constants in different dimensions are different.
    (a) Suppose that you want to make a coordinate transformation such that the potential energy can be written in the new coordinates as $V = (1/2)\kappa_1(x'^2 + y'^2 + z'^2)$. What are x', y', z' in terms of x, y, z?
    (b) Now you want to express the kinetic energy in the form $[1/2m](p_1'^2 + p_2'^2 + p_3'^2)$, where $p_1' = m_1(dx'/dt)$, etc. What will be the values of the masses m_1, m_2, m_3 in terms of the mass m and the spring constants?

36. According to the equipartition theorem, the average energy per degree of freedom is $(1/2)kT$.
    (a) What is the root mean square speed of a gold atom (atomic mass number 197) vibrating around its equilibrium position in a gold wire at room temperature?
    (b) How does this compare with the root mean square speed of conduction electrons in this wire, assuming that they behave like a gas?

37. Demonstrations of the photoelectric effect require the use of ultraviolet light of maximum wavelength 264 nm in order to kick electrons off the surface of copper. From this, estimate the maximum energy of electrons in copper's conduction band. (Take the energy of an electron free in space, all by itself, and standing still as zero.)

38. A current of 1 ampere (1 coulomb per second) is flowing through a copper
    wire of 4 mm diameter. There are 8.4×10^{28} atoms per cubic meter, and each
    contributes one electron to the conduction band.
    (a) How many conduction electrons are there in one centimeter of this wire?
    (b) Roughly what is the net drift velocity of the conduction electrons in this
        wire?
    (c) How does the answer to part (b) compare with the root mean square ther-
        mal velocity of free conduction electrons in the wire at room temperature
        (295 K)?

39. The amplitude of vibration for atoms in a typical solid at room temperature
    is about 10^{-11} m.
    (a) How much energy is there in each degree of freedom, on average?
    (b) What is the elastic (spring) constant, κ?
    (c) Roughly what is the vibrational frequency for an atom of aluminum (mass
        number 27)? Hint: For simple harmonic motion, do you remember how
        the frequency, mass, and spring constant are related?

Section F
40. Calculate the isothermal compressibility of 47 liters of CO_2 gas at room tem-
    perature (295 K) and atmospheric pressure (1.013×10^5 N/m^2), assuming
    that it obeys the ideal gas law reasonably well.

41. In deriving the difference $C_p - C_V = p(\partial v / \partial T)_p$ we assumed that
    $(\partial u_0 / \partial T)_p = (\partial u_0 / \partial T)_V = 0$. Suppose that this assumption were not true.
    Use equation 10.13 to write out the exact expression for $C_p - C_V$. Then use
    equation 10.8 and the chain rule to show that

    $$C_p - C_V = (R/B)(1 + a/pv^2)$$

    for a van der Waals gas.

42. Steam is under a very high pressure $p = 100$ atm, so that its molar vol-
    ume is small ($v = 0.3$ liters). The van der Waals constants for steam are
    $a = 5.5$ liters^2 atm/mole^2 and $b = 0.030$ liters/mole. What are the difference
    $C_p - C_V$ and the temperature T for this gas?

43. The molar heat capacity at constant volume for a rarefied gas is measured to
    be $C_V = 33.3$ J/(mole K).
    (a) Does it have monatomic or polyatomic molecules?
    (b) How many degrees of freedom does each molecule have?

44. The molar volume of copper is 7.1 cm^3 and its coefficient of thermal vol-
    ume expansion is $\beta = 4.2 \times 10^{-5}$/K. Each copper atom has six degrees of
    freedom. At room temperature and atmospheric pressure, by what percent
    do you expect C_p and C_V to differ? Which is bigger?

45. The equation of state for some system is $p^2 T^{-1/3} e^{aV} = b$, where a and b are constants.
    (a) Write this in differential form, expressing dV in terms of dT and dp.
    (b) Express the isothermal compressibility of this system in terms of T, V, and p.
    (c) Express the coefficient of volume expansion for this system in terms of T, V, and p.

46. Calculate the coefficient of volume expansion, β, for a van der Waals gas in terms of p, v, a, and b.

47. Consider an ideal gas whose molecules have ν degrees of freedom and which is undergoing some process that we are tracing on a p–V diagram. In terms of p, V, and ν, what is the slope of this path at the point (p, V) on the diagram if the process is (a) isobaric, (b) isothermal, (c) adiabatic (pV^γ = constant, where $\gamma = (\nu + 2)/\nu$)?

48. Consider a system that behaves according to the van der Waals equation of state.
    (a) Write this in differential form, expressing dv in terms of dT and dp.
    (b) Show that the isothermal compressibility of the system is given correctly by our result A/pB (see Table 10.4).
    (c) What is $(\partial p/\partial T)_V$?

49. An equation of state for a certain material is found to be $pV^2 - aTV = bT$, where a and b are constants. In terms of p, V, T, find (a) the coefficient of thermal expansion, (b) the isothermal compressibility.

50. We are going to examine how well the van der Waals model works for liquids if we use the constants for the gas phase from Table 10.1. The molar volume of liquid water is about 0.018 liters $= 0.59b$.
    (a) Using the van der Waals constants for water (Table 10.1), estimate the difference $C_p - C_V$. How does this compare with the actual value at 20 °C listed in Table 10.2?
    (b) Do the same for the coefficient of volume expansion, β.

51. Using the van der Waals model and Table 10.1, answer the following:
    (a) What is the pressure exerted upon nitrogen gas in a container if at a temperature of 500 K its molar volume is 0.5 liter?
    (b) What is its coefficient of volume expansion under these conditions?

52. A system behaves according to the equation $[p + \frac{a}{(v-c)^{1/3}}](v - \frac{b}{p}) = RT$, with appropriate values of the constants a, b, c. Express the following as a function of T, v, p:
    (a) the coefficient of volume expansion, β,
    (b) The isothermal compressibility, κ,
    (c) The difference in molar heat capacities, $C_p - C_V$.

53. The equation of state relating pressure, temperature, and molar volume for a particular material is given by $(pv + A)e^{Bv} = RT$, where A, B, and R are constants. Express the following as a function of T, v, and p: (a) β, (b) κ, (c) $C_p - C_V$.

54. (a) Starting with the equation preceding 10.13, show that for an ideal gas,
$C_p = [(v + 2)/2]R$.
   (b) Using the specific heats listed in Table 8.1, find the effective number of degrees of freedom per molecule for helium gas and liquid water (the molar masses are 4.00 and 18.0 grams, respectively).

Chapter 11
Choice of variables

The interdependence of so many different thermodynamical variables means that there are several different but equivalent ways of writing any expression. So we have many options, and we can choose whichever variables we wish. For example, the first law is normally written with S, V, N as the independent variables:[1]

$$dE(S, V, N) = T(S, V, N)dS - p(S, V, N)dV + \mu(S, V, N)dN \qquad (11.1)$$

But we may prefer other variables. For example, if we have a thermometer, pressure gauge, and meter stick, we may prefer to work with T, p, V and write the first law as

$$dE = AdT + Bdp + CdV,$$

where E, A, B, C are all functions of T, p, V. Indeed, we can do this if we want to.

In short, the redundancy among the various parameters may at first seem like a great deal of trouble. But in this chapter we learn ways to use this interdependence to our advantage.

---

[1] Remember that if there is more than one type of mechanical or diffusive interaction, then there are correspondingly more independent variables.

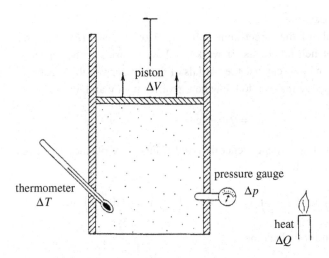

Figure 11.1 We are studying a system that is undergoing nondiffusive interactions, and it is convenient for us to measure temperature and pressure. How can we express the changes in other properties, such as internal energy, volume, or entropy (ΔE, ΔV, or ΔS) in terms of our measured changes in temperature and pressure ΔT and Δp?

A   Changing variables

Suppose that we want to determine changes in some property X for a *closed* system. The constraint $dN = 0$ reduces the number of independent variables from three to two. If we have a thermometer and a pressure gauge, we would like to obtain an expression of the form

$$dX = AdT + Bdp,$$

so that measurements of ΔT and Δp can be directly converted into the desired change ΔX (Figure 11.1). How do we get such an expression? There are many ways to do it.

A.1   The general case

Imagine we have several variables $v, w, x, y, \ldots$, of which only two are independent. Suppose that we wish to know how one varies with two others – for example, how x varies with y and z. How do we answer this question?

The direct approach
The most direct way of attacking this problem is to write

$$dx = \left(\frac{\partial x}{\partial y}\right)_z dy + \left(\frac{\partial x}{\partial z}\right)_y dz. \tag{11.2}$$

The partial derivatives may not be very useful as they stand, but if we are either lucky or clever, we can convert them into tabulated or easily measured properties of the system, such as pressure, temperature, heat capacities, etc. There are many ways to do this, as illustrated below in subsection 11A.3 and as you will find in the homework problems.

Alternative approach

Suppose the partials of the direct approach, e.g., $(\partial x/\partial y)_z$ and $(\partial x/\partial z)_y$, are inconvenient or not helpful for us. If we know how x varies with respect to any other variables then we can use the partials of these other variables instead.

For example, suppose that we do know how x varies with v and w:

$$dx = f\,dv + g\,dw.$$

How do we use this to find its dependence on y and z? First we write the derivatives dv, dw each in terms of dy, dz:

$$dv = \left(\frac{\partial v}{\partial y}\right)_z dy + \left(\frac{\partial v}{\partial z}\right)_y dz, \qquad dw = \left(\frac{\partial w}{\partial y}\right)_z dy + \left(\frac{\partial w}{\partial z}\right)_y dz.$$

Then our expression for dx becomes

$$dx = f\,dv + g\,dw = f\left[\left(\frac{\partial v}{\partial y}\right)_z dy + \left(\frac{\partial v}{\partial z}\right)_y dz\right] + g\left[\left(\frac{\partial w}{\partial y}\right)_z dy + \left(\frac{\partial w}{\partial z}\right)_y dz\right].$$

Collecting the coefficients of dy and dz, we have the form we wanted,

$$dx = A\,dy + B\,dz, \tag{11.3}$$

with

$$A = f\left(\frac{\partial v}{\partial y}\right)_z + g\left(\frac{\partial w}{\partial y}\right)_z, \qquad B = f\left(\frac{\partial v}{\partial z}\right)_y + g\left(\frac{\partial w}{\partial z}\right)_y. \tag{11.3'}$$

Although we now have the desired variables, y, z, the coefficients are a mess. Don't be discouraged. In a few pages we will learn how to simplify these ugly partial derivatives.

Example 11.1 Suppose that we want to know how the internal energy changes with temperature and pressure for a *closed* system ($dN = 0$), i.e., we want an expression like $dE = (\cdots)dT + (\cdots)dp$.

We can start with the first law for nondiffusive interactions ($dN = 0$),

$$dE = T\,dS - p\,dV,$$

but we must write the changes dS, dV in terms of dT, dp to get the form we want:

$$dS = \left(\frac{\partial S}{\partial T}\right)_p dT + \left(\frac{\partial S}{\partial p}\right)_T dp, \qquad dV = \left(\frac{\partial V}{\partial T}\right)_p dT + \left(\frac{\partial V}{\partial p}\right)_T dp.$$

Putting these into the above expression for dE and collecting terms, we have our desired expression for the dependence of E on T and p, i.e.,

$$dE = A\,dT + B\,dp,$$

with

$$A = T\left(\frac{\partial S}{\partial T}\right)_p - p\left(\frac{\partial V}{\partial T}\right)_p \quad \text{and} \quad B = T\left(\frac{\partial S}{\partial p}\right)_T - p\left(\frac{\partial V}{\partial p}\right)_T. \quad (11.4)$$

A.2 Why switch?

It is clear that, through transformations such as the above, we can write thermo-dynamic expressions in terms of virtually whichever set of variables we wish. But why would we prefer one set of variables over another? This question is easy to answer. It is clearly to our advantage to express the properties of a system in terms of parameters that are either (a) constrained, or (b) easy to measure.

For example, suppose that we are interested in some property X of a system undergoing a nondiffusive process at atmospheric pressure ($dN = dp = 0$), and we have a thermometer at our disposal. Then it is advantageous to express this property in terms of T, p, and N, because we are left simply with

$$dX = AdT + Bdp + CdN = AdT, \quad (dp = dN = 0),$$

and dT is easy to measure.

Of course, the coefficient A has some partial derivatives in it but, using the techniques that follow, we can write these in terms of easily measured properties of the system.

As another example, consider the adiabatic expansion ($dS = 0$) and cooling of a rising air mass that remains intact ($dN = 0$). These two constraints leave just one independent variable. We should be able to express the change in some property X in terms of just one convenient variable, such as the pressure:

$$dX = Adp + BdS + CdN = Adp, \quad (ds = dN = 0).$$

A.3 The partial derivatives

Through the above approaches, we can express the property of interest, X, in terms of whichever variables we wish. The resulting expressions contain partial derivatives, some of which are readily dealt with, and some of which are not.

Many partial derivatives are easily measured properties of the system (Figure 11.2). For example, we have already encountered (equations 8.1, 8.7, and 10.7–10.12)

$$\left(\frac{\partial E}{\partial S}\right)_V = T, \quad \left(\frac{\partial S}{\partial V}\right)_E = \frac{p}{T}, \quad \left(\frac{\partial S}{\partial T}\right)_V = \frac{C_V}{T},$$

$$\left(\frac{\partial S}{\partial T}\right)_p = \frac{C_p}{T}, \quad \left(\frac{\partial V}{\partial T}\right)_p = V\beta, \quad \left(\frac{\partial V}{\partial p}\right)_T = -V\kappa. \quad (11.5)$$

where T is the temperature, p is the pressure, C_y is the heat capacity at constant y, β is the coefficient of thermal expansion, and κ is the isothermal compressibility.

Table 11.1. *The interrelationships between S, T, p, V for nondiffusive processes, written in terms of the easily measured properties of the system, C_p, C_V, β, κ, T, p, V. The reciprocal of each relationship is also valid*

$$\left(\frac{\partial S}{\partial T}\right)_p = \frac{C_p}{T} \qquad \left(\frac{\partial S}{\partial V}\right)_T = \frac{\beta}{\kappa} \qquad \left(\frac{\partial T}{\partial V}\right)_S = -\frac{T\beta}{C_V \kappa}$$

$$\left(\frac{\partial S}{\partial T}\right)_V = \frac{C_V}{T} \qquad \left(\frac{\partial S}{\partial V}\right)_p = \frac{C_p}{TV\beta} \qquad \left(\frac{\partial T}{\partial V}\right)_p = \frac{1}{V\beta}$$

$$\left(\frac{\partial S}{\partial p}\right)_T = -V\beta \qquad \left(\frac{\partial T}{\partial p}\right)_S = \frac{TV\beta}{C_p} \qquad \left(\frac{\partial p}{\partial V}\right)_S = -\frac{C_p}{VC_V \kappa}$$

$$\left(\frac{\partial S}{\partial p}\right)_V = \frac{C_V \kappa}{T\beta} \qquad \left(\frac{\partial T}{\partial p}\right)_V = \frac{\kappa}{\beta} \qquad \left(\frac{\partial p}{\partial V}\right)_T = -\frac{1}{V\kappa}$$

The definitions of the heat capacities C_p and C_V, the coefficient of thermal expansion β, and the isothermal compressibility κ are as follows:

$$C_p = \left(\frac{\partial Q}{\partial T}\right)_p = T\left(\frac{\partial S}{\partial T}\right)_p, \qquad C_V = \left(\frac{\partial Q}{\partial T}\right)_V = T\left(\frac{\partial S}{\partial T}\right)_V, \qquad \beta = \left(\frac{1}{V}\frac{\partial V}{\partial T}\right)_p, \qquad \kappa = -\left(\frac{1}{V}\frac{\partial V}{\partial p}\right)_T$$

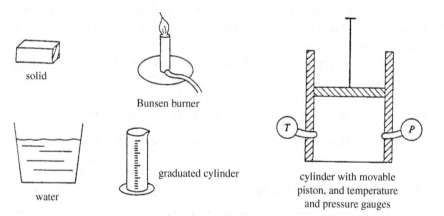

Figure 11.2 For a gas, the water, or a solid, how would you measure the coefficient of thermal expansion (β), isothermal compressibility (κ), or heat capacity (C_p or C_V), using the equipment in this figure? (Assume that you can measure heat transfer, $\Delta Q = C_{(p\,or\,V)}\Delta T$, and volume change, ΔV.)

solid

Bunsen burner

graduated cylinder

water

cylinder with movable piston, and temperature and pressure gauges

Other partial derivatives may also be converted into measurable properties using methods outlined below. These techniques are applied to the commonly encountered partial derivatives for nondiffusive interactions (i.e., all those involving S, T, p, V) in homework problems 1–12, with results listed in Table 11.1.

1 Maxwell's relations

Maxwell's relations (Table 9.3) constitute a valuable resource. If you see a partial derivative you don't like, you can use the appropriate Maxwell's relation to substitute another for it. Of the twelve, those in the first column, M1, M4, M7, and M10, apply exclusively to nondiffusive interactions.

2   Chain rule

Second, you can use the chain rule. For example, suppose you have a derivative $(\partial s/\partial t)_{u,v}$ that you don't like. You can use the chain rule to express this in the form

$$\left(\frac{\partial s}{\partial t}\right)_{u,v} = \left(\frac{\partial s}{\partial x}\right)_{u,v}\left(\frac{\partial x}{\partial t}\right)_{u,v}. \tag{11.6}$$

What should you choose for the intermediate variable x? Anything you want! Since u and v are both held constant, there is only one independent variable, and you can choose it to be any other variable that is convenient for you.

Example 11.2 Suppose that we are working on some nondiffusive process ($dN = 0$) where there are two independent variables and we run into the derivative $(\partial S/\partial V)_p$. This could be determined by measuring the change in volume ΔV as heat $T\Delta S$ is added at constant pressure, giving us the needed $(\Delta S/\Delta V)_p$. But there are other alternatives.

We could use Maxwell's relation M7 to turn it into $(\partial p/\partial T)_S = (\Delta p/\Delta T)_S$, which we could measure by compressing the system adiabatically ($S = $ constant) and measuring the changes in pressure Δp and temperature ΔT. But maybe we don't care to do that either.

So let's try using the chain rule. Since p and N are being held constant, there is only one independent variable. What should we choose as an intermediate variable?

Any would work. But the temperature T would be a good choice, because changes in S, V, T are related through heat capacities and coefficients of thermal expansion. The chain rule with T as the intermediate variable gives

$$\left(\frac{\partial S}{\partial V}\right)_p = \left(\frac{\partial S}{\partial T}\right)_p\left(\frac{\partial T}{\partial V}\right)_p.$$

To transform these partials we could use equations 11.5 or Table 11.1, or we could work them out from first principles:

$$\left(\frac{\partial S}{\partial T}\right)_p = \frac{1}{T}\left(\frac{\partial Q}{\partial T}\right)_p = \frac{C_p}{T},$$

$$\left(\frac{\partial T}{\partial V}\right)_p = \left[\left(\frac{\partial V}{\partial T}\right)_p\right]^{-1} = \left[V\frac{1}{V}\left(\frac{\partial V}{\partial T}\right)_p\right]^{-1} = (V\beta)^{-1}.$$

With this, we have for our partial derivative

$$\left(\frac{\partial S}{\partial V}\right)_p = \left(\frac{\partial S}{\partial T}\right)_p\left(\frac{\partial T}{\partial V}\right)_p = \frac{C_p}{TV\beta}.$$

That is, we have transformed our undesirable partial derivative into easily measurable and tabulated properties of the system.

3 Ratios

A third way of transforming a partial derivative into something more desirable goes as follows. Suppose that you wish to transform $(\partial x/\partial y)_z$ into something else. Noting that z is constant, we write dz in terms of dx and dy:

$$0 = dz = \left(\frac{\partial z}{\partial x}\right)_y dx + \left(\frac{\partial z}{\partial y}\right)_x dy.$$

From this, we can solve for the ratio dx/dy at constant z, which is the definition of the partial derivative $(\partial x/\partial y)_z$:

$$\left(\frac{\partial x}{\partial y}\right)_z = -\frac{(\partial z/\partial y)_x}{(\partial z/\partial x)_y} = -\left(\frac{\partial z}{\partial y}\right)_x \left(\frac{\partial x}{\partial z}\right)_y. \tag{11.7}$$

Exercise caution when using this relationship, because it is tricky. The right side looks deceptively similar to the chain rule, but in fact it is quite different. Notice in particular the negative sign, and the variables that are being held constant.

Example 11.3 Use the above technique to write the derivative $(\partial T/\partial p)_V$ as something more desirable.

Since V is held constant and T and p are the variables of interest, we write

$$0 = dV = \left(\frac{\partial V}{\partial T}\right)_p dT + \left(\frac{\partial V}{\partial p}\right)_T dp.$$

This gives

$$\left(\frac{\partial T}{\partial p}\right)_V = -\frac{(\partial V/\partial p)_T}{(\partial V/\partial T)_p} = \frac{V\kappa}{V\beta} = \frac{\kappa}{\beta}.$$

Summary of Section A

There is a great deal of interdependence among thermodynamical variables. It is usually most convenient to use as independent variables those that are either (a) constrained, or (b) easy to measure.

To express how property x varies with the variables y and z we could write (equation 11.2)

$$dx = \left(\frac{\partial x}{\partial y}\right)_z dy + \left(\frac{\partial x}{\partial z}\right)_y dz,$$

or we could start with some already known relationship between variables (e.g. the first law) and then use the technique of equation 11.2 to eliminate the variables we don't want by expressing them in terms of more convenient variables.

The resulting expressions contain partial derivatives of the form $(\partial x/\partial y)_z$, which we may wish to avoid. We would prefer to write them in terms of easily measured properties of the system such as temperature, pressure, volume, heat capacities, compressibilities, and coefficients of thermal expansion. There are three ways of doing this.

1. Use Maxwell's relations.
2. Use the chain rule with an intermediate variable t chosen for convenience (equation 11.6):

$$\left(\frac{\partial x}{\partial y}\right)_z = \left(\frac{\partial x}{\partial t}\right)_z \left(\frac{\partial t}{\partial y}\right)_z.$$

3. For a variable z that is being held constant, express dz in terms of dx and dy, which gives (equation 11.7)

$$\left(\frac{\partial x}{\partial y}\right)_z = -\frac{(\partial z/\partial y)_x}{(\partial z/\partial x)_y} = -\left(\frac{\partial z}{\partial y}\right)_x \left(\frac{\partial x}{\partial z}\right)_y.$$

Table 11.1 lists the results of these approaches for all the commonly encountered partial derivatives for nondiffusive interactions.

B Examples

We now use a few more examples to illustrate the above techniques for expressing a given property in terms of whichever variables we wish.

B.1 Testing equipartition

We derived the equipartition theorem (Section 4E and subsection 8A.3) using the assumptions that the potential energy reference level u_0 is constant and that the energy in each degree of freedom can be expressed in the form $b\xi^2$. A consequence of these assumptions is that changes in the internal energy of a system are related to changes in temperature according to

$$dE = \frac{N\nu}{2}k\,dT, \tag{11.8}$$

We can test the accuracy of this result experimentally by rewriting the first law for closed systems in terms of T, V instead of S, V:

$$dE = T\,dS - p\,dV = T\left[\left(\frac{\partial S}{\partial T}\right)_V dT + \left(\frac{\partial S}{\partial V}\right)_T dV\right] - p\,dV.$$

Using Table 11.1 for the partial derivatives, this becomes

$$dE = C_V\,dT + \left(\frac{T\beta}{p\kappa} - 1\right)p\,dV.$$

This agrees with the prediction of the equipartition theorem (11.8) if

$$C_V = \frac{N\nu}{2}k \quad \text{and} \quad \frac{T\beta}{p\kappa} - 1 = 0.$$

So either of these could be used to test for violation of the equipartition theorem; we say that the "degree of violation" is given by

$$\left| \frac{2C_V}{Nvk} - 1 \right| \quad \text{or} \quad \left| \frac{T\beta}{p\kappa} - 1 \right|. \tag{11.9}$$

Example 11.4   Let us test the equipartition theorem for a van der Waals gas, for which the equation of state is (equation 10.7):

$$\left(p + \frac{a}{v^2} \right)(v - b) = RT.$$

In subsection 10F.3 we derived the isothermal compressibility κ and coefficient of thermal expansion β for a van der Waals gas (Table 10.4):

$$\kappa = \frac{1}{p}\frac{A}{B}, \qquad \beta = \frac{R}{pv}\frac{1}{B},$$

where the two factors A and B are given by (equation 10.19):

$$A = \left(1 - \frac{b}{v} \right), \qquad B = \left(1 - \frac{a}{pv^2} + \frac{2ab}{pv^3} \right),$$

and go to unity in the limit of rarefied gases ($v \gg b$). If we insert these expressions into the second of the violation tests given in equation 11.9 and use the van der Waals equation to write the product RT in terms of p and v, we get (homework)

$$\text{degree of violation} = \left| \frac{T\beta}{p\kappa} - 1 \right| = \frac{a}{pv^2}. \tag{11.10}$$

The physical reason for this violation is the weak intermolecular attraction ($p \to p + a/v^2$), which gives a small amount of potential energy that is *not* of the appropriate $b\xi^2$ form. Therefore, the degree of violation of the equipartition theorem is determined by how this assumption-violating a/v^2 term compares with the unmodified ideal gas pressure, p.

B.2 $\Delta E(T, p)$

Suppose that we wish to measure changes in internal energy using a thermometer and pressure gauge. In Example 11.1 we saw that for a closed system ($dN = 0$) we could write the change in internal energy as a function of T and p. For small but finite changes, the expression for dE becomes

$$\Delta E = A\Delta T + B\Delta p,$$

with (11.4)

$$A = T\left(\frac{\partial S}{\partial T} \right)_p - p\left(\frac{\partial V}{\partial T} \right)_p \quad \text{and} \quad B = T\left(\frac{\partial S}{\partial p} \right)_T - p\left(\frac{\partial V}{\partial p} \right)_T.$$

Although this is the form we want, the partial derivatives are scary. Fortunately, using the relationships in Table 11.1 we can transform all four partial derivatives into easily measured properties of the system, getting

$$\Delta E = (C_p - pV\beta)\Delta T + (p\kappa - T\beta)V\Delta p. \qquad (11.11)$$

B.3 $\Delta S(p, V)$ and $\Delta S(T, p)$

Next, suppose that we wish to know the entropy change ΔS with changes in pressure and volume $\Delta p, \Delta V$ for a closed system ($\Delta N = 0$). We could begin by writing

$$\Delta S = \left(\frac{\partial S}{\partial p}\right)_V \Delta p + \left(\frac{\partial S}{\partial V}\right)_p \Delta V.$$

We can then express these partial derivatives in terms of more familiar parameters by either consulting Table 11.1 or using Maxwell's relations and the chain rule. Either way, this expression for the change in entropy becomes (homework)

$$\Delta S = \frac{C_V \kappa}{\beta T}\Delta p + \frac{C_p}{TV\beta}\Delta V. \qquad (11.12)$$

This formula will allow us to determine the small change in entropy ΔS during some process by measuring the corresponding small changes in pressure and volume $\Delta p, \Delta V$.

In the homework problems a similar method is used to show that changes in entropy can be determined from changes in temperature and pressure, according to[2]

$$\Delta S = \frac{C_p}{T}\Delta T - V\beta\Delta p. \qquad (11.13)$$

B.4 Variations in the heat capacities

Suppose that we know the heat capacity C_V at temperature T for a system at volume V_1, and we wish to know what it would be at the same temperature but a different volume, V_2. We need not make another measurement, providing that we know the equation of state relating T, p, V.

At volume V_2, the heat capacity is

$$C_{V_2} = C_{V_1} + \int_{V_1}^{V_2} dC_V = C_{V_1} + \int_{V_1}^{V_2} \left(\frac{\partial C_V}{\partial V}\right)_T dV.$$

From the definition of heat capacity, and using $\partial Q/\partial T = T\partial S/\partial T$,

$$\frac{\partial C_V}{\partial V} = \frac{\partial}{\partial V}\frac{\partial Q}{\partial T} = T\frac{\partial^2 S}{\partial V\partial T} = T\frac{\partial}{\partial T}\frac{\partial S}{\partial V},$$

---

[2] If $\Delta S = \Delta Q/T$, why is there the second term? (Hint: $C_p\Delta T = \Delta Q$ only if the pressure is constant.)

where the independent variables are T, V. That is, the partial with respect to one implies that the other is held constant. Using Maxwell's equation M4, this becomes (homework)

$$\left(\frac{\partial C_V}{\partial V}\right)_T = T\left(\frac{\partial^2 p}{\partial T^2}\right)_V,$$

and therefore the heat capacity at another volume but the same temperature is given by

$$C_{V_2} = C_{V_1} + T\int_{V_1}^{V_2}\left(\frac{\partial^2 p}{\partial T^2}\right)_V dV. \tag{11.14}$$

The integrand can be calculated from the equation of state relating T, p, V.

In the homework problems the same procedure can be followed to derive a similar result for C_p:

$$C_{p_2} = C_{p_1} - T\int_{p_1}^{p_2}\left(\frac{\partial^2 V}{\partial T^2}\right)_p dp. \tag{11.15}$$

Again, the integrand can be calculated from the equation of state.

Summary of Section B

In this section, we have used some specific examples to illustrate how to express the change in any property in terms of the changes in any other two, for nondiffusive processes.

The equipartition theorem assumes that the energy in each degree of freedom can be written as $\varepsilon = b\xi^2$, where b is a constant and ξ is a coordinate, and that the potential energy reference level u_0 is constant. Under these assumptions, it predicts that the change in internal energy of a system is given by (equation 11.8)

$$dE = \frac{N\nu}{2}k dT.$$

Applying the techniques of the previous section to partial derivatives, we can write the first law as

$$dE = C_V dT + \left(\frac{T\beta}{p\kappa} - 1\right)pdV.$$

Comparing these two equations, we can test the degree to which real systems violate the equipartition theorem by either of the following (equation 11.9):

$$\text{degree of violation} = \left|\frac{2C_V}{N\nu k} - 1\right| \quad \text{or} \quad \left|\frac{T\beta}{p\kappa} - 1\right|.$$

Applying this test to the van der Waals model for gases reveals that its violation of equipartition is due to the long-range attraction between molecules.

Further examples give the change in internal energy in terms of changes in temperature and pressure (equation 11.11)

$$\Delta E = (C_p - pV\beta)\Delta T + (p\kappa - T\beta)V\Delta p,$$

the change in entropy in terms of changes in pressure and volume or changes in temperature and pressure (equations 11.12, 11.13)

$$\Delta S = \frac{C_V \kappa}{\beta T}\Delta p + \frac{C_p}{TV\beta}\Delta V, \qquad \Delta S = \frac{C_p}{T}\Delta T - V\beta\Delta p,$$

and the value of the heat capacities at one pressure or volume in terms of their value at another (equations 11.14, 11.15),

$$C_{V_2} = C_{V_1} + T\int_{V_1}^{V_2}\left(\frac{\partial^2 p}{\partial T^2}\right)_V dV,$$

$$C_{p_2} = C_{p_1} - T\int_{p_1}^{p_2}\left(\frac{\partial^2 V}{\partial T^2}\right)_p dp.$$

Problems

Section A

In problems 1–12 we examine all the various possible partial derivative relationships between the variables S, T, p, V for nondiffusive interactions ($dN = 0$), in order to write them in terms of easily measurable properties of the system, $C_p, C_V, \beta, \kappa, T, p, V$. For each, you should demonstrate that the given conversion is correct. (In some cases, suggestions for a possible approach are provided in parentheses.)

1. $(\partial S/\partial T)_p = C_p/T$

2. $(\partial S/\partial T)_V = C_V/T$

3. $(\partial S/\partial p)_T = -V\beta$ (M10)

4. $(\partial S/\partial p)_V = \kappa C_V/\beta T$ (chain rule, and the answer to problem 8 below.)

5. $(\partial S/\partial V)_T = \beta/\kappa$ (chain rule, and the answer to problem 3 above.)

6. $(\partial S/\partial V)_p = C_p/TV\beta$ (chain rule)

7. $(\partial T/\partial p)_S = TV\beta/C_p$ (M7, and the answer to problem 6 above.)

8. $(\partial T/\partial p)_V = \kappa/\beta$ (ratios, write $dV = 0$ in terms of dT and dp.)

9. $(\partial T/\partial V)_S = -\beta T/\kappa C_V$ (M1, and the answer to problem 4 above.)

10. $(\partial T/\partial V)_p = 1/V\beta$

11. $(\partial p/\partial V)_S = -C_p/(C_V V\kappa)$ (ratios, write $dS = 0$ in terms of dp and dV, and use the results of problems 4 and 6 above. Or use the chain rule and the results from problems 7 and 9 above.)

12. $(\partial p/\partial V)_T = -1/V\kappa$

For problems 13 through 20, you will demonstrate how one differential depends on two others for nondiffusive interactions. For each case derive an expression of the form $dx(y, z) = A dy + B dz$. Each problem will have two parts.

(a) First, the coefficients A and B are expressed as the appropriate partial derivatives.

(b) Then these partial derivatives are converted into easily measurable properties of the system, $C_p, C_V, \kappa, \beta, T, p, V$, using the results to problems 1–12 above or Table 11.1.

13. $dV(p, T)$

14. $dE(p, T)$ (Hint: Start with $dE = TdS - pdV$. Rewrite dS and dV in terms of dp and dT. Then use M10.)

15. $dE(T, V)$ (Hint: For any expression that involves E it is easiest to start with the first law.)

16. $dV(S, T)$

17. $dE(p, V)$

18. $dE(S, T)$

19. $dT(p, V)$

20. $dS(T, p)$

21. $dS(E, p)$

22. You wish to know how the pressure varies with temperature for a system that has a fixed number of particles and is confined to a fixed volume.
    (a) What are the constraints?
    (b) What is the number of independent variables?
    (c) Write down an expression for dp in terms of dT, using the appropriate partial derivative.
    (d) Now replace the partial derivative with the properties $C_p, C_V, \kappa, \beta, T, p, V$ as appropriate.

23. You have a system that has a fixed number of particles held at fixed volume, and you wish to know how the entropy varies with the temperature.
    (a) What are the constraints?
    (b) Write down an expression for dS in terms of dT, using the appropriate partial derivative.
    (c) Now replace the partial derivative with the properties $C_p, C_V, \kappa, \beta, T, p, V$ as appropriate.

24. Find an expression with appropriate partial derivatives for each of the following and then replace the partial derivatives with the properties $C_p, C_V, \kappa, \beta, T, p, V$ as appropriate (for part (a), you might have to use the first law to find the partials of V with respect to E and S):
(a) $dV(E, S)$, (b) $dV(p, S)$, (c) $dV(p, T)$, (d) $dS(p, T)$.

25. We are studying a system that is undergoing diffusive interactions at constant pressure and temperature.
(a) How many independent variables are there?
(b) Write down an expression for the change in entropy dS with respect to the number of particles dN.
(c) See if you can convert the above partial derivative into another, using any of Maxwell's relations.

26. We are studying a system that is undergoing diffusive interactions at constant pressure and temperature.
(a) Starting with the first law, $dE = TdS - pdV + \mu dN$, write down an expression for dE in terms of dN.
(b) Of the two partial derivatives appearing in this expression, convert both to other partial derivatives, using the appropriate Maxwell's relations.

27. A certain solid is not isotropic, having different coefficients of linear expansion in each dimension: $\alpha_x = 0.5 \times 10^{-5}/K$, $\alpha_y = 1.5 \times 10^{-5}/K$, $\alpha_z = 2.0 \times 10^{-5}/K$. What is its coefficient of volume expansion, β? (Hint: $X' = X_0(1 + \alpha_x \Delta T)$, etc., and $V' = X'Y'Z'$. So what is V' in terms of V_0?)

28. Suppose you have available the equipment and materials of Figure 11.2. How would you measure the coefficient of volume expansion for (a) the gas, (b) the water, (c) the solid? (Hint: You might need to use the liquid to help you with the solid. Assume that you can measure the volume of the solid by the volume of water displaced.)

29. Suppose you have available the equipment and materials of Figure 11.2. How would you measure:
(a) the heat capacity at constant volume for the gas,
(b) the heat capacity at constant pressure for the liquid,
(c) the isothermal compressibility of the liquid,
(d) the isothermal compressibility of the solid?

30. Prove that $(\partial x/\partial y)_z = -(\partial x/\partial z)_y (\partial z/\partial y)_x$.

Section B

31. Derive equation 11.9, starting from the first law for nondiffusive interactions.

32. Derive equation 11.10 for a van der Waals gas from equation 11.9, and the expressions for κ and β given in Example 11.4.

33. For water vapor, the van der Waals constants are $a = 5.46$ liter^2 atm/mole^2 and $b = 0.0305$ liter/mole.
   (a) At a temperature of 500 K, the molar volume is about 7 liters when the pressure is 5.8 atm. For this particular system, estimate the degree of violation of equipartition.
   (b) Repeat for water vapor when its molar volume is 0.1 liters under a pressure of 43 atm.

34. (a) Starting from the first law for nondiffusive interactions, derive equations 11.4.
   (b) Show that equation 11.11 follows from equation 11.4.

35. Derive the result 11.12 for $\Delta S(p, V)$.

36. Show that $(\partial C_V / \partial V)_T = T(\partial^2 p / \partial T^2)_V$.

37. Derive the result 11.15 using the definition of the heat capacity C_p and one of Maxwell's relations.

38. Use result 11.14 to show that C_V does not depend on the volume for (a) an ideal gas, (b) a van der Waals gas.

39. (a) Using the method of ratios, show that $(\partial p / \partial T)_V = \beta / \kappa$.
   (b) The coefficient of thermal expansion and the isothermal compressibility for steel are 3.5×10^{-5}/K and 7.1×10^{-12}/Pa, respectively, and for water are 2.1×10^{-4}/K and 5.0×10^{-10}/Pa, respectively. For each, what pressure increase would be required to keep the volume constant if the temperature were raised by $1\,^\circ$C?

40. (a) Show that for a closed system $\Delta S = (C_p / T)\Delta T - V\beta\Delta p$ (equation 11.13).
   (b) For water, the molar heat capacity is 75.3 J/(K mole), the molar volume is 1.8×10^{-5} m^3/mole, and the coefficient of thermal expansion is 2.1×10^{-4}/K. What is the change in molar entropy of liquid water if you heat it up from $0\,^\circ$C to $100\,^\circ$C at atmospheric pressure, at the same time increasing the pressure by 10 000 atm? (1 atm $= 1.013 \times 10^5$ Pa. You need to integrate.)

41. Use the result 11.15 to show that C_p:
   (a) does not depend on pressure, for an ideal gas;
   (b) does depend on pressure, for a van der Waals gas, the integrand being equal to

$$\left(\frac{\partial^2 v}{\partial T^2}\right)_p = -\frac{2aR^2}{(pvB)^3}\left(1 - \frac{3b}{v}\right) \qquad \text{with} \qquad B = 1 - \frac{a}{pv^2} + \frac{2ab}{pv^3}.$$

42. Give the Maxwell's relations corresponding to M1, M4, M7, and M10 for a system for which there is magnetic interaction rather than volume expansion.

The external magnetic field B is in the z direction and the z-component of the system's magnetic moment is $\mathcal{M}$. (Since $dW = -Bd\mathcal{M}$, you replace pdV by $-Bd\mathcal{M}$ in the first law.)

43. Find an expression for $\Delta E(T, \mathcal{M})$ for a system involved in thermal and magnetic interaction only. The external magnetic field B is in the z-direction and the z-component of the system's magnetic moment is $\mathcal{M}$. (Use $dW = -Bd\mathcal{M}$; the expression will have some partial derivatives in it.)

Chapter 12
Special processes

In many important processes one or more of the thermodynamic variables are constrained (Figure 12.1). Isobaric processes take place at constant pressure, isothermal processes at constant temperature, and isochoric processes at constant volume. If there is no heat exchange with the surroundings, the process is adiabatic. If there is no exchange of particles, the process is nondiffusive. Each constraint simplifies our studies by reducing the number of independent variables.

A Isobaric processes

A considerable number of processes occur in closed systems at a constant pressure (Figure 12.1a). The two constraints $dp = 0$ and $dN = 0$ ensure that all properties of the system depend on just one independent variable. Using the techniques of the preceding chapter, we can choose this variable to be whatever we wish.

Example 12.1 How does internal energy depend on temperature for isobaric, nondiffusive, processes?

Figure 12.1 (a) Processes throughout vast regions of space are isobaric and/or isothermal. The gravitational collapse of interstellar clouds, which forms stars, is nearly adiabatic. (Courtesy of NASA) (b) A thunderhead is the result of the rapid adiabatic cooling of air as it rises and expands. (Plymouth State University photograph, courtesy of Bill Schmitz) (c) The physical and chemical processes carried out through marine plankton are isobaric and isothermal. (Courtesy of Mark Moline, California Polytechnic State University)

Figure 12.1   (*cont.*)

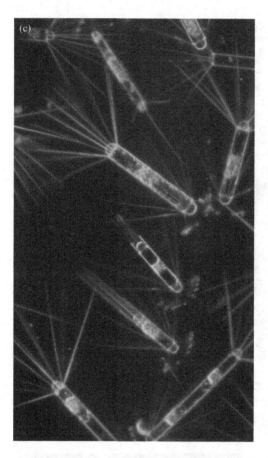

We choose our variables to be N, p, T because N and p are constant and we are measuring changes in T. The change in internal energy is given by the first law,[1]

$$\mathrm{d}E = T\mathrm{d}S - p\mathrm{d}V = T\left(\frac{\partial S}{\partial T}\right)_p \mathrm{d}T - p\left(\frac{\partial V}{\partial T}\right)_p \mathrm{d}T.$$

With the help of Table 11.1 we can convert these partial derivatives to give

$$\mathrm{d}E = (C_p - pV\beta)\mathrm{d}T \qquad \text{(isobaric processes)}. \tag{12.1}$$

This is what we wanted. If we know how the properties C_p, p, V, and β depend on T then we can integrate equation 12.1 to find the relationship for finite changes.

B   Isothermal processes

Isothermal nondiffusive processes also operate under two constraints, $\mathrm{d}N = \mathrm{d}T = 0$. So again, there is just one remaining independent variable, which we are free to choose.

---

[1] Note the convention that we usually display only two variables for nondiffusive ($\mathrm{d}N = 0$) processes.

Example 12.2 How does a system's internal energy vary with volume for isothermal, nondiffusive, processes ($dT = dN = 0$)?

We choose our variables to be T, N, V because T and N are constant and we are measuring changes in V. The change in internal energy is given by the first law:

$$dE = T dS - p dV = T \left(\frac{\partial S}{\partial V} \right)_T dV - p dV.$$

Again, we use Table 11.1 to convert the partial derivative to easily measured properties:

$$dE = \left(\frac{T\beta}{\kappa} - p \right) dV \qquad \text{(isothermal processes)}. \qquad (12.2)$$

In the homework problems, this result is used to show that the internal energy of an ideal gas depends only on its temperature, not on its volume.

C Adiabatic processes

A process during which there is no heat transfer is called adiabatic ($dQ = 0$). Many processes occur sufficiently rapidly that there can be no significant heat transfer with the surroundings yet sufficiently slowly that the system itself remains in equilibrium. For these adiabatic and "quasistatic" processes, we can write

$$dQ = T dS = 0 \qquad \text{(adiabatic and quasistatic)}.$$

(For a nonequilibrium adiabatic process, such as an adiabatic free expansion, entropy is not conserved: during such a process $dQ = T dS$ does *not* hold.)

C.1 Examples

Very little heat is transferred during the rapid compression of gases in the cylinder of an engine. Yet the speed of the piston is much slower than the thermal motion of the gas molecules, so the gas remains in equilibrium throughout this adiabatic process. Convection in many fluids involves quasistatic adiabatic expansion and compression as the materials rise and fall. Examples include convection in the Earth's mantle, in stellar interiors, in the ocean, and in planetary atmospheres. Much of our weather is caused by adiabatic processes in air. As air moves across the Earth or changes elevations, it encounters different pressures that produce adiabatic changes in volume and temperature.

As a fluid rises, it encounters lower pressure, so it expands and cools adiabatically. To determine whether a fluid is stable against vertical convection, we compare the adiabatic cooling rate with the existing vertical temperature gradient in the fluid, called the "lapse rate" (Figure 12.2). If the adiabatic cooling rate is greater than the lapse rate then the rising fluid becomes cooler than its surroundings and stops rising. In our atmosphere, such stability against vertical convection

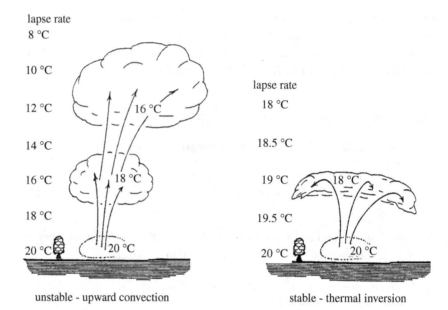

Figure 12.2 Rising air expands and cools as it encounters the lower pressure at higher elevation. Whether it continues to rise depends on whether it cools faster or slower with elevation than the lapse rate for the air around it. (Left) If it cools more slowly than the lapse rate, it will remain warmer than its surroundings and continue to rise. (Right) If it cools more rapidly than the lapse rate, it will become cooler than its surroundings and stop rising. Such a condition is described as a "thermal inversion."

is known as "thermal inversion" (Figure 12.2). We often see this condition in the early morning when the air near the ground is cooler than that above. Smoke and car fumes are trapped near the ground, unable to rise, and so "smog" forms in urban areas. Conversely, if the adiabatic cooling rate, is smaller than the lapse rate then the rising fluid remains warm compared with its surroundings and continues to rise.

The adiabatic expansion and cooling of rising moist air masses causes moisture to condense within the air mass, and the release of latent heat warms the air mass and fuels further rising. This is why thunderheads billow upwards so violently, and it is the fuel for violent storms, including hurricanes and typhoons. The adiabatic cooling of rising air is also the reason why cool air and snow are found at the top of a mountain rather than at its base. (After all, wouldn't you otherwise expect the colder air be denser and sink to the *base* of the mountain?)

C.2   Interrelationships between T, p, and V

For quasistatic adiabatic nondiffusive processes, the two constraints $dN = dS = 0$ ensure that there is only one independent variable. Therefore, any one property can be expressed in terms of any one other. For example,

$$dT = \left(\frac{\partial T}{\partial p}\right)_S dp, \qquad dT = \left(\frac{\partial T}{\partial V}\right)_S dV, \qquad dV = \left(\frac{\partial V}{\partial p}\right)_S dp.$$

Using the techniques of the last chapter, or Table 11.1, we can convert the above partial derivatives to get the following results for quasistatic adiabatic processes:

$$\frac{dT}{T} = \frac{V\beta}{C_p} dp, \qquad \frac{dT}{T} = -\frac{\beta}{\kappa C_V} dV, \qquad \frac{dV}{V} = -\frac{\kappa C_V}{C_p} dp. \qquad (12.3)$$

The properties T, V, C_p, C_V, κ are always positive, and the coefficient of thermal expansion, β, is usually positive.[2] Consequently, the equations 12.3 tell us that the following must be true for isentropic nondiffusive processes:

- increasing pressure causes increased temperature (provided β is positive);
- increasing volume causes decreased temperature (provided β is positive);
- increasing pressure causes decreased volume.

C.3 Adiabatic processes in ideal gases

For quasistatic adiabatic processes in ideal gases,

$$dE = -p dV \qquad \text{(first law)},$$
$$p dV + V dp = Nk dT \qquad \text{(differential form of ideal gas law)},$$
$$dE = \frac{Nv}{2} k dT \qquad \text{(equipartition)}.$$

$$(12.4)$$

These three equations, which relate the four differentials dE, dT, dV, dp, allow us to find the relationship between any two of these by eliminating the other two. We could also get the same relationships from equations 12.3 above, using the values of the parameters C_p, C_V, β, κ for ideal gases obtained in subsection 10F.3:[3]

$$C_V = \frac{v}{2} R, \qquad C_p = C_V + R, \qquad \beta = \frac{1}{T}, \qquad \kappa = \frac{1}{p}.$$

Either approach gives us the following relationships for ideal gases:

$$\frac{dT}{T} = \left(1 - \frac{1}{\gamma}\right) \frac{dp}{p}, \qquad \frac{dT}{T} = (1 - \gamma) \frac{dV}{V}, \qquad \gamma \frac{dV}{V} = -\frac{dp}{p}, \qquad (12.5)$$

where

$$\gamma = \frac{C_p}{C_V} = \frac{v + 2}{v}. \qquad (12.6)$$

---

[2] An important exception is water between $0\,°C$ and $4\,°C$.
[3] Use $Nk = nR$ for converting between number of molecules N and number of moles n.

By integrating the three equations 12.5, we find that for quasistatic adiabatic processes in ideal gases,[4]

$$TV^{\gamma-1} = \text{constant}, \qquad Tp^{1/\gamma-1} = \text{constant}, \qquad pV^{\gamma} = \text{constant}. \qquad (12.7)$$

Summary of Sections A–C

Any two constraints on a system leave only one independent variable, so any one property can be expressed as a function of any other. For example, we can write the change in internal energy of a system for isobaric nondiffusive processes as (equation 12.1)

$$dE = (C_p - pV\beta)dT$$

and for isothermal nondiffusive processes as (equation 12.2)

$$dE\left(\frac{T\beta}{\kappa} - p\right)dV.$$

Also common and important are those processes that occur sufficiently rapidly that essentially no heat enters or leaves the system during the process. Such isentropic (adiabatic and quasistatic) nondiffusive processes have only one independent variable, so any one variable can be written in terms of any other. For these processes (equation 12.3),

$$\frac{dT}{T} = \frac{V\beta}{C_p}dp, \qquad \frac{dT}{T} = -\frac{\beta}{\kappa C_V}dV, \qquad \frac{dV}{V} = -\frac{\kappa C_V}{C_p}dp.$$

For the particular case of ideal gases, the factors in the coefficients can be evaluated from the equation of state, and so equations 12.3 can be integrated, giving (equations 12.7, 12.6)

$$TV^{\gamma-1} = \text{constant}, \qquad Tp^{1/\gamma-1} = \text{constant}, \qquad pV^{\gamma} = \text{constant},$$

with

$$\gamma = \frac{C_p}{C_V} = \frac{\nu+2}{\nu}.$$

These same equations may also be derived from the differential forms of the first law, the ideal gas law, and equipartition.

D   Reversibility

The second law demands that, during any process, the total entropy of all the interacting systems must either increase or remain constant:

$$\Delta S_0 = \Delta S_1 + \Delta S_2 + \Delta S_3 + \cdots \geq 0.$$

---

[4] From any one of these we can get the other two simply by using the ideal gas law to change variables.

If the total entropy increases then the process cannot be reversed, because the reversed process would require a decrease in entropy, in violation of the second law. Therefore, for a process to be reversible, there must be no change in total entropy ($\Delta S_0 = 0$).

In Section 9B we found that when two systems are interacting thermally, mechanically, and/or diffusively, the change in entropy of the combined system is (equation 9.7)

$$
\begin{aligned}
\mathrm{d}S_0 &= \mathrm{d}S_1 + \mathrm{d}S_2 \\
&= \frac{1}{T_2}\left[\left(\frac{T_2 - T_1}{T_1}\right)\mathrm{d}Q_1 + (p_1 - p_2)\,\mathrm{d}V_1 - (\mu_1 - \mu_2)\,\mathrm{d}N_1\right] \\
&\geq 0.
\end{aligned}
$$

We can see immediately that in order for a process to be reversible (i.e., $\mathrm{d}S_0 = 0$) then:

- if heat is transferred, the temperatures must be equal;
- if volume is transferred, the pressures must be equal;
- if particles are transferred, the chemical potentials must be equal.

Although we have been thinking of two separate interacting systems, these results apply equally to interactions between different parts of the same system. If the system initially has some uneven distribution of temperatures, pressures, or chemical potentials within it, then as it moves toward equilibrium, the changes that occur cannot be reversed.

If processes cannot be reversed when entropy increases, then how can a refrigerator remove heat that has leaked inside it (the first term in the equation) or a fallen book be put back on the shelf (the second term), or a pump put air back into a tire once it has leaked out (the third term)? For these reverse processes to occur, another system must be involved, such as the fuels consumed to produce the electricity or to power the muscles. Even though you may reverse these processes in your kitchen or library, or in the tire, you burn fuels to do this. So in each case at least one of the systems cannot be brought back to its starting point. The world's total entropy has increased and the full process cannot be reversed.

Notice that reversibility involves considerations of entropy, not heat transfer alone. Because the two are closely related, we sometimes confuse them. With the help of equation 9.7 above, you should be able to think of adiabatic processes ($\mathrm{d}Q = 0$) that are *not* reversible. Examples would be mechanical interactions where the two pressures differ, or diffusive interactions where the two chemical potentials differ. Likewise, you should be able to think of nonadiabatic processes ($\mathrm{d}Q \neq 0$) that *are* reversible. If the temperatures of two systems are the same, then the entropy does not increase when heat is transferred between them.

Example 12.3 A flying ball of putty smashes into a wall (Figure 12.3). Is the process reversible?

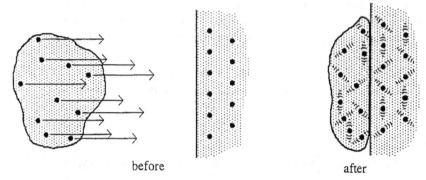

before   after

Figure 12.3 A ball of putty before and after its collision with a wall. The motion of some representative molecules is indicated. During the collision the coherent component of their motion (translation) is transformed into random thermal motion. Since there are far more states available for the molecules to oscillate in random directions than for them to all move in the same direction, the entropy increases during the collision.

Consider the motion of the putty molecules before the collision. In addition to their thermal motion, there is an overall translation. During the collision, the translational kinetic energy gets transformed into additional random motion of the molecules (hence the putty and the wall become hotter). There are far more states available for the molecules to move in random directions at various speeds than for them all to move the same direction at the same speed. That is, the entropy of the states of random motion is far greater than that of the coherent collective motion.

Since the entropy has increased, the process is irreversible. We would never see the putty later cool down and fly back along the path it came. This would require all the molecules to be moving in the same direction at the same instant, which is very improbable – a state of very low entropy. The entropy simply does not decrease once it has increased.

This last example can be made into a more general statement regarding friction. In addition to their thermal motion, the molecules in a moving object are going in the same direction at the same speed, involving states of rather low entropy. Friction transforms this coherent motion into more random motions, for which there are far more available states; hence the entropy is higher entropy. Consequently, friction always increases the total entropy, and so whenever friction is present, the process is irreversible. For reasons like these, increased thermal motion is the end product or "waste basket" of all other forms of energy.

E   Nonequilibrium processes

Most of the tools we develop in this course are for systems in equilibrium, that is, for quasistatic processes. In general, nonequilibrium processes are more difficult

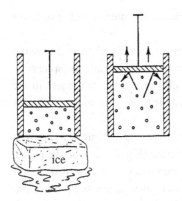

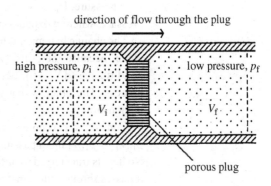

to study, because we do not have these powerful tools to work with. Nonetheless, insights gained from equilibrium studies help us understand some nonequilibrium processes as well.[5]

In most materials mechanical equilibrium is attained quickly and thermal equilibrium rather slowly, with relaxation times related to the speed of sound and thermal conductivity, respectively. Heat transfer proceeds particularly slowly in gases because they are such poor conductors of heat.

Figure 12.4 (Left) One way to cool a gas is to compress it, cool it with something cold, and then insulate it and let it expand, doing work and cooling further as it does. (Right) The throttling process involves an insulated tube with a constriction, such as a porous plug. The higher pressure on the left forces the gas through the constriction, which expands from volume V_i to volume V_f as it enters the region of lower pressure on the right. It does work $p_f V_f$ on the gas ahead of it as it pushes it on down the tube and likewise receives work $p_i V_i$ from the gas behind it.

E.1 Joule–Thompson process

The cooling of gases is important both in refrigeration and liquefaction. To cool a gas, we could do the following (Figure 12.4, on the left):

* compress it,
* then hold it against something cold to remove as much heat as possible,
* finally, let it expand, cooling further as it does.

This last step is tricky. Containers that are sufficiently strong to withstand large pressure changes tend to have large heat capacities. So much of the energy lost to expansion would be regained from the container walls, thereby defeating our purpose.

This problem can be avoided in the Joule–Thompson, or "throttling," process, for which the expanded cooled gas moves to a different part of the container (Figure 12.4, on the right). A gas is forced through a tube in which there is a constriction of some kind. Before getting to the constriction the gas is under

---

[5] For a system that is not in equilibrium, the temperature, pressure, and chemical potential may vary from one region to another, or we may not be able to define them at all. Consequently, we must be clever in using familiar tools. For example, we often think of a nonequilibrium system as being composed of subsystems, each of which has a well-defined temperature, pressure, and chemical potential. Then we imagine that changes happen incrementally, so that each subsystem is close to equilibrium at all times.

high pressure. Upon passing through the constriction, it enters a region of lower pressure, where it expands and cools.

The change in temperature can be expressed in terms of the change in pressure alone, because there are two constraints. One of these is that the process is nondiffusive ($dN = 0$), because no other particles join or leave the gas in the process. The other is that the enthalpy is constant ($dH = 0$), as we will now demonstrate.

Consider a certain amount of gas whose pressure and volume change from p_i, V_i to p_f, V_f upon passing through the constriction (Figure 12.4, on the right). No heat is transferred to or from the gas during the process, but work is done. As it passes through the constriction, it does work $p_f V_f$ as it pushes the gas in front of it on down the tube. Likewise, it receives work $p_i V_i$ from the gas behind it as it gets pushed through the constriction. From the first law, we can write the change in internal energy of this gas as

$$\Delta E = \Delta Q - \Delta W \qquad \Rightarrow \qquad E_f - E_i = 0 - (p_f V_f - p_i V_i).$$

Rearranging terms shows that the enthalpy $H = E + pV$ is unchanged during this process:

$$E_f + p_f V_f = E_i + p_i V_i \qquad \Rightarrow \qquad H_f = H_i.$$

So we have two constraints ($dH = dN = 0$), leaving only one independent variable. To take advantage of the constraint $dH = 0$, we reexpress the change in enthalpy:

$$\Delta H = 0 = T\Delta S + V\Delta p \qquad \text{(throttling).} \tag{12.8}$$

This equation relates ΔS to Δp. But we want to relate ΔT to Δp. So we must write ΔS in terms of ΔT and Δp and then use Table 11.1 for the partial derivatives:

$$\Delta S = \left(\frac{\partial S}{\partial T}\right)_p \Delta T + \left(\frac{\partial S}{\partial p}\right)_T \Delta p = \frac{C_p}{T}\Delta T - V\beta\Delta p.$$

Putting this expression for ΔS back into equation 12.8 and solving for ΔT gives

$$\Delta T = \frac{(T\beta - 1)V}{C_p}\Delta p \qquad \text{(throttling).} \tag{12.9}$$

This is what we were looking for. It allows us to calculate the drop in temperature by integrating over the change in pressure from one side of the constriction to the other. To integrate, we would have to know how β, V, C_p depend on T, p. Either an equation of state or careful measurements could give us that information.

Throttling would not cool an ideal gas at all because, for an ideal gas, $T\beta - 1 = 0$ (can you explain why?). Furthermore, the enthalpy of an ideal gas is

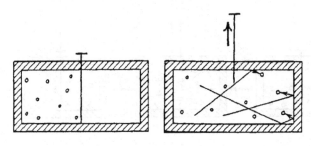

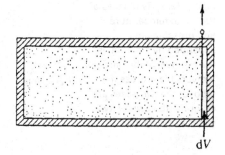

dV

Figure 12.5 (Left) In free expansion, a partition is removed and the gas rushes into the empty part of the container. Because nothing is moved by the expanding gas no work is done and, if the container is insulated, its internal energy remains unchanged. (Right) To use *equilibrium* thermodynamics (e.g., equation 12.11) in this process, we would have to let the gas freely expand in infinitesimal increments, and integrate.

proportional to its temperature:

$$H = E + pV = \frac{N\nu}{2}kT + NkT = \left(\frac{\nu}{2} + 1\right)NkT \qquad \text{(ideal gas)}.$$

So for an ideal gas, constant enthalpy means constant temperature.

In most real gases, however, there are weak long-range forces of attraction between molecules, and they move *against* these forces as they spread out. So under expansion, they slow down and their temperature falls. (Rising potential energy means decreasing kinetic energy.) Since these interactions are stronger at closer distances, we expect throttling to be more effective if it takes place at higher pressures and higher gas densities. Indeed, this is usually true, as can be seen in the homework problems.

Hydrogen and helium are different, however, because their long-range mutual attraction is so weak. Under pressure, collisional repulsion dominates at normal temperatures, making the potential energy reference level u_0 positive. So when expanded, the potential energy decreases, causing an increase in thermal energy and temperature. Highly compressed hydrogen could self-ignite while expanding in air! So, to cool these gases via throttling we must start at colder temperatures, where the repulsive forces due to intermolecular collisions are not so prominent.

E.2 Free expansion

In a "free expansion," a gas expands without doing any work at all. Consider a gas in one section of a rigid insulated container, as in Figure 12.5. The other section is completely empty. As the partition is removed and the gas rushes into the empty section, the molecules collide with stationary rigid walls. Because the walls don't move, the gas does no work on them. The combination of no heat transfer and no work means no change in internal energy. So free expansion operates under the two constraints $dN = dE = 0$, and we can express the changes in any property in terms of just one independent variable.

To take advantage of the constraint $dE = 0$, we start with the first law,

$$\Delta E = 0 = T\Delta S - p\Delta V \qquad \text{(free expansion)}. \qquad (12.10)$$

Figure 12.6 Two gases of the same temperature and pressure are separated by a barrier. Is the removal of the barrier a reversible process if (above) the two gases are different, (below) the two gases are identical?

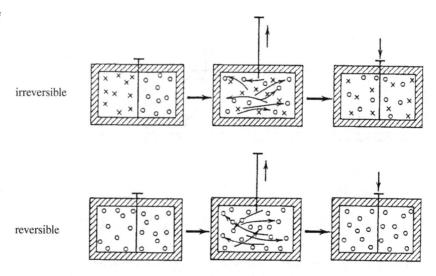

irreversible

reversible

If we wish to relate changes in temperature, ΔT, to changes in volume, ΔV, we must write ΔS in terms of these two variables, using Table 11.1 to convert the partial derivatives:[6]

$$\Delta S = \left(\frac{\partial S}{\partial T}\right)_V \Delta T + \left(\frac{\partial S}{\partial V}\right)_T \Delta V = \frac{C_V}{T} \Delta T + \frac{\beta}{\kappa} \Delta V.$$

Inserting this into equation 12.10 and solving for ΔT gives

$$\Delta T = \left(\frac{p\kappa - T\beta}{\kappa C_V}\right) \Delta V \qquad \text{(free expansion).} \qquad (12.11)$$

Free expansion, like throttling, would not cool an ideal gas at all, because for an ideal gas, $p\kappa - T\beta = 0$ (homework). Another way of seeing this is to refer to our model equation for the internal energy of an ideal gas:

$$E = \frac{N\nu}{2}kT.$$

Since the internal energy remains unchanged, so must the temperature. By contrast, most real gases do cool under free expansion. As we have seen, the long-range intermolecular attraction makes them slow down as they spread out.

E.3 Mixing

Free expansion is one example of an irreversible process initiated by the removal of barrier constraints. Another example is that illustrated in Figure 12.6, where two gases of equal temperature and pressure are separated by a barrier. If the two

---

[6] Can you explain why entropy increases even though no heat is transferred? (Think about the number of states in coordinate space.)

Table 12.1. *Typical thermal Conductivities of some materials*

Material	k (W/m K)	Material	k (W/(m K))
silver	410	masonry	0.62
copper	380	water	0.56
aluminum	210	sheet rock and plaster	0.50
steel	45	wood	0.10
rock	2.2	insulation	0.042
glass	0.95	air	0.023

gases have different compositions, the removal of the barrier allows them to mix, which is an irreversible process. The thoroughly mixed state has higher entropy – there is more volume in coordinate space for each molecule. No matter how long you wait, you will never see the two gases separate again. The entropy has increased and cannot go back.

An important variation of the above experiment is to do it for two gases that have the *same* initial composition. When the partition is removed, nothing changes. The original state can be recovered simply by reinserting the partition. In this case, the removal of the barrier causes no change in entropy. The process is reversible.

The above two processes are the same except for the identities of the particles. In the first case the entropy increases, and in the second case it does not. This illustrates that the way in which we measure entropy, or equivalently the way in which we count accessible states, depends on whether the particles are identical. We have seen this before and will see it again.

E.4   Thermal conduction

The second law demands that heat flows from hot to cold, but it does not dictate the *rate* of flow. Experimentally, we find that the heat flux or "thermal current density", measured in watts/m², depends on the temperature gradient:

$$J_Q = -k \frac{\partial T}{\partial x}, \tag{12.12}$$

where the constant of proportionality k is the "thermal conductivity" of the material (Table 12.1), and the minus sign indicates that the flow is *backwards* to the gradient. This is analogous to the electrical current density, which is backwards to the voltage gradient:

$$J_{\text{electrical}} = -\sigma \frac{\partial V}{\partial x},$$

the constant σ being the "electrical conductivity" of the material.

The total rate of heat flow, $\dot{Q}$, is the product of the flow rate per unit area times the area A. For a temperature drop ΔT over a distance Δx, we have

$$\dot{Q} = J_Q A = -k\frac{\Delta T}{\Delta x} A.$$

The collection of factors $kA/\Delta x$ is the reciprocal of the "thermal resistance," R, so the above equation can be written as

$$\dot{Q} = -\frac{1}{R}\Delta T \quad \text{where} \quad R = \frac{\Delta x}{kA}. \tag{12.13}$$

Again, this is the same as the corresponding relation for electrical resistance,

$$I_{\text{electrical}} = -\frac{1}{R}\Delta V \quad \text{where} \quad R = \frac{\Delta x}{\sigma A} :$$

thus, the resistance increases with thickness and decreases with cross sectional area.

By analogy with the flow of electricity, you can see that the laws of thermal resistors in series and parallel must be the same as for electrical resistors:[7]

$$R_{\text{series}} = R_1 + R_2 + \cdots, \qquad \frac{1}{R_{\text{parallel}}} = \frac{1}{R_1} + \frac{1}{R_2} + \cdots. \tag{12.14}$$

For example, for a wall with windows (resistance R_w), wall space with interior insulation (resistance R_i), and walls space with interior studs (resistance R_s), the total thermal resistance would be given by

$$\frac{1}{R_{\text{total}}} = \frac{1}{R_w} + \frac{1}{R_i} + \frac{1}{R_s},$$

because these thermal resistors are side by side, in parallel. Each consists of several materials in series, however, so you would have to use the series formula for each. For example, if the window consists of two sheets of glass (resistance R_g) with an air space (resistance R_a) between, then the window's thermal resistance is

$$R_w = R_g + R_a + R_g.$$

The insulated wall might have some plaster (resistance R_p) followed by sheet rock (resistance R_s) followed by insulation (resistance R_i) followed by exterior siding (resistance R_e), giving a total thermal resistance

$$R_w = R_p + R_s + R_i + R_e.$$

Each individual resistance depends on area, thickness, and thermal conductivity, according to equation 12.13 above.

The thermal resistances per square foot or square meter are called R-values and they are now given for all building materials, so you don't have to calculate

---

[7] The first is derived from the fact that the total temperature change is the sum of the changes across the individual layers, and the second is derived from the fact that the total flow is the sum of the flows through neighboring parts.

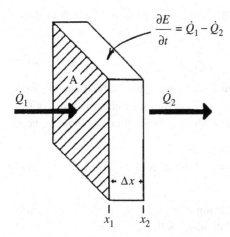

$$\frac{\partial E}{\partial t} = \dot{Q}_1 - \dot{Q}_2$$

$\dot{Q}_1$

$\dot{Q}_2$

A

Δx

x_1 x_2

Figure 12.7 The rate of change of the energy stored in the region indicated is the difference between the rate $\dot{Q}_1$ at which heat flows in and the rate $\dot{Q}_2$ at which it flows out.

them from scratch (i.e., from equation 12.13). But still you will have to put them together using the parallel and series formulas to calculate the total thermal resistance of walls and ceilings. Most building codes now place restrictions on thermal resistances, and as our energy resources become more depleted these restrictions will become tighter. You will also become increasingly interested as your energy bills rise.

E.5 The heat equation

In the previous subsection, we were interested in heat flow through a material across which there is a uniform temperature gradient. But what if the temperature gradient is not uniform? No problem – we have the machinery to calculate flow rates in this case too.

We begin by noting that the rate of increase in internal energy E within the region (x_1, x_2) in Figure 12.7 is the difference between the rates at which heat flows in and out:

$$\dot{Q}_1 - \dot{Q}_2 = \frac{\partial E}{\partial t}. \tag{12.15}$$

If the region's cross sectional area is A and thickness is $\Delta x = x_2 - x_1$, we can convert the left-hand side of this equation to the following form (using $\dot{Q} = J_Q A$):

$$\dot{Q}_1 - \dot{Q}_2 = -\frac{\partial \dot{Q}}{\partial x} \Delta x = -\frac{\partial J_Q}{\partial x} A \Delta x.$$

Substituting this expression into the left of equation 12.15, we divide both sides by the volume of the region ($V = A \Delta x$) and get

$$-\frac{\partial J_Q}{\partial x} = \frac{\partial \rho_E}{\partial t}, \quad \text{where} \quad \rho_E = \frac{E}{V} = \text{energy density}. \tag{12.16}$$

This is the one-dimensional form of the "continuity equation," which in three dimensions is

$$-\nabla \cdot \mathbf{J} = \frac{\partial \rho}{\partial t}.$$ (12.17)

It applies to anything that is conserved – mass, energy, electrical charge, etc., so you will encounter it again and again in your various fields of study. In the homework problems, this equation is integrated over an arbitrary volume to show that it simply means

$$\text{rate in} - \text{rate out} = \text{rate of change inside.}$$ (12.18)

We can rewrite equation 12.16 by using expression 12.12 for the thermal current density, $J_Q = -k(\partial T/\partial x)$, and dividing both sides by k:

$$\frac{\partial^2 T}{\partial x^2} = \frac{1}{k}\frac{\partial \rho_E}{\partial t}.$$ (12.19)

This tells us that energy flows out of regions (ρ_E decreases) where the temperature has a relative maximum (i.e., a negative second derivative) and into regions (ρ_E increases) where there is a relative minimum (i.e., a positive second derivative).

Now working on the right-hand side of the equation, the change in internal energy, dE, is related to the temperature change dT by (equations 8.13 and 8.14):

$$dE = mcdT.$$

Dividing by the volume, we obtain

$$d\rho_E = \rho_m cdT,$$

where ρ_m is the mass density and c the specific heat. With this replacement, equation 12.19 becomes

$$\frac{\partial^2 T}{\partial x^2} = \frac{1}{K}\frac{\partial T}{\partial t}, \qquad \text{where} \quad K = \frac{k}{\rho_m c}.$$ (12.20)

This is called the "heat equation." As mentioned above, you will encounter it many different fields of science, because it applies to the flow of anything that is driven by gradients and is conserved. The symbols will change depending on whether you are dealing with energy, electrical charge, mass, probabilities, etc., but it will always be of the form

$$\frac{\partial^2 Z}{\partial x^2} = A\frac{\partial Z}{\partial t}.$$ (12.21)

In three dimensions this becomes

$$\nabla^2 Z = C\frac{\partial Z}{\partial t},$$

where C is any constant.

In the homework problems you can show that a particular solution to the heat equation 12.20 is

$$T(x, t) = \frac{1}{\sqrt{4\pi K t}} e^{-(x-a)^2/4Kt}. \tag{12.22a}$$

You should recognize this as a Gaussian, beneath which the integrated area on $x = (-\infty, \infty)$ is 1 and which has standard deviation (subsection 3B.2)

$$\sigma = \sqrt{2Kt}. \tag{12.22b}$$

Notice that this Gaussian starts out (at $t = 0$) as an infinitely narrow spike ($\sigma = 0$), which spreads out as time progresses. This spike is actually a delta function, $\delta(x - a)$, since the total area beneath $T(x, t)$ at any time (including $t = 0$) is 1. In the homework problems, you can use this fact to show that if the initial temperature distribution is given by $T(x, t = 0) = f(x)$ then the temperature distribution at any later time is given by

$$T(x, t) = \int_{-\infty}^{\infty} f(x') \left(\frac{1}{\sqrt{4\pi K t}} e^{-(x-x')^2/4Kt} \right) dx', \tag{12.23}$$

where

$$f(x) = T(x, t = 0).$$

Summary of Sections D and E

A process is reversible only if the total entropy of the interacting systems remains constant. Such a process may or may not involve the transfer of heat energy between systems. Thermal, mechanical, and diffusive interactions may be reversible only if the interacting systems have the same temperature, pressure, or chemical potential, respectively. Friction transforms the coherent motion of molecules in one direction into random thermal motions in all directions, for which there are more accessible states and therefore higher entropy.

During the throttling process in a gas, the number of particles and the enthalpy remain constant. The change in temperature of the gas is related to the change in pressure through (equation 12.9)

$$\Delta T = \frac{(T\beta - 1) V}{C_p} \Delta p \qquad \text{(throttling)},$$

where β is the coefficient of thermal expansion. Such temperature changes are the result of intermolecular forces, so there would be none in an ideal gas.

The free expansion of a gas results in no change in internal energy, because there is no work and no heat transfer. Hence, any change in potential energy is compensated by a corresponding change in kinetic energy, and hence temperature. Such changes depend on the intermolecular forces. Changes in temperature and volume are related through (equation 12.11)

$$\Delta T = \left(\frac{p\kappa - T\beta}{\kappa C_V} \right) \Delta V \qquad \text{(free expansion)}.$$

Mixing systems of equal temperature and pressure may or may not be reversible, depending on whether the particles of the two systems are identical. Correspondingly, the counting of states for systems of identical particles is different from that for non-identical particles.

For heat conduction the thermal current density depends on the temperature gradient (equation 12.12):

$$J_Q = -k\frac{\partial T}{\partial x}.$$

The rate of heat flow through a material is (equation 12.13)

$$\dot{Q} = -\frac{1}{R}\Delta T, \qquad \text{where } R = \frac{\Delta x}{kA},$$

R is the "thermal resistance", A the cross sectional area, and Δx its thickness.

For materials in series or parallel the total thermal resistance is given by (equation 12.14)

$$R_{\text{series}} = R_1 + R_2 + \cdots, \qquad \frac{1}{R_{\text{parallel}}} = \frac{1}{R_1} + \frac{1}{R_2} + \cdots$$

If the temperature gradient is not uniform, we can use the following expression for the rate of heat flow per unit area (equation 12.16):

$$-\frac{\partial J_Q}{\partial x} = \frac{\partial \rho_E}{\partial t},$$

where ρ_E is the energy density. This is the one-dimensional form of the continuity equation and applies to the flow of all conserved quantities. It means (equation 12.18)

$$\text{rate in} - \text{rate out} = \text{rate of change inside.}$$

When written in terms of the temperature, equation 12.16 becomes the heat equation (12.20),

$$\frac{\partial^2 T}{\partial x^2} = \frac{1}{K}\frac{\partial T}{\partial t}, \qquad \text{where} \qquad K = \frac{k}{\rho_m c}.$$

If the initial temperature distribution is given by $T(x, t = 0) = f(x)$ then the temperature distribution at any later time is given by (equation 12.23)

$$T(x, t) = \int_{-\infty}^{\infty} f(x') \left(\frac{1}{\sqrt{4\pi K t}} e^{-(x-x')^2/4Kt} \right) dx'.$$

Problems

Section A

Problems 1–5 deal with *isobaric nondiffusive* interactions. In each case, find the required expression by first using the appropriate partial derivatives and then using Table 11.1 to express them as easily measured properties of the system.

1. Find the expressions that relate changes in entropy to changes in (a) volume, (b) temperature.

2. Starting with the first law, find the expressions that relates changes in internal energy to changes in (a) volume, (b) temperature.

3. Find the expression that relates changes in volume to changes in temperature.

4. Find the expressions that relate changes in each of the following to changes in temperature: (a) enthalpy, (b) Helmholtz free energy, (c) Gibbs free energy.

5. Repeat the above problem for changes in volume.

6. The overwhelming mass of life in the oceans is microscopic and single-celled. For the first 90% of the history of Earth, all life was in the oceans. Under what constraints do the corresponding biophysical processes operate? (Assume that an organism stays relatively stationary in space during any one process.) Such processes are of great scientific importance.

7. Consider some isobaric process. This constraint $dp = 0$ means that there are two independent variables. Starting from the first law,
   (a) write down an expression for dE in terms of dV and dN. (Hint: First write dS in terms of dV and dN.)
   (b) write down an expression for dE in terms of $d\mu$ and dN.

Section B
Problems 8–12 deal with *isothermal nondiffusive* interactions. Again, in each case find the required expression by first using the appropriate partial derivatives and then using Table 11.1 to convert them into easily measured properties of the system.

8. Find the expressions that relate changes in entropy to changes in (a) volume, (b) pressure.

9. Starting with the first law, find the expressions that relate changes in internal energy to changes in (a) volume, (b) pressure.

10. Find the expression that relates changes in volume to changes in pressure.

11. Find the expressions that relate changes in each of the following to changes in pressure: (a) enthalpy, (b) Helmholtz free energy, (c) Gibbs free energy.

12. Repeat the above problem for changes in volume.

13. In Example 12.2 we showed that $dE = (T\beta/\kappa - p)dV$ for an isothermal nondiffusive process. Show that dE is independent of dV for an ideal gas, by showing the coefficient of dV to be zero.

14. What is the coefficient of dV in the above problem for a van der Waals gas? Express your answer in terms of T, v, and the van der Waals coefficients a and b.

15. Consider a mole of steam under very high pressure, so that the van der Waals model is the correct equation of state. The constants for steam are $a = 5.5 \, \text{liter}^2 \, \text{atm/mole}^2$, $b = 0.030$ liter/mole. Suppose that the steam is initially at a pressure of 100 atm with a volume of 0.3 liter and then is expanded to twice this volume. Find (a) the initial temperature of the steam, (b) the final temperature if the expansion is isobaric, (c) the final pressure if the expansion is isothermal.

16. Consider the isothermal expansion of an ideal gas of n moles at temperature T. If it expands from volume V_i to V_f, what is (a) the work done by the gas in terms of n, T, V_i, and V_f, (b) the heat added to the gas in terms of n, T, V_i, and V_f?

17. For a fluid in hydrostatic equilibrium, the pressure varies with altitude according to $dp = -(M/V)g\,dz$, where M/V is the mass density.
    (a) Show that if the atmosphere were an ideal gas at constant temperature, the pressure would fall off exponentially with altitude according to $p = p_0 e^{-\alpha z}$, with $\alpha = mg/kT$ and m the average mass of an air molecule (4.84×10^{-26} kg for dry air).
    (b) If the atmosphere were at $0\,^\circ$C, what change in altitude would be required for the pressure to decrease by $1/2$?

Section C

18. Starting with the first law and the differential forms of $E = (Nv/2)kT$ and $pV = NkT$, prove that, for adiabatic processes in an ideal gas (where $\gamma = (v+2)/v$),
    (a) $pV^\gamma = $ constant,
    (b) $TV^{\gamma-1} = $ constant,
    (c) $Tp^{1/\gamma-1} = $ constant,
    (d) $\gamma = C_p/C_V$ (refer to equations 10.14 and 10.15).

19. A certain gas is compressed adiabatically. The initial and final values of the pressure and volume are $p_i = 1$ atm, $V_i = 2$ liters, $p_f = 1.1$ atm, $V_f = 1.85$ liters. What is the number of degrees of freedom per molecule for this gas?

20. Consider a gas, initially at pressure p_i and volume V_i, that is expanded adiabatically to final volume V_f. Find an expression for the work done by the gas in terms of p_i, V_i, V_f, and γ.

21. For an ideal gas undergoing a nondiffusive process, show that

(a) the heat capacities are $C_V = (v/2)Nk$ and $C_p = [(v+2)/2]Nk$,

(b) the change in enthalpy is $\Delta H = C_p \Delta T$.

22. Consider 0.446 mole of an ideal gas, initially at temperature 0 °C, pressure 1 atm, volume 10.0 liters, that has five degrees of freedom per molecule. Suppose that this gas is expanded to a volume of 15 liters. What would be the new values of its temperature and pressure if this expansion were (a) isobaric, (b) isothermal, (c) adiabatic?

23. For each of the three processes in problem 22, calculate (a) the work done by the gas, (b) the heat added to the gas, (c) the change in internal energy, (d) the change in enthalpy.

24. For most solids and liquids, C_p and C_V are nearly the same, $C_p \approx C_V = C$. Furthermore C, β, κ, and the volume V remain reasonably constant over fairly wide ranges in T and p. Assuming that they are thus constant, integrate the results 12.3, $dT/T = (V\beta/C_p)dp$, $dT/T = -(\beta/C_V\kappa)dV$, $dV/V = -(\kappa C_V/C_p)dp$, to find the relationships between the initial and final values of T, p, and V for adiabatic processes in solids and liquids.

25. Starting with the first law for quasistatic adiabatic (i.e., isentropic) non-diffusive processes and expressing all partial derivatives in terms of easily measured parameters, using Table 11.1, show how dE varies with (a) dp, (b) dV, (c) dT.

26. Pressure decreases with height y in the atmosphere at a rate given by $p = p_0 e^{-\alpha y}$, where $\alpha = 0.116$ km^{-1}. For air, $\gamma = C_p/C_V = 1.4$.
    (a) If the temperature of the air on the ground is 293 K, find what the temperature at an altitude of 1 km would be if temperature changes at an adiabatic rate.
    (b) If the actual temperature at that altitude is 285 K, would there be upward convection or would there be an inversion layer?

27. Near the bottom of a certain ocean basin (about 5 km down) the temperature of the water is 3 °C. Even in the high pressures at that depth, the following parameters are roughly the same as at the ocean's surface: the coefficient of volume expansion $\beta = 2.1 \times 10^{-4}$/K, the isothermal compressibility $\kappa = 4.6 \times 10^{-10}$ m^2/N, the molar heat capacity $C_V = 75$ J/K, the molar volume $v = 1.8 \times 10^{-5}$ m^3, the density $\rho = 1.03 \times 10^3$ kg/m^3.
    (a) What is the change in pressure with each additional one meter depth?
    (b) Measurements taken show that down there the temperature increases with depth at a rate of 0.7×10^{-4} K/m. Is the water stable against vertical convection?

28. Calculate the parameters β and κ in terms of p, V and T for a gas obeying the ideal gas law.

29. Sometimes we have "conditional stability" in our atmosphere, which means that the lower atmosphere is stable against vertical convection (i.e., thermal inversion) if the air is dry but not if it is moist. How can this be?

30. With the help of Table 11.1, derive equations 12.3 for isentropic nondiffusive interactions.

31. One liter of air ($v = 5$, so $\gamma = 1.4$) at 1 atm pressure and 290 K is compressed adiabatically until the final pressure is 8 atm. What are the final volume and temperature, and how much work has been done on the air?

32. Because mechanical disturbances propagate through materials much faster than the speed of thermal conduction, the compressions and rarefactions associated with the propagation of sound are adiabatic. The wave equation for sound is

$$\partial^2 y/\partial x^2 = (1/v^2)\partial^2 y/\partial t^2,$$

where y is the displacement, $v^2 = B/\rho$ (ρ is the mass density and v the wave speed), and B is the *adiabatic* bulk modulus, $B = -V(\partial p/\partial V)_S$.
(a) For an ideal gas, what is B in terms of p and γ?
(b) For dry air at standard pressure (1 atm) and 290 K, $\gamma = 1.4$ and $\rho = 1.22\,\text{kg/m}^3$. What is the speed of sound?
(c) Using this result, write down the correct expression for the speed of sound in dry air at any temperature T.

Section D

33. Consider a large rotating flywheel connected to a piston, as in Figure 13.2. As the flywheel rotates, the gas in the cylinder is alternately compressed and expanded quasistatically by the moving piston. If the container and piston are thermally insulated and there is no friction, is the process reversible? (Hint: Is the entire system back to its starting point after one complete rotation of the flywheel?)

34. Consider an insulated container holding fresh water and sea water, which are separated by a partition. If they are initially at the same temperature and pressure and the partition is removed, is the process reversible? That is, can the original system be regained by reinserting the partition?

35. Think about friction. A book sliding across a level table comes to a stop. Because systems always go towards configurations of higher entropy, the final stopped state must have higher entropy than the preceding moving state. Can you explain the reason for this in terms of molecular motion? Explain why it would be extremely improbable for the book to subsequently slide back to where it started from.

36. When a book slides across a table and comes to a stop, both systems (i.e., the book and the table top) get slightly warmer as a result of the friction, so the

internal energy of both systems increases. How can this happen if energy is conserved?

Section E

37. A certain gas undergoing a throttling process is initially at a pressure of 100 atm and a temperature of $0\,^\circ$C, and the molar volume is 0.25 liters. If the molar heat capacity of this gas is $C_p = 29$ J/(K mole) and the coefficient of volume expansion is $\beta = 5 \times 10^{-3}$/K, what change in the temperature of the gas do you expect if the pressure is reduced by 1 atm (1 atm $= 1.013 \times 10^5$ Pa)?

38. One mole of an ideal gas with $\nu = 5$ ($\gamma = 1.4$) is initially at a temperature, pressure, and volume of $0\,^\circ$C, 100 atm and 0.224 liters, respectively. It is allowed to expand until its pressure is reduced to 10 atm. What is its final temperature if this expansion is (a) adiabatic, (b) carried out through a throttling process?

39. The van der Waals constants for oxygen gas, O_2, are $a = 1.36$ liters^2 atm/mole^2, $b = 0.0318$ liters/mole. Some oxygen gas is under a pressure of 100 atm and has a molar volume of 0.25 liters, and there are five degrees of freedom per molecule.
    (a) What is its temperature?
    (b) What is its coefficient of thermal expansion, β?
    (c) If this gas were undergoing a throttling process, what would be the rate of temperature decrease with pressure, dT/dp, in units of $^\circ$C per atmosphere?

40. The molar heat capacity at constant pressure for a gas whose molecules have ν degrees of freedom apiece is $C_p = [(\nu + 2)/2]R$. Using the van der Waals model and expressing your answers in terms of ν, p, v, and the van der Waals coefficients a, b, calculate the coefficient in the following formulas:
    (a) $dT = (\cdots)dp$ for the throttling process.
    (b) $dT = (\cdots)dv$ for free expansion.

41. In part (b) of the previous problem, we obtained $dT = (-2a/v^2\nu R)dv$ for the free expansion of a van der Waals gas. Integrate to find the relationship between T_i, T_f, v_i, and v_f for this process.

42. A mole of steam ($\nu = 6$) under an initial pressure of 300 atm is allowed to expand freely from volume 0.15 liters to a volume twice that size. The van der Waals constants for steam are $a = 5.5$ liters^2/(atm mole)^2, $b = 0.030$ liter/mole.
    (a) What is the initial temperature of the steam?
    (b) Using the results of part (a) and problem 41, find the final temperature of the steam.

43. An ideal gas expands from initial volume V_i to final volume V_f. The initial temperature is T_i. Calculate the changes ΔE, ΔQ, and ΔW for the gas, in terms of N, v, T_i, V_i, and V_f, for (a) free expansion, (b) adiabatic expansion (hint: $pV^\gamma = $ constant; the value of the constant in terms of T_i and V_i can be obtained from $p_i V_i = NkT_i$), (c) isothermal expansion.

44. Is the Joule–Thomson throttling process adiabatic? Is it isentropic? Explain each answer.

45. (a) Suppose that you have two sets of four distinguishable coins, and you flip all four coins in each system. How many different possible states are there?
    (b) Now suppose that the two sets each contain four *identical* coins, and you flip all four coins in each set, keeping the two sets separate from each other. How many different possible states are there?
    (c) Now you combine the two sets into one set of eight identical coins, all of which you flip. How many different possible states are there now?

46. You have a container divided into two equal volumes, V. On each side, you have a gas of N particles. The corrected number of states per particle, ω_c, is the same for the particles on both sides:

$$\omega_c = c(V/N)(E_{\text{therm}}/Nv)^{v/2}$$

where c is a constant. You remove the partition between the two volumes. In terms of N and ω_c, find the number of states available to the combined system
    (a) before you remove the partition,
    (b) after you remove the partition if the two gases are different,
    (c) after you remove the partition if the two gases are the same.
    (d) How is this last result explained? If the volume available to each particle doubles, shouldn't ω_c double too?

47. Prove the formulas 12.14 for thermal resistors in series and parallel. (In series the total temperature drop ΔT is the sum of the individual drops, but $\dot{Q}$ is the same for all. In parallel, the two conditions are reversed.)

48. As you go down through the Earth's crust, the temperature increases at a rate of about 30 °C per kilometer of depth (the rate is more beneath oceans and less beneath continents).
    (a) Using this temperature gradient and the thermal conductivity of rock from Table 12.1, calculate the rate at which heat reaches each square meter of the Earth's surface from its interior.
    (b) Averaged over the entire Earth, day and night and all seasons, solar energy reaches the Earth's surface at a rate of 175 W/m^2. How does this compare with the energy arriving from the Earth's interior?

49. You can estimate the amount of energy leaking through one $3\,\mathrm{m} \times 20\,\mathrm{m}$ exterior wall during a winter, as follows. Assume $15\,\mathrm{m}^2$ of window space, $37\,\mathrm{m}^2$ of insulated wall, and $8\,\mathrm{m}^2$ of wall studs and other solid wood. Use the thermal conductivities given in Table 12.1 to find the thermal resistances of each of these parts.

   (a) *Windows* What is the thermal resistance R_w (the R-value) of the windows? Assume double-pane glass, each pane being 3 mm thick and with an air space 1 cm thick. Also, to be realistic, assume two relatively immobile air spaces 1 mm thick on the extreme interior and exterior surfaces of these double-glazed windows (that is, there are five layers: 1 mm air, 3 mm glass, 1 cm air, 3 mm glass, and another 1 mm air).

   (b) *Insulated walls* What is the thermal resistance R_i of the insulated walls? Assume 2 cm of sheet rock, followed by 13 cm of insulation, followed by 3 cm of wood. Again assume two immobile air spaces 1 mm thick on the extreme interior and exterior surfaces, giving a total of five layers altogether.

   (c) *Wall studs* What is the thermal resistance R_s of the wall-stud area? Assume 16 cm of solid wood and 2 cm of sheet rock, but again with two immobile air spaces 1 mm thick on the extreme interior and exterior surfaces, a total of four layers altogether.

   (d) Now calculate the thermal resistance of the entire wall.

   (e) If the inside–outside temperature differential averages $20\,^{\circ}\mathrm{C}$ during the winter, how much energy flows out through this wall during the three winter months?

   (f) Electrical heating costs about 15 cents per kilowatt hour, and gas heating costs about one third of that. How much would this lost energy cost you, for each kind of heating?

50. Show that equation 12.22a is a solution to the heat equation 12.20.

51. Show that equation 12.23 is a solution to the heat equation 12.20 satisfying the initial condition $T(x, t = 0) = f(x)$.

Chapter 13
Engines

A The general idea

Engines convert heat into work. Thermodynamics owes both its name, "heat-motion", and much of its early development to the study of engines. The working system for most engines interacts both thermally and mechanically with other systems, so its properties depend on *two* independent variables. Most engines are cyclical, so that the working system goes through the following stages:

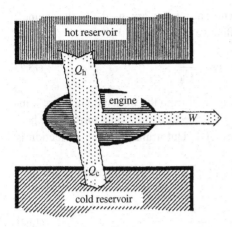

Figure 13.1 Energy flow for an engine. It takes heat Q_h from a hot reservoir, does work W with it, and exhausts the remaining heat Q_c into a cold reservoir.

- it is heated;
- it expands and does work, pushing a piston or turbine blades;
- it is cooled further;
- it is compressed back into its original state, ready to begin the cycle again.

The expansion occurs when the working system is hot and is under high pressure or has a larger volume, and the compression occurs when it is cooler and is under lower pressure or has a smaller volume. Therefore, the work done *by* the engine while expanding is greater than the work done *on* the engine while being compressed. So there is a net output of work by the engine during each cycle. This is what makes engines useful. If you can understand this paragraph, you understand nearly all engines.

The details vary from one engine to the next. The working system could be any of a large variety of gases or volatile liquids. The source of heating could be such things as a flame, a chemical explosion, heating coils, steam pipes, sunlight, or a nuclear reactor. The cooling could be provided by such things as air, water, ice, evaporation, or radiative coils. Whatever the source of heat we call this source the "hot reservoir," and whatever the source of cooling we call this source the "cold reservoir." These generic terms allow us to analyze all engines together, irrespective of the particular sources of heating and cooling and of the particular nature of the working system.

Figure 13.1 illustrates the basic process for all engines. The engine takes heat Q_h from the hot reservoir at high temperatures, does work W with it, and then "exhausts" the remaining heat Q_c into the cold reservoir at a lower temperature. Conservation of energy demands that the energy input equals the energy output:

$$Q_h = W + Q_c. \tag{13.1}$$

Since the purpose of an engine is to turn heat into work, we define the engine's efficiency e to be a measure of how effective it is at doing this:

$$\text{efficiency } e = \frac{W}{Q_h}. \tag{13.2}$$

Since we pay for the heat input Q_h and benefit from the work output W, we want to make this ratio as large as possible. A perfect engine would turn all the heat input into work, giving it an efficiency of 1. Unfortunately, such perfection is impossible, as we will see.

Combining equations 13.1 and 13.2 provides an equivalent way of defining efficiency:

$$e = \frac{Q_h - Q_c}{Q_h} \tag{13.3}$$

You can see that maximizing the efficiency means minimizing the heat exhausted Q_c. Heat exhausted is heat that has not been turned into work and is therefore wasted.

The operation of real engines involves turbulence, temperature gradients, friction, and other losses, as well as varying constraints. Unfortunately, the thermodynamical tools that we have developed are most easily applied to systems in equilibrium and subject to well-defined constraints. Therefore our study of engines often involves approximating the cycles of real engines with "model cycles," which are sequences of different stages, each stage having an appropriate constraint such as being purely isobaric, isothermal, adiabatic, or isochoric. We also need to assume that each process is quasistatic, so that the engine is in equilibrium at all times, and that we can ignore frictional losses. This allows us to calculate the work done or the heat added during any stage of the cycle by

$$\Delta W = \int p \, dV, \qquad \Delta Q = \int T \, dS.$$

To describe the working system, we may choose whichever two independent variables we wish. We often find it convenient to work with the pressure and volume (p, V), and we represent the changes in these two variables on a p–V diagram. This representation has the advantage that since the work done equals $\int p \, dV$ it can be read directly off the diagram as the area under the curve.

Sometimes we choose the two independent variables to be the temperature and entropy (T, S), and we can represent the changes in these variables on a T–S diagram. This representation has the advantage that the heat input is the integral $\int T \, dS$ and therefore can be read directly from the T–S diagram as the area under the curve. Since heat and work are of central importance in the analysis of engines, we use the (p, V) and (T, S) representations extensively.

 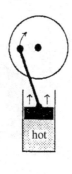 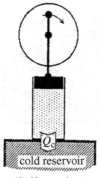 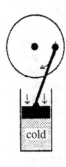

(1) Heat addition   (2) Adiabatic expansion   (3) Heat exhaust   (4) Adiabatic compression
                          (hot, high pressure)   (cool, low pressure)

B   Examples

B.1   Gas piston engines

Consider a gas piston engine that stores energy in a heavy flywheel, as in Figure 13.2. The engine goes through the following four-stage cycle, illustrated in Figures 13.2 and 13.3.

1 *Heat addition* (ignition stage)   When the gas is fully compressed we add heat, causing the pressure to increase.

2 *Adiabatic expansion* (power stage)   The heat source is removed. The hot high-pressure gas expands and pushes the piston, making the flywheel turn.

3 *Heat exhaust* (exhaust stage)   When the gas is fully expanded we extract heat from it, causing the pressure to fall further.

Figure 13.2 Elements of a simple piston engine. (1) Heat Q_h is added to the compressed gas, causing its temperature and pressure to rise. (2) The hot high-pressure gas drives the piston, making the flywheel rotate. (3) Heat Q_c is exhausted, causing the temperature and pressure to fall. (4) The flywheel compresses the cooled low-pressure gas with ease. This brings it back to the starting point.

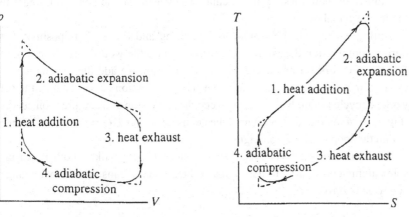

Figure 13.3 *p–V* and *T–S* diagrams for the four stages of the simple piston engine given above. The solid lines indicate the actual cycle and the broken lines indicate our model cycle, which has pure and abruptly changing constraints for each of the four stages.

Figure 13.4 (Left) The work done is the integral $\int p\, dV$, which is the area under the curve in a *p–V* plot; dV is positive during expansion and negative during compression. The area under the expansion curve minus that under the compression curve is simply the area between the two curves. (On the right) Similarly, heat transfer is the integral $\int T\, dS$, which is the area under the curve in a *T–S* plot; dS is positive as heat is added and negative as it is removed. So the net heat added during any cycle is equal to the area between the two curves.

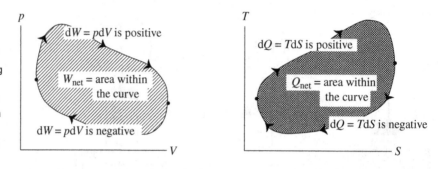

4 *Adiabatic compression* (compression stage) We stop the heat removal. The turning flywheel compresses the gas back to its original pressure and volume.

Notice that during the expansion stage the gas is hotter, and so it exerts greater pressure than in the compression stage. For this reason, it does more work on the piston while expanding than the piston does on it when it is being compressed. Therefore, the *net* work done by the engine for each cycle is positive. Heat has been converted into work.

Changes in pressure and volume for the cycle are shown on the *p–V* diagram of Figure 13.3 (on the left). The corresponding changes in temperature and entropy are shown in the *T–S* diagram on the right and go as follows. As heat is added to the gas (stage 1) the temperature and entropy both increase, and as heat is removed (stage 3) they both decrease. The entropy is constant for the adiabatic stages (i.e., the two stages that have no heat exchange), the temperature falling during expansion (stage 2) and rising during compression (stage 4). Figure 13.3 also illustrates how we simplify calculations by using single-constraint stages to model the actual cycle.

During expansion, the volume is increasing and the work is positive, but during compression the volume is decreasing and the work is negative. So, to find the net work for any cycle on a *p–V* diagram, we add the area under the expansion curve and subtract that under the compression curve. That is, the net work per cycle is simply the area between the expansion and compression curves (Figure 13.4, on the left). Similarly, the net heat transfer $\Delta Q = \int T\, dS$ is the area within the loop on the *T–S* diagram.

After any complete cycle, the system's internal energy (like all other measurable parameters) returns to its original value, which means that the net change per cycle is zero:

$$\Delta E = 0 = \Delta Q - \Delta W \qquad \Rightarrow \qquad \Delta Q_{net} = \Delta W_{net}. \qquad (13.4)$$

So the area of the cycle on the *p–V* diagram must be equal to its area on the *T–S* diagram.

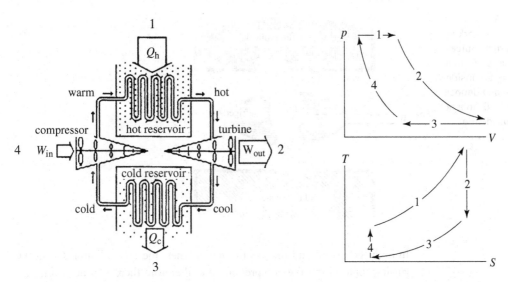

Figure 13.5 Elements of a gas turbine engine, and the model cycle on a *p–V* diagram and on a *T–S* diagram. The cycle includes (1) isobaric heating from the hot reservoir, (2) adiabatic expansion through the turbine, (3) isobaric cooling in the cold reservoir, (4) adiabatic compression in the compressor.

B.2   Gas turbines

A turbine is a series of fan blades that are forced to spin as hot pressurized gas shoots through. The gas expands and cools as it passes through the spinning blades, thereby turning heat into work. In an engine, some of the work output of the turbine is used to power the compressor (a turbine running backwards), which compresses the gas on the return cycle. The gases in the compressor are relatively cool and easily compressed. So the compressor does less work on the cooled gas than is done by the heated gas on the turbine. This means that there is a net work output, which can be used to power a car, an airplane propeller, an electrical generator, or a variety of other machines.

In a turbine engine the gas flows continuously and is heated and cooled under conditions of constant pressure instead of constant volume. That is, in contrast with piston engines, the turbine engine's hot and cold reservoirs cause changes in volume rather than pressure. (The changes in pressure occur in the turbine and compressor, which are *outside* the hot and cold reservoirs.) The cycle is illustrated in Figure 13.5 and goes as follows.

1 *Isobaric heating* (ignition stage)   Initially warm and under high pressure, the gas is heated further as it flows through the hot reservoir.
2 *Adiabatic expansion* (power stage)   The hot pressurized gas then expands and cools as it shoots through the turbine and into the region of lower pressure beyond. It turns the turbine blades as it goes.
3 *Isobaric cooling* (exhaust stage)   The warm gas is then cooled isobarically while passing through the cold reservoir.
4 *Adiabatic compression* (compression stage)   The cold low-pressure gas is then forced by the compressor back into the high-pressure region, becoming compressed and warmed. It is now back to its initial state and is ready to repeat the cycle.

Figure 13.6 A
refrigerator uses work W
from an external source
to extract heat Q_c from a
cold reservoir (the inside
of the fridge) and deposit
heat $Q_h = W + Q_c$ in a hot
reservoir (outside the
fridge).

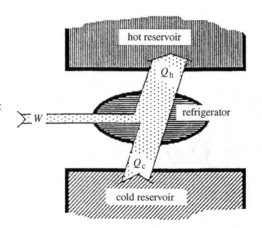

Figure 13.6 A refrigerator uses work W from an external source to extract heat Q_c from a cold reservoir (the inside of the fridge) and deposit heat $Q_h = W + Q_c$ in a hot reservoir (outside the fridge).

In both the piston and the gas turbine engines, the gas is hotter during the expansion than during the compression and therefore there is a net work output. The basic difference between them is that in the piston engine the higher temperature produces a greater *pressure* whereas in the gas turbine engine the higher temperature produces a larger *volume*. Either way, the work done ($\int p dV$) is larger when the gas is hotter.

C   Refrigerators

Refrigerators are engines running backwards (compare Figure 13.6 with 13.1). In an engine the gas expands when it is hot and is compressed when it is cold, so there is a net output of work. By contrast, the gas in a refrigerator is compressed when it is hot and expands when it is cold, so that there is a net input of work. In this reversed cycle, the heat is transferred from the cold to the hot reservoir, which is the opposite of what engines do.

A good refrigerator will extract as much heat as possible from the cold reservoir (the inside of the fridge) with as little work as possible, thereby minimizing your utility bill. Therefore the appropriate performance measure is the amount of heat that can be extracted per unit of work expended. This is called the coefficient of performance:

$$\text{coefficient of performance} = \frac{Q_c}{W}. \tag{13.5}$$

The diagram of a typical kitchen refrigerator (Figure 13.7) is like that of a turbine engine but a simple expansion chamber replaces the turbine (because you simply want to cool the gas and don't want it to push or turn anything). The cold coils are inside the refrigerator, and the hot coils are outside. Each set of coils is known as a "heat exchanger." The basic cycle goes as follows.

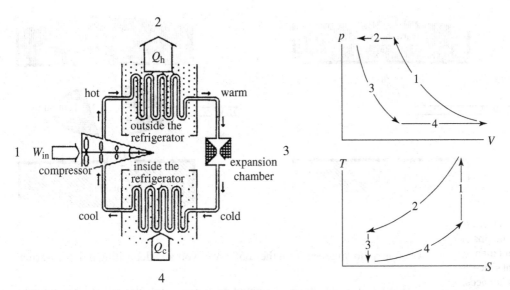

Figure 13.7 A refrigerator's cycle. (1) Adiabatic compression, (2) heat exhaust, (3) adiabatic expansion, (4) heat absorption.

1 *Adiabatic compression*   Work from the outside runs the compressor; compressing the gas makes it hot. (Notice that work has been done on the system.)

2 *Heat exhaust*   The hot pressurized gas is now hotter than the ambient outside temperature. So, as this gas passes through the outside coils, it releases heat into the air of the room (the hot reservoir) and cools down to room temperature. (Notice that heat has been exhausted into the hot reservoir.)

3 *Adiabatic expansion*   The pressurized gas at room temperature then flows through the expansion chamber, expanding and cooling as it enters the low-pressure region beyond. It may even cool sufficiently to condense into a liquid.

4 *Heat absorption*   The cold low-pressure fluid is now colder than the inside of the refrigerator, so as it flows through the cold coils it absorbs heat from the inside of the refrigerator, becoming warmer and perhaps vaporizing if it is a liquid at this stage. (Notice that heat has been absorbed from the cold reservoir.) Now the gas is back to its original state and is ready to start the cycle again.

D   Heat pumps

If the compressor on your refrigerator could be run in either direction, you would have a "heat pump." Run in one direction it would remove heat from the inside of the refrigerator and deposit it on the outside. When put in reverse it would do the opposite.

These devices are used to heat and cool buildings. One set of coils is outside the building, and one is inside. In the winter, you run the compressor in the direction that transfers heat from the outside to the inside. In the summer you run it in the opposite direction. Either way it is acting as a refrigerator, because either

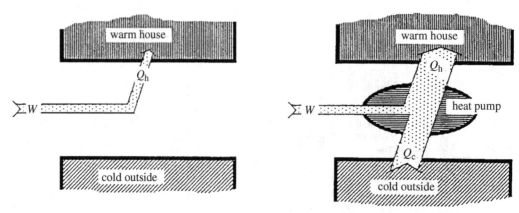

Figure 13.8 Illustration of why heat pumps are more efficient than conventional space heaters and furnaces, which simply release energy W and deposit it inside the house. Heat pumps use this much energy to extract additional heat from the outside air, increasing the total heat deposited inside the house.

way it is removing heat from the cooler reservoir and depositing it in the warmer one.

Because a relatively small amount of work can transfer rather large amounts of heat, heat pumps are more efficient than furnaces for heating buildings (Figure 13.8). For example, suppose that 1 unit of energy can either be burned in a furnace or used to power a heat pump. If the heat pump uses this 1 unit to transfer 4 units of heat from the cold reservoir, then it would provide $1 + 4 = 5$ times more heating than the furnace. Unfortunately, the initial equipment expense is greater. So, compared with conventional heaters, heat pumps generally have greater initial equipment costs but smaller long-term operating costs.

E Types of cycle

If the engine's or refrigerator's working fluid remains a gas throughout the cycle, this is a "gas cycle." However, if the gas becomes a liquid at any time, it is a "vapor cycle." The latent heat absorbed during vaporization and released during condensation permits relatively large heat transfer with relatively small temperature differentials. This gives vapor cycles an advantage in certain applications, such as refrigeration and steam turbines.

In engines with "closed cycles" the same fluid goes through every cycle. In "open cycles," a new fluid is used each time. The old fluid is driven out, carrying the exhaust heat with it. It may also carry combustion products if the engine uses internal combustion. Common examples of open cycles include automobile and jet airplane engines.

Whether the same old fluid is recirculated or new fluid is taken in as the old fluid is released makes no difference to either the engine or the environment. Both cycles release heat and combustion products. In closed cycles they are released by the furnaces and heat exchangers and in open cycles they are released in the exhaust gases. The net effect is the same.

Summary of Sections A–E

An engine transforms heat into work. In each cycle it takes heat Q_h from a hot reservoir, does work W with it, and exhausts the remaining heat, Q_c, into a cold reservoir (equation 13.1):

$$Q_h = W + Q_c.$$

In general, the working system in an engine is a fluid that goes through the following four-stage cycle: heating, expansion, cooling, compression. It expands when hot, and is compressed when cold. Therefore, the work done *by* it during expansion is larger than the work done *on* it when it is compressed. So the net work done per cycle is positive.

The efficiency of an engine measures the fraction of the incoming heat that gets transformed into work (equations 13.2, 13.3):

$$\text{efficiency } e = \frac{W}{Q_h} = \frac{Q_h - Q_c}{Q_h}.$$

Most engines interact both thermally and mechanically with other systems, so all properties depend on two variables. We assume that the working system is in equilibrium at all times, so the work done and heat added during any part of the cycle are the integrals $\int p\, dV$ and $\int T\, dS$, respectively. The net work done and net heat added during an entire cycle are the areas within the cycle on a p–V and a T–S diagram, respectively. During any cycle, the net work done by the engine is equal to the net heat gained by it (equation 13.4):

$$\Delta Q_{net} = \Delta W_{net} \qquad \text{(one complete cycle)}.$$

In a gas piston engine, the hot expanding gas pushes a piston. Some of this energy is stored (e.g., in a flywheel) so that it can be used to compress the gas later in the cycle when the gas is cooler. In the gas turbine engine, the heated expanding gases do work by turning a turbine, and the cooled gases have work done on them by a compressor.

A refrigerator is an engine running backwards. With external work W coming in, it removes Q_c from the cold reservoir and deposits $Q_h = W + Q_c$ in the hot reservoir. The "coefficient of performance" is a measure of how efficiently the refrigerator removes heat from the cold reservoir (equation 13.5):

$$\text{coefficient of performance} = \frac{Q_c}{W}.$$

A heat pump is a refrigerator that can run in either direction, so it can absorb and release heat through either set of coils.

If the working system remains a gas throughout the cycle it is called a gas cycle. If it condenses into a liquid for any part of the cycle, it is called a vapor cycle. In a closed cycle, the same fluid gets recycled. In an open cycle, the old fluid is released and new replacement fluid taken in during each cycle.

Figure 13.9 Illustration of the slopes of the lines for isobaric, isothermal, adiabatic, and isochoric processes on a p–V diagram (on the left), and a T–S diagram (on the right).

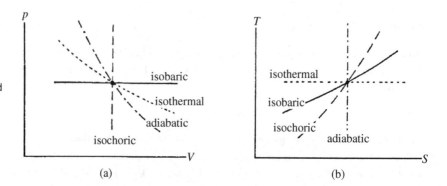

(a)   (b)

F Performance analysis

F.1 Model cycles and constraints

We now examine the behaviors of engine operating systems under the various constraints that we use in our model cycles. Figure 13.9a illustrates these qualitatively on a p–V diagram. Obviously the isobaric ($\Delta p = 0$) line must be horizontal and the isochoric ($\Delta V = 0$) line must be vertical. The lines representing isothermal and adiabatic processes are sloping, because for both processes the pressure decreases as the systems are expanded. The adiabatic line slopes more steeply because the system cools off during adiabatic expansion, so the particles exert even lower pressure than those of a system expanded isothermally.

Figure 13.9b illustrates these processes on a T–S diagram. Of course the isothermal line ($\Delta T = 0$) must be horizontal and the adiabatic line ($\Delta Q = T\Delta S = 0$) vertical. The isochoric and isobaric lines are sloping because for both processes the temperature increases as you add heat. But the temperature rises faster in the isochoric case. Can you explain why?

Each stage of a model cycle operates under one constraint. This constraint reduces the number of independent variables from two to one, so we can express any property in terms of just one variable. To do this we can use models, equations of state, or the techniques of Chapter 11.

For the particular case of an ideal gas, Table 13.1 gives the change in internal energy ΔE, the work done ΔW, the heat added ΔQ, and the change in entropy ΔS for various types of process. In this particular table, most changes are expressed in terms of the change in temperature. In the homework problems, you can derive these equations and express some of them in terms of changes in pressure or volume.

F.2 Enthalpy

In addition to theoretical analyses, in which we often break up an engine's cycle into a sequence of single-constraint stages, there is also a simple way of doing performance analysis experimentally that requires only the measurement of temperatures and pressures at various points within the engine.

Table 13.1. ΔE, ΔW, ΔQ ($\Delta E = \Delta Q - \Delta W$), and ΔS *for an ideal gas operating under various constraints*

Quantity	Isobaric	Isochoric	Adiabatic*	Isothermal
ΔE	$nC_V\Delta T$	$nC_V\Delta T$	$nC_V\Delta T$	0
ΔW	$nR\Delta T$	0	$-nC_V\Delta T$	$nRT\ln\left(\dfrac{V_f}{V_i}\right)$
ΔQ	$nC_p\Delta T$	$nC_V\Delta T$	0	$nRT\ln\left(\dfrac{V_f}{V_i}\right)$
ΔS	$nC_p\ln\left(\dfrac{T_f}{T_i}\right)$	$nC_V\ln\left(\dfrac{T_f}{T_i}\right)$	0	$nR\ln\left(\dfrac{V_f}{V_i}\right)$

Helpful relationships (v = number of degrees of freedom per molecule, $C_{V,p}$ are the molar heat capacities)

$$pV = nRT, \quad E = nC_vT, \quad C_V = (v/2)R, \quad C_p = C_V + R, \quad \gamma = C_p/C_V = (v+2)/v$$

Other variables The following relationships may be used to exchange variables in the above expressions:

isobaric	isochoric	adiabatic	isothermal
$T = \left(\dfrac{p}{nR}\right)V$	$T = \left(\dfrac{V}{nR}\right)p$	$pV^\gamma = \text{const.}$	$V = \dfrac{nRT}{p}$
$\dfrac{T_f}{T_i} = \dfrac{V_f}{V_i}$	$\dfrac{T_f}{T_i} = \dfrac{p_f}{p_i}$	$TV^{\gamma-1} = \text{const.}$	$\dfrac{V_f}{V_i} = \dfrac{p_i}{p_f}$

* For adiabatic processes, $\Delta W = -\Delta E = -nC_V\Delta T = \int p\,dV = [1/(\gamma-1)][p_iV_i - p_fV_f]$.

Consider the change in internal energy of a fluid flowing from region 1 to region 2, as in Figure 13.10, on the left. In region 1 it has pressure and volume p_1, V_1 and in region 2 its pressure and volume are p_2, V_2. As we have seen, it receives work p_1V_1 from the fluid behind as it gets pushed out of region 1, and it does work p_2V_2 on the fluid ahead as it pushes into region 2. The change in its internal energy is equal to the net work done on it,

$$E_2 - E_1 = p_1V_1 - p_2V_2.$$

Now suppose that the fluid also interacts thermally and/or mechanically with some external system. Then the change in its internal energy is given by (Figure 13.10, on the right)

$$E_2 - E_1 = p_1V_1 - p_2V_2 + Q_{ext} - W_{ext}.$$

If we now move the pV terms to the other side of the equation, we have

$$(E_2 + p_2V_2) - (E_1 + p_1V_1) = Q_{ext} - W_{ext}.$$

Figure 13.10 Consider a fluid flowing through a tube. (On the left) As it flows through the resistive barrier, it does work p_2V_2 in pushing the fluid ahead of it and receives work p_1V_1 from the fluid behind. So the net change in enthalpy is zero. (On the right) If it also interacts with an external system, receiving heat Q_{ext} and/or doing work W_{ext}, the net change in enthalpy is equal to this energy exchange.

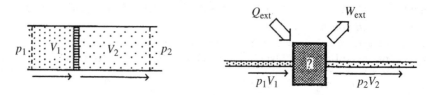

As we saw in subsection 12E.1, the combination $E + pV$ is the enthalpy of the fluid. Therefore, we can determine the work done or heat transferred during any process by the change in the fluid's enthalpy:

$$H_2 - H_1 = Q_{ext} - W_{ext}. \tag{13.6}$$

Note that, as usual, Q_{ext} is the heat added to the fluid and W_{ext} is the work done by the fluid.

For most gases, the enthalpy can be determined simply by measuring the temperature. With the help of the ideal gas law, we have for n moles of the gas

$$H = E + pV = \frac{\nu}{2}nRT + nRT = \frac{\nu + 2}{2}nRT = nC_pT \qquad \text{(ideal gas)}. \tag{13.7}$$

With this, equation 13.6 becomes

$$H_2 - H_1 = nC_p(T_2 - T_1) = Q_{ext} - W_{ext} \qquad \text{(ideal gas)}. \tag{13.8}$$

For liquids and denser gases the enthalpy depends on both temperature and pressure, and this dependence varies from one material to the next. The molar enthalpies of most common working fluids, such as water, various refrigerants, ammonia, etc., are listed in tables as functions of their temperature and pressure. Therefore, by measuring the temperature and pressure at various points in the cycle, we can find the change in enthalpy, which tells us the work done or heat added during each of the various stages.

If you did not have an appropriate table of enthalpies, you could make one yourself. Start in the rarefied "ideal" gas phase at low pressure and high temperature T, where you can use equation 13.7 ($H = nC_pT$). Then you can determine the enthalpy for any other temperature and pressure, including the liquid phase, simply by transferring the amount of heat and/or work (Q_{ext}, W_{ext}) needed to get to that point. Equation 13.8 would give you the enthalpy H_2 at the new point in terms of the enthalpy H_1 at the first point. This can be done for water in the homework problems.

The important point here is that to study any engine component experimentally, you do not need to go inside the component. If you simply measure the fluid's

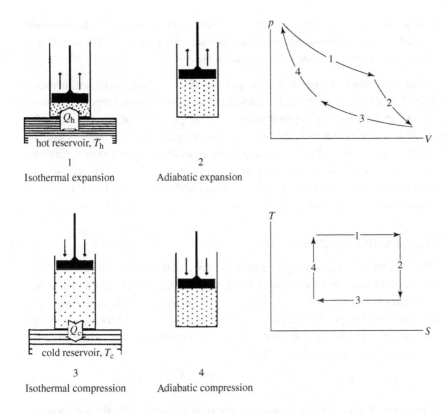

Figure 13.11 Illustration of the four stages of a Carnot cycle. (1) Isothermal expansion at temperature T_h, (2) adiabatic expansion until the temperature falls to T_c, (3) isothermal compression at temperature T_c, (4) adiabatic compression until the temperature rises back to T_h.

1
Isothermal expansion

2
Adiabatic expansion

3
Isothermal compression

4
Adiabatic compression

temperature and pressure before it enters and after it leaves that component, then you will know what has happened inside.

G The Carnot engine

G.1 The cycle

We now study an imaginary engine first proposed by Sadi Carnot in 1824 and illustrated in Figure 13.11. It is reversible and, as we will show, this means that it has the highest possible efficiency for any engine operating between two given temperatures. For this reason, it serves as a model for real engines to emulate. The cycle goes as follows.

1 *Isothermal expansion* (ignition) at temperature T_h The compressed gas expands slowly while in thermal equilibrium with the hot reservoir. Expanding gases tend to cool off so, to keep the temperature constant, heat Q_h must flow into it from the reservoir.

2 *Adiabatic expansion* (power)   The gas is then removed from the hot reservoir and continues to expand and cool as it pushes the piston outward.

3 *Isothermal compression* (exhaust)   When it reaches temperature T_c, the system is put on the cold reservoir and the piston reverses direction. Compression normally heats a gas so, to keep the temperature constant, heat Q_c must be exhausted into the reservoir.

4 *Adiabatic compression* (compression)   The gas is then removed from the cold reservoir and the compression continues until its temperature has risen to T_h. At this point it is back to where it started and is ready to repeat the cycle.

For a process to be reversible the entropy of the universe must not increase. Therefore heat transfer must be carried out at equal temperatures and volume transfer at equal pressures (equation 9.7). Although heat flows from hot to cold, the flow *rate* goes to zero in the limit of equal temperatures. The same is true for volume transfer in the limit of equal pressures. So for an engine to be reversible, the heat and work transfers must proceed at infinitesimal rates. Hence, infinite time is required to complete a cycle. So a reversible engine, such as Carnot's, is purely theoretical and not at all practical.

G.2   Carnot efficiency

The efficiency of a Carnot engine can be determined from the temperatures of the hot and cold reservoirs alone. After one complete cycle, the entropy and all other properties return to their original values. The entropy changes only during the two isothermal stages, because those are the only times when heat enters or leaves the engine. It receives Q_h at temperature T_h and exhausts Q_c at temperature T_c. Therefore, for one complete cycle,

$$\Delta S_{cycle} = 0 = \frac{Q_h}{T_h} - \frac{Q_c}{T_c} \quad \Rightarrow \quad \frac{Q_h}{T_h} = \frac{Q_c}{T_c} \qquad \text{(Carnot)}. \qquad (13.9)$$

This shows that the amount of heat transferred is proportional to the temperature, so we can write the efficiency (equation 13.3) of the Carnot engine in terms of the temperatures:

$$e_{Carnot} = \frac{Q_h - Q_c}{Q_h} = \frac{T_h - T_c}{T_h} = 1 - \frac{T_c}{T_h}. \qquad (13.10)$$

G.3   The efficiency of other engines

To show that a reversible engine has the maximum possible efficiency, we consider a reversible engine r and any other engine a. We use the entire work output of the second engine to drive the reversible engine backwards, making it operate as a refrigerator.

As is illustrated in Figure 13.12, engine a takes heat Q_h^a from the hot reservoir, does work W with it, and exhausts the remaining Q_c^a into the cold reservoir. The work output W goes entirely into the reversible engine. With this work, the reversible engine takes heat Q_c^r from the cold reservoir and deposits Q_h^r into the hot reservoir. If we put both engines inside a "black box" (the broken-line box in in Figure 13.12) then all we see is a net transfer of energy from the hot to the cold reservoir:

$$Q_{net} = Q_h^a - Q_h^r.$$

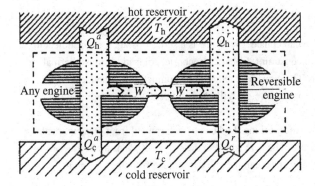

The second law requires that heat flows from hot to cold and not vice versa, so $Q_{net} \geq 0$. This has implications as follows:

$$Q_h^a \geq Q_h^r \quad \Rightarrow \quad \frac{W}{Q_h^a} \leq \frac{W}{Q_h^r} \quad \Rightarrow \quad e_a \leq e_r. \tag{13.11}$$

So we have proven that no engine a operating between these two reservoirs may have efficiency greater than that of a reversible engine.

Figure 13.12 Proof that a reversible engine r is more efficient than any other engine a. The work from engine a drives the reversible engine backwards. The combined system (inside the broken-line box) simply transfers heat $Q_h^a - Q_h^r$ from the hot reservoir to the cold one. Because the second law demands that heat flows from hot to cold, $Q_h^a \geq Q_h^r$, and so $(W/Q_h^a) \leq (W/Q_h^r)$. That is, engine a cannot be more efficient than the reversible engine r.

G.4 Lessons for other engines

The Carnot engine gives us insight into how to make real engines more efficient. Equation 13.10 tells us that a smaller temperature ratio T_c/T_h should make a more efficient engine. In practice, the low temperature, T_c, is restricted by the environment. To make it colder would require refrigeration. As you can show in the homework problems, the work needed to refrigerate the cold reservoir is greater than the work gained by the greater temperature differential. So you lose more than you gain. The upper limit on temperature T_h is usually decided by materials. The engine must be sturdy and durable, and higher temperatures take a toll in these areas. Stationary engines, such as those in power plants, can usually be made of more durable materials than portable engines, such as those in automobiles and airplanes. Consequently stationary engines can operate at higher temperatures and therefore are usually more efficient.

In real engines friction, turbulence, temperature gradients, and other dissipative and nonequilibrium effects cause the entropy to increase, making the actual efficiency fall short of the maximum possible, i.e., the efficiency of a reversible engine. The "coefficient of utility" is a measure of how close a real engine comes to achieving the maximum possible efficiency for the temperatures between which it operates:

$$\text{coefficient of utility} = \frac{\text{actual efficiency}}{\text{Carnot efficiency}}. \tag{13.12}$$

Summary of Sections F and G

To analyze an engine, we can consider its cycle as a series of stages, each subject to the one most appropriate constraint. Isobaric, isochoric, adiabatic, and isothermal constraints are common. The constraint reduces the number of independent variables to one, so all properties (e.g., ΔW, ΔQ, ΔS, ΔE) can be expressed in terms of just one variable.

The molar enthalpy of the working fluid can be determined from measurements of temperature and pressure. The heat added to the system, Q_{ext}, or the work done by the system, W_{ext}, during any stage can then be determined from the change in enthalpy (equation 13.6):

$$H_2 - H_1 = Q_{ext} - W_{ext} \qquad \text{(any fluid)}.$$

If the working fluid is an ideal gas, this becomes (equation 13.8)

$$H_2 - H_1 = nC_p(T_2 - T_1) = Q_{ext} - W_{ext} \qquad \text{(ideal gas)}$$

The Carnot cycle is reversible and consists of four stages:

1 isothermal expansion in contact with the hot reservoir at T_h;
2 adiabatic expansion during which the temperature drops from T_h to T_c;
3 isothermal compression in contact with the cold reservoir at T_c;
4 adiabatic compression during which the temperature rises from T_c back to T_h.

For an engine to be reversible requires there to be no change in total entropy and thus no temperature or pressure difference between the engine and the systems with which it interacts. Consequently, both heat transfer and work must proceed infinitely slowly.

In the isothermal stages, the heat transfer is in proportion to the reservoir's temperature. Therefore, the efficiency of the Carnot engine is given by (equation 13.10)

$$e_{Carnot} = \frac{Q_h - Q_c}{Q_h} = \frac{T_h - T_c}{T_h} = 1 - \frac{T_c}{T_h}.$$

Because it is reversible, the Carnot engine is the most efficient engine possible operating between any two given temperatures.

Study of the Carnot engine suggests that we can make real engines more efficient if we (a) minimize the ratio T_c / T_h and (b) minimize the entropy loss due to temperature and pressure differentials, friction, turbulence, etc. The coefficient of utility is a measure of how close an engine comes to achieving its maximum possible efficiency (equation 13.12):

$$\text{coefficient of utility} = \frac{\text{actual efficiency}}{\text{Carnot efficiency}}.$$

H   Some common internal combustion engines

We now look more closely at some of the more popular engines. We begin with internal combustion engines, where the heat source is chemical explosions within the engine itself. These engines tend to be relatively small and portable and are used in cars, trucks, power tools, jets, and airplanes. Whether they use pistons or turbines, internal combustion engines have open cycles; the combustion products are exhausted along with the unused heat each cycle and the fuel is drawn in with the fresh supply of air. The values of W and Q_h can be determined from measurements of enthalpies, as discussed earlier this chapter, or alternatively W can be determined from mechanical measurements and Q_h from the heat content of the fuels consumed.

H.1   Four-stroke gasoline engines

Although some small gasoline piston engines have just two strokes per cycle, those that are perhaps more familiar to you have two extra strokes: one to exhaust the combustion products and one to draw in the fuel mixture for the next ignition. The stages of these four-stroke gasoline engines are illustrated in Figure 13.13, along with actual and model "Otto" cycles on p–V and T–S diagrams. These stages are as follows.

1 Power stroke
  (a) *Ignition*   With the fuel–air mixture near maximum compression, a spark ignites them. The pressure and temperature rise quickly during this explosion.
  (b) *Expansion*   The hot gas then pushes forcefully against the piston, the pressure and temperature falling as the gas expands adiabatically.
2 Exhaust stroke
  (a) *Pressure release*   At the end of the expansion stroke, the exhaust valve opens, the exhaust escapes, and the pressure drops to nearly atmospheric.
  (b) *Purge*   With the exhaust valve still open, the moving piston continues to force out the waste heat and combustion products.
3 Intake stroke
  The exhaust valve closes and the intake valve opens. As the piston moves back down, the fuel–air mixture enters from the carburetor.
4 Compression stroke
  The intake valve closes. The moving piston compresses and heats the fuel–air mixture until it is ready for ignition, and the cycle repeats.

H.2   Four-stroke diesel engines

The four-stroke diesel and gasoline engines are similar. The diesel fuel burns more slowly (Figure 13.14a), so the timing of the onset of ignition (i.e., the spark) is not as crucial. In fact, many diesel engines do not even use a spark. Rather, during

Figure 13.13 (Left) The four strokes of an internal combustion piston engine, the first and second strokes each being divided into two stages. (Right) p–V and T–S plots for the cycle, the first representing the actual cycle and the other two representing the single-constraint model Otto cycle.

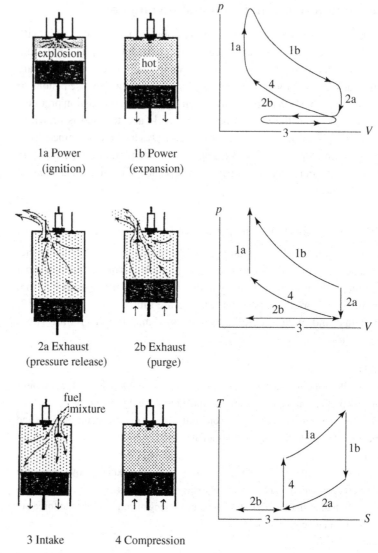

1a Power (ignition)   1b Power (expansion)

2a Exhaust (pressure release)   2b Exhaust (purge)

3 Intake   4 Compression

compression the fuel mixture's temperature rises to the point where it can ignite easily with the help of a hot "glow plug."

In gasoline engines the piston moves relatively little during the fast-burning ignition stage, so we model this part of the cycle as heat input under *isochoric* conditions. By contrast, the diesel fuel burns so slowly that the piston moves considerably as the fuel is burning. Therefore it is more appropriate to model this stage of the engine's operation as being heat addition under *isobaric* conditions, as indicated in Figures 13.14b, c.

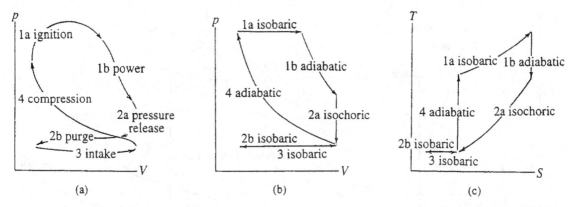

Figure 13.14 *p–V* and *T–S* diagrams for a four-stroke diesel piston engine. Plot (a) represents a real engine whereas plots (b) and (c) represent the single-constraint model cycle used to simplify calculations. The main difference from the gasoline engine is that the diesel fuel burns more slowly.

H.3   Jets and other gas turbines

We now look at gas turbine engines that use internal combustion. The cycle is described below and illustrated in Figure 13.15.

1 *Intake and compression*   (adiabatic compression)   Air is drawn in and compressed. The work required for compression is taken from the work output of the turbine (often on the same axle). Fuel is sprayed into it.

2 *Ignition*   (heat input)   The fuel–air mixture is ignited.

3 *Turbine*   (adiabatic expansion)   The hot gases then shoot out through the turbine. Although the pressure of the gas entering the turbine is essentially the same as that leaving the compressor, the much higher temperature means that the volume is correspondingly much larger and therefore more work is done in the turbine.

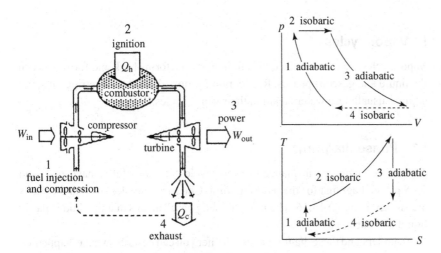

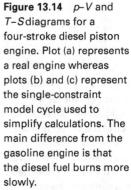

Figure 13.15   Illustration of an internal combustion gas turbine engine, such as that used in jet aircraft.

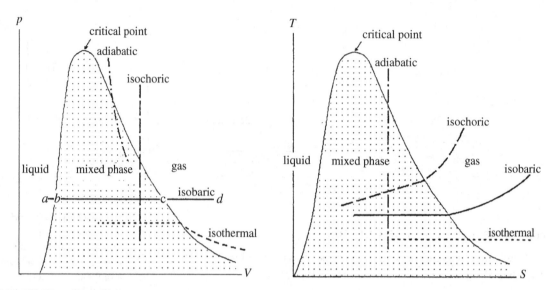

Figure 13.16 p–V and T–S diagrams for a typical fluid (not drawn to scale) indicating regions where it is a liquid, a gas, and a mixture of the two. See if you can understand the slopes of the lines for the processes marked adiabatic, isothermal, isobaric, and isochoric. Above the critical point a distinction between gas and liquid can no longer be made.

In the case of jet engines, the turbine blades draw off from the escaping gases only enough work to run the compressor and whatever generators and ancillary equipment are needed for the airplane. The rest of the work remains in the motion of the escaping gases. From Newton's third law, it is the momentum of these escaping gases that provides the forward thrust to the airplane.

If you compare this with our earlier description of a gas turbine cycle, you may think that the fourth stage, the exhausting of the remaining heat, is missing. It isn't. Instead of being exhausted into the atmosphere via a heat exchanger, it is exhausted directly with the escaping gases, as fresh cool gases are drawn in through the intake scoop.

I Vapor cycles

Vapor cycles are characterized by a fluid that transforms back and forth between the liquid and gaseous phases. Refrigerators employ a variety of refrigerants, but engines usually use water/steam in their vapor cycles.

I.1 Phase diagrams

To analyze engine performance in vapor cycles, it is helpful to have p–V and T–S phase diagrams for the working fluid. Generic examples of such diagrams are shown in Figure 13.16. The term "mixed phase" means that the fluid is partly liquid and partly gas.

To understand these diagrams, see whether you can describe what is happening as you move along each of the lines. In particular, answer each of the following questions for each line, one process at a time.

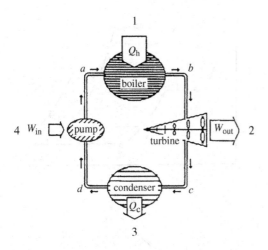

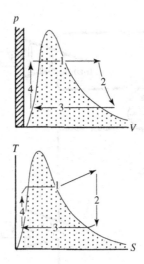

Figure 13.17 Elements of a steam turbine engine. The water goes through the following cycle: (1) the boiler adds heat Q_h; (2) the turbine extracts work W_{out}; (3) the condenser removes heat Q_c; (4) the pump pushes the water back into the boiler, doing work W_{in}. With temperature and pressure gauges, we can determine the enthalpy at the points a, b, c, d, which tells us the heat transfer and work done during each of the four stages.

- Is heat being added, removed, or neither?
- Is the volume increasing, decreasing, or neither?
- Within the mixed phase, is liquid evaporating or gas condensing, and why?
- As the system crosses from the gaseous to the mixed phase, what is happening and why?

I.2 Rankine cycle

The Rankine cycle is widely used in power plants. It resembles a gas turbine cycle but includes liquid–vapor transitions. It is illustrated in Figure 13.17 and goes as follows.

1) *Boiler* (hot reservoir)   Liquid water enters the boiler, where it is heated by some external source such as contact with pipes containing superheated steam from a nuclear reactor or fossil fuel furnace. The heat vaporizes the water, producing very hot and highly pressurized steam in the boiler.
2) *Turbine* (adiabatic expansion)   The hot steam then shoots through a turbine, turning the blades as it expands and cools. In the particular case of electrical power production, the turbine shaft is connected to an electrical generator, so the work done by the steam on the turbine blades is turned into electrical energy.
3) *Condenser* (cold reservoir)   The cooled steam then passes through a heat exchanger such as a large radiator immersed in the cold waters from a nearby river. This cools it sufficiently for it to condense completely into the liquid state again.
4) *Pump* (adiabatic compression)   The liquid water then gets pumped back into the boiler under high pressure, where it is ready to repeat the cycle.

As illustrated in Figure 13.17, an analysis of the cycle can be accomplished by the placement of temperature and pressure gauges at the points a, b, c, d; these measurements enable us, using equation 13.6, to obtain the enthalpy of the steam

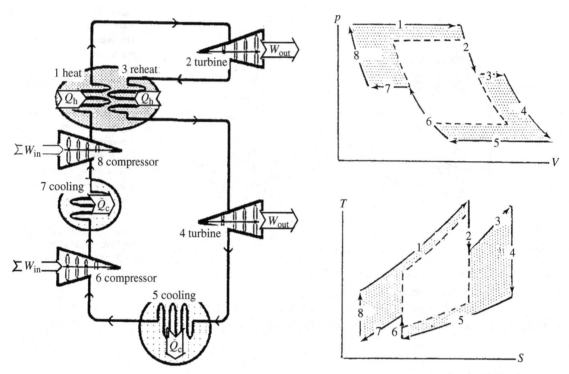

Figure 13.18 Illustration of a gas turbine engine with a second-stage compressor and a "reheat cycle." The shaded areas in the p-V and T-S diagrams are added by the second compressor 7–8 and by the reheat cycle 3–4.

or water at these four points, from which such things as the engine's efficiency can be determined: we have

$$e = \frac{W_{\text{net}}}{Q_h} = \frac{W_{\text{out}} - W_{\text{in}}}{Q_h}, \quad \text{or} = \frac{Q_h - Q_c}{Q_h}$$

with

$$Q_h = H_b - H_a, \quad Q_c = H_c - H_d, \quad W_{\text{in}} = H_a - H_d, \quad \text{and} \quad W_{\text{out}} = H_b - H_c.$$

J   Increasing the efficiency

Particularly in view of our dependence on rapidly diminishing non-renewable energy resources, there is considerable interest in increasing the efficiency of our engines. One way to do this is to reduce the entropy loss to temperature and pressure variations within the engine's working fluid and to friction in the moving parts. But other ways to increase the efficiency are revealed by our analysis of Carnot engines:

- increase the average temperature at which heat enters the engine;
- decrease the average temperature at which heat is exhausted.

For the gas cycle, we can create higher temperatures by greater compression before ignition and lower temperatures by greater expansion during the power stroke. So a larger compression ratio, the ratio of the gas's maximum and minimum volumes,

$$\text{compression ratio} = \frac{V_{max}}{V_{min}}, \tag{13.13}$$

promotes higher efficiency. Piston engines achieve larger compression ratios through either greater compression before ignition or longer power strokes. Gas turbine engines achieve larger compression ratios by using staged compressors and turbines, such as the cycle illustrated in Figure 13.18.

Summary of Sections F–I

Internal combustion engines use open cycles, and this allows them to purge the combustion products and receive a fresh fuel supply each cycle. Piston engines often have two extra strokes to accomplish this. One pair of compression and expansion strokes dispenses the normal four stages of a gas piston engine (ignition, power, exhaust, compression), and another pair of strokes purges the combustion products and draws in the fresh fuel mixture. The fuel ignites at the end of the compression stroke. Gasoline burns quickly, so the combustion is modeled as isochoric heat addition. Diesel fuel burns more slowly and the combustion is modeled as isobaric heat addition. Otherwise the two engines work similarly.

Internal combustion gas turbine engines continuously exhaust the combustion products, along with the excess heat leaving the turbine, and also continuously bring a fresh fuel mixture into the compressor.

The p–V and T–S phase diagrams for any fluid used in vapor cycles include a substantial region of mixed phase, partly liquid and partly gas. A Rankine cycle is a vapor cycle commonly used in power production, usually with water as the working fluid. The Rankine cycle is similar to a gas turbine, but in such a cycle the vapor becomes liquid in the condenser (by heat removal) after leaving the turbine, and then is pumped to the boiler, where it is vaporized again (by heat addition) before shooting through the turbine.

The efficiency of engines can be increased by decreasing the ratio T_c / T_h, that is, by increasing the differential between the temperature at which heat enters the engine and the temperature at which it is exhausted. One way of accomplishing this is to increase the compression ratio, defined as the ratio of the gas's maximum and minimum volumes during the cycle (equation 13.13):

$$\text{compression ratio} = \frac{V_{max}}{V_{min}}$$

For piston engines, this could be accomplished by compressing the gas into a smaller volume before ignition and/or by having a longer expansion stroke before exhaust. For gas turbines, we could use staged compressors to compress the gas more before ignition, and staged turbines to allow the gases to expand further.

Problems

Section A

1. Consider n moles of an ideal gas with v degrees of freedom per molecule, initially at temperature T_i, which expands from volume V_i to V_f. Calculate the heat added ΔQ and work done ΔW in terms of n, v, T_i, V_i, and V_f if the expansion is (a) isobaric, (b) isothermal, (c) adiabatic.

2. Show that for adiabatic processes in an ideal gas the work done is given by $\Delta W = (p_i V_i - p_f V_f)/(\gamma - 1)$.

3. One mole of an ideal gas at 300 K is expanded isothermally until its volume is doubled. How much heat energy is absorbed by the gas during this process?

Section B

4. Heating a stretched rubber band increases its tension, making it tend to contract, and a contracting rubber band tends to cool off. You are invited to design an engine that does work by pulling rather than pushing. You will need a hot reservoir, a cold reservoir, and a rubber band with one end fixed and one attached to a movable piston. Suppose that the heat addition and removal are done isothermally (e.g., the system is immersed in reservoirs as appropriate). Sketch and label the four parts of the cycle for this engine, and also sketch and label the four parts on a F–L diagram, which is like a p–V diagram, except using tension $(-F)$ and length (L). (Note that the work done is $dW = -FdL$, because the system is pulling, not pushing.)

5. Repeat the above problem for the case where the heat is added and removed under conditions of constant length.

6. The cylinder in an automobile engine has a radius of about 5 cm. The top of the piston begins its compression stroke at about 23 cm from the head end and travels about 18 cm during the stroke. The gas in the cylinder behaves as an ideal gas with five degrees of freedom per molecule. It begins at 310 K and 1 atm pressure when the piston is at the bottom of its stroke (i.e., fully expanded).
   (a) How many moles of gas are in the cylinder?
   (b) What are the temperature and pressure of the gas at the end of the adiabatic compression stroke?
   (c) How much work is done on the gas by the piston during the compression stroke?
   (d) When the gas is completely compressed, the combustion of the gasoline increases the temperature by another 800 K. What is the temperature now?
   (e) How much heat energy has been added to the gas?

(f) How much work is done by the gas as it expands adiabatically back to its starting point?

(g) What is the efficiency of this engine?

7. Suppose that the gas piston engine of Figures 13.2 and 13.3 starts at pressure and volume p_1, V_1. It is heated isochorically until its pressure is p_2, then expanded adiabatically until its volume is V_3, and next undergoes isochoric cooling and adiabatic compression to complete the cycle. In terms of γ, p_1, V_1, p_2, and V_3, find

(a) the pressure and volume at the end of each stage,

(b) the heat added ΔQ and the work done ΔW for each of the four stages. Hint: $\Delta Q = nC_V\Delta T$ for isochoric processes, with $C_V = (v/2)R$ and $v/2 = [1/(\gamma - 1)]$, $pV = nRT$.

8. Consider the gas turbine engine of Figure 13.5. Suppose that the gas starts at pressure and volume p_1, V_1. It is heated isobarically until its volume is V_2 and then expanded adiabatically until its volume is V_3. The cycle is then completed with isobaric and then adiabatic compression. In terms of γ, p_1, V_1, V_2, and V_3, find

(a) the pressure and volume at the end of each stage,

(b) the heat added ΔQ and work done ΔW for each of the four stages. Hint: $\Delta Q = nC_p\Delta T$ for isobaric processes, with $C_p = [(v+2)/2]R$ and $(v+2)/2 = [\gamma/(\gamma - 1)]$, $pV = nRT$.

9. Redraw the two diagrams in Figure 13.3 for the case where the piston engine receives and exhausts heat under isobaric, rather than isochoric, conditions.

10. You are interested in finding the heat added ΔQ, the work done ΔW, and the change in internal energy ΔE for each stage of the engine model cycle depicted in Figures 13.2 and 13.3. Complete the following table for the values of ΔQ, ΔW, ΔE for each of the four stages, giving the energies in joules. What is the efficiency of this engine?

	ΔQ	ΔW	ΔE
Stage 1	5		
Stage 2		4	
Stage 3	-2		
Stage 4			

11. Consider a four-stage gas engine whose cycle goes as follows: isobaric expansion, isothermal expansion, isochoric heat removal, adiabatic compression. Sketch this cycle (a) on a p–V diagram, (b) on a T–S diagram. (c) Make a table showing the signs $+$, $-$, 0 for ΔQ, ΔW, ΔE for each of the four stages.

12. Repeat problem 11 for a four-stage gas engine whose cycle goes as follows: isothermal expansion, isochoric heat removal, isobaric compression, adiabatic compression.

13. Repeat problem 11 for a four-stage gas engine whose cycle goes as follows: isothermal expansion, adiabatic expansion, isobaric compression, isochoric heat addition.

14. A certain engine contains n moles of an ideal gas whose molecules each have ν degrees of freedom. It begins its cycle at pressure, volume, and temperature p_0, V_0, T_0. In stage 1 it expands isobarically to volume $2V_0$. In stage 2 its pressure drops isochorically to $p_0/2$. In stage 3 it is compressed isobarically back to its original volume, V_0. In stage 4 it is heated isochorically back to its starting point.
    (a) Sketch this cycle on a p–V diagram.
    (b) Express T_0 in terms of $(p_0$, V_0, and $n)$.
    (c) Make a table expressing the work done, the heat added, and the change in internal energy for the engine in each of the four stages, expressing them in terms of nRT_0.
    (d) What is the efficiency of this engine?
    (e) The maximum efficiency of an engine operating between the temperature extremes T_h and T_c is given by $e = (T_h - T_c)/T_h$. What would be the maximum efficiency of an engine operating between the temperature extremes of the engine considered above?

15. A 1 gigawatt electrical power plant is 40% efficient.
    (a) If coolant water flows through it at a rate of 2.16×10^5 m^3/ hour, by how many degrees Celsius is the coolant water heated? (Water's specific heat is 4186 J/kg and its density is 10^3 kg/m^3.)
    (b) If the power plant uses evaporative cooling, how much water is evaporated per second? (It takes about 2.5×10^6 J to heat up and evaporate 1 kg of water.)

Section C

16. Is the integral $\int p dV$ around the cycle in Figure 13.7 positive or negative?

17. The cycles of the refrigerator of Figure 13.7 and the gas turbine of Figure 13.5 are quite similar. In particular, both send the working fluid to the hot reservoir after passing through the compressor. But in one you would expect the hot reservoir to be slightly hotter than the coils, and the other the hot reservoir would be slightly cooler. Which is which, and why?

18. What are the lower and upper limits on:
    (a) the efficiency of an engine,
    (b) the coefficient of performance for a refrigerator?

Sections D, E, and F

19. Consider one mole of an ideal gas with v degrees of freedom per molecule and temperature T. In terms of v and T, what would be the slope of the curve on a T–S plot for (a) isochoric heat addition, (b) isobaric heat addition, (c) isothermal heat addition? (d) Do these have curvature and, if so, are they concave upwards or downwards?

20. Consider one mole of an ideal gas with v degrees of freedom per molecule and pressure and volume p, V. In terms of v, p and V, what would be the slope of the curve on a p–V plot for (a) isobaric expansion, (b) isothermal expansion, (c) adiabatic expansion? (d) Do these have curvature and, if so, are they concave upwards or downwards?

21. Show that for the isothermal expansion of an ideal gas at temperature T (see Table 13.1):
    (a) $\Delta Q = nRT \ln(V_2/V_1)$;
    (b) $\Delta S = nR \ln(V_2/V_1)$.
    (c) Express these in terms of p_1, p_2, and T.

22. Show that for isobaric processes in an ideal gas (see Table 13.1):
    (a) $\Delta S = nC_p \ln(T_2/T_1)$;
    (b) $\Delta W = nR\Delta T$.
    (c) Express these in terms of p, V_2, and V_1.

23. Consider an ideal gas that has initial pressure and volume p_i, V_i and undergoes isothermal expansion to volume V_f. Find the following in terms of p_i, V_i, and V_f:
    (a) the work done by the gas;
    (b) the final pressure, p_f.

24. An ideal gas whose molecules each have five degrees of freedom has initial pressure and volume p_i, V_i. If this gas undergoes adiabatic expansion to volume V_f, find the following in terms of p_i, V_i, and V_f:
    (a) the work done by the gas;
    (b) the final pressure, p_f.

25. Consider an ideal gas whose molecules each have five degrees of freedom. It starts out at pressure and volume p_1, V_1, is isothermally expanded to volume V_2, and then adiabatically expanded further, to volume V_3. In terms of p_1, V_1, V_2, and V_3, find the following:
    (a) the total work done by the gas during the expansion,
    (b) the final pressure, p_3.
    (If necessary refer to the previous two problems.)

26. Consider an ideal gas having initial pressure and volume p_1, V_1. In terms of p_1, V_1, and γ, write down an expression for the variation of p with V during

(a) isothermal expansion, (b) adiabatic expansion. (c) Show that the slope dp/dV of the path on a p–V diagram is steeper for adiabatic processes than for isothermal processes, given that $\gamma = C_p/C_V > 1$.

27. The equation of state for a certain material is $pV^{10}T^5 = $ constant. This material starts out at $(p, V, T) = (1 \text{ atm}, 0.5 \text{ liters}, 300 \text{ K})$. Its heat capacity is $C_p = 3.5$ J/K. It is compressed until its volume is reduced by 1%. Calculate the heat added to the system if the compression is (a) isothermal (use Maxwell's relations, writing $p = \text{constant}/V^{10}T^5$, and taking $\partial p/\partial T_V$), (b) isobaric. Hint: $\Delta Q = (\partial Q/\partial V)\Delta V$.

28. A system whose equation of state is $p^6T^{-1}e^{aV} = b$, where a and b are constants, expands isobarically at pressure p from volume V_1 to volume V_2. Find expressions for $\Delta W, \Delta Q, \Delta S$, and ΔE in terms of p, C_p, V_1, V_2, a, and b.

29. Consider a refrigerator or heat pump, as in Figure 13.7, for which the fluid temperatures just before entering stages $1, 2, 3, 4$ are given by $40\,°C, 20\,°C, -10\,°C, 5°C$, respectively. The operating fluid is a gas with five degrees of freedom per molecule. For each mole of fluid that passes through the cycle, find
    (a) the work done on the gas by the compressor,
    (b) the heat energy released by the gas into the hot reservoir,
    (c) the energy lost to work in the expansion chamber,
    (d) the heat absorbed in the cold coils,
    (e) the coefficient of performance for this refrigerator.

30. Assuming that water vapor is an ideal gas with $\nu = 6$, calculate the following for one mole:
    (a) The enthalpy at $400\,°C$ and 3 atm pressure;
    (b) the enthalpy at $100\,°C$ and 1 atm pressure;
    (c) the enthalpy of one mole of liquid water at $100\,°C$ and 1 atm pressure.
        Hint: $\Delta H = \Delta E + \Delta(pV)$, with $\Delta E = \Delta Q - \Delta W$; ΔQ can be obtained from the latent heat and $\Delta W = p\Delta V$ is the work done on the water vapor by the atmosphere as its volume decreases from the gaseous to the liquid phase.
    (d) Use the answer to part (c) and the specific heat of water to find the enthalpy at $50\,°C$ and 1 atm.
    (e) Use the answer to part (c) and the specific heat of water to find the enthalpy at $10\,°C$ and 8 atm. (Liquid water is essentially incompressible.)

31. The cylinder of a piston engine contains an ideal gas whose molecules have five degrees of freedom apiece. The gas initially has volume 0.2×10^{-3} m^3, pressure 5×10^5 Pa, and temperature $600\,°C$. It is then expanded to a volume

1.0×10^{-3} m^3. Calculate the heat added ΔQ and the work done ΔW if the expansion is (a) isothermal, (b) isobaric, (c) adiabatic.

32. Consider a three-stage engine whose working fluid is one mole of a gas whose molecules have five degrees of freedom each. Each cycle starts at pressure and volume p_1, V_1 and expands to volume V_2. In terms of p_1, V_1, and V_2, find the heat added ΔQ and work done ΔW for each of the three stages if the cycle is:
    (a) adiabatic expansion, followed by isobaric compression, followed by isochoric heat addition,
    (b) isothermal expansion, followed by isobaric compression, followed by isochoric heat addition,
    (c) adiabatic expansion, followed by isothermal compression, followed by isochoric heat addition,
    (d) isothermal expansion, followed by isochoric heat removal, followed by adiabatic compression.
    Sketch a p–V diagram for each of the above processes.
    Sketch a T–S diagram for each of the above processes.

33. Consider a three-stage engine whose working fluid is one mole of a gas whose molecules have five degrees of freedom each. At the first stage the gas starts at temperature T_1 and expands until the temperature falls to T_2. For each of the two cycles described below, sketch the p–V diagram. Also, find the heat added ΔQ and the work done ΔW for each of the three stages in terms of T_1, and T_2.
    (a) Adiabatic expansion, followed by isothermal compression, followed by isochoric heat addition.
    (b) Adiabatic expansion, followed by isobaric compression, followed by isochoric heat addition.

34. Consider a four-stage engine whose working fluid is one mole of a gas whose molecules have five degrees of freedom each. It starts at temperature T_1, expands isobarically until the temperature rises to T_2, expands adiabatically until the temperature falls to T_3 ($> T_1$), and undergoes isochoric heat removal until the temperature falls back to T_1 and then isothermal compression back to the starting point.
    (a) Calculate the heat added ΔQ and the work done ΔW during each stage of this cycle interms of T_1, T_2, T_3.
    (b) Sketch the cycle on a p–V diagram.
    (c) Sketch the cycle on a T–S diagram.

35. An ideal diatomic gas of n moles at initial pressure and volume p_0, V_0 expands to final volume $2V_0$.

(a) Calculate the heat added, the work done, and the change in internal energy for each of three cases: the expansion is isobaric, isothermal, or adiabatic. Express your answers in terms of $p_0 V_0$.

(b) Calculate the final temperature in terms of n, p_0, and V_0 for each of these cases.

36. You have a material (not unlike water) whose molar specific heat C_p at atmospheric pressure is 30J/(mole K) in the gaseous phase, 75 J/(mole K) in the liquid phase, and 35 J/(mole K) in the solid phase. Furthermore, the material condenses at 373 K with a latent heat of 41 000 J/mole and freezes at 273 K with a latent heat of 6000 J/mole. Assuming that it behaves as an ideal gas at 600 K, find (a) its enthalpy at 600 K, (b) its molar enthalpy at 500 K, 374 K, 372 K, 323 K, 274 K, 272 K, and 200 K.

(At constant pressure, $\Delta H = \Delta E + p\Delta V = \Delta Q$, so you don't have to worry about changes in volume at phase transitions.)

37. Show that the change in molar enthalpy of an ideal gas that is expanded or compressed adiabatically is given by $\Delta h = \int v dp = \left(\frac{\gamma}{\gamma-1}\right)(p_f v_f - p_i v_i)$.

Section G

38. Consider the Carnot cycle, shown in Figure 13.11. Suppose that the temperatures of the hot and cold reservoirs are 800 K and 400 K, respectively. Complete the following table for the values of ΔQ, ΔW, ΔE for each of the four stages, where the energies are in joules. (The system is an ideal gas.) What is the efficiency of this cycle?

	ΔQ	ΔW	ΔE
Stage 1	4		
Stage 2		1	
Stage 3		-1	
Stage 4		-1	

39. You are interested in finding the heat added ΔQ, the work done ΔW, and the change in internal energy ΔE for each stage of the Carnot cycle depicted in Figure 13.11. The working fluid is an ideal gas with five degrees of freedom per molecule ($\gamma = 1.4$). It starts out at the beginning of stage 1 with pressure $p_1 = 4 \times 10^5$ Pa, volume $V_1 = 10^{-3}$ m³, and temperature $T_1 = 600$ K. It expands isothermally in stage 1 to volume $2V_1$ and then adiabatically in stage 2 to volume $3V_1$.

(a) Make a table showing the pressure, volume, and temperature at the end of each of the four stages.

(b) Make a table showing the values of ΔQ, ΔW, and ΔE in joules for each of the four stages.

(c) What is the efficiency of this cycle?

40. The outside coils of a certain electric heat pump operate at $10\,^{\circ}\text{C}$ below the ambient temperature, and those on the inside operate at $10\,^{\circ}\text{C}$ above the ambient temperature. Otherwise, there are no losses at all. Suppose that the outside temperature is freezing and the indoor temperature is kept at $17\,^{\circ}\text{C}$.

(a) How many joules of work are required for your heat pump to deposit one joule of heat energy inside?

(b) Now suppose that the combined power plant and transmission line efficiency is 30%. That is, of the energy of the original primary fuels, 30% reaches your heat pump as electrical energy. How many joules of energy in the primary fuels are required per joule of energy delivered to your house?

(c) Is it a more efficient use of primary fuels to run your electric heat pump or to burn the primary fuels in your space heater?

41. Consider a system that is interacting thermally, but not mechanically or diffusively, with outside systems. (An example is the combined system in the broken-line box of Figure 13.12.) Show that if heat were to flow from cold to hot then the entropy would decrease, in violation of the second law.

42. You can prove in the following way that it is overall more efficient for an engine to exhaust at the ambient environmental temperature than to exhaust at a cooler temperature that requires refrigeration. Suppose that the hot reservoir is at T_h, the ambient temperature is T_a, and the refrigerated temperature is T_c. The engine cannot put out more work than would a Carnot engine operating between T_h and T_c, so $W_{\text{eng}} = eQ_h = a[(T_h - T_c)/T_h]\,Q_h$, where $a \le 1$. Likewise the refrigerator cannot require less work input than a Carnot refrigerator, so $W_{\text{frig}} = bQ_a[(T_a - T_c)/T_a]$, where $b \ge 1$. The net work output is that from the engine minus that needed to run the refrigerator, $W_{\text{net}} = W_{\text{eng}} - W_{\text{frig}}$. Show that this quantity is less than the work you would get out of the engine if it simply ran between T_h and T_a without any refrigeration, which would be $W_{\text{eng}} = aQ_h[(T_h - T_a)/T_h]$. (For the Carnot engine, $Q_h/T_h = Q_a/T_a$.)

43. For a Carnot cycle, prove that $Q_h/T_h = Q_c/T_c$ using the following approach. First calculate Q_h and Q_c in terms of the volumes during isothermal expansion and contraction of an ideal gas. Then use $TV^{\gamma-1} = \text{constant}$ for adiabatic processes to relate the ratios of the volumes appearing in this result. This should give you the result you need. (To avoid confusion, sketch out the four stages on a p–V diagram and label the four end points.)

44. A motorcycle engine does work at a rate of about 6 kW while burning a liter of gasoline every 20 minutes.
    (a) Given that the mass of one gasoline molecule is about 114 atomic mass units (1 amu $= 1.66 \times 10^{-27}$ kg), and the mass of a liter of gasoline is 0.7 kg, how many gasoline molecules are there in a liter?
    (b) If each gasoline molecule releases 57 eV of energy upon oxidation, how much heat energy (in joules) is provided by burning one liter of gasoline?
    (c) How much work (in joules) is done altogether by the engine while it burns 1 liter of gasoline?
    (d) What is the efficiency of this motorcycle engine?

45. The ignition temperature for the gasoline in the motorcycle in problem 44 is 1100 K and the exhaust temperature is 570 K.
    (a) What is the Carnot efficiency for an engine operating between these temperatures?
    (b) What is the coefficient of utility for the motorcycle engine?

46. Most electrical power is produced by burning fossil fuels. The heat produced is used to run steam turbines that drive generators. The combustion temperature can be controlled to some extent through dilution of the combustion gases with extra air passing through the furnaces. To make our limited fossil fuel resources stretch as far as possible, should we make the combustion temperature as high or as low as possible? Why?

Sections H

47. Why do we model the heat transfer for gasoline piston engines as isochoric and that for turbine engines as isobaric?

48. For a typical engine operating speed of 4000 rpm, estimate the time needed for the gasoline and diesel fuel mixtures to burn, respectively.

49. Consider an engine whose working system is n moles of a gas whose molecules have five degrees of freedom apiece. The engine undergoes a three-stage cycle, starting at pressure and volume p_1, V_1. The first stage is isobaric expansion to volume V_2, the second stage is adiabatic expansion, and the third stage is isothermal compression back to the starting point (p_1, V_1). In terms of n, p_1, V_1, and V_2, find
    (a) the pressure, volume, and temperature at the end of each stage,
    (b) the heat added, the work done, and the change in internal energy during each stage.

50. Consider an engine whose working system is n moles of a gas whose molecules have five degrees of freedom apiece. It has a three-stage cycle, starting at pressure and volume p_1, V_1. The first stage is isochoric heat addition until the pressure reaches p_2, the second stage is adiabatic expansion, and

the third stage is isothermal compression back to the starting point (p_1, V_1). In terms of n, p_1, V_1, and p_2, find

(a) the pressure, volume, and temperature at the end of each stage,

(b) the heat added, the work done, and the change in internal energy during each stage.

51. Consider the Otto cycle, shown in the p–V diagrams of Figure 13.13, and assume that the gas is a diatomic ideal gas. Imagine that after the intake stroke the temperature and pressure of the gas are 300 K and 10^5 Pa.

(a) If it is then compressed adiabatically to $1/10$ its initial volume, what is its new pressure and temperature?

(b) Suppose that the sudden ignition of the fuel then doubles the temperature of the gas (isochorically). What is this new temperature of the gas?

(c) What is the maximum possible efficiency of an engine that operates between the highest and lowest temperatures in this Otto cycle?

52. A car that gets 20 miles to the gallon is travelling down the highway at 60 miles per hour. The engine is 20% efficient. The heat content of a gallon of gas is about 1.3×10^8 J. What is the rate (in kilowatts) at which the engine is doing work?

Section I

53. On a p–V phase diagram such as that in Figure 13.16, sketch the path followed during isothermal expansion for a fluid that starts out as a liquid.

54. On a T–S phase diagram such as that in Figure 13.16, sketch the path followed during isobaric heat addition for a fluid that starts out as a liquid.

55. The tropical ocean is a giant collector of solar energy. Surface waters have temperatures around 24 °C, and deeper waters have temperatures around 4 °C. This temperature differential can be used to run engines that drive electrical generators.

(a) Estimate the amount of solar energy $Q_h - Q_c$ stored in the ocean by means of this temperature differential; the warm surface layer is typically 200 m thick and the tropical ocean covers about $1/3$ of the Earth's surface area. (Earth's radius is 6400 km.)

(b) The world consumes energy at a rate of about 9 billion kilowatts. How many joules of energy is this per year?

(c) What fraction of the solar energy stored in the oceans would have to be harvested annually to meet the world's energy needs?

(d) Answer part (c) for the case where 90% of the energy is lost during conversion.

(e) What is the Carnot efficiency for engines running between these two temperatures?

(f) What is the actual efficiency for engines running between these two temperatures if the coefficient of utility is 0.1?

56. The solar power incident on the Earth's surface, averaged over all latitudes, seasons, weather conditions, and times of day and night, is about $175 \, \text{watts/m}^2$. How does the rate at which the Earth receives solar energy compare with the rate at which humans consume energy? (The Earth's human population consumes energy at a rate of 9 billion kilowatts, and the Earth's radius is 6400 km.)

Chapter 14
Diffusive interactions

In Chapter 9 we showed that temperature governs thermal interactions, pressure governs mechanical interactions, and chemical potential governs diffusive interactions. They do this in ways that are so familiar to us that we call them "common sense":

- thermal interaction. Heat flows towards lower temperature.
- mechanical interaction. Boundaries move towards lower pressure.
- diffusive interaction. Particles go towards lower chemical potential.

In this chapter we examine diffusive interactions, working closely with the chemical potential μ and the Gibbs free energy $N\mu$.

Figure 14.1 Diffusion into regions of lower concentration is demanded by the second law, because increasing the volume per particle also increases the entropy. In this example, when the partition is removed the diffusion into previously unoccupied regions doubles the number of states accessible to each molecule, so the number of states available to the system increases by a factor 2^N. Does the chemical potential increase or decrease? (see problem 2)

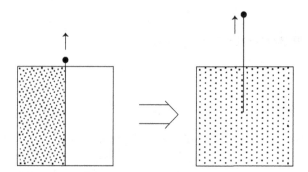

A   The chemical potential

In Chapter 5 we learned that the equilibrium distribution of particles is determined by the fact that particles seek configurations of

- lower potential energy,
- lower particle concentration.

Although the first of these is familiar in our macroscopic world (e.g., balls roll downhill), the second is due to thermal motions, which are significant only in the microscopic world (Figure 14.1).

Both factors trace their influence to the second law. The number of states per particle, and hence the entropy of the system, increases with increased volume in either momentum space or position space. Deeper potential wells release kinetic energy, making available more volume in momentum space, V_p. And lower particle concentrations mean more volume per particle in position space, V_r.[1]

The two factors are interdependent. The preference for regions of lower potential energy affects particle concentrations, and vice versa. There is a trade-off. The reduction in one must more than offset the gain in the other (Figure 14.2). We are now in a position to evaluate these interrelationships more precisely and make quantitative predictions.

A.1   Dependence on temperature and pressure

Suppose that two systems are initially in diffusive equilibrium. What would happen if we heated or compressed one of them? According to equation 9.13 the resulting change in chemical potential would be

$$d\mu = -\left(\frac{S}{N}\right)dT + \left(\frac{V}{N}\right)dp,\qquad(14.1)$$

where S, V, N are all positive. So the preference for lower chemical potential means that particles would flow towards regions of higher temperature ($dT > 0$)

---

[1] There is a subtlety here involving identical particles The corrected number of states per particle (equation 6.8) becomes $\omega_c = -e\omega/N = eV_r V_p/Nh^3$, so the volume per particle becomes V_r/N.

Figure 14.2 In the microscopic world, thermal energy allows some particles to diffuse into regions of higher potential energy and lower concentration. Macroscopic analogies would be the mist and spray rising from the base of Niagara Falls. Most of the water remains in the river at the base, but some mist rises back up into regions of higher potential energy and lower concentration. (New York State Office of Parks and Recreation; Historic Preservation, courtesy of Allen James)

and lower pressure ($dp < 0$). The underlying reason is that higher temperature corresponds to more room in momentum space, V_p, and lower pressure increases the volume in position space, V_r.

A.2 Dependence on potential energy and particle concentration

The tendency of particles to seek regions of lower potential energy and lower concentration can be quantified by manipulating the first law. We begin with the integrated form (equation 9.12) and solve for μ, as we did in equation 10.3c:

$$\mu = \frac{E + pV - TS}{N} = \overline{\varepsilon} + p\overline{\upsilon} - kT \ln \omega_c, \qquad (14.2)$$

where $\overline{\varepsilon}, \overline{\upsilon}, k \ln \omega_c$ are the average energy, volume, and entropy per particle, respectively. For further insight, we rewrite each term on the right-hand side, as follows:

$$\overline{\varepsilon} = u_0 + \frac{\upsilon}{2}kT, \qquad (14.3a)$$

$$p\overline{\upsilon} \approx kT \text{ for gases} \qquad \text{and} \qquad p\overline{\upsilon} \approx 0 \text{ for liquids and solids,}^2 \qquad (14.3b)$$

$$\omega_c = \frac{e\omega}{N} = \frac{eV_p V_r}{h^3 N} = \text{constant} \times \frac{T^{\upsilon/2}}{\rho}. \qquad (14.3c)$$

---

2 More precisely, it is typically 10^3 to 10^5 times smaller than the other terms for solids and liquids. Beware that the symbol we use for volume per particle "υ" is very similar to that which we have used for molar volume. Although a discerning eye might notice that one is slightly wider and has a line over the top to indicate average value, context is probably the best way to distinguish between them.

For the last expression we have used $\rho = N/V_r$ for particle density together with Table 6.2 for the dependence of ω_c on T. Substituting these formulas into equation 14.2 gives

$$\mu = u_0 + kT \ln \rho + f(T),\qquad(14.4)$$

where the function $f(T)$ depends on the system.

This is the expression we wanted. It displays the dependence of the chemical potential on the potential energy reference level u_0 and the particle concentration ρ for systems held at any given temperature T. It confirms our previous assertion: the preference for lower chemical potential means that particles diffuse towards regions of lower potential energy (lower u_0) and lower particle concentration (lower ρ).

A.3   Equilibrium concentrations

Expression 14.2 can also be used to predict equilibrium concentrations in diffusively interacting systems. Multiplying both sides by $(-1/kT)$ and exponentiating, it gives

$$e^{-\mu/kT} = \frac{e\omega}{N}\, e^{-(\bar{\varepsilon}+p\bar{v})/kT} \qquad \left(\text{using } \omega_c = \frac{e\omega}{N}\right).$$

If two systems are in diffusive and thermal equilibrium ($\mu_1 = \mu_2$, $T_1 = T_2$) then $e^{-\mu/kT}$ is the same for both, which tells us that

$$\frac{\omega}{N}\, e^{-(\bar{\varepsilon}+p\bar{v})/kT} \text{ is the same for both,}\qquad\text{hence}\qquad N \propto \omega e^{-(\bar{\varepsilon}+p\bar{v})/kT}.\quad(14.5)$$

In the next chapter we will derive this expression in a different way, using

$$\begin{pmatrix}\text{number of}\\\text{particles }(N)\end{pmatrix} = \begin{pmatrix}\text{number of}\\\text{states }(\omega)\end{pmatrix} \times \begin{pmatrix}\text{probability that a state has}\\\text{a particle in it }(\propto e^{-(\bar{\varepsilon}+p\bar{v})/kT})\end{pmatrix}.$$

Using the expressions for $\bar{\varepsilon}$ and $p\bar{v}$ from equations 14.3 ($(\bar{\varepsilon} - p\bar{v})/kT = u_0 + \text{constant}$) and absorbing e^{constant} into the proportionality constant, we can write equation 14.5 as

$$N \propto \omega e^{-\frac{u_0}{kT}},\qquad(14.5')$$

Or, if we prefer to work with particle densities we can divide both sides by the volume:

$$\rho \propto \rho_\omega e^{-u_0/kT} \qquad \left(\rho = \frac{N}{V},\quad \rho_\omega = \frac{\omega}{V}\right).\qquad(14.5'')$$

These equations also confirm our expectation that more particles will be found where the number of available states ω is larger and the potential energies u_0 are lower. Applying 14.5″ to our atmosphere, for example, where a molecule's potential energy at altitude h is given by $u_0 = mgh$, we find that the atmosphere thins by $1/2$ for roughly every 5.6 km of elevation.

Summary of Section A

Temperature rules thermal interactions, pressure rules mechanical interactions, and chemical potential rules diffusive interactions. Particles move towards regions of lower chemical potential because that increases the number of accessible states, as required by the second law.

Chemical potential varies with temperature and pressure according to (equation 14.1)

$$d\mu = -\left(\frac{S}{N}\right)dT + \left(\frac{V}{N}\right)dp.$$

Its explicit dependence on potential energy u_0 and particle concentration ρ is revealed by first rewriting the integrated form of the first law (equation 14.2), so that

$$\mu = \frac{E + pV - TS}{N} = \bar{\varepsilon} + p\bar{v} - kT\ln\omega_c,$$

where $\bar{\varepsilon}, \bar{v}, k\ln\omega_c$ are the average energy, volume, and entropy per particle, respectively, and then rearranging the terms on the right to give (equation 14.4)

$$\mu = u_0 + kT\ln\rho + f(T).$$

Starting with equation 14.2, multiplying by $(-1/kT)$, and exponentiating both sides, we can also show that for two systems in equilibrium, the number of particles (N) is related to their potential energies u_0 and the number of states ω through (equations 14.5, 14.5′, 14.5″)

$$N \propto \omega e^{-(\bar{\varepsilon}+p\bar{v})/kT} \propto \omega e^{-u_0/kT}, \quad \text{or} \quad \rho \propto \rho_\omega\, e^{-u_0/kT}.$$

B Colligative properties of solutions

We now use the tools of the previous section to study the "colligative" properties of solutions – those that depend on the concentration of the solute but not on its nature. We will frequently use sea water as an example, which on average is about 3.5% salt by weight or 2% salt by number of particles (mostly ions).

B.1 Changes in the chemical potential

Of all the particles in solution, a fraction f are solute particles, which may or may not have solvent molecules attached. We focus on the remaining fraction,

$1 - f$, which are unattached solvent molecules. Because they are not interacting with the solute, they experience a negligible change in potential energy (u_0 is unchanged).[3] If the original pure solvent's density was ρ_0, addition of the solute reduces its density to $(1 - f)\rho_0$. And, according to equation 14.4, the solvent's chemical potential changes by[4]

$$\Delta\mu = \mu - \mu_0 = kT \ln(\rho/\rho_0) = kT \ln(1 - f) \approx -fkT, \tag{14.6}$$

where we have assumed that $f \ll 1$.

We first wish to study the effects of solutes on phase equilibrium, using the liquid and vapor phases for illustration. Imagine that a liquid of pure solvent is initially in diffusive equilibrium with its vapor phase ($\mu_{\text{liq}} = \mu_{\text{vap}}$). We then add solutes (initially holding the temperature and pressure constant), which decreases the chemical potential of the liquid solvent in accordance with equation 14.6.

The change in the solvent's chemical potential means that it is no longer in diffusive equilibrium with its vapor phase. So there will be a net transfer between vapor and liquid until equilibrium is reestablished. Since the two phases begin and end in equilibrium, their two chemical potentials must change by the same amount. According to equations 14.1 and 14.6, this readjustment requires changes in temperature and/or pressure according to

$$-\left(\frac{S}{N}\right)_{\text{vap}} \Delta T + \left(\frac{V}{N}\right)_{\text{vap}} \Delta p = -\left(\frac{S}{N}\right)_{\text{liq}} \Delta T + \left(\frac{V}{N}\right)_{\text{liq}} \Delta p - fkT,$$

where the left-hand side is the change in the chemical potential of the vapor, $\Delta\mu_{\text{vap}}$, and the right-hand side is the change in the chemical potential of the liquid, $\Delta\mu_{\text{liq}}$. Rearranging terms gives

$$-\left[\left(\frac{S}{N}\right)_{\text{vap}} - \left(\frac{S}{N}\right)_{\text{liq}}\right] \Delta T + \left[\left(\frac{V}{N}\right)_{\text{vap}} - \left(\frac{V}{N}\right)_{\text{liq}}\right] \Delta p = -fkT. \tag{14.7}$$

We now use this result to look at changes in pressure at constant temperature ($\Delta T = 0$) and then at changes in temperature at constant pressure ($\Delta p = 0$).

B.2 Vapor pressure

For changes in pressure at a fixed temperature ($\Delta T = 0$), we note that the vapor volume can be approximated by the ideal gas law value ($V/N = kT/p$), and that of the liquid is negligible in comparison. So equation 14.7 becomes

$$\frac{\Delta p}{p} = -f \qquad (\text{if } \Delta T = 0). \tag{14.8}$$

That is, the vapor pressure decreases in proportion to the decrease in solvent concentration.

---

[3] True, this picture may be an oversimplified one, but it works. Another simplified approach that also works is to ignore the interactions altogether and assume constant total particle density, so that the solvent particle density still decreases by the factor $1 - f$.

[4] Use $\ln a - \ln b = \ln(a/b)$ and assume that the temperature is held constant as the solute is added.

This makes sense. In diffusive equilibrium, the rate of molecules going from the liquid to the vapor phase is equal to the rate going the other way. If there are fewer molecules in the liquid trying to escape into the vapor then there must be correspondingly fewer molecules in the vapor trying to go back. That is, the vapor pressure is correspondingly reduced.

B.3 Freezing and boiling points

We next look at the effect of solutes on the freezing and boiling points of liquid solutions held at fixed pressure ($\Delta p = 0$). In this case, equation 14.7 becomes

$$\frac{1}{N}\left(S_{\text{vap}} - S_{\text{liq}}\right)\Delta T = fkT.$$

The change in entropy $\Delta S = S_{\text{vap}} - S_{\text{liq}}$ for N particles going between the two phases is determined from the latent heat released upon condensation, $\Delta S = \Delta Q/T = \mathcal{L}/T$, so that the above expression becomes

$$\frac{1}{N}\frac{\mathcal{L}}{T}\Delta T = fkT, \qquad \text{or} \qquad \Delta T = f\frac{NkT^2}{\mathcal{L}} \quad \text{(if } \Delta p = 0). \tag{14.9}$$

In the homework problems, this equation can be used to show that the salts dissolved in sea water raise its boiling point by about 0.6 °C and depress its freezing point by about 2.0 °C. ($\mathcal{L}$ is positive for vaporization and negative for freezing, as latent heat is added for the one and removed for the other.)

B.4 Osmosis

Many advanced organisms regulate their internal body fluids in such a way that their cells are continually bathed in an environment that ensures optimum performance. This regulation is accomplished by special membranes, such as lungs, gills, skins, and guts, that separate their internal body fluids from the external environment. These membranes are semi-permeable, meaning that they allow water molecules to pass through, but not the dissolved salts.[5] The diffusion of water molecules through these membranes is called "osmosis."

For vertebrates like ourselves, the body fluid salinity is between those of fresh and sea water. If we drink fresh water, the water diffuses through our gut into the more saline body fluids. But if we drink sea water then osmosis goes the other way, causing dehydration. Swelling due to injury can be reduced by soaking the swollen body part in salt water, thus drawing the excess fluids out through the skin and into the saltier solution.

The underlying reason for diffusion through these semi-permeable membranes is that the water on the salt side has a slightly lower chemical potential. According to equation 14.6 the difference for typical sea water ($f = 0.02$) at normal temperatures is about (homework)

$$\Delta\mu = -fkT \approx -0.0005 \text{ eV}$$

---

[5] This is statement is oversimplified but basically correct.

The amount of back pressure that would be required to oppose osmotic diffusion is called the osmotic pressure. It has to be sufficient to offset the 0.0005 eV that the chemical potential loses owing to the difference in concentration. Assuming the same temperature on both sides, equation 14.1 tells us that the required back pressure would be

$$\Delta p = \frac{N}{V}\Delta\mu.$$

Knowing that the molar volume of water is 18 cm^3, we find that the osmotic pressure is

$$\Delta p \approx 2.7 \times 10^6 \text{ Pa} \approx 27 \text{ atm.}$$

Pressures such as this would rupture most organic membranes. Thus most fresh water organisms cannot survive prolonged periods in sea water, and vice versa.

Summary of Section B

When a solute is added, the chemical potential of a solvent is reduced by (equation 14.6)

$$\Delta\mu = kT \ln(1-f) \qquad \approx \qquad -fkT,$$

for $f \ll 1$, where f is the fraction of all particles that belong to the solute. This change in the solvent's chemical potential means that corresponding changes in the temperature and/or pressure are required to bring it back into diffusive equilibrium with another phase. For liquid-vapor equilibrium, these changes are given by (equation 14.7)

$$-\left[\left(\frac{S}{N}\right)_{\text{vap}} - \left(\frac{S}{N}\right)_{\text{liq}}\right]\Delta T + \left[\left(\frac{V}{N}\right)_{\text{vap}} - \left(\frac{V}{N}\right)_{\text{liq}}\right]\Delta p = -fkT.$$

Such properties, whose changes depend on the concentration f of the solute but not its particular nature, are called colligative.

Application of equation 14.7 reveals that a liquid solvent's vapor pressure is reduced by (equation 14.8)

$$\frac{\Delta p}{p} = -f,$$

and its freezing point is lowered and its boiling point raised by (equation 14.9)

$$\Delta T = f\frac{NkT^2}{\mathcal{L}},$$

where $\mathcal{L}/N$ is the latent heat per particle.

Owing to the difference in chemical potential, fresh water molecules tend to diffuse through semi-permeable membranes towards the saltier side, a process called osmosis. The reverse pressure needed to oppose this diffusion is called the osmotic pressure. The drop in chemical potential between fresh and sea water amounts to about 0.0005 eV, which corresponds to an osmotic pressure of about 27 atmospheres.

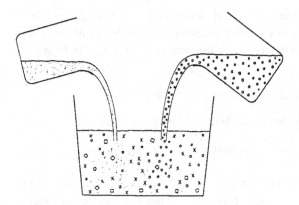

Figure 14.3 After the reactants are mixed and equilibrium is reached, how do we know what the equilibrium concentrations of the various interacting chemicals will be?

C   Chemical equilibrium

We now examine chemical reactions to see how the equilibrium concentrations of the various reactants are interrelated (Figure 14.3). The reactants could be any type of particle, such as atoms, ions, or molecules. We think of all the particles of one type as being one subsystem of the larger group. For example, the following would be the subsystems in the dissociation of water or nitric acid molecules:

$$H^+ + OH^- \leftrightarrow H_2O, \quad \text{all} \quad H^+, OH^-, \text{ and } H_2O \text{ particles};$$

$$H^+ + NO_3^- \leftrightarrow HNO_3, \quad \text{all} \quad H^+, NO_3^-, \text{ and } HNO_3 \text{ particles}.$$

C.1   Gibbs free energy and the law of mass action

Consider a chemical reaction that has reached equilibrium under conditions of constant temperature and pressure (so that $\Delta\mu_i = 0$, by equation 14.1). As we learned in subsection 9F.4, the Gibbs free energy must be a minimum, and therefore its derivatives are zero. So to first order we have (equation 9.17)

$$\Delta G = \sum_i \mu_i \Delta N_i = 0 \qquad \text{at equilibrium} \quad (T, p \text{ constant}).$$

The changes in the numbers of particles of the various reactants, ΔN_i, are in proportion to their "stoichiometric coefficients". For example, in the burning of hydrogen,

$$2H_2 + O_2 \leftrightarrow 2H_2O,$$

for every two hydrogen (H_2) molecules consumed, one oxygen molecule (O_2) is consumed, and two water (H_2O) molecules are produced. So, in this example, the ratios of the changes in numbers are given by

$$\Delta N_{H_2} : \Delta N_{O_2} : \Delta N_{H_2O} = -2 : -1 : +2.$$

The numbers $-2, -1, +2$ are the stoichiometric coefficients for this reaction.[6]

In general, if b_i represents the stoichiometric coefficient for the ith reactant in a process then the ratios of the changes in the numbers of particles for the various reactants are given by

$$\Delta N_1: \Delta N_2: \Delta N_3: \cdots = b_1 : b_2 : b_3 : \cdots,$$

and we can write the equilibrium condition 9.17 in the more convenient form

$$\sum_i \mu_i b_i = 0 \qquad \text{at equilibrium} \quad (T, p \text{ constant}). \tag{14.11}$$

Because we are interested in particle concentrations, we use equation 14.4 to identify the dependence of μ on ρ, combining all other factors into one messy function $\zeta(T)$, which does not interest us except that it depends only on T.[7] Then

$$\mu = kT[\ln \rho - \ln \zeta(T)]. \tag{14.12}$$

We insert this expression for the chemical potentials of the reactants into equation 14.11 and divide out the common factor kT to get

$$\sum_i b_i (\ln \rho_i - \ln \zeta_i) = 0.$$

Moving $b_i \ln \zeta_i$ to the other side of the equation and using $y \ln x = \ln x^y$ gives

$$\sum_i \ln \rho_i^{b_i} = \sum_i \ln \zeta_i^{b_i}.$$

Taking the antilogarithm of both sides gives

$$\rho_1^{b_1} \rho_2^{b_2} \rho_3^{b_3} \cdots = \zeta_1^{b_1} \zeta_2^{b_2} \zeta_3^{b_3} \cdots.$$

This "law of mass action" describes how the equilibrium concentrations of the various reactants are related through their stoichiometric coefficients. Because the functions on the right-hand side depend only on the temperature, the law can be written as

$$\rho_1^{b_1} \rho_2^{b_2} \rho_3^{b_3} \cdots = A(T) \qquad \text{(law of mass action)}. \tag{14.13}$$

The function $A(T)$ is called the "equilibrium constant" for the particular reaction. In principle it can be calculated from the functions $\zeta(T)$ for the various reactants, but it is usually determined experimentally and is listed in tables for most common reactions at various temperatures. Reactant concentrations may be measured in

[6] If the reaction proceeds in the other direction – from right to left – the stoichiometric coefficients would all reverse sign. This does not matter – all that does matter is their *relative* signs.

[7] In the language of equation 14.4, $\zeta(T) = \exp\{-[u_0 + f(T)]/kT\}$. Strictly speaking, this function may also depend on the pressure, and we could include this dependence if we wished. For example, the depth of the potential well, u_0, depends on particle spacing. But liquids are nearly incompressible, and the potential energies of gas molecules are minuscule, so we can safely ignore the pressure dependence for most cases.

whichever units are convenient, providing that the appropriate unit conversion is made in the equilibrium constant.

C.2 Reaction rates

The law of mass action relates the concentrations of reactants after equilibrium has been achieved. But it does not tell us the rate of the reaction nor whether it will happen in the first place. A reactant may have to surmount a potential barrier, called the "activation energy", in order to leave its present state for one of lower chemical potential. For example, it may have to be torn from one molecule before it can interact with another. The required activation energy might severely retard the rate of reaction, or it might prevent it from happening at all. Chemical intermediaries called "catalysts" are sometimes used to provide an alternate path that reduces or avoids the potential barrier and speeds up the reaction. Thermal motions of the reactants usually help to speed up a reaction, so higher temperatures may also help.

Summary of Section C

When chemical reactants are in equilibrium at a given temperature and pressure, the second law requires that the Gibbs free energy be a minimum (equation 9.17):

$$\Delta G = \sum_i \mu_i \Delta N_i = 0 \quad \text{at equilibrium} \quad (T, p \text{ constant}).$$

The changes ΔN_i in the numbers of particles of the various reactants are in proportion to their stoichiometric coefficients b_i, which appear in the chemical equation, so the above condition becomes (equation 14.11)

$$\sum_i \mu_i b_i = 0 \quad \text{at equilibrium} \quad (T, p \text{ constant}).$$

For gases and liquid solutions we are able to write the chemical potential as (equation 14.12)

$$\mu = kT[\ln \rho - \ln \zeta(T)],$$

where $\zeta(T)$ is a function of the temperature only. If we insert this expression for the chemical potential of each reactant into equation 14.11, divide by kT, and take the antilogarithm we get

$$\rho_1^{b_1} \rho_2^{b_2} \rho_3^{b_3} \cdots = \zeta_1^{b_1} \zeta_2^{b_2} \zeta_3^{b_3} \cdots$$

or (equation 14.13)

$$\rho_1^{b_1} \rho_2^{b_2} \rho_3^{b_3} \cdots = A(T),$$

where the function $A(T)$ is called the equilibrium constant. This law of mass action shows how the equilibrium concentrations ρ_i of the reactants are related through their stoichiometric coefficients b_i.

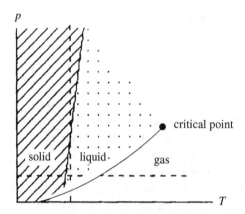

Figure 14.4 Phase diagram for the solid–liquid–gas phases of a typical material. The vertical and horizontal broken lines indicate what would happen if we changed the pressure at constant temperature or the temperature at constant pressure, respectively.

D   Phase equilibrium

We now turn our attention from equilibrium among chemical reactants to equilibrium between physical phases. For illustrative purposes we will often refer to the familiar solid–liquid–gas phases, which are characterized by differences in densities and rigidity. But distinctive phases can also be identified in other properties, such as electrical, magnetic, thermal, acoustic, and fluid behaviors.

The diffusion of particles from one phase to another is governed by the chemical potential, which depends on two intrinsic variables, usually chosen to be the temperature and pressure (or another mechanical parameter). From equation 14.1 you can see that changes in either variable must cause changes in the chemical potential, and such changes may make one phase favored over another.

D.1   Phase diagrams

A plot depicting the range of variables over which the various phases are stable is called a phase diagram. At any point on the line separating two phases, their chemical potentials are equal and they are in diffusive equilibrium. Figure 14.4 illustrates a phase diagram for the solid, liquid, and gaseous phases of a typical material. The melting and boiling points of most materials rise with increasing pressure, because greater pressure favors the denser phase and so more thermal motion is required to change it.

Beyond a certain temperature and pressure, called the "critical point," it is no longer possible to distinguish between liquid and gas. The thermal motion of the molecules prevents them from sticking together. We could gradually compress this gas until it has attained the density of a liquid without it ever condensing. Critical points for some familiar gases are listed in Table 14.1. The strong electrical polarization of water molecules (Table 10.1) makes them particularly "sticky" and gives them some remarkable properties, such as expansion upon freezing, as discussed in the next section. This "stickiness" also means that more thermal

Table 14.1. *Critical points of some gases, listed in order of increasing molecular weight. Notice that the general pattern according to which T_c, p_c increase with molecular weight is broken by helium and water, owing to their respective exceptionally weak and exceptionally strong intermolecular interactions*

Gas	T_c (K)	p_c (10^5 Pa)	Gas	T_c (K)	p_c (10^5 Pa)
H_2	33	13	O_2	154	50
He	5.3	2.3	CH_4	191	46
H_2O	647	221	CO_2	304	74
N_2	126	34			

energy is required to break them apart, as is reflected in higher melting and boiling points and a higher critical temperature than other light molecules.

D.2   Clausius–Clapeyron equation

We now look more closely at how changes in pressure affect the temperature of phase transitions. For two phases to remain in equilibrium as we change T and p, their two chemical potentials must remain equal. That is, they must both change by the same amount (Figure 14.5):

$$d\mu_1 = d\mu_2 \quad \text{(to remain in equilibrium)}.$$

It is convenient to multiply both sides by Avogadro's number and use equations 9.13 or 14.1 to write the change $d\mu$ in terms of the changes in temperature and pressure:

$$N_A d\mu = -s dT + v dp, \tag{14.14}$$

where s and v are the molar entropy and molar volume, respectively. With this, the preceding statement that the two chemical potentials must change by the same amount becomes

$$-s_1 dT + v_1 dp = -s_2 dT + v_2 dp.$$

After collecting terms, we have

$$\frac{dp}{dT} = \frac{s_2 - s_1}{v_2 - v_1} = \frac{\Delta s}{\Delta v}$$

where Δs and Δv are the changes in molar entropy and volume as the system goes across the equilibrium line from one phase to the other. If L is the molar

Figure 14.5 A small portion of a line separating two phases on a phase diagram. The two phases are in equilibrium along this line, so their chemical potentials must be equal. For them to remain equal as we move along this line, they must both change by the same amount: $\Delta\mu_1 = \Delta\mu_2$.

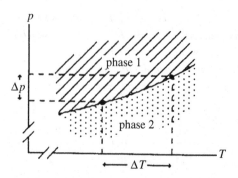

latent heat for this phase transition then we can write the change in molar entropy as $\Delta s = L/T$, and the above result becomes the Clausius–Clapeyron equation,

$$\frac{\mathrm{d}p}{\mathrm{d}T} = \frac{L}{T\Delta v} \qquad \text{(Clausius–Clapeyron equation).} \qquad (14.15)$$

It tells us how changing the pressure affects the temperature of a phase transition, or vice versa.

For most materials, the molar volume v increases as latent heat is added. So the ratio $L/T\Delta v$ is positive, and increased pressure raises the temperature of phase transitions. However, the molar volume *decreases* for melting ice. So the ratio is negative and increased pressure *decreases* the melting point! This is why ice is slippery. When you step on ice, the pressure reduces the melting point and it melts (homework). That layer of water under your foot makes it slippery. This should make sense. Increased pressure should force or favor the denser phase, and the above result says that it does.

In the case of vaporization, the molar volume of the liquid phase is nearly negligible compared with that of the gas, and the latter can be approximated by the ideal gas law:

$$\Delta v = v_{\text{gas}} - v_{\text{liq}} \approx v_{\text{gas}} \approx \frac{RT}{p}.$$

Putting this into our result 14.15 gives

$$\frac{\mathrm{d}p}{\mathrm{d}T} \approx \frac{Lp}{RT^2} \qquad \Rightarrow \qquad \frac{\mathrm{d}p}{p} \approx \frac{L}{R}\frac{\mathrm{d}T}{T^2}. \qquad (14.16)$$

This can be integrated to give (homework)

$$pe^{L/RT} = \text{constant} \qquad \text{(liquid–gas),} \qquad (14.17)$$

providing that L is reasonably constant over the range of integration. If we know one point (T, p) where the phases are in equilibrium, this relationship will tell us the others.

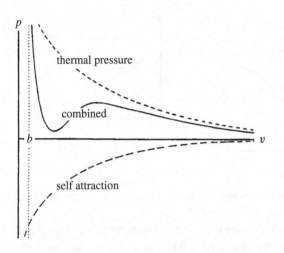

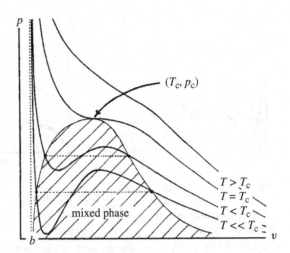

Figure 14.6 p-v plots for the van der Waals model (not to scale). (Left) Illustration of the thermal pressure, self attraction, and combined terms for a given temperature. (Right) Plots of p versus v for various temperatures. When the system is hotter than the critical temperature T_c, the thermal pressure dominates over the self-attraction for all values of the molar volume, and so there is no phase transition. The reason for the broken lines and the term "mixed phase" in the shaded region will soon become apparent.

D.3   Mean field models and liquid–gas transitions

For a better understanding of phase transitions, we often employ models such as those in Chapter 10. A system's properties reflect the interactions among its constituent particles, which are changing continuously owing to their thermal motions. So most models represent a time average and are therefore called "mean field" models.

As an example, we consider what the van der Waals model tells us about the gas–liquid phase transition. We rearrange the van der Waals equation (10.7) to express the pressure as a sum of two terms,

$$p = \frac{RT}{v - b} - \frac{a}{v^2} \tag{14.18}$$

or

$$\text{pressure} = \text{thermal pressure} - \text{self-attraction.}$$

The thermal pressure tends to keep the molecules dispersed. The self-attraction tends to pull them together and is responsible for condensation into the liquid phase.

As you can see, the thermal pressure dominates at extremely large and small volumes,

$$p \approx \frac{RT}{v - b} \quad \text{for } v \to \infty \text{ and } v \to b.$$

The self-attraction term a/v^2 can be influential at intermediate volumes, if T is not very large. The two terms in 14.18 are illustrated on the left-hand side of Figure 14.6. On the right-hand side of this figure are p-v curves for several different temperatures. You can see that above the critical temperature T_c the dip due to the attractive a/v^2 term disappears. The thermal pressure dominates over the self-attraction everywhere, so there is no condensation (i.e., phase transition) above

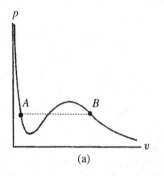

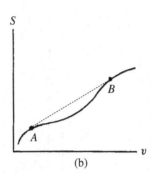

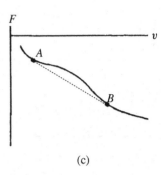

(a)   (b)   (c)

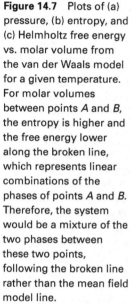

Figure 14.7 Plots of (a) pressure, (b) entropy, and (c) Helmholtz free energy vs. molar volume from the van der Waals model for a given temperature. For molar volumes between points A and B, the entropy is higher and the free energy lower along the broken line, which represents linear combinations of the phases of points A and B. Therefore, the system would be a mixture of the two phases between these two points, following the broken line rather than the mean field model line.

T_c. This disappearance of the dip identifies the critical point (T_c, p_c) indicated on the phase diagram of Figure 14.4.

We now use these p-v curves to study how the Helmholtz free energy F and the entropy S vary with volume at a fixed temperature.[8] From equation 9.15a $(dF = -SdT - pdV)$, you can see that

$$\left(\frac{\partial F}{\partial V}\right)_T = -p. \tag{14.19}$$

That is, on a plot of the Helmholtz free energy vs. volume, the slope at any point is the negative of the pressure. According to the van der Waals model for $T < T_c$, p first decreases, then increases, and next decreases again as the volume expands (Figure 14.7a), so there is a corresponding increase–decrease–increase variation in the slope of the Helmholtz free energy curve (Figure 14.7c), causing a "double hump" (at A and B). Equation 9.14 $(F = E - TS)$ implies that changes in entropy and Helmholtz free energy are (negatively) related, as is reflected in the entropy curve of Figure 14.7b.

The broken lines on the curves of Figure 14.7 represent a linear combination of the phases at points A and B, A being the liquid phase (with smaller volume) and B the vapor phase. Because the points along the broken lines have higher entropy and lower free energy than points on the solid line (which corresponds to the mean field model), the system will follow the broken line. That is, in this region the system will be partly liquid and partly vapor, the vapor content increasing as added heat causes more liquid to evaporate and the molar volume to increase.

To determine the point (T, p) at which a phase transition occurs, we use equation 14.19 to write the change in Helmholtz free energy from point A to point B at any given temperature as

$$\Delta F_{AB} = -\int_A^B pdV = \text{minus the area under the } p\text{–}v \text{ curve.} \tag{14.20}$$

But, as seen in Figure 14.7c, the change in free energy from A to B is the same whether we follow the solid or the broken curve. Therefore, the areas $(\int pdv)$

---

[8] Remember that the second law requires S to be a maximum and F a minimum for fixed temperature.

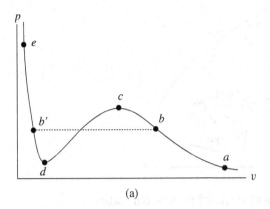

(a)

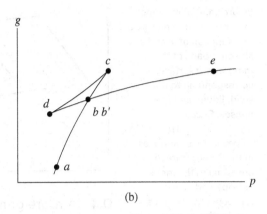

(b)

Figure 14.8 (a) Plot of p vs. v for the van der Waals model for $T < T_c$. (b) Integrating $g = g_0 + \int v\,dp$ from points a to e (i.e., backwards along the curve). The slope is the molar volume v and decreases as we go along. The pressure decreases from c to d.

under the broken and solid curves on the p-v diagram (Figure 14.7a) from A to B must be the same. This requirement determines the values (p, v) for which the phase transition begins and ends.

Instead of studying how the Helmholtz free energy F varies with molar volume, as above, we could just as well study how the molar Gibbs free energy, $g = N_A \mu$, varies with pressure. Equation 14.14, $dg = -s\,dT + v\,dp$, tells us that on a plot of g vs. p for any given temperature the slope is equal to the molar volume v. Furthermore, we can find the value of g at any pressure by integrating:

$$\left(\frac{\partial g}{\partial p}\right)_T = v \qquad \Rightarrow \qquad g(p) = g_a + \int_a^p v\,dp'.$$

For example, suppose we start at point a on the van der Waals p-v plot of Figure 14.8a and work backwards. In each region we can see what the slope of g vs. p should be, and whether g is increasing or decreasing.

- From a to c, g is increasing ($dp = +$); the slope v is large and decreasing.
- From c to d, g is decreasing ($dp = -$); the slope v is smaller and decreasing.
- From d to e, g is increasing ($dp = +$); the slope v is still smaller and decreasing.

Consequently, our plot of g vs. p looks like that in Figure 14.8b.

The second-law constraint that g be minimized means that the system would normally follow the lowest curve in Figure 14.8b, the phase transition from gas to liquid occurring at the point bb'. Notice that there is a discontinuous change in the slope at this point, which means that there is a discontinuous change in the molar volume.

Taking care to eliminate disturbances and condensation or vaporization nuclei, it is sometimes possible to create a "supercooled gas" or a "superheated liquid". These would correspond to the "metastable states" along the line segments bc or $b'd$ in Figure 14.8.

Figure 14.9 Plot of the molar Gibbs free energy g as a function of T and p, showing parts of the minimum-g surfaces corresponding to the solid, liquid, and gas phases. Because $dg = -sdT + vdp$, the slopes of these surfaces in the T and p directions are equal to the molar entropy and volume, respectively. The lines of intersection of the surfaces identify the phase boundaries. The critical point c for the liquid–gas transition is also indicated.

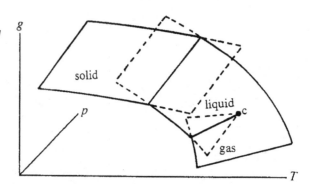

D.4   A more general treatment of phase transitions

With insights from the preceding section, we now broaden our study of the molar Gibbs free energy ($g = N_A\mu$) and phase transitions. As Figure 14.9 illustrates, $g(T, p)$ is a surface in (g, T, p) space. Since both the molar entropy s and the molar volume v are positive and since (equation 14.14)

$$dg = -sdT + vdp,$$

g must slope upward with increasing p and downward with increasing T. Furthermore, since both s and v vary with temperature and pressure, so does the slope. The minimum-g surfaces representing the solid, liquid, and gas phases have distinctively different slopes, owing to their distinctively different molar entropies and molar volumes. The intersections of these surfaces mark the phase boundaries, and the extensions of the surfaces past the phase boundary identify the supercooled or superheated phases in unstable equilibrium. A conventional phase diagram results from the projection of these phase boundaries onto the p-T plane, as in Figure 14.4. Figure 14.8b corresponds to a constant-T slice through the g-surface.

The surface of minimum-g changes slope abruptly at phase boundaries. From equation 14.14 above, you can see that the change of slope in the T direction ($\partial g/\partial T = -s$) is due to the addition or removal of latent heat, which changes the molar entropy s. And the change of slope in the p direction ($\partial g/\partial p = v$) is due to the change in molar volume v. Because of the discontinuity in the first derivatives, we often call these "first order" phase transitions.

For other phase transitions the first derivatives are continuous but the second or higher derivatives are not. Such transitions are often called "continuous" or "higher order". Because the molar heat capacity is proportional to the second derivative of g,

$$\left(\frac{\partial^2 g}{\partial T^2}\right)_p = -\left(\frac{\partial s}{\partial T}\right)_p = -\frac{1}{T}\left(\frac{\partial Q}{\partial T}\right)_p = -\frac{C_p}{T},$$

a discontinuity in the molar heat capacity could mark a higher order transition (Figure 14.10). Notice that since only first order transitions have an abrupt change in slope (and hence in molar entropy s), only first order transitions involve the

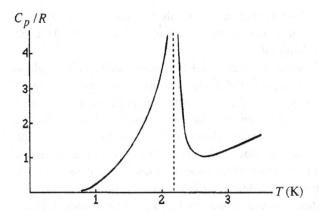

Figure 14.10 Plot of the molar heat capacity c_p for liquid helium. The discontinuity at 2.18 K (and atmospheric pressure) is called the "lambda point" because of the shape of this curve, and it marks a higher order transition.

transfer of latent heat. Also, only first order transitions have intersecting and overlapping sheets, so only first order transitions can display hysteresis.[9]

All phase transitions involve some change in the ordering of particles, so it is customary to generalize the treatment of phase transitions using a generic "order parameter," ξ. For solid–liquid or liquid–gas transitions, it might be related to the molar volumes or the interparticle spacing. And for superconducting transitions it might be related to the fraction of electrons in the superconducting state. It is often defined in such a way that it is zero in the higher-temperature phase and becomes nonzero when the transition occurs.[10] As we lower a system's temperature, the onset of a phase transition at temperature T_0 is identified by a change in the order parameter.

For the particular case of the liquid–gas transitions discussed in the preceding section, the three variables T, p, v can be related in the van der Waals model, which reduces the number of independent variables to two. In a more general study of phase transitions, we initially employ three independent variables and then apply the second law constraint that g be minimized to reduce the number of variables to two. The three initial variables are usually the temperature and pressure (or an equivalent mechanical variable) and an order parameter.

To see how this works, consider the van der Waals gas of subsection D.3 the three variables being T, p, V. According to their definitions (equations 9.14) the Helmholtz and Gibbs free energies differ by a factor pV:

$$G = F + pV$$

So, when plotting the *molar* Gibbs free energy as a function of molar volume for fixed T, p[11], we have to add a factor pv to the plot of Figure 14.7c. This changes

---

[9] "Hysteresis" means that the present state depends on its history. For example, in order to get a supercooled gas you must begin with a gas and cool it. You can't begin with a liquid.

[10] Although ξ is a continuous variable in our mean field models, one particular value is selected when we apply the second law requirement that the Gibbs free energy be minimized for any particular (T, p).

[11] Remember, v, T and p are *independent* variables until we apply the second law constraint that g be minimized.

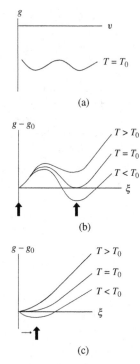

Figure 14.11 Plots of the molar Gibbs free energy g vs. molar volume v or order parameter ξ. The thick arrows indicating the equilibrium points (minimum g). (a) g as a function of v for a van der Waals liquid–gas transition (obtained by adding pv to the Helmholtz free energy curve of Figure 14.7c). (b) and (c) More generally, g as a function of the order parameter for (b) first order and (c) higher order transitions. The order parameter is defined to be zero in the higher temperature phase. As the temperature is lowered past T_0, the second law requirement for the minimization of g means that the order parameter undergoes a change in value that is discontinuous for a first order transition (b) and *continuous* for a higher order transition (c).

the plot from sloping downward to being roughly horizontal, as illustrated in Figure 14.11a. The molar volume is then determined by the point where the Gibbs free energy is minimized.

From our study of the van der Waals model, we might infer that if the phase transition is due to an interplay of attractive and dispersive forces, the mean field model's prediction for the Gibbs free energy might be a double-humped function of the order parameter. In this case the system would undergo a first order transition, with a *discontinuous* change in the order parameter ξ as the temperature falls below some value T_0 (Figure 14.11b). If the model's entropy is *not* a double-humped function of the order parameter (i.e., it has no point of inflection, unlike the curve in Figure 14.7b), then the system might undergo a higher order transition, marked by the onset of a *continuous* change in the order parameter as the temperature falls below some T_0 (Figure 14.11c). An important example is the condensation of particles into the lowest possible quantum state, which begins at a single temperature and continues as more particles fall into this "ground" state at still lower temperatures. More on this in a later chapter.

The curves of Figures 14.11b, c describe what the Gibbs free energy would be if we could change the order parameter under conditions of constant temperature and pressure. We can't really do that, of course, because Nature will always choose the one point on each curve that maximizes the entropy. But if we understand our systems well, we might be able to construct mean field models that describe our systems under such nonequilibrium conditions. One test of these models, then, would be to predict correctly the observed equilibrium values for the order parameters.

Summary of Section D

Variations in temperature and pressure cause changes in the chemical potential. The phase with the lower chemical potential is favored. Phase diagrams identify the range of variables over which each phase is in equilibrium. Two phases are in diffusive equilibrium along phase boundaries.

Phases are in diffusive equilibrium if their chemical potentials are equal. If two phases remain in equilibrium after changes in the temperature and pressure, their two chemical potentials must have changed by the same amount. This observation, along with equation 9.13 ($d\mu = -s\,dT + v\,dp$), relates changes in p and T along phase boundaries to changes in s and v across phase boundaries:

$$\frac{\Delta p}{\Delta T} = \frac{\Delta s}{\Delta v}$$

This equation can also be written as (equation 14.15)

$$\frac{\Delta p}{\Delta T} = \frac{L}{T\Delta v} \qquad \text{(Clausius–Clapeyron equation)},$$

where L is the molar latent heat for the phase transition.

For the specific case of liquid–gas phase transitions for which the ideal gas law is valid, we can ignore the volume of the liquid phase in comparison with that of the gas and integrate the above expression, assuming L is constant, to give (equation 14.17)

$$pe^{L/RT} = \text{constant} \qquad \text{(liquid–gas phase equilibrium)}.$$

This tells us how various values of (T, p) for phase equilibrium are related.

The van der Waals mean field model gives us insight into the liquid–gas phase transition. It can be written in the form (equation 14.18)

$$p = \frac{RT}{v - b} - \frac{a}{v^2}$$

or

$$\text{pressure} = \text{thermal pressure} - \text{self-attraction}.$$

At the extremes of large and small volumes, the thermal pressure dominates, but at intermediate volumes the self-attraction may take over.

In a plot of the Helmholtz free energy versus volume, the slope is given by (equation 14.19)

$$\left(\frac{\partial F}{\partial V} \right)_T = -p,$$

so that changes in the pressure are reflected in changes in the slope of F vs. V. The van der Waals model gives double humps in the plots of various properties as a function of molar volume. A linear combination of the phases at the two "humps" has lower free energy and higher entropy than the mean field model prediction for molar volumes in this range. So the system must be a mixture of the two phases in this region.

A system's molar Gibbs free energy varies with temperature and pressure according to (equation 14.14)

$$dg = -s dT + v dp,$$

so the free energy is a two-dimensional surface that slopes up with increasing p and down with increasing T. First order phase transitions are marked by discontinuous changes in molar entropy (with accompanying latent heat) and in molar volume.

For a more general treatment of phase transitions, we initially consider three independent variables, the temperature and pressure (or equivalent mechanical variable) and an order parameter. We then require that the Gibbs free energy be minimized to determine the order parameter's equilibrium value. A phase transition is identified by a change in this value of the order parameter as the temperature is lowered below some point T_0. A discontinuous change in the order parameter identifies a first order phase transition, and the onset of a continuous change identifies a continuous or higher order transition.

E   Binary mixtures

Now we wish to investigate systems that contain two or more different kinds of particles, such as you would find in solutions, fluid mixtures, minerals, or alloys. In most cases, the mixing of these components is controlled primarily by competition between the following two opposing effects.

1 On the one hand, mixing increases the system's entropy by offering each particle more volume in position space, V_r (Figure 12.6 top). This would lower the chemical potential and is the reason why particles tend to diffuse towards regions of lower concentration.

2 On the other hand, a (usually) stronger attraction between like than between unlike particles favors separation of the components, because then the potential wells are deeper. So thermal energy is released, affording the particles more volume in momentum space, V_p.

It is the competition between these two contributions to the system's entropy that is most crucial in determining whether the mixture will be homogenous or will separate into its components. We refer to them as the "mixing entropy" and the "interaction entropy," respectively.

E.1   Position space and mixing entropy

Consider two miscible fluids A and B, of equal temperatures, pressures, and volumes, which are separated by a partition. When the partition is removed, the volume available to each molecule doubles, which means more accessible states and higher entropy. Each particle's entropy increases by $k \ln 2$ and the entropy of the entire system increases by

$$\Delta S = Nk \ln 2.$$

As can be shown in the homework problems, we can easily generalize this to the case where system B has a fraction f of the total number N of particles and occupies the corresponding fraction of the total volume before the partition is removed. System A has the remaining fraction $1 - f$ of particles and volume. In this case, the increase in total entropy on removal of the partition is

$$\Delta S = fNk \ln \left(\frac{1}{f}\right) + (1 - f)Nk \ln \left(\frac{1}{1-f}\right). \tag{14.21}$$

Both terms are large and positive. We conclude that, when looking at the volume in coordinate space alone, the entropy of the mixed state is much higher than that when the two substances are separated. So the mixed state is favored.

In many materials, components are mixed in one phase but not in another. For example, liquid water separates from dissolved salts when freezing or boiling. Because entropy favors the mixed-liquid state, water's boiling point is raised and its freezing point is depressed when salts are dissolved in it (subsection 14B.3).

A similar thing happens to magmas and alloys, which are homogenous mixtures in the molten state but tend to separate into tiny crystals of distinctly different compositions when they freeze. Owing to the entropy of mixing the homogenous molten state is preferred, so the freezing point is correspondingly depressed. Solder is a familiar example of an alloy with a lowered freezing point.

E.2 Momentum space and interaction entropy

Next, we examine volumes in momentum space. The molecules of many materials experience virtually no shift in potential energy upon mixing and hence no shift in kinetic energy and no change in accessible volume in momentum space. Examples include normal gases, because their molecules are so widely separated that they have virtually no potential energy at all. Also included in this category would be many liquids whose molecules have similar structure, so that the interactions between like and unlike neighboring molecules are similar. Such fluids are said to be completely miscible.

But for many mixtures the attraction between like particles is significantly stronger than the attraction between unlike particles, so the potential wells are shallower when they are mixed. Shallower potential wells means that their potential energy rises upon mixing, resulting in reduced kinetic energy and therefore a smaller accessible volume in momentum space, V_p. The loss in kinetic energy also means that the system cools off. (One practical application is the portable cold packs used for athletic injuries.)

So when two fluids mix, we can divide the change in entropy into two parts,[12]

$$\Delta S = \Delta S_m + \Delta S_i, \tag{14.22}$$

where ΔS_m is the mixing entropy, caused by the increase in accessible volume in position space V_r, and ΔS_i is the interaction entropy, caused by changes in potential and kinetic energies and hence changes in the accessible volume in momentum space V_p. In general, upon mixing,

$$
\begin{array}{lll}
V_r \text{ increases} & \Rightarrow & \Delta S_m \text{ is positive,} \\
V_p \text{ decreases} & \Rightarrow & \Delta S_i \text{ is negative.}
\end{array}
\tag{14.23}
$$

If the gain ΔS_m from the increase in position space is larger than the loss ΔS_i from the decrease in momentum space then the second law favors mixing. If you understand this statement, you understand the behavior of binary systems.

Many people prefer to work with the Gibbs free energy rather than the entropy, because $G(G = N\mu)$ is proportional to μ, which governs diffusive interactions.

---

[12] These are the two main contributions to the change in entropy. Some (usually) smaller contributions include the effect of mixing on the phonon modes and on the states accessible to conduction electrons.

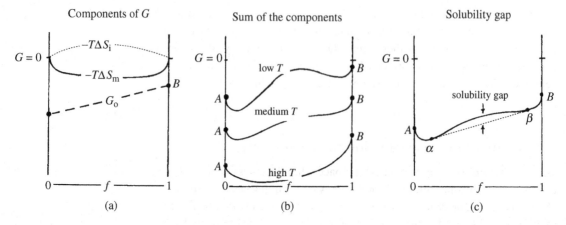

Figure 14.12 Plots of the Gibbs free energy G of a mixture of two substances, A and B, as a function of the mixing fraction f. (a) The contributions from the unmixed components, G_0 (broken line), from the entropy of mixing, $-T\Delta S_m$ (solid), and from and the interaction entropy, $-T\Delta S_i$ (dotted line). (b) As the temperature increases, G decreases and $-T\Delta S_m$ becomes more dominant. (c) The contribution $-T\Delta S_i$ from the interaction entropy may dominate for intermediate mixing ratios, resulting in a solubility gap. A linear combination (small-broken line) of the separate phases, for example α (mostly A) or β (mostly B), has lower free energy than a homogenous mixture of A and B (solid line).

Furthermore, when two phases are in equilibrium their chemical potentials are equal whereas their entropies differ owing to the latent heat that must be added or removed for transition between the two phases. According to equation 9.14c,

$$G = E - TS + pV.$$

If we assume that the total internal energy of the combined system is constant, that it is held at constant pressure and temperature[13] and that the change in volume between mixed and unmixed states is negligible, then

$$\Delta G = -T\Delta S = -T(\Delta S_m + \Delta S_i). \tag{14.24}$$

From this it is obvious that the second law requirement of maximizing the entropy S corresponds to minimizing the Gibbs free energy G under conditions of constant temperature and pressure. (This was previously proven in subsection 9F.4.)

E.3 Gibbs free energy and the solubility gap

As is illustrated in Figure 14.12a, we can write the total Gibbs free energy $(G = N\mu)$ of a mixture in two parts: that of the unmixed components, G_0, and that due to the changes brought about by the mixing process, ΔG:[14]

$$G = \underbrace{N_A\mu_{A0} + N_B\mu_{B0}}_{G_0} \quad \underbrace{-T\Delta S_i - T\Delta S_m}_{\Delta G}, \tag{14.25}$$

We are particularly interested in ΔG. Since the second law demands that G be minimized (i.e., S be maximized), the two fluids mix if the increase in $T\Delta S_m$ dominates over the decrease in $T\Delta S_i$ (i.e., if there is more gain in V_r than loss in

---

[13] We have previously stated that there are as many independent variables as there are interactions. So if ΔE, ΔT, Δp are all zero, how can anything change? The answer is that we have two interacting systems – two kinds of particles, each with their own energy and chemical potential. Furthermore, they are not in equilibrium until the mixing has stabilized.

[14] Some people prefer to call the "interaction entropy" contribution the "energy of mixing," defined as $\Delta E_m = -T\Delta S_i$. This can be thought of as an "energy barrier" that opposes mixing.

V_p). If not, they will not mix. Thus

$$\Delta S_m + \Delta S_i > 0 \qquad \text{for mixing to occur.} \qquad (14.26)$$

For insights into which of the two terms in 14.26 dominates, we make the following observations.

1   At higher temperatures, materials become more miscible

The greater thermal motion at higher temperatures favors particle mixing, and it also weakens the attractive interactions among neighboring particles by freeing particles from entrapment and breaking up favored molecular orientations or aggregates. So many mixtures tend to be homogenous at higher temperatures (where ΔS_m dominates) but separate into components at lower temperatures (where they become entrapped in potential wells and ΔS_i dominates). For example, many alloys and magmas are homogenous mixtures in the hotter molten state but separate out into tiny crystals of different compositions as they cool.

2   At low mixing fractions $T\Delta S_m$ always dominates

The interaction entropy ($\Delta S_i = \Delta E_i / T$) is linear in f for small mixing fractions. That is, if the thermal energy lost when one particle transfers to a shallower potential well is ε_0 then that for two particles is $2\varepsilon_0$, etc. In contrast, the entropy of mixing ΔS_m is *logarithmic* in the mixing fraction, so its slope is infinite in the limit of small mixing fractions, as can be shown by taking the derivative of equation 14.21:

$$\frac{d}{df} \Delta S_m \rightarrow \begin{cases} +\infty & \text{as } f \rightarrow 0, \\ -\infty & \text{as } f \rightarrow 1. \end{cases} \qquad (14.25)$$

Hence, for substances where the fraction f is nearly 0 (pure material A) or 1 (pure material B), the entropy of mixing always dominates.

In fact, since the derivative becomes nearly infinite, the addition of just a few impurities increases the entropy immensely. One practical consequence is that it is almost impossible to find pure substances in Nature and it is very difficult to remove the last few impurities from a nearly pure substance. The huge increase in entropy greatly favors the presence of at least a few impurities. (So oil and water *do* mix, but only a little.)

We have encountered this before. We found that particles diffuse into regions of higher potential energy if the decrease in concentration is sufficiently large. Here we have produced the same result using a different tool (but a tool still based on the second law).

3   The energy of mixing may lead to a solubility gap

Although $T\Delta S_m$ (where ΔS_m is the entropy of mixing) is positive and *always* dominates for small mixing fractions, $T\Delta S_i$ (where ΔS_i is the interaction entropy) may dominate at intermediate mixing fractions. This is illustrated in Figures 14.12. Because the entropy of interaction is usually negative, it causes an upward bulge in the Gibbs free energy (equation 14.24), especially at lower temperatures, as illustrated in Figure 14.12b. In Figure 14.12c, the dotted line indicates that a linear combination of phases α and β has lower Gibbs free energy than a homogenous mixture (solid line). The gap between these two lines is called the "solubility gap."

The physical reason for the solubility gap is that, as mentioned earlier, the stronger attraction between like particles may cause them to separate out in two distinct phases, α which is mostly A and β which is mostly B. (Neither phase can be pure A or pure B, because the entropy of mixing always dominates for small mixing fractions.) The stronger attraction for their own kind (i.e., the deeper potential well) releases enough kinetic energy for the increased volume in momentum space V_p to overcomes the loss in volume in coordinate space V_r caused by their separation. That is, the total volume in six-dimensional phase space (hence the entropy) increases.

4 Overall, G becomes more negative at higher temperatures

According to equation 9.14c', $G = N\mu$, and according to equation 9.15c, $dG = -S\,dT + V\,dp$. So at any given pressure, G is negative (at least for systems with attractive inter particle forces) and it decreases with increased temperature. This behavior is seen in Figure 14.12.

E.4 Phase transitions in miscible fluids

We now examine the phase transitions of fluid mixtures. We will rely heavily on graphical representations of the Gibbs free energy as a function of the mixing fraction. You should understand the features illustrated in Figure 14.12 as well as the following observations.

1 Attractive forces between particles, which are responsible for ΔS_i, are generally very strong in solids and minuscule in gases. In addition, we have seen that the influence of ΔS_m increases with temperature. Consequently, the solubility gap (where $T\Delta S_i$ dominates $T\Delta S_m$ see Figure 14.12c) may be very pronounced in the solid phase but disappear in liquids and gases (see Figure 14.12b).

2 According to equation 9.15 ($dG = -S\,dT$ for constant p, N), G decreases as temperature increases, and the rate of decrease depends on the entropy. Because $S_{gas} > S_{liq} > S_{sol}$ (entropies increase when we add latent heat), G_{gas} decreases faster than G_{liq}, and G_{liq} decreases faster than G_{sol}. So, as you heat a liquid up, for example, G_{gas} decreases fastest and will eventually become lower than G_{liq}. So the liquid vaporizes.

We will now study a phase transition in a mixture of two fluids A and B. Suppose that the boiling points for the two pure fluids are T_A, T_B, respectively, with $T_A < T_B$. As in the example shown in Figure 14.13 (1), $G_{liq} < G_{gas}$ at low temperatures, so the mixture is a liquid for all mixing fractions. But as the

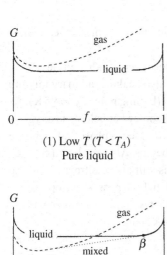

(1) Low T $(T < T_A)$
Pure liquid

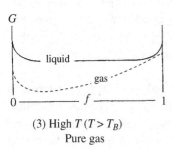

(2) Intermediate T $(T_A < T < T_B)$
Mixture of liquid and gas possible

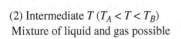

(3) High T $(T > T_B)$
Pure gas

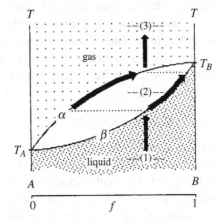

Figure 14.13 (1)–(3) As temperature *increases* the Gibbs free energy decreases, but that of the gas phase decreases faster. For temperatures between T_A and T_B, the mixture could be a gas, a liquid, or both, depending on the mixing fraction *f*. That is, a linear combination of phases α (a gas rich in *A* molecules) and β (a liquid rich in *B* molecules) could have a lower Gibbs free energy than either the gas or the liquid, so the system will be a combination of these two phases until all is vaporized. (Right) The heavy arrows show the path followed as temperature is increased for a certain mixing ratio *f*. Horizontal slices through the numbers 1, 2, 3 correspond to the diagrams at the three temperatures illustrated on the left, and the labels α, β correspond to the phases in diagram (2).

temperature is increased (Figure 14.13b) G_{gas} decreases faster than G_{liq}, and so the two curves cross for temperatures in the range between T_A and T_B. At these intermediate temperatures, the mixture is *homogenous* for small mixing ratios (where $T \Delta S_m$ dominates), being a gas for *f* near 0 and a liquid for *f* near 1. For intermediate values of *f*, however, a linear combination of the two phases α (a gas that is mostly *A*) and β (a liquid that is mostly *B*) has lower free energy, so the mixture will be a combination of these two phases α and β.

Now consider what happens if we begin in the liquid mixture phase and heat it up, as is illustrated by the lower vertical arrow in Figure 14.13 (right). When the temperature reaches a certain point, the onset of vaporization into the gaseous α phase begins and the mixing ratio *f* bifurcates, as indicated by the curved arrows. Because *A* tends to vaporize first, the vapor phase, α, is mostly *A* (smaller *f*), leaving more *B* (larger *f*) in the liquid β phase. This continues until all is vaporized and so the mixing ratio is back to its original value; now the upper vertical arrow is followed. Notice that although the phase transition for a pure substance happens at just one temperature (T_A or T_B), that for the mixture takes place over a range of intermediate temperatures.

The cooling and liquefaction of gases follows the reverse path of the arrows in Figure 14.13(right). Let's look at air. Air is, roughly speaking, a binary mixture of 79% N_2 (boiling point $T_A = 79$ K) and 21% O_2 (boiling point $T_B = 95$ K). As air is cooled, liquefaction begins at a temperature somewhat below the oxygen's boiling point (T_B). Again, entropy favors the mixed gaseous phase and therefore depresses the temperature at which condensation begins. Also, the liquid condensate is *not* pure oxygen, even while the mixture is still above nitrogen's boiling point. Nor is the remaining gaseous nitrogen pure. In both phases, entropy favors mixing. So to purify oxygen or nitrogen requires that we repeat the process many times, depending on the level of purity we wish to obtain.

E.5   Minerals and alloys

We now examine the liquid–solid transition as we cool a molten mixture and it freezes. In the solid phase the particle interactions are stronger and the temperature is lower. Both factors tend to increase $T\Delta S_i$ relative to $T\Delta S_m$ and thereby favor separation. For simplicity, we again deal with just two components, which could be different elements or materials with different chemical compositions. Many of these processes are similar to the liquid–gas transition described in the preceding section, but here we face a larger solubility gap in the solid phase.

If we begin in the homogenous molten phase and lower the temperature then both G_{liq} and G_{sol} rise, but G_{liq} rises faster.[15] When G_{liq} becomes higher than G_{sol}, the material begins to solidify, the solid phase being the state of lowest G. This process can produce many different results, depending on the mixing fraction f (Figure 14.14), as we now explore.

As the melt is solidifying

- The entropy of mixing favors the more homogenous molten phase. So freezing begins at a temperature below that of either pure substance.
- For small mixing fractions f (i.e., the mixture is mostly A), the cooling system will go through a stage where it is partly solid (phase α_{sol}, mostly A and very depleted in B) and partly liquid (phase β_{liq}, still mostly A, but enriched in B). But the composition of both these phases changes over time, relatively more B freezing out later.
- The corresponding process happens for large mixing fractions f (i.e., the mixture is mostly B). The system as it cools will go through a stage where it is partly solid (phase β_{sol}, mostly B and very depleted in A) and partly liquid (phase α_{liq}, still mostly B, but

---

[15] Again, $dG = -SdT$, with $S_{\text{liq}} > S_{\text{sol}}$. As we lower the temperature dT is negative so dG is positive.

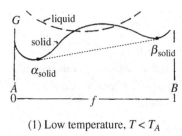

(1) Low temperature, $T < T_A$

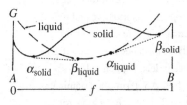

(2) Intermediate temperature,
$T_A < T < T_B$

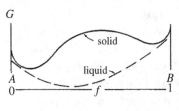

(3) High temperature, $T > T_B$

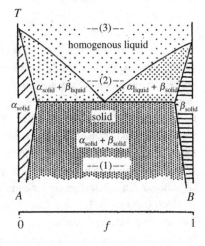

Figure 14.14 (1)–(3) Plots of Gibbs free energy vs. mixing fraction f at three temperatures for the solid and liquid phases of a binary mixture. The stable phase or mixture of phases is the one with the lowest G. Note that G decreases with increasing temperature, G_{liq} decreasing faster than G_{sol}. (Right) Phase diagram for the binary mixture, indicating the stable phase as a function of temperature and mixing fraction. A horizontal slice through each of the positions 1, 2, 3 gives the stable phase (minimum G) of the corresponding diagrams on the left. Following any vertical line downwards shows what happens as the melt freezes for that particular value of the mixing fraction.

enriched in *A*). Again, the compositions change over time, relatively more *A* freezing out later.

The final outcome, after the material is completely solidified:

- For very small mixing fraction f ($f \approx 0$), the final solid will be phase α_{sol}, i.e., mostly *A* with a little *B* impurity mixed in.
- For very large mixing fraction ($f \approx 1$), the final solid will be phase β_{sol}, i.e., mostly *B* with a little *A* impurity mixed in.
- For intermediate mixing fractions, the final solid will be a heterogeneous mixture of phases α_{sol} and β_{sol} (hence tiny crystals in igneous rocks).
- Even after solidification, the microscopic crystalline structure may slowly change with decreasing temperature, as the influence of the entropy of mixing term $T\Delta S_m$ continues to decrease in importance and heterogeneous mixtures $\alpha_{sol} + \beta_{sol}$ become favored over more homogenous mixtures, e.g., α_{sol} with *B* impurity.

Summary of Section E

The mixing of two substances A and B causes changes in the volume accessible to the particles in both position and momentum space, and these changes in entropy are reflected in the system's properties. The changes in the Gibbs free energy can be written in the form (equation 14.24)

$$\Delta G = -T\Delta S_m - T\Delta S_i,$$

where the entropy of mixing ΔS_m is positive and reflects the additional accessible volume in position space and the interaction entropy ΔS_i is negative and reflects changes in momentum space due to particle interactions. The second law requires that the Gibbs free energy be minimized, so the two substances mix only if ΔG is negative, i.e., if $T\Delta S_m$ dominates $T\Delta S_i$. This does indeed happen at high temperatures and at small mixing fractions. But the $T\Delta S_i$ term often dominates at lower temperatures, especially in solids where particle interactions are stronger. Where this happens, there is a solubility gap. That is, the two materials tend to separate out into a heterogeneous mixture of two phases: α, which is mostly A, and β, which is mostly B. The separated phases will never be completely pure, because the entropy of mixing dominates at low mixing fractions.

Owing to the entropy of mixing, the boiling and freezing points of mixtures change to favor a more homogeneously mixed phase. For example, in the cooling and liquefaction of mixed gases, the gas with the higher boiling point begins to condense first, but at a temperature below its normal boiling point. And the condensate is not pure, again because the entropy of mixing favors mixing in the liquid as well. As condensation occurs, the liquid phase contains mostly the substance with the higher boiling point, and the gas contains mostly the substance with the lower boiling point. But the composition of both phases changes as more and more of the mixture condenses.

A similar thing happens in a liquid–solid transition: the liquid state is more homogeneously mixed, and the solid phase may have a sizable solubility gap. The entropy of mixing favors the homogenous liquid phase, so the freezing point is depressed by an amount that depends on the mixing fraction. For very small mixing fractions, the frozen solid will be either α phase (mostly A with a little B impurity) or β phase (mostly B with a little A impurity). For intermediate mixing fractions, the solubility gap will ensure that the final solid is a heterogeneous mixture of the α and β phases.

Problems

Section A

1. For water vapor, $u_0 = 0$. Each molecule has six degrees of freedom in both the vapor and liquid phases. The latent heat of vaporization for water is

40 700 J/mole and as it vaporizes, it expands. The average work done per molecule in this expansion is $p\bar{v} = kT$. The specific heat of liquid water is 75.3 J/(mole K). The molar mass of water is 18 g/mole. Use this information to calculate the average energy of a water molecule (in eV) for (a) water vapor at 100 °C, (b) liquid water at 100 °C, (c) liquid water at 0 °C, (d) What is the potential energy reference level u_0 (in eV) for liquid water at 100 °C? (e) What is u_0 for liquid water at 0 °C? (f) In raising the temperature of liquid water, how much of the heat goes into thermal energy, and how much into raising u_0?

2. You are invited to answer the question in the caption to Figure 14.1.

3. At constant pressure, chemical potential μ decreases with increasing temperature (equation 14.1). Near 0 °C, which chemical potential falls faster with increasing temperature, μ_{water} or μ_{ice}? How do you know?

4. Rearrange the terms of the first law so that $-\mu dN$ is on one side of the equation and all the other terms on the other. Now show that $-\mu/T$ is a measure of the increase in entropy when a single particle enters a system with no gain in either energy or volume.

5. In a problem at the end of Chapter 13 you showed that the entropy of a system increases when a single particle is added, even when there is no increase in either thermal energy or volume. Now demonstrate this with a few examples of very small systems with thermal energy that comes in single units. The different states are identified by the different ways in which the units of energy are distributed among the particles. In each case, calculate the number of such states Ω_1 for the given system, the number of states Ω_2 if a single energyless particle is added to it, and the factor Ω_2/Ω_1 by which the number of states has increased. The system are as follows:
   (a) two particles and one unit of energy;
   (b) two particles and two units of energy;
   (c) three particles and one unit of energy;
   (d) three particles and two units of energy;
   (e) three particles and three units of energy;
   (f) Does the number of states increase with one added energyless particle in each case?
   (g) Consider the factor by which the number of states increases when one energyless particle is added. Does this factor increase when the energy per particle in the original system increases?

6. We are going to improve on the above problem by showing that the entropy of a system increases as the energy and the number of particles increase. Suppose that we distribute n units of energy among the N particles of a system. There

are $\frac{(N+n-1)!}{n!(N-1)!}$ different ways in which this can be done. According to Stirling's approximation, $m! \approx (m/e)^m$. Show that:

(a) $\frac{(N+n-1)!}{n!(N-1)!} \approx \frac{(N+n)^{N+n}}{n^n N^N}$ for large N;

(b) the number of states is an increasing function of the number of particles, N;

(c) the number of states is an increasing function of the total energy, n.

7. For an ideal gas, $pV = NkT$, $E = (N\nu/2)kT$, and the number of accessible states is given by $\Omega = \omega_c^N$, where $\omega_c = \text{constant} \times (V/N)(E/N\nu)^{\nu/2}$. Choosing E, V, N to be the independent variables, and writing

$$\Delta \ln \Omega = \frac{\partial \ln \Omega}{\partial E} \Delta E + \frac{\partial \ln \Omega}{\partial V} \Delta V + \frac{\partial \ln \Omega}{\partial N} \Delta N,$$

work out these partial derivatives to show that $T\Delta S = \Delta E + p\Delta V - \mu\Delta N$. What is the expression for μ that you get?

8. Show that (a) equation 14.4 follows from 14.2 and 14.3; (b) equation 14.5″ follows from 14.2 and 14.5.

9. Estimate ω_c for water molecules in liquid water at $17\,°C$, given that $m = 3.0 \times 10^{-26}$ kg. Remember that the number of states is $V_r V_p/h^3$. For an estimate of the accessible volume in momentum space, assume that the maximum magnitude of the momentum for any molecule is roughly equal to the root mean square value; most accessible states reside within a sphere of that radius in momentum space. Assume that any one molecule can go anywhere within the volume V occupied by the liquid, and don't forget to correct for the number of identical particles within that volume.

10. Consider the ideal monatomic gas argon, whose atomic mass is 6.68×10^{-26} kg at standard temperature and pressure (0 °C, 1 atm).

(a) Estimate ω_c for the particles of this gas (see the previous problem).

(b) What is a typical value for the average entropy per particle in this gas?

(c) From your answer to the above, estimate the chemical potential (equation 14.2).

11. Suppose that we transfer 2000 particles from system A at 270 K to system B at 800 K, both of which are at atmospheric pressure (1.013×10^5 Pa). The volume required by each particle in either system is 10^{-29} m^3. In system A each particle has $\nu_A = 6$, $u_{0,A} = -0.10\,\text{eV}$, and $\mu_A = -0.20\,\text{eV}$. In system B each particle has $\nu_B = 3$, $u_{0,B} = -0.14\,\text{eV}$, and $\mu_B = -0.36\,\text{eV}$. Calculate the following quantities in eV:

(a) $\Delta E, \Delta Q, \Delta W, \mu\Delta N$ for system A,

(b) $\Delta E, \Delta Q, \Delta W, \mu\Delta N$ for system B (hint: energy and particles are conserved, so your two ΔE's and ΔN's must be equal and opposite, also, the transferred particles are in equilibrium in system A but not upon entry into system B),

(c) the amount of thermal energy released into system B by these entering particles,

(d) the number of states per particle, ω_c, in each system.

12. Consider a system initially at $17\ ^\circ$C whose particles have six degrees of freedom apiece and for which $u_0 = -0.4\ $eV. Suppose that the particles find a second configuration, in which the potential energy reference level u_0 shifts downward to $-0.5\ $eV. By what factor do you expect the number of states per particle, ω_c, to increase? Would the second law favor the first configuration or the second?

13. The depth of the potential well for a water molecule in liquid water at $17\ ^\circ$C is about $-0.43\ $eV and in the gaseous phase is $0\ $eV. If the density of states ρ_ω in the vapor phase is about 80 times greater than that in the liquid phase, roughly what would be the absolute humidity (in grams of water vapor per cubic meter of air) for saturated air at $17\ ^\circ$C?

14. Assuming that the temperature of air is constant at $0\ ^\circ$C at all altitudes, the density of air should decrease by a factor $1/2$ for every increase of z kilometers in altitude. Find z. (The mass of an air molecule is about $4.9 \times 10^{-26}\ $kg.)

15. The density and atomic mass numbers for various substances are listed below in the following table.

Substance	Density ($10^3\ $kg/m^3)	Molecular weight
aluminum	2.7	27
mercury	13.6	200
ethyl alcohol	0.79	46
water	1.00	18

For each of these, calculate. (a) the volume per particle, V/N, (b) the value of $p\bar{v}$ in electron volts for atmospheric pressure, (c) the pressure in atmospheres at which $p\bar{v}$ would be equal to kT at 295 K.

16. Two solutions are in diffusive equilibrium at $37\ ^\circ$C and atmospheric pressure. A certain salt's concentration is 1000 times greater in solution A than in solution B. The density of single-particle states, ρ_ω, is the same for both solutions, as is the number of degrees of freedom per particle.
    (a) In which solution is the potential well u_0 deeper?
    (b) How much deeper?

17. Water molecules have six degrees of freedom apiece in all phases. At $0\ ^\circ$C, $u_{0,\ \text{water}} = -0.443\ $eV, the latent heat of fusion for ice is 6006 J/mole, and ice is 0.917 times as dense as liquid water. Using this information, calculate
    (a) the potential energy reference level for the water molecule in ice, $u_{0,\ \text{ice}}$,
    (b) the ratio $\rho_{\omega,\text{water}}/\rho_{\omega,\text{ice}}$ for the water molecules in the two phases.

18. Consider the equilibrium between water vapor and ice at $-10\,°C$. The water molecules have six degrees of freedom in both the solid and gaseous phases, and the density of ice is 917 kg/m^3. The depth of the potential well for the water molecule in ice is -0.505 eV. The ratio $\rho_{\omega,\text{vapor}}/\rho_{\omega\text{ice}}$ is about $40\,000$. With this information, calculate the density of water vapor in air when in equilibrium with ice at $-10\,°C$. (Hint: Use equation 14.5, because you cannot ignore the $p\bar{v}$ term for the vapor.)

19. The molecules of a certain salt have six degrees of freedom in the salt crystals and three degrees of freedom when dissolved in water. The density of the crystalline salt is 5 g/cm^3, and that of salt ions in the saturated salt solution at $17\,°C$ is 0.03 g/cm^3. The density of single-particle quantum states is 1000 times greater in the dissolved state than in the crystals. What is the difference between the depths of the potential wells that the salt molecule experiences in the crystalline form and in solution, $u_{0,\text{solution}} - u_{0,\text{crystal}}$?

Section B

20. Using equation 14.6, estimate how much lower sea water's chemical potential is than that of fresh water at $17\,°C$.

21. At $10\,°C$, when water vapor is in diffusive equilibrium with liquid fresh water, the air above the liquid water is about 2% water vapor by number of molecules. What is the water's vapor pressure? By how much would it decrease if the water vapor were in diffusive equilibrium with sea water rather than fresh water?

22. Use equation 14.9 to estimate by how much sea water's freezing point is depressed, and its boiling point raised, relative to fresh water. (The latent heats for water are $40\,700$ J/mole for vaporization and -6000 J/mole for freezing.)

23. Imagine an organism that lives in a hydrothermal vent where the temperature is $80\,°C$ and the salinity is 4% by number of ions. If the organism's body fluids have a salinity of 1%, what is the osmotic pressure across membranes separating this organism's body fluids from its environment?

24. Use the fact that a mole of water has a mass of 18 grams, and therefore occupies a volume 18 cm^3, to calculate the volume per water molecule, $\bar{v}$. Use this to calculate the osmotic pressure corresponding to a decrease in 0.001 eV in chemical potential experienced by water as it goes from pure water to a salt solution. How far beneath the surface of the ocean would you have to be for the pressure to be this large? (Pressure increases with depth at a rate of 1 atm per 10 meters of depth.)

25. For water at room temperature, the entropy and volume per molecule are very roughly 3×10^{-22} J/K and 3×10^{-29} m^3, respectively.

(a) Using this and equation 14.1, estimate roughly the increase in pressure that would be needed to counteract a difference of 0.01 °C in temperature across a water-permeable membrane.

(b) Without this pressure, would water diffuse from the hot toward the cold or vice versa?

(c) If the temperature gradient were such that the temperature dropped by 0.01 °C across a distance of one or two molecular widths, by how many degrees would it change over a distance of 1 cm?

Section C

26. In the minimization of the Gibbs free energy, $G = \sum_i \mu_i N_i$, we wrote $\Delta G = \sum_i \mu_i \Delta N_i = 0$. What has happened to the $\Delta \mu_i N_i$ term?

27. Suppose that the chemical potential of particle type is A is -1.3 eV, that of particle type is -0.4 eV, and that of particle type C is -1.5 eV. Particles of A and B could possibly combine to form particles of C. Would they? Why, or why not?

28. Write out equation 14.11 explicitly for the reaction of H_2SO_4 and $NaOH$ to form H_2O and Na_2SO_4, with the correct stoichiometric coefficients.

29. Particles of types A, B, and C could interact to form particles of types D and E according to the chemical equation $3A + B + 4C \leftrightarrow 2D + 3E$. If the respective chemical potentials are $\mu_A = -0.4$ eV, $\mu_B = -0.1$ eV, $\mu_C = +0.2$ eV, $\mu_D = -0.1$ eV, $\mu_E = -0.2$ eV, which way would this reaction tend to go (to the right or to the left)? Why?

30. A chemical reaction between reactant types A, B, and C is proceeding according to $3A + B \rightarrow 2C$.

(a) From this fact, can you write down a relationship between the chemical potentials μ_A, μ_B, and μ_C?

(b) Chemical potential increases as concentration increases. Explain how this fact could stop the above chemical reaction from continuing even before the particles of types A or B are all used up.

31. The entropy of water at 25 °C and one atmosphere of pressure is 188.8 joules/(mole K). Given that the molecular weight of water is 18 and that its specific heat is 4.186 J/(g K), find (a) the entropy of water at 27 °C (hint: use $\Delta Q = T \Delta S$), (b) the entropy per molecule for water at 25 °C and at 27 °C. (c) When you change the temperature of water by 2 °C from 25 to 27 °C at atmospheric pressure, by how much does the chemical potential change?

32. At 25 °C and one atmosphere of pressure, a mole of water has entropy 188.8 J/(mole K) and volume 18 cm^3. If you raise the temperature by 1 °C, by how much would you have to increase the pressure in order to keep the chemical potential of a water molecule unchanged?

33. (a) The equilibrium constant for the dissociation of water, $H_2O \rightarrow H^+ + OH^-$, is $10^{-15.745}$ mole/l at 24 °C. What is the concentration of the H^+ ions?

  (b) The negative exponent of the H^+ ion concentration (in moles per liter) is called the pH value. For example, if $\rho(H^+) = 10^{-2}$ moles per liter, the pH value is +2. What is the pH value of pure water at 24 °C?

  (c) The equilibrium constant for the dissociation of water at 60 °C is $10^{-14.762}$ moles per liter. What is the pH value of pure water at this temperature?

34. A chemical reaction $3A + B + 2C \rightarrow D + 2E$ is conducted at standard temperature and pressure for which the equilibrium constant is 10^3 (mole/liter)$^{-3}$.

  (a) If in equilibrium the concentrations in moles per liter are $\rho_A = 0.2$, $\rho_B = 0.5$, $\rho_C = 0.1$, and $\rho_D = 0.2$, what is the concentration of the reactant E?

  (b) Suppose that you now remove some reactant E. The reaction then proceeds to a new equilibrium, the new concentration of A being 0.17 moles per liter. What are the new concentrations of B, C, and D? By how much has the concentration of E decreased?

35. In Chapter 6 (Table 6.2) we showed that the corrected number of states per particle for a monatomic ideal gas is $w_c = e^{5/2}(2\pi mkT)^{3/2}/h^3\rho$.

  (a) Show that for a monatomic ideal gas, $e^{-\mu/kT} = [(2\pi mkT)^{3/2}/h^3\rho]e^{-u_0/kT}$.

  (b) What would be the function $\zeta(T)$ in equation 14.12?

  (c) For the ionization of hydrogen atoms in the photosphere of the Sun $(H \leftrightarrow H^+ + e^-)$, show that the equilibrium constant is given by

$$\frac{[\rho_{H^+}][\rho_{e^-}]}{[\rho_H]} = \frac{(2\pi m_e kT)^{3/2}}{h^3}e^{-I/kT},$$

  where $[\rho_i]$ denotes the equilibrium concentration of species i and I is the ionization potential (13.6 eV).

  (d) Show that the concentration of H^+ ions is proportional to $e^{-I/2kT}$.

Section D

36. Water's latent heat of vaporization is 40 700 J/mole. Roughly how much energy (in eV per molecule) separates the band of the liquid-phase states from the free continuum for water molecules? (Don't forget that the latent heat includes the energy used per molecule for expansion into the gaseous phase, $p\bar{v} \approx kT$.)

37. Make a qualitative sketch of the phase diagram of a substance for which the solid and liquid phases are equally dense.

38. Water and methane molecules both have about the same mass, yet water has a much higher boiling point at any given pressure. Why do you suppose this is?

39. The mutual attraction between water molecules is much greater than that between ammonia molecules. For which material do you expect the critical temperature to be higher, and why?

40. The molar entropy of ice at 0 °C is 160.2 J/(mole K). Using $L_{vaporization} = 40680$ J/mole, $L_{melting} = 6006$J/mole, find the molar entropy of water (a) as a liquid at 0 °C, (b) as a liquid at 100 °C, (c) as a vapor at 100 °C, (d) in ice melt at 0 °C that is 40% ice and 60% liquid.

41. Given that water boils at 100 °C at atmospheric pressure and that its latent heat is 40 680 J/mole, under roughly what pressure will water boil at: 140 °C, (b) 200 °C, (c) 300 °C, (d) 40 °C, (e) 0 °C? (f) The exact answer to part (e) is 0.00603 atm. What is the % error in your answer to part (e), and what approximation(s) went into the derivation of equation 14.17 that might account for this error?

42. What approximation went into the derivation of equation 14.17 that would make it especially inaccurate near the critical point?

43. Under atmospheric pressure iron melts at 1530 °C, and the latent heat of fusion is 1.49×10^4 J/mole. At the melting point, the density of the solid phase is 7.80 g/cm^3 and that of the liquid phase is 7.06 g/cm^3. The atomic mass for iron is 55.8.
   (a) What are the molar volumes (in m^3) for iron in the solid and liquid phases at the melting point?
   (b) What is the change in molar volume of the iron when the solid is melted?
   (c) To force iron to melt at a higher temperature, would the pressure on it have to increase or decrease?
   (d) At what pressure would the melting point be 1600 °C?

44. Under atmospheric pressure, water freezes at 0 °C. The latent heat of fusion is 6.01×10^3 J/mole. At the melting point, the density of ice is 0.917 g/cm^3, and that of liquid water is 0.9999 g/cm^3. The molecular weight of water is 18.
   (a) What are the molar volumes (in m^3) for water in the solid and liquid phases at the melting point?
   (b) What is the change in molar volume of the water when the solid is melted?
   (c) To force water to melt at a lower temperature, would the pressure on it have to increase or decrease?
   (d) At what pressure would the melting point be −30 °C?

45. If a 700 N man stands on ice skates for which the area of contact with the ice is 10 mm², by how many degrees is the melting point of the ice lowered? (The latent heat of fusion is 6.01×10^3 J/mole. At the melting point, the density of ice is 0.917 g/cm³, and that of liquid water is 0.9999g/cm³. The molecular weight of water is 18.)

46. Suppose that the surface of some ice is at -1 °C, and you want to know how much pressure you must apply to melt it. The latent heat of fusion is 6.01×10^3 J/mole. At the melting point, the density of ice is 0.917 g/cm³ and that of liquid water is 0.9999 g/cm³. The molecular weight of water is 18.
    (a) How much pressure will you have to apply to melt it?
    (b) If you weigh 700 N, the area of contact between your shoes and the ice must be less than what?
    (c) With soft-soled shoes on very flat ice, will your pressure be sufficient to melt the ice?
    (d) If you are wearing hard-soled shoes, and the ice is bumpy, so that the area of contact between your shoes and the ice is only 20 mm² (the tops of the bumps), will you melt the ice?
    (e) On flat ice, you slide much farther with ice skates than with shoes. Why?

47. For a certain liquid–gas phase transition, the molar latent heat changes with temperature according to $L = A + BT$, where A and B are constants. What equation of the form $f(p, T) = $ constant describes the line separating the two phases on a phase diagram? Assume that the gas is an ideal gas and that the molar volume of the liquid phase is negligible in comparison with that of the gas.

48. In Figure 14.6, you can see that for the $T = T_c$ curve, and only for that curve, there is a point of inflection where both the first and second derivatives are zero.
    (a) Use this information to find T_c, p_c in terms of the van der Waals constants a, b.
    (b) With this result and the data of Table 10.1, estimate the values T_c, p_c for water and methane. (The true values are 647 K, 221 atm and 191 K, 46 atm, respectively.)

49. Answer the following by going back to fundamentals regarding accessible volumes in position and momentum space. Figure 14.7b is an isothermal curve, and it shows that as volume increases, entropy increases. Why? If volume increases during an adiabatic expansion, why doesn't the entropy increase?

50. In the isothermal plot of g vs. p of Figure 14.8b there is a discontinuous change in the slope of the line of minimum g at the point bb'. Why does this imply a sudden change in the volume?

Section E

51. Initially systems A and B have different types of particle and are separated by a removable partition. System A initially has a fraction f of the total volume and of the total number of particles. The partition is removed and the particles mix.

    (a) Show that the increase in the total entropy due to changes in the accessible volume is given by equation 14.21.

    (b) Show that $-T\Delta S_m$ is concave upwards everywhere.

52. Solder is a mixture of tin and lead. The melting point of tin is 232 °C and that of lead is 327 °C. The melting point for solder (depending on the mixing fractions) generally lies in the range 180 °C to 200 °C. Why is the melting point of the mixture lower than that of either ingredient?

53. When a certain two miscible fluids mix their temperature rises. Why is this?

54. When two systems mix at 290 K, they absorb 10 000 J of heat from their surroundings. What is the change in interaction entropy of the system upon mixing?

55. When plotting the entropy of mixing, ΔS_m, vs. the mixing fraction f, the slope becomes infinite in the limits $f \to 0$ and $f \to 1$ (equation 14.25). Prove this by taking the derivative with respect to the mixing fraction (d/df) of the entropy of mixing (see equation 14.21).

56. Imagine that particles interact in pairs. If the interaction energy of two particles is 2ε, then half of that (ε) can be associated with each particle. Let $\varepsilon_A, \varepsilon_B, \varepsilon_X$ represent the interaction energy per particle for pairs of As, Bs, and unlike particles, respectively. (For attractive interactions, these energies are negative.) There are Nf particles of type B and $N(1-f)$ of type A.

    (a) Show that the change in potential energy (equal and opposite to the change in thermal energy) is given by

    $$\Delta U = Nf(1-f)(2\varepsilon_X - \varepsilon_A - \varepsilon_B).$$

    (b) In the common case where attraction between like particles is stronger than that between unlike particles, $2\varepsilon_X - \varepsilon_A - \varepsilon_B > 0$, so the answer to part (a) is positive. Potential energy rises and thermal energy falls, and the answer to (a) represents energy that must be added to the system to keep its temperature constant. As can be seen in Figure 14.12, $\Delta G = -T\Delta S_m - T\Delta S_i$ must be concave upward everywhere to avoid a solubility gap. What restriction does this place on the factor $2\varepsilon_X - \varepsilon_A - \varepsilon_B$ for any value of f?

    (c) Suppose that $\varepsilon_A = \varepsilon_B = -0.25\,\text{eV}$ and $\varepsilon_X = -0.20\,\text{eV}$. At what temperature would A and B be miscible for all f?

57. Referring to Figure 14.14, consider the cooling and solidification of a molten magma with a mixing fraction of 0.25 (the process is represented by a vertical line at $f = 0.25$). See whether you can explain the various phases through which the cooling magma goes in the phase diagram (going downward along your line) by what you see on the corresponding plots of Gibbs free energy in the diagrams on the left.

Part VI
Classical statistics

Chapter 15
Probabilities and microscopic behaviors

Early in this book we used statistics to study the behavior of small systems. Expansion of this statistical approach enabled us to develop more powerful methods for larger systems, as we have been doing in the past 11 chapters. Now we return our attention to small systems, this time armed with our arsenal of statistical tools.

A myriad of small systems is begging to be studied, and the particular tools we use depend on the type of information that we wish to obtain. Often we must be creative in developing an appropriate approach to the problem at hand. But some statistical tools can be used in broad classes of studies and are therefore particularly valuable. The rest of this book is devoted to developing and illustrating some of the more popular standard tools.

A The ensembles

If we flip two coins, the probabilities for the possible heads–tails outcomes would be

$$P_{hh} = 1/4, \qquad P_{ht\,or\,th} = 1/2, \qquad P_{tt} = 1/4.$$

If we had a very large number of identically prepared systems, each consisting of two flipped coins, then we would expect 1/4 of them to have two heads, 1/2 of them to have one head and one tail, and 1/4 to have two tails, because these are the respective probabilities. A very large number of identically prepared systems (real or imagined) is called an "ensemble."

If a property x has the value x_i when the system is in state i, and the probability that it is in this state is P_i, then the mean value of the property x is given by (Chapter 2)

$$\overline{x} = \sum_i P_i x_i.$$

Because P_i is also the fraction of the members of an ensemble that would be in this state, this definition of mean value is also sometimes called the "ensemble average."

We classify ensembles according to how their members are interacting with outside systems. If there is no interaction at all, it is called a "microcanonical ensemble". In Chapter 7 we considered a completely isolated system A_0 that consisted of two interacting subsystems A_1 and A_2, and we calculated the probabilities P_i for the various possible energy distributions between them. If instead of one such isolated system A_0 we considered a huge number, all identically prepared, it would be a "microcanonical ensemble." The probability P_i for any particular distribution of energies between the two subsystems would then be reflected in the fraction of the ensemble's members that had this particular energy distribution.

In a "canonical ensemble" the members are interacting *thermally and/or mechanically* with an outside system. An example is the oxygen molecules in our atmosphere. The probability for any molecule to be in a certain quantum state is reflected in the fraction of all oxygen molecules that are in that state. Because each molecule is interacting thermally and mechanically, but not diffusively ($N =$ one molecule for each member) with the atmosphere, altogether they form a canonical ensemble.

In addition to possible thermal and mechanical interactions, the members of a "grand canonical ensemble" interact diffusively with their environment. An example would be a large number of identical quantum states, each of which has particles entering and leaving it (see below). Another example would be a large number of identical ice crystals interacting diffusively with the moisture in the air.

B   Probability that a system is in a certain state

The most powerful tools of thermodynamics are based on the statistics of large numbers. How can we use these tools to study the behaviors of a small component of a system, such as a single molecule or a single quantum state? The task is

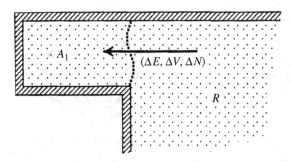

Figure 15.1 A small system, A_1, is interacting thermally, mechanically, and diffusively with a huge reservoir, R. To move into state s the small system must take energy, volume, and particles ΔE, ΔV, ΔN from the reservoir. What then is the probability that it is in this state?

accomplished through a clever trick. If the whole system is isolated, then any gain in the energy, volume, or particles possessed by the small component is at the expense of the rest of the system. Because the "rest of the system" is very large, we can apply the powerful statistical tools to this "rest of the system" in order to infer the behavior of the small component.

We begin by considering a tiny system A_1 interacting with a large reservoir R, as in Figure 15.1. The number of states for the combined system is the product of the number of states for the two subsystems (Sections 2B or 6D). The number of states for the reservoir, Ω_R, can be written in terms of its entropy S_R: $\Omega_R = e^{S_R/k}$ (equation 7.11). So

$$\Omega_0 = \Omega_1 \Omega_R \quad \longrightarrow \quad \Omega_0 = \Omega_1 e^{S_R/k}. \tag{15.1}$$

Now suppose that the small system is in one particular state ($\Omega_1 = 1$), which has taken energy ΔE, volume ΔV, and particles ΔN from the reservoir. This reduces the entropy of the reservoir by an amount given by the first law (equation 8.6).

$$S_R = S_{R0} - \frac{\Delta E + p\Delta V - \mu\Delta N}{T},$$

where S_{R0} was the original entropy of the reservoir before it lost ΔE, ΔV, ΔN to the small system. If we insert this and $\Omega_1 = 1$ (one specified state) into equation 15.1, we have

$$\Omega_0 = 1 \times \exp\left(\frac{S_{R0}}{k} - \frac{\Delta E + p\Delta V - \mu\Delta N}{kT}\right).$$

Finally, we recall that the probability for any particular configuration (i.e., any distribution of energy, volume, or particles) is proportional to the number of states Ω_0 corresponding to that configuration (using the fundamental

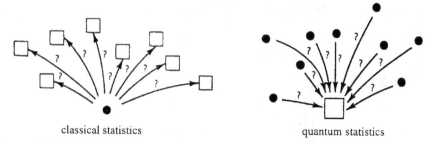

classical statistics   quantum statistics

Figure 15.2 Figurative illustration of the difference between classical (Boltzmann) and quantum statistics. In classical statistics, attention is typically focused on the individual small system or particle, which could occupy any of several different states. In quantum statistics, attention is focused on the individual quantum state, which could be occupied by various numbers of particles.

postulate, Section 6B). Incorporating the factor $e^{S_{R0}/k}$ into the constant of proportionality, we have the following very important result.

The probability that a system is in a state s

When a small system is interacting with a reservoir of temperature T, the probability that it is in a state s which takes energy, volume, and particles ΔE, ΔV, ΔN from the reservoir is given by

$$P_s = C \exp\left(-\frac{\Delta E + p\Delta V - \mu \Delta N}{kT}\right),$$ (15.2)

where C is a constant of proportionality determined by the requirement that the sum over all possible configurations must give a total probability of 1,

$$\sum_s P_s = 1.$$ (15.2′)

Notice that we have managed to write the probability that the small system is in a certain state purely in terms of its influence on the reservoir R. States that cause a greater reduction in the reservoir's entropy are less probable.

Since the factor $1/kT$ is always present in the exponent, it is customary and convenient to use the symbol β for this factor:

$$\beta \equiv \frac{1}{kT}.$$ (15.3)

C   Two approaches

The result 15.2 is often applied in one of two different ways that correspond to two different ways of viewing the microscopic system (Figure 15.2):

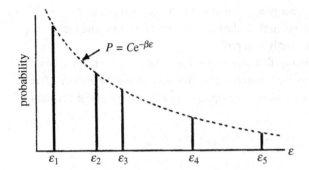

Figure 15.3 Classical statistics: a plot of the probability that a small system is in a state s with energy ε_s, when interacting with a reservoir. The relative probabilities are indicated by the heights of the lines. Probabilities are smaller for states of higher energies, because the removal of energy from the reservoir reduces its entropy.

1 as a particle or other small system that can occupy any of several different states;
2 as a quantum state that could be occupied by various numbers of particles.

The first approach is often referred to as "Boltzmann" or "classical statistics" and the second as "quantum statistics."

As we will soon learn, the approach that we use depends on the nature of the problem. The classical approach is the most convenient when either:

• the small systems are distinguishable, so that we can identify the one of interest, or
• the probability that two or more small systems might attempt to occupy the same quantum state is very small.

The quantum approach is most convenient when neither of these applies, that is, when the small systems are identical and the probability that two or more might attempt to occupy the same state cannot be ignored.

C.1 Classical statistics

In the classical approach the small system is identified. For example, it could be a certain set of genes in a chromosome or a single nitrogen molecule in a room full of air. It has a fixed number of particles and exchanges none with the reservoir.[1] So $\Delta N = 0$, and the probability of finding it in state s is (Figure 15.3)

$$P_s = Ce^{-\beta(\varepsilon_s + pv_s)} \qquad \text{(classical statistics)} \qquad (15.4)$$

---

[1] So what kind of ensemble would it be if you had a large number of these small systems?

where ε_s and υ_s are the energy and volume (previously written as ΔE and ΔV) that the small system has taken from the reservoir. States that require more energy or volume are correspondingly less probable.

The energy and volume that are taken from the reservoir are customarily measured relative to the ground state. But they could be measured relative to any level $\varepsilon_0, \upsilon_0$ by making an appropriate adjustment in the constant of proportionality:

$$P_s = Ce^{-\beta[(\varepsilon_s - \varepsilon_0) + p(\upsilon_s - \upsilon_0)]} = \underbrace{Ce^{\beta(\varepsilon_0 + p\upsilon_0)}}_{C'} e^{-\beta(\varepsilon_s + p\upsilon_s)}.$$

As can be shown in the homework problems, the $p\upsilon_s$ term is usually negligible in comparison with ε_s. Consequently, the probability that a small system is in a state s is normally written as

$$P_s = Ce^{-\beta\varepsilon_s} \qquad \text{(classical statistics, } p\upsilon_s \ll \varepsilon_s\text{).} \qquad (15.5)$$

C.2 Quantum statistics

In the second approach, we consider the "small system" to be a single quantum state. Its volume is fixed,[2] but it may contain various numbers of particles. So although it exchanges no volume with the reservoir, it may take particles ($\Delta V = 0$ but $\Delta N \neq 0$).[3] Consequently, the probability for it to be in a certain configuration is, from equation 15.2,

$$P = Ce^{-\beta(\Delta E - \mu \Delta N)}.$$

If the energy of a particle in this quantum state is ε then the probability that there are n particles in the state is (with $\Delta N = n$ and $\Delta E = n\varepsilon$)

$$P_n = Ce^{-n\beta(\varepsilon - \mu)} \qquad \text{(quantum statistics).} \qquad (15.6)$$

In both the classical and the quantum approaches, the constant of proportionality C is determined by the condition that the system must certainly have one of the possible configurations. For classical statistics, the small system must certainly be in one of the states, and for quantum statistics, the state must certainly contain some number of particles. Thus we have

$$C \sum_s e^{-\beta\varepsilon_s} = 1$$

$$\Rightarrow C = \frac{1}{\sum_s e^{-\beta\varepsilon_s}} \qquad \text{(classical statistics),} \qquad (15.7)$$

---

[2] Remember, a quantum state identifies a certain volume in phase space, which means that its volumes in both position and momentum space are fixed.

[3] So what kind of ensemble would a large number of identical quantum states be?

$$C \sum_n e^{-n\beta(\varepsilon-\mu)} = 1$$

$$\Rightarrow C = \frac{1}{\sum_n e^{-n\beta(\varepsilon-\mu)}} \quad \text{(quantum statistics)}.$$

Summary of Sections A–C

When a small system interacts with a large reservoir at temperature T, the probability that it is in a particular state s that takes energy, volume and particles $(\Delta E, \Delta V, \Delta N)$ from the reservoir is given by equation 15.2

$$P_s = C \exp\left(-\frac{\Delta E + p\Delta V - \mu\Delta N}{kT}\right)$$

There are two common ways of applying this, depending on the nature of the small system. In the one approach, called Boltzmann or classical statistics, we consider the small system to be a certain particle or group of particles that can occupy various states. The probability that it occupies state s with energy ε_s is given by (equation 15.5)

$$P_s = C e^{-\beta\varepsilon_s} \quad \text{(classical statistics, } p\upsilon_s \ll \varepsilon_s),$$

where (equation 15.3)

$$\beta \equiv \frac{1}{kT}.$$

In the second approach, called quantum statistics, we consider the system to be a certain quantum state that may be occupied by various numbers of particles. If the energy of a single particle in this state is ε then the probability that there are n particles in this state is given by (equation 15.6)

$$P_n = C e^{-n\beta(\varepsilon-\mu)} \quad \text{(quantum statistics)}.$$

The constant of proportionality is determined by requiring that the sum over all possibilities gives unity (equation 15.7). In classical statistics,

$$C = \frac{1}{\sum_s e^{-\beta\varepsilon_s}}; \quad \text{in quantum statistics } C = \frac{1}{\sum_n e^{-n\beta(\varepsilon-\mu)}}.$$

D   Applications of quantum statistics

We now use two examples to illustrate the use of quantum statistics. In the first, we find the probability that a certain quantum state in an oven has a photon in it, and in the second we find the probability that a certain quantum state in a metal has an electron in it.

Example 15.1 Consider a photon quantum state of energy 0.1 eV in an oven at 500 K. The chemical potential of a photon is zero. How much more likely are we to find 0 photons than 1 in this particular state?

For $T = 500$K and $\varepsilon - \mu = 0.1$eV, we have

$$\beta(\varepsilon - \mu) = \frac{1}{kT}(\varepsilon - \mu) = 2.32.$$

The ratio of the two probabilities is

$$\frac{P_0}{P_1} = \frac{Ce^{-0}}{Ce^{-1\beta(\varepsilon-\mu)}} = e^{2.32} = 10.$$

It is 10 times more likely that this particular state is empty.

Example 15.2 In a particular metal at 300 K, the electrons each have chemical potential $\mu = -2.03$ eV. A certain quantum state has energy -2.00 eV, and it can contain no more than one electron. What are the probability that it is empty and the probability that it is full?

Using $T = 300$ K and $\varepsilon - \mu = 0.03$ eV, we have

$$\beta(\varepsilon - \mu) = \frac{1}{kT}(\varepsilon - \mu) = 1.16.$$

According to equation 15.7, the constant C in equation 15.6 for the probability is

$$C = \frac{1}{\sum_{n=0,1} e^{-n\beta(\varepsilon-\mu)}} = \frac{1}{e^0 + e^{-1.16}} = \frac{1}{1 + 0.31} = 0.76.$$

So the required probabilities are

$$P_0 = Ce^0 = (0.76)(1) = 0.76,$$
$$P_1 = Ce^{-1.16} = (0.76)(0.31) = 0.24.$$

E Application of classical statistics

We now look at applications of classical statistics. If we are interested in *relative* probabilities we don't need the constant C, because it appears in both the numerator and denominator of these ratios:

$$\frac{P_i}{P_j} = \frac{Ce^{-\beta\varepsilon_i}}{Ce^{-\beta\varepsilon_j}} = e^{-\beta(\varepsilon_i - \varepsilon_j)}. \tag{15.8}$$

If we are interested in the *absolute* probability for a state, however, the constant C must be calculated from equation 15.7.

E.1 Examples

To illustrate the above, we examine the excitation of a hydrogen atom at various temperatures. The two lowest energy levels of a hydrogen atom are $\varepsilon_0 = -13.6$ eV and $\varepsilon_1 = -3.4$ eV.

Example 15.3 At room temperature, what would be the ratio of hydrogen atoms in the first excited state to the number in the ground state? Ignore degeneracies (Section 6C).

We use equation 15.8 with $\varepsilon_1 - \varepsilon_0 = 10.2\,\text{eV}$ and $T = 295\,\text{K}$:

$$\beta(\varepsilon_1 - \varepsilon_0) = \frac{10.2\,\text{eV}}{kT} = 401.$$

So the ratio of the two probabilities is

$$\frac{P_1}{P_0} = e^{-401} = 10^{-174}.$$

Clearly, no hydrogen atoms will be in excited states at room temperature.

Example 15.4 At what temperature would we find about half as many hydrogen atoms in the first excited state as in the ground state? (Ignore degeneracies.)

From equation 15.8 we have

$$\frac{P_1}{P_0} = e^{-\beta(\varepsilon_1 - \varepsilon_0)} = \frac{1}{2} \qquad \Rightarrow \qquad -\frac{\varepsilon_1 - \varepsilon_0}{kT} = \ln\frac{1}{2}.$$

With $\varepsilon_1 - \varepsilon_0 = 10.2\,\text{eV}$, we can solve for T, getting

$$T = 1.7 \times 10^5\,\text{K}.$$

E.2 Excitation temperature

It is convenient to define the "excitation temperature" for the particles of a system:

$$T_e = (\varepsilon_1 - \varepsilon_0)/k. \tag{15.9}$$

If we look at the relative probability for the small system to reach the first excited state,

$$\frac{P_1}{P_0} = e^{-(\varepsilon_1 - \varepsilon_0)/kT},$$

we can see that for temperatures small compared with T_e the exponent is large and negative, so that there is little probability of excitation to even the first excited state. That is, for $T \ll T_e$, the system is pretty much confined to its ground state.

F Heat capacities

The molar heat capacity of a system is given by equation 10.14 as

$$C_V = \frac{\nu}{2}R,$$

where ν is the number of degrees of freedom per particle. Each degree of freedom requires a certain minimum energy for excitation, corresponding to a certain

Figure 15.4 Plot of molar heat capacity vs. temperature for a hypothetical non-condensing diatomic gas in a typical room. The number of degrees of freedom per molecule goes from 0 to 3 to 5 to 7, as the excitation temperatures for translational, rotational, and vibrational motions are passed.

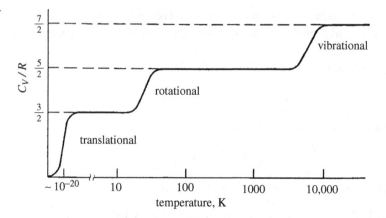

excitation temperature (equation 15.9). When the system is well below this temperature, excitations will not occur and so that degree of freedom disappears. No energy can be stored in it.

F.1 Diatomic gases

The molecules of gases each have three translational degrees of freedom. Polyatomic molecules may also rotate and vibrate. As we saw in section 4C (Figure 4.6), the common diatomic gases, such as N_2, O_2, and H_2, can rotate around two of the three rotational axes. They may also have two vibrational degrees of freedom (corresponding to the kinetic and potential energies in terms of relative coordinates) for vibrations along the molecular axis.

The excitation temperatures for a diatomic molecule in a room are typically about 10^{-20} K for translational motion, 5 K to 50 K for rotational motion, and 2000 K to 10 000 K for vibrations (homework). Consequently, the number of degrees of freedom per molecule will go from 0 to 3 to 5 to 7 as these thermal barriers are passed, and these changes are reflected in the molar heat capacities, as depicted in Figure 15.4. Although 10^{-20} K is far below what can be observed, even this excitation energy is sensitive to the size of the container. For molecules confined to dimensions measured in nanometers or less, the translational excitation temperature rises to the point where we can observe this transition in the laboratory (homework).

F.2 Solids

The atoms in solids are in three-dimensional potential wells, which give them six degrees of freedom (three kinetic and three potential). These wells are a bit wider than those for the atoms in diatomic gas molecules, so their wavelengths are longer and the excitation energies correspondingly lower. The excitation temperature for

Table 15.1. *Molar heat capacities of some common metals at room temperature, in terms of the Dulong–Petit value 3R*

Metal	$C_V/3R$	Metal	$C_V/3R$
aluminum	0.97	silicon	0.85
copper	0.98	silver	1.01
lead	1.04	zinc	1.02

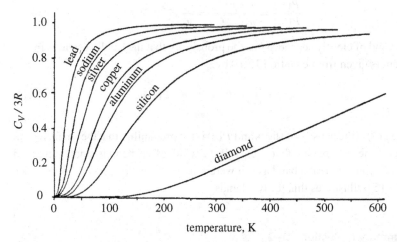

Figure 15.5 Plot of molar heat capacity vs. temperature for several common solids. At sufficiently high temperatures the molar heat capacities equal 3R, as we would expect. But at lower temperatures, there is insufficient thermal energy for many of the atoms to reach even the first excited state, so they become entrapped in the ground state and lose their degrees of freedom. Note that the excitation temperatures tend to be higher for the more rigid solids in which the atoms are bound more tightly. (The molar heat capacity of diamond does not reach 3R until around 2000 K.) Can you explain why this is? (see problem 38.)

vibrations in solids is typically measured in tens or hundreds of kelvins, which may be compared with the thousands of kelvins needed for the excitation of vibrations in common diatomic gases. Consequently, for most solids at room temperature and above, the molar heat capacities (Table 15.1) are in close agreement with the "Dulong–Petit law,".

$$C_V = 3R.$$

As the temperature is lowered, however, these degrees of freedom begin to disappear and the heat capacity drops correspondingly (Figure 15.5).

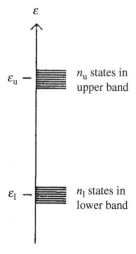

Figure 15.6 The probability for the energy of a small system to be within a band of closely spaced states is given by $P_{band} = \sum_s n_s P_s = \int g(\varepsilon) P(\varepsilon) d\varepsilon$, where $g(\varepsilon)$ is the density of states and $P(\varepsilon) = Ce^{-\beta\varepsilon}$. If the band widths are small compared with kT, it is often convenient to treat each band as a single degenerate level. Here we have two bands, with energies ε_u, ε_l and degeneracies n_u, n_l, respectively.

G   Closely spaced states

Suppose now that several different states have the same energy ε_s. If the level ε_s is n_s times degenerate, then the probability $P(\varepsilon_s)$ that a small system has this energy is equal to the number of such states times the probability that it is in any one of them.

$$P(\varepsilon_s) = n_s P_s. \tag{15.10}$$

Furthermore, the relative probability that the system is in level i rather than level j is then

$$\frac{P(\varepsilon_i)}{P(\varepsilon_j)} = \frac{n_i C e^{-\beta\varepsilon_i}}{n_j C e^{-\beta\varepsilon_j}} = \frac{n_i}{n_j} e^{-\beta(\varepsilon_i - \varepsilon_j)}. \tag{15.11}$$

For a band of closely spaced states the probability that the small system is in one of them is given from equation 15.10 by

$$P_{band} = \sum_s n_s P_s = \int_{band} g(\varepsilon) P(\varepsilon) d\varepsilon, \tag{15.12}$$

where $g(\varepsilon)$ is the density of states and $P(\varepsilon)$ is the probability that the small system is in any one state having that energy. If the width of the band is small compared with kT, you can treat a band as if it were a single degenerate level (homework). Figure 15.6 illustrates this for two bands.

Summary of Sections D–G

The relative probability for a small system to be in levels i and j is given by (equation 15.8)

$$\frac{P_i}{P_j} = e^{-\beta(\varepsilon_i - \varepsilon_j)}.$$

If level s is n_s times degenerate, the probability that the system has energy ε_s is (equation 15.10)

$$P(\varepsilon_s) = n_s P_s,$$

and the relative probability for any two such degenerate levels is (equation 15.11)

$$\frac{P(\varepsilon_i)}{P(\varepsilon_j)} = \frac{n_i}{n_j} e^{-\beta(\varepsilon_i - \varepsilon_j)}.$$

The probability that the system's energy lies within a band of closely spaced states is (equation 15.12)

$$P_{band} = \sum_s n_s P_s = \int_{band} g(\varepsilon) P(\varepsilon) d\varepsilon.$$

The excitation temperature is defined in terms of the energy separation between the ground and first excited states (equation 15.9):

$$T_e = \frac{\varepsilon_1 - \varepsilon_0}{k}.$$

For temperatures small compared with this, the system is confined to the ground state.

The molar heat capacity of a system is given by

$$C_V = \frac{v}{2} R$$

where v is the number of degrees of freedom per particle. As the temperature is lowered below the excitation temperature, the corresponding degrees of freedom disappear. These changes in the degrees of freedom per molecule are reflected in the molar heat capacities.

H Equipartition

H.1 Average energy per degree of freedom

Imagine that a small system is a single degree of freedom and that the energy stored in this degree of freedom is of the usual form $\varepsilon = b\xi^2$, where b is a constant and ξ is a position or momentum coordinate. According to the definition of the mean value (2.1), the average energy stored in this degree of freedom is

$$\bar\varepsilon = \sum_s P_s \varepsilon_s = C \sum_s e^{-\beta \varepsilon_s} \varepsilon_s, \qquad \text{where} \quad C = \frac{1}{\sum_s e^{-\beta \varepsilon_s}}. \tag{15.13}$$

As we learned in Chapter 1 (equation 1.6), the number of quantum states within a coordinate interval $d\xi$ is proportional to the length of this interval. Therefore, the sum over states can be replaced by an integral (treating ξ as a continuous variable):

$$\sum_s \rightarrow \text{constant} \times \int d\xi.$$

With this replacement, equation 15.13 becomes

$$\bar\varepsilon = \frac{\text{constant} \times \int_{-\infty}^{+\infty} d\xi \, e^{-\beta b\xi^2} b\xi^2}{\text{constant} \times \int_{-\infty}^{+\infty} d\xi \, e^{-\beta b\xi^2}}. \tag{15.13'}$$

The constants cancel, and the numerator can be integrated by parts (homework) to get

$$\int_{-\infty}^{+\infty} d\xi \, e^{-\beta b\xi^2} b\xi^2 = 0 + \frac{1}{2\beta} \int_{-\infty}^{+\infty} d\xi \, e^{-\beta b\xi^2}.$$

This last integral is identical to that in the denominator so they cancel, giving

$$\bar{\varepsilon} = \frac{1}{2\beta} = \frac{1}{2}kT. \tag{15.14}$$

This shows that with each degree of freedom is associated an average energy $(1/2)kT$.[4]

H.2 Brownian motion

Consider a small system with three translational degrees of freedom. According to the above result its average translational kinetic energy is

$$\bar{\varepsilon}_{\text{trans}} = \frac{1}{2}m\overline{v^2} = \frac{3}{2}kT,$$

and so its root mean square speed is

$$v_{\text{rms}} = \sqrt{\overline{v^2}} = \sqrt{\frac{3kT}{m}}. \tag{15.15}$$

This gives us a measure of the thermal motion of the particles of a system, which may be extremely erratic owing to their rapid and random collisions. It applies to all particles, ranging in size from the subatomic to the gigantic. But as you will show in your homework, this thermal motion is only appreciable for particles that are very small. It was first noticed in 1827 by a Scottish botanist, Robert Brown, who was using a microscope to observe the jittering of tiny spores suspended in water. Consequently, for particles large enough to be seen with a microscope, it is often called "Brownian motion."

Summary of Section H

Using classical statistics, we can show that if the energy in a degree of freedom is of the form $b\xi^2$, where b is a constant and ξ is a position or momentum coordinate, then its average energy is given by (equation 15.14)

$$\bar{\varepsilon} = \frac{1}{2}kT.$$

Consequently, the average translational kinetic energy for any particle that moves in three dimensions is

$$\bar{\varepsilon}_{\text{trans}} = \frac{1}{2}m\overline{v^2} = \frac{3}{2}kT,$$

and so its root mean square speed is (equation 15.15)

$$v_{\text{rms}} = \sqrt{\overline{v^2}} = \sqrt{\frac{3kT}{m}}$$

---

[4] More generally, if the energy per degree of freedom is $b\xi^\alpha$ then its average is $(1/\alpha)kT$ (homework).

Problems

Sections A and B

1. Consider a reservoir with temperature, pressure, and chemical potential T, p, μ, respectively. Initially, its entropy is S_0. Then a small system interacts with it, removing energy, volume, and particles $\Delta E, \Delta V, \Delta N$. In terms of the parameters $S_0, T, p, \mu, \Delta E, \Delta V, \Delta N$, find
   (a) the reservoir's new entropy S_R,
   (b) the number of states accessible to the reservoir now,
   (c) the probability, to an overall multiplicative constant factor, that the small system is in the one state that takes $\Delta E, \Delta V, \Delta N$ from the reservoir

2. Consider a small system that is interacting with a reservoir at temperature 400 K, pressure 10^8 Pa, and chemical potential -0.3 eV.
   (a) To go from state 1 to state 2, the small system must take an additional 0.03 eV of energy and 10^{-29} m^3 of volume from the reservoir. How many times more probable is it for the small system to be in state 1 than state 2?
   (b) To go from state 1 to state 3 the small system must take 0.4 eV of energy and one particle (but no extra volume) from the reservoir. How many times more probable is it that the small system is in state 1 than state 3?
   (c) To go from state 1 to state 4, the small system must take no energy but one particle and 10^{-27}m^3 of volume from the reservoir. How many times more probable is it that the small system is in state 1 than state 4?

Section C

3. A certain very delicate organic molecule requires only 0.04 eV of energy to be excited from the ground state. Upon excitation, its volume increases by 2×10^{-31} m^3. Under what pressure would $p\Delta V$ be equal to ΔE for this excitation? To how many atmospheres would this correspond?

4. There are a few exceptional cases of extremely high pressure where the $p\Delta V$ term cannot be ignored in the probability expression $P = Ce^{-\beta(\Delta E + p\Delta V)}$. One of these is a neutron star. The radius of a hydrogen atom is about 0.53×10^{-10} m, and that of a neutron is negligible by comparison. The electron can combine with the proton to form a neutron (releasing a neutrino), but this state is 0.84 MeV higher in energy than the non-combined state. Assuming that classical statistics gives roughly the right answer, calculate the minimum pressure that must exist within a neutron star.

5. From the Bohr model, the radius of a hydrogen atom in the ground state is 0.53×10^{-10} m and the radius of the first excited state is four times larger.

The energy of the ground state is $-13.6\,\text{eV}$, and that of the first excited state is $-3.4\,\text{eV}$.

(a) What are the values of ΔE and $p\Delta V$ for excitation into the first excited state at atmospheric pressure?

(b) At what pressure would the $p\Delta V$ term be comparable to ΔE? How many atmospheres is this?

6. Consider a small system for which the energies of the states, measured relative to the ground state, are $\varepsilon_n = n(0.02\,\text{eV})$, $n = 0, 1, 2, \ldots$ The temperature of the system is 273 K. (Use $1 + x + x^2 + x^3 + \cdots = 1/(1 - x)$.)

(a) To three-decimal-place accuracy, what is the value of the constant C in $P_s = Ce^{-\beta\varepsilon_s}$?

(b) What is the probability that this system is in the ground ($n = 0$) state?

(c) What is the probability that this system is in the first excited state?

7. Repeat the above problem for the case where the temperature is 500 K.

8. In a system at 290 K, a certain quantum state has energy that is 0.02 eV above the chemical potential. Any number of particles $(0, 1, 2, 3, \ldots)$ may occupy this state. Find the probability that at any instant the state contains (a) 0 particles, (b) one particle, (c) two particles.
$(1 + x + x^2 + x^3 + \cdots = 1/(1 - x)$.)

Section D

9. A system has temperature 290 K and chemical potential $-0.2\,\text{eV}$. For a certain quantum state in this system, the energy per particle is $-0.16\,\text{eV}$, and any number of particles (bosons) may occupy it

(a) What is the value of C in the probability formula $P_n = Ce^{-n\beta(\varepsilon-\mu)}$ (to three decimal places)?

(b) What is the probability that this quantum state contains zero particles?

(c) What is the probability that this quantum state contains one particle?
$(1 + x + x^2 + x^3 + \cdots = 1/(1 - x)$.)

10. Repeat the above problem for the case of identical fermions, for which no more than one particle may occupy the state.

11. Consider the photons inside an oven at 500 K. Photons are bosons, so any number of photons may occupy one state. If the chemical potential of a photon is zero and the energy of a certain state is 0.2 eV, find

(a) the factor $\beta(\varepsilon - \mu)$ for this state,

(b) the constant C in the formula $P_n = Ce^{-n\beta(\varepsilon-\mu)}$, accurate to three decimal places,

(c) the probability of there being no photons in this state at any particular moment,

(d) the probability of there being two photons in this state at any particular moment.

12. Consider the photons inside an oven. The chemical potential of a photon is zero. Photons are bosons, so any number of photons may occupy one state. A certain state has energy 0.1 eV. At what temperature would the ratio $P_{n=1}/P_{n=0}$ be equal to e^{-1}?

13. The chemical potential for a conduction electron in a certain metal is -0.3 eV. Electrons are fermions, so no more than one electron may occupy a given state ($n = 0$ or 1 only). At room temperature (295 K), what is the probability that a state of energy -0.27 eV is (a) unoccupied, (b) Occupied?

14. Helium atoms are bosons, so any number may occupy a given state. In very cold liquid helium, suppose that there is a gap of about 3×10^{-4} eV between the ground state and the first excited state. Below about what temperature would helium atoms be mostly confined to the ground state?

15. Consider a system of particles at 1000 K for which the chemical potential is -0.4 eV. Suppose that we are interested in a certain quantum state, in which a particle would have an energy of -0.2 eV. How many times more likely is it for this state to be unoccupied than for it to have one particle in it?

Section E

16. A certain kind of molecule has four different possible electronic configurations. These have two different energies, as follows: one state of energy -0.34 eV (the ground state), and three states of energy -0.30 eV (the excited states).
    (a) If a system of such molecules is at a temperature of $17\,^\circ$C, what fraction of them will be in the ground state at any time?
    (b) At what temperature would half of them be in the excited states?

17. For a certain molecule, the first excited state lies 0.2 eV above the ground state.
    (a) At what temperature would the number of molecules in the ground state be exactly 10 times the number in the first excited state?
    (b) What is the excitation temperature for this molecule?

18. For a certain molecule in a system at 500 K, the energies of the various quantum states, measured relative to the ground state, are given by $\varepsilon = n(0.1\,\text{eV})$, $n = 0, 1, 2, \ldots$.
    (a) To three significant figures, what is the value of the constant C in the formula $P_s = Ce^{-\beta \varepsilon_s}$?
    (b) What is the probability that the molecule is in the level $n = 1$?
    (c) The probability that it is in the level $n = 2$?
    $(1 + x + x^2 + x^3 + \cdots = 1/(1-x))$.)

19. For an energy $\varepsilon = 1$ eV, at what temperature would the ratio ε/kT be equal to 1?

20. A certain type of molecule requires 0.06 eV for excitation to the first excited level, which is doubly degenerate. The other states are much higher, and we can safely ignore them.
    (a) What fraction of the molecules will be in the first excited level at 17 °C?
    (b) At what temperature would we find 1% of these molecules at the first excited level?

21. If 0.03 eV is required to excite the molecules in a certain system into their first excited state, roughly what is the temperature below which there is a rather low probability of excitation?

22. The mass of an average air molecule is about 5×10^{-26} kg. Assuming the atmospheric temperature to be a constant 275 K at all altitudes and assuming the gravitational acceleration to be a constant 9.8 m/s², at what altitude h would you expect to find the air density to be exactly half that at sea level? (Hint: $P_h/P_0 = 1/2$.)

23. A certain liquid at room temperature (295 K) has Avogadro's number of molecules, each having a certain magnetic moment. Owing to quantum effects, there can be only three possible orientations of this magnetic moment, the component along any one direction being equal to $(1, 0, -1)\mu_B$ (μ_B is the Bohr magneton and equals 9.3×10^{-24} J/T.) The interaction energy between a magnetic moment and an external magnetic field is given by $E = -\mu \cdot \mathbf{B}$. In an external field of 2 tesla, find
    (a) the value of $\beta \mu_B B$ ($\beta = 1/kT$),
    (b) the respective probabilities that a molecule is in each of the three possible alignments,
    (c) the average magnetic moment per molecule,
    (d) the magnetic moment of the entire liquid.

24. The probability that a system is in a state s is given by $P_s = Ce^{-\beta \varepsilon_s}$, where $C = (\sum e^{-\beta \varepsilon_s})^{-1}$. Prove that you may measure energies relative to any level you wish, by multiplying both numerator and denominator by $e^{\beta \varepsilon_0}$ where ε_0 is any arbitrary energy. (Remember, T is the temperature of the reservoir, which is constant.)

25. An air molecule has a mass of about 5×10^{-26} kg. Consider two different possible states in which an air molecule might be found. In state 1 the molecule moves in a certain direction at a speed of 400 m/s, and in state #2 the molecule is standing still. What is the ratio of the probabilities for the air molecule to be in these two states, P_1/P_2, if the air temperature is (a) 17 °C, (b) −40 °C?

26. The rotational inertia of a certain molecule around any axis is 10^{-48} kg m². The rotational angular momentum is quantized in the form $L = \sqrt{l(l+1)}\, \hbar$,

where $l = 0, 1, 2, \ldots$ What is the ratio of the probabilities for the molecule to be in the $l = 1$ rotational state and the $l = 0$ state (P_1/P_0) if the temperature is (a) $200\,^\circ\text{C}$, (b) $-100\,^\circ\text{C}$?

27. The first excited state for the electrons of a certain molecule is 0.08 eV above the ground state. When excited, the volume of the electronic cloud increases by $10^{-28}\,\text{m}^3$. At 295 K and 1 atm, find
   (a) $\Delta E + p\Delta V$ for this excitation,
   (b) the ratio of probabilities P_1/P_0.

28. The first vibrational state of a certain molecule requires an excitation energy of 0.2 eV. At what temperature are there one quarter as many molecules in the first excited vibrational state as in the ground state?

Section F

29. Starting from the first law for nondiffusive interactions, $dE = dQ - p\,dV$, and equipartition, prove that the molar heat capacity of a substance is given by $C_V = (v/2)R$. (Assume that the potential energy reference level, u_0, is constant.)

30. The atomic mass number of a single nitrogen atom is 14, and the separation between the two atoms in a nitrogen molecule (N_2) is $1.098 \times 10^{-10}\text{m}$. The rotational kinetic energy is given by $\varepsilon_{\text{rot}} = (1/2I)L^2$, where I is the rotational inertia about the particular axis and the angular momentum is quantized in the form $L^2 = l(l+1)\hbar^2$, where $l = 0, 1, 2, \ldots$ Find
   (a) the rotational inertia of the nitrogen molecule about a perpendicular bisector of the molecular axis,
   (b) the energy required for excitation from the non-rotating $l = 0$ state to the first excited rotational state,
   (c) the excitation temperature for this excitation.

31. Repeat the above problem for a hydrogen (H_2) molecule. The atomic mass number for hydrogen is 1 and the separation between the two atoms is 0.373×10^{-10} m.

32. The rotational inertia of a diatomic molecule around the molecular axis is typically about 10^{-7} times the rotational inertia around a perpendicular bisector of the molecular axis. With this information and the answers to the two preceding problems, estimate the excitation temperature for rotations about the molecular axis for nitrogen and for hydrogen.

33. Suppose that you wish to know the molar heat capacity of a certain gas composed of tetrahedral-shaped molecules. This depends on the temperature, because if certain excited states are inaccessible then the molecule will not be able to store energy in those particular degrees of freedom.

(a) If the first excited vibrational level of this molecule requires 0.2 eV of energy, roughly what is the temperature below which you can ignore the vibrational degrees of freedom?

(b) If the rotational inertia about any axis is 10^{-46} kg m^2, roughly what is the temperature below which you can ignore the rotational degrees of freedom? (See problem 30 for the quantized form of the rotational energy.)

(c) At room temperature, what would you expect the molar heat capacity C_V of this gas to be?

34. Consider an air molecule, of mass 5×10^{-26} kg, in a cubical room that measures 3 m each side.

(a) What are the lowest and next lowest possible kinetic energies for this molecule? (Hint: What are the longest possible wavelengths in each dimensions? From these, get the momentum components and thus the minimum kinetic energy.)

(b) What is the excitation temperature for translational motion?

35. What is the excitation temperature for an air molecule confined to a tiny cubical container of side 1 nm? (See the previous problem.)

36. An atom of mass 10^{-25} kg is confined in a very tiny box (e.g., within a crystal lattice), so that it can only move about 2×10^{-11} meters in any dimension. Roughly what is the excitation temperature for translational motion for this atom?

37. A system of N particles has total energy E, so that the average energy per particle is given by $\bar{\varepsilon} = E/N$. If the states available to any particle have energies $\varepsilon_1, \varepsilon_2, \varepsilon_3, \ldots$, write down an expression that implicitly determines the temperature of the reservoir in terms of $\varepsilon_1, \varepsilon_2, \varepsilon_3, \ldots$ (Hint: Write down the expression that gives the average energy per particle in terms of the probabilities of being in the various states. This should do it.)

38. Answer the question in the caption of Figure 15.5.

Section G

39. A certain particle is interacting with a reservoir at 500 K and can be in any of four possible states. The ground state has energy -3.1 eV, and the three excited states all have the same energy, -3.0 eV. What is the probability that it is in (a) the ground state, (b) a particular excited state, (c) any state of energy -3.0 eV?

40. Consider a system whose molecules each have five accessible states – the ground state and four excited states, all lying 0.1 eV above the ground state.

At what temperature would as many molecules be in excited states as are in the ground state?

41. About 0.4 eV is required to dissociate a single water molecule in liquid water at room temperature into H^+ and OH^- ions. If the dissociated molecule has twice as many different states accessible to it as does the non-dissociated molecule, what fraction of the water molecules in a glass of water at room temperature (295 K) would be dissociated at any instant?

42. In the hydrogen atom, the ground state has energy -13.6 eV and degeneracy 2. The first excited state has energy -3.4 eV and degeneracy 8.
    (a) At 295 K, what is the ratio of the number of atoms in the first excited state and the number in the ground state?
    (b) At what temperature would this ratio be equal to $1/2$?

43. Consider a system for which the energies of particles come in two bands. In the lower band there are n_1 states, evenly spaced, extending from energy $\varepsilon = a$ to $\varepsilon = b$. In the upper band there are n_u evenly spaced states, extending from energy $\varepsilon = c$ to $\varepsilon = d$. Answers should be in terms of $a, b, c, d, n_1, n_u, \beta$).
    (a) What is the density of states, $g(\varepsilon)$, (the number of states per unit energy increment) in each of the bands?
    (b) Using $g(\varepsilon)$ from part (a) find the ratio of the number of all particles in excited states to the number in the ground states.
    (c) Now take the result from part (b) and apply it to the case where the band widths are very small compared with kT i.e., $(b - a)/kT \ll 1$, $(d - c)/kT \ll 1$. Factor out $e^{-\beta(c-a)}$ and use the expansion $e^x \approx 1 + x$ for $x \ll 1$. Your answer should be the same as that for two levels, of energy c and a, that are n_u and n_1 times degenerate, respectively.
    (d) Repeat part (c) but applied to the other extreme, where the band widths are very large compared with kT. Again, factor out $e^{-\beta(c-a)}$ but use the fact that $e^{-x} \approx 0$ for $x \gg 1$.

44. In a certain system at $17\,°C$, there are n states in both the lower band and the upper band.
    (a) If the lower states all have energy -0.31 eV and the upper states all have energy -0.29 eV, what is the ratio of the numbers of particles in the two bands?
    (b) Now suppose that the states in the upper band are evenly spread between -0.29 and -0.27 eV. You will have to sum over all the upper states, using a density of states given by $g(\varepsilon) = n/(0.02 \text{ eV})$ in the integral. In this case, what is the ratio of the numbers of particles in the upper and lower states?
    (c) Repeat the above, but assume that the states in the lower band are evenly spread between -0.32 and -0.31 eV and those in the upper band are evenly spread between -0.29 and -0.27 eV.

Section H

45. A nitrogen molecule (N_2) has mass 4.7×10^{-26} kg. In air at 20 °C:
    (a) How many translational degrees of freedom does a single nitrogen molecule have?
    (b) What is the average translational kinetic energy of a single nitrogen molecule?
    (c) What is the root mean square speed, v_{rms}, of a single nitrogen molecule?
    (d) What is the root mean square speed of a hydrogen molecule (H_2), with mass 3.3×10^{-27} kg?

46. The mass of a single nucleon is 1.7×10^{-27} kg and that of an electron is 9.1×10^{-31} kg. In violent nuclear reactions, such as those occurring in exploding warheads or in the interiors of stars, the temperatures are roughly 10^7 K. In this environment, what is the root mean square speed of (a) a nucleon, (b) an electron?

47. In Section H of the main text we calculated the average energy per degree of freedom, $\bar{\varepsilon} = (\int e^{-\beta\varepsilon}\varepsilon d\xi)/(\int e^{-\beta\varepsilon}d\xi)$, for the case where the energy term has the form $\varepsilon = b\xi^2$. Show that when you integrate the numerator by parts, you get $(1/2)kT$ times the denominator.

48. Suppose that the energy stored in a degree of freedom is given by $\varepsilon = b|\xi|^\alpha$, where ξ is a coordinate, b is a constant, and α is any positive exponent. Following a development parallel to that in Section H, prove that the average energy stored in this degree of freedom is given by $(1/\alpha)kT$. (Hint: If the integrand is symmetric then the integral from $-\infty$ to $+\infty$ can be written as twice the integral from 0 to $+\infty$. This allows you to eliminate the absolute-value symbol in $|\xi|$.)

49. A small dust particle has mass 10^{-8} g. It falls onto a glass of ice-cold water where it is supported by the surface tension, and moves freely in only *two* dimensions. What is the root mean square speed of its Brownian motion there?

50. A certain kind of plankton lives in the ocean at temperatures around 8 °C. It is unicellular and has approximate dimensions 8 μm by 5 μm by 3 μm and the same density as water. It relies entirely on thermal energy to propel it through the ocean water and bring it into contact with nutrients. What is
    (a) its root mean square speed, (b) the total distance it travels in a day (roughly)?

51. A certain grandfather clock was driven by a pendulum, made of a 0.5 kg weight on the end of a 2 m wire of negligible mass. But it hasn't been

running for years. The pendulum just hangs there inside the glass case at room temperature (295 K). It has two degrees of freedom (kinetic and potential), and there is no air convection inside the case.

(a) What is the mean thermal energy of the pendulum?
(b) What is the root mean square amplitude of the thermally induced swings of the weight?

Chapter 16
Kinetic theory and transport processes in gases

In this chapter we will use classical statistics to analyze gases. Each individual gas molecule is a "small system" that interacts thermally and mechanically with the "large reservoir" of the rest of the gas. We first find the "Maxwell distribution"[1] of particles as a function of their momenta or velocities and then apply this result to various processes.

A Probability distributions

A.1 One dimension

Consider a single particle moving in the x direction. The number of quantum states in any momentum interval is proportional to the length of that interval, dp_x (equation 1.6), and the probability for the particle to be in any one of these states is proportional to $e^{-\beta\varepsilon}$ (classical statistics). We combine these two factors to write the probability for the momentum to lie in the range dp_x as

$$P(p_x)d\,p_x = Ce^{-\beta p_x^2/2m}d\,p_x.$$

The value of the constant C is determined by demanding that the x-momentum

---

[1] Named after James Clerk Maxwell who first performed this analysis in 1859.

must have some value, so the sum over all p_x must equal unity (homework):

$$\int_{-\infty}^{+\infty} P(p_x)\mathrm{d}p_x = 1. \tag{16.1}$$

To do this integral, we could use the table of integrals in Appendix E, or we might simply notice that the distribution is Gaussian in the momentum (Section 3B), so that it has the form

$$P(p_x) = \frac{1}{\sqrt{2\pi}\sigma}e^{-p_x^2/2\sigma^2}, \qquad \text{with } \sigma^2 = \frac{m}{\beta}. \tag{16.2}$$

Either way we find that $C = (\beta/2\pi m)^{1/2}$, so the probability distribution is

$$P(p_x)\mathrm{d}p_x = \sqrt{\frac{\beta}{2\pi m}}e^{-\beta p_x^2/2m}\mathrm{d}p_x. \tag{16.3}$$

A.2   Three dimensions

The probability that a system meets two or more independent criteria is the product of the probabilities that it meets each (Section 2C). Therefore, the probability that a particle has its three momentum components in the ranges $\mathrm{d}p_x$, $\mathrm{d}p_y$, $\mathrm{d}p_z$, respectively, is the product of the three individual probabilities:

$$P(p_x, p_y, p_z)\mathrm{d}p_x\mathrm{d}p_y\mathrm{d}p_z = P(p_x)\mathrm{d}p_x\ P(p_y)\mathrm{d}p_y\ P(p_z)\mathrm{d}p_z.$$

Using the notation

$$\mathbf{p} = (p_x, p_y, p_z), \qquad p^2 = p_x^2 + p_y^2 + p_z^2,$$

and

$$\mathrm{d}^3p = \mathrm{d}p_x\mathrm{d}p_y\mathrm{d}p_{z,},$$

this becomes

$$P(\mathbf{p})\mathrm{d}^3p = \left(\frac{\beta}{2\pi m}\right)^{3/2}e^{-\beta p^2/2m}\mathrm{d}^3p. \tag{16.4}$$

If we only care about the momentum's magnitude and not its direction, we can write the three-dimensional element d^3p in spherical coordinates,

$$\mathrm{d}^3p = p^2\mathrm{d}p\sin\theta\mathrm{d}\theta\mathrm{d}\phi.$$

Then integration over the angles gives the volume of a hollow spherical shell of radius p and thickness $\mathrm{d}p$ (Figure 16.1):

$$4\pi p^2\mathrm{d}p$$

So, the probability that the momentum's magnitude lies within the range p and $p + \mathrm{d}p$ is

$$P(p)\mathrm{d}p = 4\pi\left(\frac{\beta}{2\pi m}\right)^{3/2}e^{-\beta p^2/2m}p^2\mathrm{d}p. \tag{16.5}$$

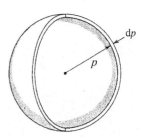

Figure 16.1   The volume of a hollow spherical shell of radius p and thickness $\mathrm{d}p$ is $4\pi p^2\mathrm{d}p$. So the number of quantum states with momenta in the range $\mathrm{d}p$ increases quadratically with p.

In this form, the probability distribution for the particle's momentum is no longer pure Gaussian. It is the product of the number of accessible states (proportional to $4\pi p^2 dp$) times the (Gaussian) probability that it is in any one of them.

These results are easily converted into velocity distributions by writing $\mathbf{p} = m\mathbf{v}$ and putting the extra factors m into the constant of proportionality.

Summary of Section A

The probability distributions for the particles of a gas are as follows. In one dimension (equation 16.3)

$$P(p_x)dp_x = \sqrt{\frac{\beta}{2\pi m}}e^{-\beta p_x^2/2m}dp_x \qquad \text{where} \quad \beta = 1/kT.$$

In three dimensions (equation 16.4)

$$P(\mathbf{p})d^3p = \left(\frac{\beta}{2\pi m}\right)^{3/2} e^{-\beta p^2/2m}d^3p,$$

or, if we care only about the momentum's magnitude and not its direction, we have (equation 16.5)

$$P(p)dp = 4\pi \left(\frac{\beta}{2\pi m}\right)^{3/2} e^{-\beta p^2/2m}p^2 dp.$$

The corresponding distributions in velocities are obtained by substituting $\mathbf{p} = m\mathbf{v}$:

$$P(v_x)dv_x = \sqrt{\frac{\beta m}{2\pi}}e^{-\beta mv_x^2/2}dv_x, \qquad (16.3')$$

$$P(\mathbf{v})d^3v = \left(\frac{\beta m}{2\pi}\right)^{3/2} e^{-\beta mv^2/2}d^3v, \qquad (16.4')$$

$$P(v)dv = 4\pi \left(\frac{\beta m}{2\pi}\right)^{3/2} e^{-\beta mv^2/2}v^2 dv. \qquad (16.5')$$

B   Mean values

We can use the above probability distributions, along with the standard integrals given in Appendix E, to calculate the mean value of any function of molecular velocities or momenta. Here are some examples.

Example 16.1 What is the average value of v_x for a molecule in a gas at temperature T?

The required mean value is given by

$$\bar{v}_x = \int v_x P(v_x)dv_x = \sqrt{\frac{\beta m}{2\pi}} \int_{-\infty}^{+\infty} e^{-\beta mv_x^2/2}v_x dv_x = 0.$$

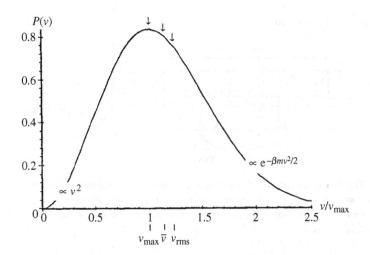

Figure 16.2 The Maxwell probability distribution for particles in a gas. The quadratic term ($\propto v^2$) dominates for small speeds, and the exponential ($\propto e^{-(\beta m/2)v^2}$) dominates at larger speeds. Values of the most probable speed v_{max}, the average speed $v = 1.128 v_{max}$, and the root mean square speed $v = 1.225\, v_{max}$ are indicated.

We might have anticipated this result. The molecule is equally likely to be moving in either direction, so the average velocity must be zero.

Example 16.2 What are the average values of the speed and the square of the speed ($\bar{v}$ and $\overline{v^2}$) for a molecule in a gas at temperature T (Figure 16.2)?

The speed is independent of direction, so we use the probability distribution 16.5′. The required mean values are

$$\bar{v} = \int v P(v) dv = 4\pi \left(\frac{\beta m}{2\pi}\right)^{3/2} \int_0^\infty e^{-\beta m v^2/2} v^3 dv = \sqrt{\frac{8kT}{\pi m}}, \tag{16.6}$$

$$\overline{v^2} = \int v^2 P(v) dv = 4\pi \left(\frac{\beta m}{2\pi}\right)^{3/2} \int_0^\infty e^{-\beta m v^2/2} v^4 dv = \frac{3kT}{m}, \tag{16.7}$$

where we refer to Appendix E to evaluate the integrals. The average speed increases with temperature and decreases with molecular mass, as we would expect.

The last result (16.7) repeats our previous proof that an average energy of $(1/2)kT$ is associated with each degree of freedom. Multiplying both sides of this equation by $m/2$, we see again that for a particle moving in three dimensions,

$$\tfrac{1}{2} m \overline{v^2} = \tfrac{3}{2} kT.$$

C Particle flux

We now look at particle flux, or "current density," which is the rate at which particles cross a unit perpendicular area. If all particles were moving in the x direction with the same velocity then the flux would be the product of the number

ρ = number of particles per unit volume

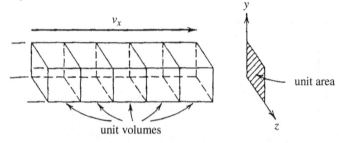

unit area

unit volumes

Figure 16.3　The flux of particles in the x direction. The number of particles crossing a unit area per second is equal to the product of the number of particles per unit volume, ρ, times the number of unit volumes that cross a unit area per second, v_x. Or in symbols, $J_x = \rho v_x$.

of particles per unit volume, ρ, times the number of unit volumes that cross unit perpendicular area per second, v_x (Figure 16.3),

$$J_x = \rho v_x.$$

If there is a distribution of particle velocities, we first look only at the part of the flux due to those particles moving with velocities in the range between v_x and $v_x + dv_x$. Since the fraction of the particles with velocities in this range is given by the probability distribution $P(v_x)dv_x$ of equation 16.3', we have

$$d\rho = \rho P(v_x)dv_x,$$

and the part of the flux due to these particles is

$$dJ_x = d\rho\, v_x = \rho P(v_x)v_x dv_x. \tag{16.8}$$

Example 16.3 What is the rate at which cabin air escapes from a punctured spaceship?

We define the cabin wall to be the y-z plane, with the cabin on the left so that particles striking it are moving in the positive x direction. We use equation 16.3' for the probability $P(v_x)dv_x$ and integrate the flux over all $v_x > 0$ to find the total flux of particles striking each square meter of the wall per second:

$$J_x = \int_{v_x > 0} dJ_x = \rho \int_0^\infty P(v_x)v_x dv_x = \rho\sqrt{\frac{kT}{2\pi m}}. \tag{16.9}$$

We need to multiply this by the area of the hole, A, to get the rate of loss in particles per second. Writing the density as $\rho = N/V$, the rate of particle loss becomes

$$\frac{dN}{dt} = -\frac{N}{V}\sqrt{\frac{kT}{2\pi m}}A. \tag{16.10}$$

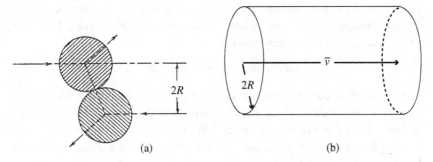

Figure 16.4 (a) For two particles of radius R to collide, their centers need come only within $2R$ of each other. (b) As a particle travels, we can think of it as cutting out a cylindrical volume each second of area $\sigma = \pi(2R)^2$, and length $\bar{v}$. The average number of collisions that this particle undergoes each second would be equal to the number of other particles whose locations were centered within this volume, if these other particles were standing still.

In the homework problems, it can be shown that, for a gas at constant temperature, integration of equation 16.10 gives exponential decay in the cabin pressure.

We could also use the particle flux to calculate the pressure exerted by a gas on the walls of its container. The momentum of each particle changes by $-2mv_x$ when it collides elastically with a wall in the yz-plane. Because pressure equals force per unit perpendicular area and force equals the rate of change of momentum, the pressure exerted on the wall would be

$$p = \int\limits_{v_x > 0} 2mv_x \, dJ_x, \tag{16.11}$$

where dJ_x is given by equation 16.8. This integral can be evaluated in a homework problem to obtain the ideal gas law, $p = NkT/V$.

D Collision frequency and mean free path

We now calculate how often gas molecules undergo collisions and how far they travel between these collisions, on average. These are called their "collision frequency" and "mean free path," respectively, and are given the symbols ν_c and $\bar{l}$.[2]

A molecule collides with any other molecule that comes within a center-to-center distance of $2R$, where R is the (effective) molecular radius (Figure 16.4a). Since a molecule moves with an average speed $\bar{v}$, we can think of it as cutting out a volume each second of length $\bar{v}$ and cross sectional area $\sigma = \pi(2R)^2$

---

[2] Beware, the symbols for velocity, v, and for frequency ν (Greek nu) look very similar. So occasionally you might have to distinguish them by context.

(Figure 16.4b). If the other molecules were sitting still, then the number of collisions it undergoes per second would be the number of other molecules whose centers are located within this volume:

$$\nu_c = \text{particle density} \times \text{volume cut out} = \rho \sigma \bar{v}.$$

But the other particles are not sitting still. Collisions involve the *relative* motions of molecules and, as we will soon show, the average relative speeds among colliding molecules are larger than their average absolute speeds by a factor $\sqrt{2}$. Consequently, the actual collision frequency of the molecules is given by

$$\nu_c = \sqrt{2}\rho\sigma\bar{v} \qquad \sigma = 4\pi R^2, \text{ where } R \text{ is the molecular radius.} \qquad (16.12)$$

The collision frequency increases for denser gases (ρ), fatter molecules (σ), and faster motion ($\bar{v}$). The average distance $\bar{l}$ traveled between collisions is simply the product of the average speed times the average time between collisions:

$$\bar{l} = \bar{v}\left(\frac{1}{\nu_c}\right) = \frac{1}{\sqrt{2}\rho\sigma}. \qquad (16.13)$$

The reason for the extra factor $\sqrt{2}$ is that collisions are much more frequent between molecules moving in opposite directions than between those moving in the same direction. (Just as you pass more cars on a highway that are coming towards you in the opposite lanes than are going with you in your lanes.) So the collision frequency is weighted in favor of those molecules with higher relative velocities. To derive this factor we write the probability that particle 1 has velocity v_1 and particle 2 has velocity v_2 as the product of the respective probabilities,

$$P(v_1, v_2) = P(v_1)P(v_2) = Ce^{-\beta m(v_1^2 + v_2^2)/2} \qquad (16.14)$$

Next we write v_1 and v_2 in terms of the center of mass and the relative velocities V, u, which are given by

$$V = \frac{v_1 + v_2}{2} \qquad \text{and} \qquad u = v_1 - v_2.$$

so that

$$v_1 = V + \frac{u}{2}, \quad v_2 = V - \frac{u}{2},$$

Putting these expressions for v_1 and v_2 into equation 16.14 gives the following distribution in the center of mass and relative velocities (see problem 25):

$$P(V, u) = Ce^{-\beta m V^2} e^{-\beta m u^2/4} \qquad (16.15)$$

For collisions, we don't care about the center of mass velocity, so we either ignore it or integrate it out. But we do care about the distribution in the *relative* velocity u. The above equation shows that it is of the same form as that for absolute velocities, but with the replacement either $m \to m/2$ or $v \to u/\sqrt{2}$:

$$Ce^{-\beta m v^2/2} \to Ce^{-\beta m u^2/4}.$$

This gives the origin of the factor $\sqrt{2}$. In particular, we see from equation 16.6 that the average relative speed would be

$$\bar{u} = \sqrt{2}\,\bar{v} = \sqrt{\frac{16kT}{\pi m}} \tag{16.16}$$

E Transport processes

A property that is unevenly distributed will become more uniform as the random thermal motions of the molecules cause mixing.[3] The rate of this diffusive transport depends on average molecular speeds and the mean free path. The faster and the farther the molecules go, the more quickly the mixing progresses.

We call Q the property's density and define the x direction as the direction in which it varies, so that $Q = Q(x)$. Here we list three familiar examples of these "transport processes," along with the corresponding property whose density varies.

- In *molecular diffusion*   the density of molecules of type i varies with x:

$$Q(x) = \rho_i(x). \tag{16.17a}$$

- In *thermal conduction*   the density of the thermal energy (i.e., the density of the particles times the average thermal energy of each) varies with x:

$$Q(x) = \rho\left[\frac{\nu}{2}kT(x)\right]. \tag{16.17b}$$

- In *viscous flow*   the momentum density of flow in the y direction varies with x:

$$Q(x) = \rho m \bar{v}_y(x). \tag{16.17c}$$

E.1 One speed and one dimension

We begin by looking at the flux for the case where all particles move with the same speed v in the x direction half going in the $+x$ direction and the other half in the $-x$ direction. Once we get this result, we will then average it over all speeds and all directions.

---

[3] If some property of the gas varies from one region to the next, then the gas is not in equilibrium. Nonetheless, we can safely use the tools of equilibrium thermodynamics as long as the relative variations are small on the scale of the molecular separations (10^{-8} to 10^{-9} m in a typical gas).

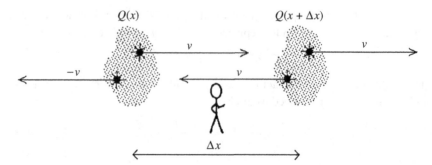

Figure 16.5 Suppose that all molecules are moving in the x direction with velocities $\pm v$, and that we are microscopic observers standing between positions x and $x + \Delta x$, watching the molecules pass by. The density of the property is $Q(x)$ to our left and $Q(x + \Delta x)$ to our right. In each of these regions, only half the particles are moving towards us, so they carry only half the property Q from that region. So the flux from our left is $Q(x)v/2$, and that from our right is $-Q(x+\Delta x)v/2$.

We will imagine that we are microscopic observers, midway between points x and $x + \Delta x$, where the property's densities are $Q(x)$ and $Q(x + \Delta x)$, respectively (Figure 16.5). In both regions only half of the particles are coming towards us (the other half are going away from us) and so the flux (density times velocity) past us of the property from each direction is given by

$$\text{flux from left} = +\frac{Q(x)}{2}v, \qquad \text{flux from right} = -\frac{Q(x + \Delta x)}{2}v.$$

The sum of these two gives the *net* flux of this property past us:

$$\text{net flux} = -\frac{v}{2}[Q(x + \Delta x) - Q(x)] = -\frac{v}{2}\frac{dQ}{dx}\Delta x.$$

But how do we decide upon the distance "Δx"? To answer this question we observe that molecules entering a new region generally require more than one collision each, on average, in order to either completely acquire or deliver the property for that region. For this reason, the distance Δx would be some small number n of mean free paths ($\Delta x = n\bar{l}$), and the preceding equation would become

$$\text{net flux} = -\frac{n\bar{l}v}{2}\frac{dQ}{dx}. \tag{16.18}$$

E.2   All speeds and all directions

In a real gas the particles are moving with a distribution of speeds and in all directions. The x-components of a particle's mean free path and velocity are given by

$$\bar{l}_x = \bar{l}\cos\theta, \qquad v_x = v\cos\theta,$$

where θ is the angle that its direction of motion makes with the x-axis. So, for real gases we need to replace the product $\bar{l}v$ in equation 16.18 by $\bar{l}_x v_x$ and average over all speeds and all directions. For those particles coming from the left[4] we would have

$$\bar{l}v \rightarrow (\bar{l}_x v_x)_{ave} = \int\limits_0^\infty P(v)dv \frac{1}{2\pi} \int\limits_{\phi=0}^{2\pi} \int\limits_{\theta=0}^{\pi/2} \sin\theta \; d\theta d\phi \; v\bar{l}\cos^2\theta,$$

Where the first integral is over speed and the second and third are over angle. Because $\bar{l}$ is independent of both the speed and the angle (equation 16.13) and v is independent of the angle, the right-hand side breaks into three factors:

$$(\bar{l}_x v_x)_{ave} = \bar{l} \left[\int\limits_0^\infty P(v)v dv \right] \left[\frac{1}{2\pi} \int\limits_{\phi=0}^{2\pi} \int\limits_{\theta=0}^{\pi/2} \sin\theta \; d\theta d\phi \; \cos^2\theta \right] = \bar{l} [\bar{v}] \left[\frac{1}{3} \right],$$

where $\bar{l}$ and $\bar{v}$ are given by equations 16.13 and 16.6, respectively. We put this result for $(\bar{l}_x v_x)_{ave}$ into equation 16.18 to get the "diffusion equation,"

$$\text{net flux of } Q = J_x = -\frac{n\bar{l}\,\bar{v}}{6} \frac{dQ}{dx}. \tag{16.19}$$

We note that the main features of this result are common sense. First, the minus sign indicates that the net transport is in the direction opposite to the gradient. Diffusion takes things from higher concentrations toward lower concentrations, and not vice versa. Second, the factors that are the particles' average speed and the mean free path tell us that the faster and farther the particles move, the faster diffusion progresses.

For each of the equations 16.17a–c, we group all constants together, and the diffusion equation takes on the following form.

- *Molecular diffusion*

$$J_x = -D\frac{d\rho_i}{dx}, \qquad \text{where the "diffusion constant" is} \quad D = \frac{n\bar{l}\bar{v}}{6}. \tag{16.20a}$$

- *Thermal conduction*

$$J_x = -K\frac{dT}{dx}, \qquad \text{where the "thermal conductivity" is} \quad K = \frac{n\bar{l}\,\bar{v}}{6}\rho\frac{v}{2}k. \tag{16.20b}$$

- *Viscous flow*

$$J_x = -\eta\frac{dv_y}{dx}, \qquad \text{where the "coefficient of viscosity" is} \quad \eta = \frac{n\bar{l}\bar{v}}{6}\rho m. \tag{16.20c}$$

In the last equation, the momentum flux is called the "stress" and it measures the sideways "drag" or "viscous" force between neighboring layers of a fluid

---

[4] For those coming from the right, the θ integral goes from $\pi/2$ to π, and we get the same answer.

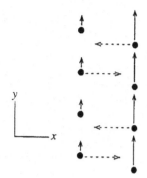

Figure 16.6 The two columns of dots represent two neighboring layers of a fluid that are flowing in the y direction at different speeds. The broken-line arrows indicate diffusion of the molecules in the x direction. Particles diffusing to the left carry more y momentum than those diffusing to the right. This mixing makes the fluid on the left speed up and that on the right slow down. That is, the diffusion causes each layer to exert a viscous drag on the other.

(Figure 16.6). It has the units of pressure (force/area), but it differs in that the force is parallel to the area rather than perpendicular to it.

E.3 Conserved properties

Generalizing from one to three dimensions, we can write all the above processes as

$$\mathbf{J} = -D\,\nabla Q \qquad \text{(diffusion equation)}, \qquad (16.21)$$

where Q is the density of the diffusing property and D is the appropriate diffusion constant. If this property is conserved (as are particles, energy, and momentum, for example) then it must satisfy the continuity equation (equation 12.17)

$$\frac{\partial Q}{\partial t} = -\nabla \cdot \mathbf{J} \qquad \text{(continuity equation)},$$

which states that the change in concentration of the property inside any volume is equal to the difference between the rate at which it enters and leaves (subsection 12E.5).

Taking the divergence of equation 16.21, using this expression for $\nabla \cdot \mathbf{J}$ in the continuity equation above, and then dividing by D gives the generalized form of the "heat equation" (equations 12.20 and 12.21):

$$\nabla^2 Q = \frac{1}{D}\frac{\partial Q}{\partial t} \qquad \text{(heat equation)}, \qquad (16.22)$$

whose solution was given in equation 12.23. Namely, if the property varies in the x direction and its concentration at time $t = 0$ is given by $Q(x, t = 0) = f(x)$ then its concentration at any later time is given by

$$Q(x, t) = \int_{-\infty}^{\infty} f(x') \left(\frac{1}{\sqrt{4\pi Dt}} e^{-(x-x')^2/4Dt} \right) dx'. \qquad (16.23)$$

The first order "diffusion equation" 16.21 states that things diffuse from higher to lower concentrations ($\mathbf{J}$ is *backwards* to the gradient), and the second order "heat equation" 16.22 states that concentrations even out, decreasing near local maxima ($\partial Q/\partial t < 0$ when $\nabla^2 Q < 0$) and increasing near local minima ($\partial Q/\partial t > 0$ when $\nabla^2 Q > 0$).

Summary of Sections B–E

Using the probability distribution for molecular speeds 16.5' we find the following. The mean values of the speed and speed squared for a particle in a gas are (equations 16.6, 16.7)

$$\bar{v} = \sqrt{\frac{8kT}{\pi m}}, \qquad \overline{v^2} = \frac{3kT}{m}.$$

The flux of particles moving in any direction and the rate at which the gas particles exit an opening of area A in their container (ρ is the particle density) are (equations 16.9, 16.10)

$$J_x = \rho \sqrt{\frac{kT}{2\pi m}}, \qquad \frac{dN}{dt} = -\frac{N}{V} \sqrt{\frac{kT}{2\pi m}} A.$$

The collision frequency and the mean free path (σ is the collisional cross section and equals $4\pi R^2$, where R is the effective molecular radius) are (equations 16.12, 16.13)

$$v_c = \sqrt{2}\rho\sigma\bar{v}, \qquad \bar{l} = \bar{v}\left(\frac{1}{v_c}\right) = \frac{1}{\sqrt{2}\rho\sigma}.$$

The average relative speed ($\bar{u}$) and absolute speed ($\bar{v}$) of a system of colliding particles are related by (equation 16.16)

$$\bar{u} = \sqrt{2}\,\bar{v}.$$

If Q is the density of some property of the gas that varies from one region to the next, it will even out as the random thermal motions of the molecules cause mixing. If we define the x direction to be the direction in which Q varies, then the net flux of this property past a point is given by the diffusion equation (equation 16.19)

$$J_x = -\frac{n\bar{l}\bar{v}}{6}\frac{dQ}{dx},$$

where the average speed and mean free path are given by equations 16.6 and 16.13, respectively, and where n is a measure of the number of collisions required to transfer the property. Applications include the following important processes (equations 16.20a–c).

- *Molecular diffusion*

$$J_x = -D\frac{d\rho_i}{dx}, \qquad \text{where the diffusion constant is} \quad D = \frac{n\bar{l}\bar{v}}{6}.$$

- *Thermal conduction*

$$J_x = -K\frac{dT}{dx}, \qquad \text{where the thermal conductivity is} \quad K = \frac{n\bar{l}\bar{v}}{6}\rho\frac{v}{2}k.$$

- *Viscous flow*

$$J_x = -\eta\frac{dv_y}{dx}, \qquad \text{where the coefficient of viscosity is} \quad \eta = \frac{n\bar{l}\bar{v}}{6}\rho m.$$

The generalization of the diffusion equation to all directions for the density Q of any property is (equation 16.21)

$$\mathbf{J} = -D\nabla Q \qquad \text{(diffusion equation)},$$

where D is the appropriate diffusion constant. When combined with the continuity equation for conserved properties, it gives (equation 16.23)

$$\nabla^2 Q = \frac{1}{D}\frac{\partial Q}{\partial t} \qquad \text{(heat equation)},$$

which we have encountered before, at the end of Chapter 12.

Problems

For many of these problems, it will be helpful to consult the table of standard integrals in Appendix E.

Section A

1. The probability that the x-component of velocity of a molecule lies in a certain range is 0.3, that the y-component lies in a certain range is 0.2, and that it the z-component lies in a certain range is 0.1. What is the probability that all three components lie in the prescribed ranges?

2. For motion in one dimension and in terms of m, k, T, for what value of the momentum will the probability be half as large as the probability for a molecule to stand still?

3. Suppose that you invest half of your money in each of two businesses. Each business has a 10% chance of failing. What is the probability that (a) both will fail, (b) neither will fail?

4. Starting with $P(\mathbf{p})d^3 p = \left(\frac{\beta}{2\pi m}\right)^{3/2} e^{-\beta p^2/2m} d^3 p$, derive $P(\mathbf{v})d^3 v$ by replacing p with mv.

5. The probability that the x-component of a molecule's velocity lies in the range dv_x is given by equation 16.3'. Check that the normalization is correct by integrating this probability distribution over all v_x.

6. According to equation 16.3', the distribution in v_x for particles in a gas is a Gaussian distribution. (See Section 3B.) The molecular mass is m and the temperature is T.
   (a) What is the standard deviation for the x-velocities of these particles?
   (b) the coefficient for a Gaussian distribution is $1/\sqrt{2\pi}\sigma$. Is the coefficient obtained in this way the same as that in equation 16.3'?
   (c) Since the distribution is centered around $v_x = 0$, the square of the standard deviation is equal to the mean value of v_x^2. What is the mean value of $(1/2)mv_x^2$?

7. Check the normalization of the expression 16.4' for $P(\mathbf{v})d^3 v$, by expressing $d^3 v$ in spherical coordinates and then integrating over all values of these coordinates.

8. In terms of the molecular mass m and the temperature T, what is the most probable speed for a molecule in a gas? (Hint: "Most probable" means that $P(v)$ is a maximum.)

9. If you have access to a computer that does numerical integration, find the fraction of the molecules in a gas that have energies above (a) $(3/2)kT$, (b) $3kT$, (c) $6kT$.

Section B

10. Do the integrations in equations 16.6 and 16.7 to show that the answers given are correct.

11. For a nitrogen (N_2) molecule at room temperature ($m = 4.7 \times 10^{-26}$ kg, $T = 295$ K), what is its (a) average velocity, (b) average speed $\bar{v}$, (c) root mean square speed, (d) most probable speed (see problem 8)?

12. You are interested in the average value of v^3 the speed cubed, for the molecules of a gas.
    (a) What is this in terms of m, k, and T?
    (b) What is the cube root of the mean cubed speed, $\sqrt[3]{\overline{v^3}}$, for helium ($m = 6.6 \times 10^{-27}$ kg) at 295 K?

13. What is the ratio of the root mean square speeds for water and carbon dioxide molecules at 295 K? (The molecular mass numbers are 18 and 38, respectively.)

14. Of all the air molecules in a room, at any instant half are going in the $+z$ direction and the other half in the $-z$ direction.
    (a) In terms of m, k, and T, what is the average z-component of velocity, v_z, of the half going in the $+z$ direction? (Hint: You might wish to write $v_z = v \cos \theta$ and use spherical coordinates, with $0 \le \theta \le \pi/2$.)
    (b) What fraction of the average speed $\bar{v}$ is your answer to (a)?

Section C

15. Do the integrations in equation 16.9 to show that the answer given is correct.

16. The density of air molecules at room temperature (295 K) and atmospheric pressure is about 2.7×10^{25} molecules/m^3. Their average mass is 4.8×10^{-26} kg. Use this and equation 16.9 to answer the following.
    (a) What is the flux of particles striking a wall of your room (in particles per square meter per second)?
    (b) If a micrometeorite punctured a hole 0.2 mm in diameter in the wall of a spaceship, at what rate would molecules leave if the air were held at atmospheric pressure and room temperature?

17. Suppose that a spaceship has a tiny hole in its side, of area 1 mm^2. The volume of the cabin is 40 m^3, and it is kept at a constant temperature of 17 °C.

Initially, the density of air molecules in the cabin is 2.7×10^{25} molecules/m^3.
The average mass of an air molecule is 4.8×10^{-26} kg.
(a) Initially, how many molecules leave the cabin per second?
(b) The rate at which the molecules leave the cabin is given by $dN/N = -Cdt$, where C is a constant. What is the value of this constant?
(c) Find the expression for the number of particles, N, as a function of the time t.
(d) How much time is required before the density of the air in the cabin falls to half its initial value?

18. In the problem 17, in which air escapes through a puncture in a spacecraft:
    (a) What is the average molecular kinetic energy of the escaping molecules? (Hint: the faster moving molecules exit more rapidly, so the escaping molecules tend to have higher than average kinetic energies. You might weight the kinetic energy of particles having x-component of velocity v_x by the fraction of the exiting flux having that kinetic energy.)
    (b) How fast will the temperature of the remaining air decrease? (Answer in terms of m, T, A, and V, i.e., the molecular mass, temperature, area of the hole, and volume of the air in the spacecraft.)

19. A light bulb manufacturer wishes to evacuate bulbs by putting a tiny 0.5 mm diameter hole in each bulb and then placing it in a vacuum. The air in the bulb is at a temperature of 290 K and has a volume of 200 cm^3. The average mass of an air molecule is 4.8×10^{-26} kg. How much time is required before the amount of air inside the bulb is reduced to 10^{-6} times its previous value? (Hint: See problem 17.)

20. In Section C in the main text, we found that the pressure exerted by a gas on the walls of its container is given by $p = \int_{v_x>0} 2mv_x \, dJ_x$, where dJ_x is the flux due to the particles whose velocities lie in the range dv_x. Do the integration, and see whether the answer gives $pV = NkT$.

Section D
21. Four cars are all equidistant from an intersection and are traveling toward it at 10 m/s. Suppose that you are in the car traveling north. What would be your speed relative to the car traveling (a) south, (b) east, (c) west, (d) another car traveling north at 10 m/s?, (e) If cars going in each of these directions were equally spaced, which would you pass most frequently?

22. If the temperature of a gas is doubled, by what factor do the following change: (a) the collision frequency, (b) the mean free path?

23. Assuming that they both have the same temperature, pressure, and molecular radius, do molecules in nitrogen gas (N_2) or water vapor (H_2O) undergo collisions more frequently? By how many times?

24. Consider two systems, one of pure helium gas and one of pure nitrogen gas, at standard temperature and pressure (0 °C, 1 atm). Under these conditions, a mole of gas occupies 22.4 liters of volume.
    (a) Given the following densities of their liquid phases, roughly what is the molecular radius of each? N_2, density $= 0.808$ g/cm^3 (liquid phase); He, density $= 0.145$ g/cm^3 (liquid phase).
    (b) In their gas phases, what is the mean free path of a molecule of each?
    (c) Also in their gas phases, what is the collision frequency for a molecule of each?

25. Show that equation 16.15 follows from 16.14.

Section E

26. Calculate the value of the following functions, averaged over the $+z$ hemisphere $\theta \leq \pi/2$: (a) $\cos\theta$, (b) $\sin\theta$, (c) $\cos^2\theta$, (d) $\cos\phi$.

27. Suppose that the particle density, ρ, increases in the $+x$ direction.
    (a) What is the sign of $\partial\rho/\partial x$?
    (b) According to equation 16.20a, in which direction will the net particle flux be? (D is positive.)

28. If the temperature of a gas is doubled, is the rate of diffusion increased or decreased? By what factor?

29. All else being equal, would the rate of transport of any property be larger or smaller in gases with fatter molecules? Can you give a physical explanation for this?

30. All else being equal, does diffusion in a gas go faster or slower, or is there no difference, if (a) the molecules are fatter, (b) the molecules are more massive, (c) the gas is denser, (d) the gas is hotter?

31. In air at 295 K, the molecules have five degrees of freedom each and an average mass of 4.8×10^{-26} kg, and there are 2.5×10^{25} molecules/m^3. Each molecule has radius 1.9×10^{-10} m. Assume that $n = 1$. With this information, estimate the value of the following for diffusive processes in air: (a) the diffusion constant, (b) the coefficient of thermal conductivity, (c) the coefficient of viscosity.

32. If you calculate the coefficient of thermal conductivity for air from equation 16.20 (see part (b) of the preceding problem), you will find that it is about five times smaller than that measured experimentally. How might you account for the difference?

33. The coefficient of thermal conductivity for air is about 0.023 W/(m K). Estimate the net flux of heat through double-glazed windows, if the air gap

is 0.8 cm thick and the temperature outside is 15 °C cooler outside than that inside.

34. For viscosity, we consider the flow speed v for various layers of a gas. If our concern is the motion of particles in the y direction, why is the flux of particles in the x direction relevant?

35. Suppose that a quantity being transported diffusively is measured in units of "#" and that its density Q in a certain region of space varies as ax^3. In terms of #, kg, m, s, what are the units of (a) Q, (b) a, (c) the diffusion constant D? (d) In terms of D and a, what is the rate of change of concentration at the point $x = 0.5\,\text{m}$?

36. Suppose that we are interested in diffusion in the x direction and that some property is suddenly injected into a gas at point x_0, in such a way that its initial density function is given by $Q(x, t = 0) = Q_0\delta(x - x_0)$. Diffusion causes it to spread out with time. In terms of the diffusion constant D:
    (a) How does the width of the spread, as measured by the standard deviation σ, vary with time?
    (b) How does the concentration at point $x = x_0$ vary with time?

Chapter 17
Magnetic properties of materials

Individual atoms have magnetic moments due to the orbits and spins of the electrically charged particles within them. Interaction with imposed external magnetic fields tends to produce some ordering of these magnetic moments. But this ordering is opposed by thermal motion, which tends to randomize their orientations. It is the balance of these two opposing influences that determines the magnetization of most materials.

A   Diamagnetism, paramagnetism, and ferromagnetism

Consider what happens when we place a material in an external magnetic field. According to Lenz's law, any change in magnetic field through a current loop produces an electromotive force that opposes the intruding field. On an atomic level, each electron orbit is a tiny current loop. The external field places an extra force on the orbiting electrons, which causes small modifications of their orbits and a slight magnetization of the material in the direction *opposite* to the external field (homework). This response is called "diamagnetism" and is displayed by all materials.

In addition, there is a tendency for the tiny atomic magnets to change their orientations to line up with an imposed external field (Figure 17.1). This response is called "paramagnetism." It gives the material a net magnetic moment in a direction *parallel* to the imposed external magnetic field. Not all materials are paramagnetic, because in some materials the atoms have no net magnetization to begin with and in others the atomic magnets cannot change their orientations.

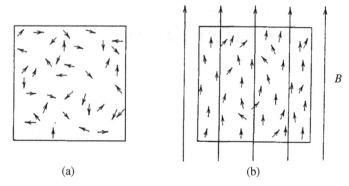

(a)   (b)

Figure 17.1 Paramagnetism. (a) Normally, the magnetic moments of the particles in a system have random alignment. (b) When the material is placed in an external magnetic field B, the magnetic moments tend to become aligned with this field, because it is the state of lower energy and higher probability. Thermal agitation prevents perfect alignment of all the particles.

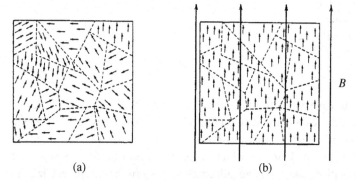

(a)   (b)

Figure 17.2 Ferromagnetism. (a) In ferromagnetic materials, the magnetic moments of neighboring atoms are strongly coupled, so that all within any one domain have nearly the same alignment. This gives each domain a large magnetic moment. (b) Here the domains are fully aligned with an imposed external field. Because of their large magnetic moments, the magnetic energy $\mu_z B$ is very large compared with kT, so the tendency for alignment dominates over the randomizing influence of thermal agitation.

But most materials are paramagnetic, and their paramagnetism dominates over diamagnetism.

In a few materials the magnetic moments of neighboring atoms are very strongly coupled, so that they form "domains" within which the atoms all have nearly the same magnetic orientation. These materials become very strongly magnetized when placed in external magnetic fields, because the domains as a whole line up with the external field and the strong coupling between neighboring atomic magnetic moments prevents any one from changing its orientation (Figure 17.2). This effect is called "ferromagnetism."

Paramagnetism provides the most appropriate challenge for our statistical tools. The alignment of the tiny atomic magnetic moments with an external field is a statistical process, the tendency for alignment being opposed by thermal agitation. For ferromagnetism, the system's basic independent elements are the magnetic moments of macroscopic domains, which are much larger than those of the individual atoms and therefore much less affected by thermal agitation. Diamagnetism can be understood from classical electricity and magnetism, without statistical tools.

B The nature of the atomic magnets

B.1 General

In subsection 1B.7 we learned that the energy of interaction between a magnetic moment μ and an external magnetic field, $\mathbf{B}$, is given by $-\mu \cdot \mathbf{B}$. So if we define the z direction to be that of the external magnetic field, the magnetic interaction energy is $-\mu_z B$ (equation 1.17). We also learned that the magnetic moment is proportional to the angular momentum component L_z (equation 1.15) and that the latter is quantized in units of $\hbar$: $L_z = l_z \hbar$, where $l_z = 0, \pm 1, \pm 2, \ldots, \pm l$. Thus we have

$$\mu_z = g \left(\frac{e\hbar}{2m} \right) l_z, \qquad g = \begin{cases} +1, & \text{proton in orbit,} \\ 0, & \text{neutron in orbit,} \\ -1, & \text{electron in orbit,} \end{cases}$$

(17.1)

for $l_z = 0, \pm 1, \pm 2, \ldots, \pm l$. We found a similar expression for the magnetic moments due to the particles' spins:

$$\mu_z = g \left(\frac{e\hbar}{2m} \right) s_z, \qquad g = \begin{cases} +5.58, & \text{proton spin} \\ -3.82, & \text{neutron spin,} \\ -2.00, & \text{electron spin.} \end{cases}$$

Here

$$s_z = \begin{cases} \pm 1/2, \pm 3/2, \ldots, \pm s, & \text{fermions,} \\ 0, \pm 1, \pm 2, \ldots, \pm s & \text{bosons,} \end{cases}$$

(17.2)

Protons, neutrons, and electrons are spin-1/2 fermions, so for these particles s_z can only have the values $\pm 1/2$.

The constant $e\hbar/2m$ in the above expressions varies inversely with the particle's mass. For the electron it is called the Bohr magneton and given the symbol μ_B. For a proton or neutron it is called the nuclear magneton and given the symbol μ_N. We have

$$\frac{e\hbar}{2m} = \begin{cases} \mu_B = 9.27 \times 10^{-24} \text{ J/T} & \text{Bohr magneton (electrons),} \\ \mu_N = 5.05 \times 10^{-27} \text{ J/T} & \text{nuclear magneton (nucleons).} \end{cases}$$

(17.3)

The less massive electrons have much larger magnetic moments and therefore dominate the magnetic moments of most atoms. In fact, we often ignore

Figure 17.3 An atomic electron makes two contributions to the magnetic moment of the atom, one due to its orbit and the other due to its spin. Both orbits and spins create tiny current loops, with corresponding magnetic moments. The gyromagnetic ratio g is -1 for the orbit and -2 for the spin.

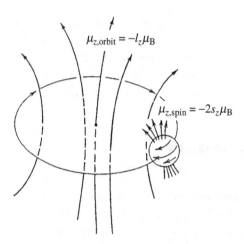

$\mu_{z,\text{orbit}} = -l_z \mu_B$

$\mu_{z,\text{spin}} = -2s_z \mu_B$

contributions from the nucleus except in the contexts of nuclear physics, nuclear magnetic resonance, high-resolution spectroscopy, and the small residual magnetism of atoms whose electrons have zero net angular momentum. Because of its dominance, we will concentrate on electron magnetism for the rest of this chapter. The treatment of nuclear magnetism would be the same, except for the replacements

$$\mu_B \to \mu_N, \qquad g_{\text{electron}} \to g_{\text{appropriate nucleon}}. \tag{17.4}$$

B.2   The electrons

An electron's total magnetic moment is the sum of the contributions from its orbit and from its spin. So we combine equations 17.1, 17.2, and 17.3 to write the total magnetic moment as (Figure 17.3)

$$\mu_z = \left(g_{\text{orbit}}l_z + g_{\text{spin}}s_z\right)\mu_B = -(l_z + 2s_z)\mu_B. \tag{17.5}$$

For an atom, l and s are the total orbital and total spin angular momentum quantum numbers for all the electrons combined, and their z-components may take on the following values:

$$l_z = 0, \pm 1, \pm 2, \ldots, \pm l,$$

$$s_z = \begin{cases} 0, \pm 1, \pm 2, \ldots \pm s, & \text{for an even number of electrons,} \\ \pm 1/2, \pm 3/2, \ldots \pm s, & \text{for an odd number of electrons.} \end{cases}$$

$$\tag{17.6}$$

The range of values for l_z and s_z is quite limited, because the electrons in an atom tend to have configurations that minimize the total angular momentum. In fact, completed electron shells have no net angular momentum ($l = 0, s = 0$), and only electrons in the outer unfilled shells make any net contribution at all. Shells that are nearly filled are best treated as being completely filled shells plus positively charged "holes" corresponding to the missing electrons.

C Paramagnetism

C.1 The general case

We now study the magnetic moment of a paramagnetic material that is placed in an external magnetic field, **B**. The magnetic moment **M** of a large number of atoms is the product of the number of atoms and the average magnetic moment of each: the z-component is given by

$$M_{z,\text{total}} = N\,\overline{\mu}_z \quad \text{and so} \quad M_{z,\text{molar}} = N_A\overline{\mu}_z. \tag{17.7}$$

To calculate the mean value $\overline{\mu}_z$ we use the definition of mean values 2.1 and the probabilities given by equations 15.5 and 15.7:

$$\overline{\mu} = \sum_s P_s\mu_s, \quad \text{where} \quad P_s = Ce^{-\beta\varepsilon_s} = \frac{e^{-\beta\varepsilon_s}}{\sum_s e^{-\beta\varepsilon_s}}. \tag{17.8}$$

According to equation 17.5, the state s is defined by the quantum numbers l_z, s_z, and its energy is given by $-\mu_z B$. So the mean value of the z-component of particle's magnetic moment is

$$\overline{\mu}_z = \sum_{l_z,s_z} P_{(l_z,s_z)}\mu_z = \frac{\sum_{l_z,s_z} e^{\beta\mu_z B}\mu_z}{\sum_{l_z,s_z} e^{\beta\mu_z B}}. \tag{17.9}$$

Writing μ_z explicitly as $-(l_z + 2s_z)\mu_B$, this becomes (problem 11)

$$\overline{\mu}_z = -\mu_B \frac{\sum_{l_z,s_z} e^{-(l_z+2s_z)x}(l_z + 2s_z)}{\sum_{l_z,s_z} e^{-(l_z+2s_z)x}}, \quad \text{where } x = \beta\mu_B B = \frac{\mu_B B}{kT}. \tag{17.10}$$

Example 17.1 A material of 10^{25} atoms at 295 K is in a magnetic field of 1 T. If the total orbital and spin angular momentum quantum numbers of each atom's electrons are $l = 0$ and $s = 1/2$, what is the average magnetic moment of a single atom, and of the entire material?

We use equation 17.10. For an external field of 1 T and a temperature of 295 K,

$$x = \mu_B B/kT = 2.3 \times 10^{-3}.$$

The quantum numbers l_z, s_z take on only the two sets of values $(0, \pm1/2)$, so $l_z + 2s_z = \pm1$ and

$$\overline{\mu}_z = -\mu_B \frac{e^{-x}(1) + e^{+x}(-1)}{e^{-x} + e^{+x}} = -\mu_B(-2.3 \times 10^{-3}) = 2.1 \times 10^{-26}\,\text{J/T}.$$

The magnetic moment of the entire system of 10^{25} atoms is

$$M_{z,\text{total}} = N\overline{\mu}_z = 0.21 \text{ J/T}.$$

Expression 17.10 for the average magnetic moment $\overline{\mu}_z$ can be simplified if the value of $x = \mu_B B/kT$ in the exponent is either much smaller or much larger than 1. We now examine these two regions more closely and show that the magnetic

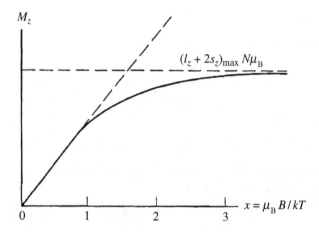

Figure 17.4 Plot of magnetic moment M_z vs. $x = \mu_B B/kT$ for a typical paramagnetic substance. For $\mu_B B/kT \ll 1$, the magnetic moment increases linearly with the external magnetic field according to the Curie law, $M_z = C(B/T)$. For $\mu_B B/kT \gg 1$, the magnetic moment reaches its maximum value, $(l_z + 2s_z)_{max} N\mu_B$. For $\mu_B B/kT$ intermediate between these two extremes, it would be necessary to use equations 17.7 and 17.10 for the magnetic moment.

moment increases linearly with x for $x \ll 1$ and approaches a constant maximum value for $x \gg 1$ (Figure 17.4).

C.2 $\mu_B B \ll kT$

In the limit $x = \mu_B B/kT \ll 1$, the magnetic energy is small compared with the thermal energy, so the tendency for magnetic moments to align with the external field is greatly reduced by thermal randomization. Example 17.1 demonstrated that even for a strong magnetic field at room temperature, x is only 0.0023. That is typical. Except for very low temperatures or magnetic fields much stronger than are attainable on Earth, we can expect x to be small.

For these small exponents, we can use the expansion $e^{-\delta} \approx 1 - \delta$ for $\delta \ll 1$ to transform equation 17.10 into

$$\bar{\mu}_z = -\mu_B \frac{\sum_{l_z, s_z} [1 - (l_z + 2s_z)x](l_z + 2s_z)}{\sum_{l_z, s_z} [1 - (l_z + 2s_z)x]}.$$

All terms linear in $l_z + 2s_z$ add up to zero, because they come in positive and negative pairs (e.g., $l_z = 0, \pm 1, \pm 2, \ldots$), which cancel. So we drop the linear terms, which gives

$$\bar{\mu}_z = \mu_B x \frac{\sum_{l_z, s_z} (l_z + 2s_z)^2}{\sum_{l_z, s_z} 1}, \qquad \text{where } x = \frac{\mu_B B}{kT}. \qquad (17.11)$$

If we explicitly factor out B/T from x and multiply by Avogadro's number, we get the following "Curie law" for the molar magnetic moment of this substance:

$$M_{z,\text{molar}} = N_A \overline{\mu}_z = C \frac{B}{T}, \qquad \text{for} \quad \frac{\mu_B B}{kT} \ll 1, \qquad (17.12)$$

where the Curie constant C is given by (problem 12)

$$C = \frac{N_A \mu_B^2}{k} \frac{\sum_{l_z,s_z}(l_z + 2s_z)^2}{\sum_{l_z,s_z} 1} = \left(3.75 \frac{\text{J K}}{\text{T}^2 \text{ mole}} \right) \frac{\sum_{l_z,s_z}(l_z + 2s_z)^2}{\sum_{l_z,s_z} 1}. \qquad (17.12')$$

This result tells us that for most materials under most ordinary conditions, the magnetic moment is directly proportional to B/T. This ratio reflects the battle between the energy of magnetic alignment ($\propto B$) and the energy of thermal randomization ($\propto T$).

The factor on the right of equation 17.12' is a ratio of positive integers that depends on the allowed values of l_z and s_z. The denominator is simply the number of different (l_z, s_z) states. In most cases, the sum includes all values of l_z from $-l$ to $+l$ and all s_z from $-s$ to $+s$. However, the allowed values of l_z and s_z may be interdependent if there is strong "spin–orbit coupling" between the spin and orbital magnetic moments.

C.3 $\mu_B B \gg kT$

The limit $x = \mu_B B/kT \gg 1$ is primarily of interest to those doing low-temperature physics or astrophysics. In this limit, magnetic energy dominates over thermal energy, so the tendency to align with the external field dominates over thermal randomization.

The sums in equation 17.10 include factors like e^{nx}, where n is an integer. When $x \gg 1$, $e^{nx} \gg e^{(n-1)x}$, so we only need consider the one term in which the exponent attains its maximum value. All other terms are negligible in comparison. This greatly simplifies the calculation of $\overline{\mu}_z$, because if we keep only the largest term then the exponential factor in the numerator cancels with that in the denominator, and using $(l_z + 2s_z)_{\text{min}} = -(l_z + 2s_z)_{\text{max}}$, we are left with simply

$$\overline{\mu}_z = \mu_{z,\text{max}} = (l_z + 2s_z)_{\text{max}}\mu_B, \qquad \text{for} \quad \frac{\mu_B B}{kT} \gg 1. \qquad (17.13)$$

That is, the average value of the magnetic moment of a particle is its maximum value, because all magnetic moments are lined up with the external field. This is what we would expect in the limit of high magnetic field and low thermal agitation.

The total magnetic moment per mole in this limit would be

$$M_{z,\text{molar}} = N_A \mu_B (l_z + 2s_z)_{\text{max}}$$
$$= \left(5.58 \frac{\text{J}}{\text{T mole}} \right) (l_z + 2s_z)_{\text{max}}, \qquad \text{for} \quad \frac{\mu_B B}{kT} \gg 1.$$

$$(17.14)$$

Summary of Chapter 17

Diamagnetism is the tendency displayed by all materials to produce a small magnetic moment that opposes the introduction of an applied external magnetic field. It is correctly described by Lenz's law in electromagnetism. Paramagnetism is a stronger effect and is displayed by most materials. It reflects the tendency for the atomic magnetic moments to align themselves with the imposed external field. Ferromagnetism is observed in a few materials where the neighboring atomic magnets interact very strongly with each other, producing domains in which the atoms have nearly the same alignments. In this chapter we have used classical statistics to study paramagnetism.

The magnetic moments caused by orbiting electrical charges are as follows (equation 17.1)

$$\mu_z = g\left(\frac{e\hbar}{2m}\right)l_z, \qquad g = \begin{cases} +1, & \text{proton in orbit,} \\ 0, & \text{neutron in orbit,} \\ -1, & \text{electron in orbit,} \end{cases}$$

$l_z = 0, \pm1, \pm2, \ldots, \pm l$, and those due to their spins are given by (equation 17.2)

$$\mu_z = g\left(\frac{e\hbar}{2m}\right)s_z, \qquad g = \begin{cases} +5.58, & \text{proton spin,} \\ -3.82, & \text{neutron spin,} \\ -2.00, & \text{electron spin,} \end{cases}$$

$$s_z = \begin{cases} \pm1/2, \pm3/2, \ldots, \pm s, & \text{fermions,} \\ 0, \pm1, \pm2, \ldots, \pm s, & \text{bosons.} \end{cases}$$

The factor $e\hbar/2m$ is given by (equation 17.3)

$$\frac{e\hbar}{2m} = \begin{cases} \mu_B = 9.27 \times 10^{-24} \text{ J/T} & \text{Bohr magneton (electrons),} \\ \mu_N = 5.05 \times 10^{-27} \text{ J/T} & \text{nuclear magneton (nucleons).} \end{cases}$$

Because the Bohr magneton is so much larger than the nuclear magneton, the magnetic properties of most materials are dominated by the electrons. The gyromagnetic ratio for the electron orbit is -1 and for the electron spin is -2, so we can write the magnetic moment for the electrons of an atom as (equation 17.5)

$$\mu_z = \left(g_{\text{orbit}}l_z + g_{\text{spin}}s_z\right)\mu_B = -(l_z + 2s_z)\mu_B,$$

where the quantum numbers l_z, s_z may take on the values (equation 17.6)

$$l_z = 0, \pm1, \pm2, \ldots, \pm l,$$
$$s_z = \begin{cases} 0, \pm1, \pm2, \ldots \pm s, & \text{for an even number of electrons,} \\ \pm1/2, \pm3/2, \ldots \pm s, & \text{for an odd number of electrons,} \end{cases}$$

and where the values of l and s depend on the particular electronic configuration. If there is spin–orbit coupling, the allowed values of l_z and s_z may not be independent of each other.

The same treatment applies to molecules, ions, free electrons, or whatever the basic elementary unit of the material is. Nuclear magnetism would be treated in the same way, except that the Bohr magneton is replaced by the nuclear magneton and the electron's gyromagnetic ratio by that of the nucleons (equation 17.4):

$$\mu_B \rightarrow \mu_N, \qquad g_{electron} \rightarrow g_{appropriate\ nucleon}.$$

We considered a material in an external magnetic field B oriented along the positive z-axis. According to classical statistics the average value of an atom's magnetic moment is (equation 17.10)

$$\overline{\mu}_z = -\mu_B \frac{\sum_{l_z, s_z} e^{-(l_z + 2s_z)x} (l_z + 2s_z)}{\sum_{l_z, s_z} e^{-(l_z + 2s_z)x}}, \qquad \text{where} \quad x = \beta \mu_B B = \frac{\mu_B B}{kT}.$$

In the common case where $\mu_B B \ll kT$, this can be written as the Curie law for the magnetic moment per mole (equations 17.12, 17.12'):

$$M_{z, molar} = C \frac{B}{T}, \qquad \text{for} \quad \frac{\mu_B B}{kT} \ll 1,$$

where

$$C = \left(3.75 \frac{\text{J K}}{\text{T}^2 \text{ mole}}\right) \frac{\sum_{l_z, s_z} (l_z + 2s_z)^2}{\sum_{l_z, s_z} 1}.$$

In the more exceptional case where $\mu_B B \gg kT$, all magnetic moments align with the magnetic field and the molar magnetic moment reaches its maximum value (equation 17.14):

$$M_{z, molar} = N_A \mu_B (l_z + 2s_z)_{max}$$

$$= \left(5.58 \frac{\text{J}}{\text{T mole}}\right) (l_z + 2s_z)_{max}, \qquad \text{for} \quad \frac{\mu_B B}{kT} \gg 1.$$

Example 17.2 Consider a material whose molecular electron clouds have orbital and spin angular momentum quantum numbers $l = 1$ and $s = 1/2$, so that the allowed values of l_z are $0, \pm 1$ and the allowed values of s_z are $\pm 1/2$. If a mole of this material at room temperature were in an external field of 3 T, what would be its magnetic moment?

For this case, the factor $x = \mu_B B/kT$ is given by

$$x = 6.8 \times 10^{-3}.$$

Because this is much less than 1, the material obeys the Curie law. The six different states are given by

$$(l_z, s_z) = (1, 1/2), (1, -1/2), (0, 1/2), (0, -1/2), (-1, 1/2), (-1, -1/2),$$

so that

$$\frac{\sum_{l_z, s_z} (l_z + 2s_z)^2}{\sum_{l_z, s_z} 1} = \frac{2^2 + 0^2 + 1^2 + (-1)^2 + 0^2 + (-2)^2}{6} = \frac{10}{6}.$$

Putting this into equation 17.12′ gives the following value for the Curie constant:

$$C = \left(3.75 \, \frac{J\,K}{T^2 \, mole}\right) \frac{10}{6} = 6.25 \, \frac{J\,K}{T^2 \, mole}.$$

Thus the molar magnetic moment is

$$M_{z,molar} = C\frac{B}{T} = \left(6.25 \times \frac{3}{295}\right) \frac{J}{T \, mole} = 0.0636 \, \frac{J}{T \, mole}.$$

Example 17.3 If the above material's temperature were 0.1 K, what would be its magnetic moment?

In this case the factor $x = \mu_B B/kT$ is given by

$$x = 20 \gg 1,$$

so all the magnetic moments are aligned along the z-axis and we use the result 17.14 with $(l_z + 2s_z)_{max} = z$, obtaining

$$M_{z,molar} = \left(5.58 \, \frac{J}{T \, mole}\right) (l_z + 2s_z)_{max} = 11.2 \, \frac{J}{T \, mole}.$$

Problems

Section A

1. Consider the magnetic moment caused by the orbit of an electron that is in the plane of this page and going clockwise as seen from above.
   (a) Does this magnetic moment point into or out of the page?
   (b) We now apply an external magnetic field that points down into the page. Does the resulting magnetic force on the orbiting electron point radially inward or outward? Does it add to or detract from the Coulomb force of attraction between the electron and the nucleus?
   (c) With this change in the centripetal force, will the electron's orbital speed increase or decrease? Will its magnetic moment increase or decrease? Does this change in its magnetic moment parallel or oppose the applied external field?
   (d) Repeat the above for an applied magnetic field pointing upward out of the page.

2. (a) Explain physically why ferromagnetic materials normally become much more strongly magnetized than paramagnetic materials. (Hint: Why would thermal agitation be less effective in randomizing the atomic alignments?)
   (b) Under what conditions might you expect paramagnetism to be as strong as ferromagnetism?

Section B

3. The spin gyromagnetic ratio of the neutron is negative. What does that imply about the distribution of charge within it?

4. Suppose that a positively charged spin-1/2 particle with a negative spin gyromagnetic ratio were found. What would you conclude about the net charge and charge distribution within this particle?

5. The magnitude of the total angular momentum of an atom is given by $\sqrt{l(l+1)}\hbar$, and the maximum z-component is $l\hbar$. What is the minimum angle that the angular momentum vector can make with the positive z-axis for the following values of l: (a) 1, (b) 2, (c) 3?

6. If all electrons are spins-1/2 particles ($s = 1/2$, $s_z = \pm 1/2$), how is it possible for the z-component of the spin angular momentum of some atoms or molecules to be whole integers ($s_z = 0, \pm 1$, etc.)?

7. A spin-1/2. negatively charged, particle has spin gyromagnetic ratio $g = -2$ (something like an electron, but with a different mass). If its magnetic moment is $\mu_z = \pm 4.5 \times 10^{-26}$ J/T, what is its mass?

8. A certain atom has two electrons outside the last closed shell, and these are each in an $l = 1$ orbit. Given this information only, what are the possible values of l and s for this atom's electrons? List all the combinations (l_z, s_z) that might be possible.

9. Calculate the spin magnetic moment for a meson that has $g_{spin} = +1$ and mass 1.37×10^{-27} kg.

10. A charmed quark is a spin-1/2 particle with $g_{spin} = +2/3$. Its precise mass is unknown but is estimated to be around 3.5×10^{-27} kg. Estimate the magnetic moment for this quark.

Section C

11. Show that equation 17.10 follows from 17.9.

12. Show that equation 17.11 follows from 17.10 (for $x \ll 1$) and that equations 17.12 and 17.12' follow from 17.11.

13. Show that equation 17.13 follows from 17.10 (for $x \gg 1$).

14. For a material in a magnetic field of 5 T, for what temperature would $x = \mu_B B/kT$ be equal to unity?

15. The tendency for particles' magnetic moments to line up in an imposed magnetic field is opposed by their random thermal motions. Thermal energies are typically kT and magnetic energies are $\mu_z B$.

(a) Calculate these two energies for an electron in a magnetic field of 1 T at 290 K. Which dominates?

(b) At what temperature would the two be comparable for a proton in a field of 1 T?

16. In sodium metal, each atom has two complete electronic shells for which $l = 0, s = 0$, and one electron left over in a state for which $l = 0, s = 1/2$. If a mole of sodium is sitting in an external field of one tesla at room temperature (295 K), find (a) its Curie constant, (b) its magnetic moment.

17. Consider a mole of a certain material for which there is strong spin–orbit coupling in the electron clouds, such that the only allowed sets of values for (l_z, s_z) are $(1, -1/2), (0, 1/2), (0, -1/2), (-1, 1/2)$.
(a) What is the Curie constant?
(b) In an external field of 2 T, what is its magnetic moment at a temperature of 10^3 K? At 10 K?
(c) If it is sitting in the Earth's field of about 10^{-4} T, below what temperature would you expect this material to become nearly completely magnetized?

18. Consider a mole of a material for which each molecule's electron cloud has $l = 1, s = 0$; there is no spin–orbit coupling. It is placed in an external field of 2 T. For this material, find
(a) its Curie constant,
(b) its magnetic moment at a temperature of 273 K,
(c) the value of $x = \mu_B B/kT$ at a temperature of 1.4 K,
(d) the magnetic moment at a temperature of 1.4 K.

19. Consider a mole of a material made of molecules for which the electron clouds each have $l = 0, s = 1$; there is no spin–orbit coupling. It is placed in an external magnetic field of strength 3 T, and is at a temperature of 2 K. For this material, find
(a) the value of $x = \mu_B B/kT$,
(b) the magnetic moment.

20. Consider a material whose molecular electron clouds each have $l = 2, s = 1/2$; there is no spin–orbit coupling. A mole of this material is at 295 K and in a field of strength 10^{-4} T.
(a) What is the Curie constant for this material?
(b) What is its magnetic moment?
(c) What is the average magnetic moment per molecule?

21. Consider a material for which the atomic electron clouds have no net angular momentum ($l = 0, s = 0$). The neutrons in each nucleus also have no net angular momentum ($l = 0, s = 0$), but the protons in each nucleus do have a net spin angular momentum: $l = 0, s = 1$. Therefore the magnetic properties of this material are due entirely to the protons in the nucleus.

(a) What is the value of $N_A \mu_N^2 / k$?

(b) What is the value of the Curie constant for the magnetization of this material? (Be careful! Remember that $g_{spin} \neq -2$ for the proton, and so the "ratio of integers" won't be integers.)

(c) In a field of 1 T and at room temperature (295 K), what would be the magnetic moment per mole of this material?

22. What is the Curie constant for a mole of a material whose molecular electron clouds have no spin–orbit coupling if (a) $(l, s) = (0, 1/2)$, (b) $(l, s) = (1, 3/2)$, (c) $(l, s) = (2, 0)$?

23. What is the maximum molar magnetic moment for each substance in the problem above?

24. Consider a system of 10^{24} spin-1/2 electrons at temperature 7 K. If they are in a magnetic field of 5 T, what is the total magnetic moment of this electron system?

25. Consider a system of 10^{25} electrons in an external magnetic field of 2 T. Each electron is in an $l = 1$ orbit. There is strong spin–orbit coupling, so that the only allowed combinations (l_z, s_z) are $(1, 1/2)$, $(0, 1/2)$, $(0, -1/2)$, $(-1, -1/2)$. For this material, find (a) the Curie constant, (b) the magnetic moment at 290 K, (c) the magnetic moment at 3 K.

Chapter 18
The partition function

We now examine a remarkable tool called the "partition function," which can be used to simplify many calculations. We begin by recalling that when a small system is interacting with a large reservoir, the probability for it to be in state s is $P_s = Ce^{-\beta E_s}$ and the mean value of any property f of the small system is given by

$$\overline{f} = \sum_s f_s P_s \quad \text{with} \quad P_s = Ce^{-\beta E_s}. \tag{18.1}$$

Unfortunately, the sum over states can include a very large number of terms, and the calculation must be repeated for each property f that we study.

The partition function facilitates these otherwise tedious calculations. To produce this function, we cannot avoid a sum over all states. However, once this sum is done and the partition function is known, many different properties may be calculated directly from it.

A Definition

Requiring the sum of all probabilities to equal unity determines the constant C in the expression 18.1:

$$\sum_s P_s = C \sum_s e^{-\beta E_s} = 1.$$

Thus

$$P_s = \frac{1}{Z} e^{-\beta E_s} \quad \text{where} \quad Z = \sum_s e^{-\beta E_s}. \tag{18.2}$$

The sum on the right of 18.2 is called the "partition function" and is customarily given the symbol Z, as indicated. As you can see, each term in this sum is proportional to the corresponding probability. And the probabilities, in turn, are proportional to the number of states accessible to the combined system (small system plus reservoir). It is this relationship to the entropy that gives the partition function its seemingly magical powers. It contains no new physics. It simply makes calculations of mean values easier by providing automatically the correct probability weightings.

Like all thermodynamic quantities, the partition function generally depends on three variables, usually chosen to be T, V, N, where T is the temperature of the *reservoir*, and V, N are the volume and number of particles for the *small system*. The dependence on T is explicit in the exponential factor $\beta = 1/kT$ but the dependence on V and N is more subtle, influencing both the sum over states $\sum_s$ and the energy of each state, E_s. Although the temperature does influence the probabilities, it does *not* affect the energy of any one state.[1] (For example, the outer electron of an atom has only certain states available to it. At higher temperatures there is a greater probability for the electron to be in an excited state, but the energy of each state remains the same.) So if we were to rewrite the partition function, indicating explicitly its dependence on the three variables, we would have

$$Z(T, V, N) \equiv \sum_{s(V,N)} e^{-\beta(T)E_s(V,N)}. \tag{18.3}$$

B Calculation of mean values

We now look at some examples. Assume that we have already calculated the partition function Z, so that we know its dependence on T, V, N. The partial derivative with respect to any of these variables will imply that the other two are held constant. Sometimes it will be more convenient to use $\beta (= 1/kT)$ instead of the variable T. As always, we will use the generic symbols p, V for all types of mechanical interactions and μ, N for all types of particles.

---

[1] Although there are exceptions to this statement (e.g., collisional broadening), it is generally correct.

B.1   Internal energy

The average internal energy of the small system is given by

$$\overline{E} = \sum_s P_s E_s = \frac{1}{Z} \sum_s e^{-\beta E_s} E_s.$$

Noticing that

$$e^{-\beta E_s} E_s = -\frac{\partial}{\partial \beta} e^{-\beta E_s},$$

we can write

$$\overline{E} = -\frac{1}{Z} \frac{\partial}{\partial \beta} \sum_s e^{-\beta E_s} = -\frac{1}{Z} \frac{\partial Z}{\partial \beta} = -\frac{\partial}{\partial \beta} \ln Z. \qquad (18.4)$$

In a similar fashion, we can find the mean square of the internal energy:

$$\overline{E^2} = \sum_s P_s E_s^2 = \frac{1}{Z} \sum_s e^{-\beta E_s} E_s^2$$

$$= \frac{1}{Z} \frac{\partial^2}{\partial \beta^2} \sum_s e^{-\beta E_s} = \frac{1}{Z} \frac{\partial^2 Z}{\partial \beta^2}. \qquad (18.5)$$

From these two results, 18.4 and 18.5, the standard deviation for the fluctuations of the internal energy about its mean value can be calculated through the relationship (homework)

$$\sigma^2 = \overline{E^2} - \overline{E}^2 = \frac{\partial^2}{\partial \beta^2} \ln Z, \qquad (18.6)$$

and from equation 18.4 this becomes, using $\beta = 1/kT$,

$$\sigma^2 = -\left(\frac{\partial \overline{E}}{\partial \beta}\right)_{V,N} = kT^2 \left(\frac{\partial \overline{E}}{\partial T}\right)_{V,N} = kT^2 C_V. \qquad (18.6')$$

B.2   Other properties of the system

The above examples exploit the partition function's dependence on the temperature T. Other calculations exploit its dependence on V and N. For these, we first examine the differential form of the system's Helmholtz free energy (9.15a), which also depends on T, V, N:

$$d\overline{F} = -\overline{S}dT - \overline{p}dV + \overline{\mu}dN.$$

The bars indicate mean values because, like the internal energy, these properties of the small system may fluctuate as it interacts with the reservoir. From this expression, we see that

$$\overline{S} = -\left(\frac{\partial \overline{F}}{\partial T}\right)_{V,N}, \qquad \overline{p} = -\left(\frac{\partial \overline{F}}{\partial V}\right)_{T,N}, \qquad \overline{\mu} = \left(\frac{\partial \overline{F}}{\partial N}\right)_{T,V}. \qquad (18.7)$$

We now use this expression for $\overline{S}$ in the definition 9.14a of the Helmholtz free energy:

$$\overline{F} = \overline{E} - T\overline{S} = \overline{E} + T\left(\frac{\partial \overline{F}}{\partial T}\right)_{V,N} \quad \Rightarrow \quad \overline{F} - T\left(\frac{\partial \overline{F}}{\partial T}\right)_{V,N} = \overline{E}. \tag{18.8}$$

With the help of equation 18.4 you can easily show that the solution to this differential equation is a function of the form[2]

$$\overline{F} = -kT \ln Z + CT,$$

where C is a constant. The third law requirement that the entropy must go to zero at $T = 0$,

$$\overline{S} = -\left(\frac{\partial \overline{F}}{\partial T}\right)_{V,N} \longrightarrow 0 \qquad \text{as} \qquad T \longrightarrow 0,$$

demands that the constant C is zero (homework), and we conclude that the Helmholtz free energy is related to the partition function by

$$\overline{F} = -kT \ln Z. \tag{18.9}$$

So once we know the partition function, Z, we can use 18.9 this in 18.7 to find the mean values of the entropy, pressure, and chemical potential of the small system.

Thus, with the independent variables T, V, N and equations 18.9, 18.4, and 18.7 (giving us $\overline{F}, \overline{E}, \overline{S}, \overline{p}, \overline{\mu}$) we have enough information to calculate nearly any property.

Summary of Sections A and B

The partition function for a small system interacting with a large reservoir is defined by (equation 18.2)

$$Z = \sum_s e^{-\beta E_s}.$$

The partition function is particularly useful in calculating mean values. For example, the mean value of the small system's internal energy, the square of its internal energy, and the standard deviation for fluctuations in energy are given by (equations 18.4–18.6)

$$\overline{E} = -\frac{\partial}{\partial \beta} \ln Z, \qquad \overline{E^2} = \frac{1}{Z}\frac{\partial^2 Z}{\partial \beta^2},$$

$$\sigma^2 = \overline{E^2} - \overline{E}^2 = \frac{\partial^2}{\partial \beta^2} \ln Z.$$

---

[2] For those who are well versed in differential equations, the first term, $-kT \ln Z$, is a particular solution to the complete equation 18.8 and the second term, CT, is the general solution to the corresponding homogenous equation. Since we are holding V and N constant, the constant C can be any function $f(V, N)$ that is independent of T.

From the differential form of the Helmholtz free energy (equation 9.15a)

$$d\overline{F} = -\overline{S}dT - \overline{p}dV + \overline{\mu}dN,$$

we see that (equation 18.7)

$$\overline{S} = -\left(\frac{\partial \overline{F}}{\partial T}\right)_{V,N}, \qquad \overline{p} = -\left(\frac{\partial \overline{F}}{\partial V}\right)_{T,N}, \qquad \overline{\mu} = \left(\frac{\partial \overline{F}}{\partial N}\right)_{T,V}.$$

Using the first of these in the definition of the Helmholtz free energy ($\overline{F} = \overline{E} - T\overline{S}$), we obtain a differential equation for $\overline{F}$, whose solution is (equation 18.9)

$$\overline{F} = -kT \ln Z.$$

With the independent variables T, V, N and equations 18.9, 18.4, and 18.7 (giving us $\overline{F}, \overline{E}, \overline{S}, \overline{p}, \overline{\mu}$) we have enough information to calculate virtually any property.

C Many subsystems and identical particles

C.1 Distinguishable subsystems

Suppose that a system consists of two distinct subsystems, A and B, as in Figure 18.1. The energy of the combined system is the sum of the energies of the two subsystems. Therefore,

$$Z = \sum_s e^{-\beta E_s} = \sum_{a,b} e^{-\beta(E_a + E_b)} = \sum_a e^{-\beta E_a} \sum_b e^{-\beta E_b} = Z_A Z_B.$$

Extending this result to three or more subsystems gives

$$Z_{\text{combined}} = Z_A Z_B Z_C \cdots \qquad \text{(distinguishable subsystems)}. \qquad (18.10)$$

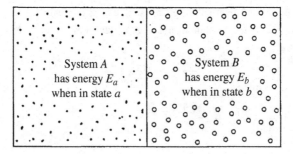

Combined system has energy $E_a + E_b$

Figure 18.1 The sum over all states of the combined system is the sum over the states *a* of system *A* and the states *b* of system *B*. Consequently, the partition function for the combined system is the product of the partition functions for the subsystems.

The partition function for the combined system is the product of those for the subsystems.

C.2 Identical subsystems

We now look at identical subsystems. Examples would include identical molecules in a gas, identical electrons in a semiconductor, identical nuclei in a reactor, etc. If the subsystems are *identical* then the result 18.10 is not correct. As we saw in subsection 6D.2, there are $N!$ different ways to arrange N distinguishable systems among any N states. But if the systems are identical, then all $N!$ arrangements are the same. For example, if systems A and B are identical, then (system A in state 1, system B in state 2) would be identical to (system A in state 2, system B in state 1). So rather than being two separate terms in our sum over states, they should be just one.

You would think, then, that we could simply divide by $N!$ to correct for this over-counting. Unfortunately, it is not that simple. The factor $N!$ applies only if all N subsystems are in *different* states. If there is some overlap (i.e., more than one system is in the same state), then this correction is too severe. For example. there is only one way (not $N!$) that N distinguishable subsystems can be in the *same* state.[3] This problem of overlapping identical subsystems is better handled using quantum statistics, as we will see in the next chapter.

Fortunately, there are large classes of important problems for which there is little overlap among identical subsystems, so the correction factor $N!$ of classical statistics is indeed correct. So we will proceed with this approach and note that the results for Z, F, S, p, μ are strictly valid for identical subsystems only if they are in different states – that is, little or no overlap. With this correction, we have for the sum over states

$$\sum_{s} \rightarrow \frac{1}{N!} \sum_{a} \sum_{b} \sum_{c} \cdots \qquad \text{(identical subsystems, no overlap)},$$

so that

$$Z = \frac{1}{N!} Z_A Z_B Z_C \cdots$$

Using Stirling's formula,

$$\frac{1}{N!} \approx \left(\frac{e}{N}\right)^N,$$

we have the following result for a system consisting of N identical subsystems:

$$Z = \left(\frac{e\zeta}{N}\right)^N \qquad \text{(identical subsystems, no overlap)} \qquad (18.11a)$$

where ζ is the partition function for any one of them:

$$\zeta = Z_A = Z_B = Z_C = \cdots = \sum_{i} e^{-\beta \varepsilon_i} \qquad \text{(one of the subsystems).} \quad (18.11b)$$

---

[3] This might happen, for example, at low temperatures, where many systems might be in the ground state.

If the energies and states accessible to each subsystem are not affected by the other subsystems, we say that the subsystems are "independent." For these, ζ does not depend on N.

Equation 18.11a expresses the partition function for a system of N identical systems or particles in terms of the partition function for just one of them. We can do the same for the Helmholtz free energy by putting the result 18.11a for Z into equation 18.9:

$$\overline{F} = -kT \ln Z = -NkT \ln \left(\frac{e\zeta}{N} \right) \qquad \text{(identical subsystems, no overlap).} \quad (18.12)$$

With this, we can use the formulas of Section B to obtain any property. For example, the third of equations 18.7 gives the following chemical potential for independent subsystems (homework):

$$\overline{\mu} = \left(\frac{\partial \overline{F}}{\partial N} \right)_{T,V} = -kT \ln \left(\frac{\zeta}{N} \right) \qquad \text{(independent identical particles, no overlap).}$$

$$(18.13)$$

Summary of Section C

The partition function for a system is the product of the partition functions for its individual distinguishable subsystems (equation 18.10):

$$Z = Z_A Z_B Z_C \cdots \qquad \text{(distinguishable subsystems).}$$

If the N subsystems are identical and there is little probability of that any of them are in exactly the same state, then we must correct for $N!$ identical permutations that are included in the partition function sum 18.2, and the partition function becomes (equation 18.11a)

$$Z = \left(\frac{e\zeta}{N} \right)^N \qquad \text{(identical subsystems, no overlap),}$$

where ζ is the partition function for any subsystem (equation 18.11b),

$$\zeta = \sum_i e^{-\beta\varepsilon_i} \qquad \text{(one of the subsystems).}$$

Similarly, the Helmholtz free energy for N identical subsystems is (equation 18.12)

$$\overline{F} = -kT \ln Z = -NkT \ln \left(\frac{e\zeta}{N} \right).$$

For independent subsystems, ζ is independent of N and (equation 18.13)

$$\overline{\mu} = \left(\frac{\partial \overline{F}}{\partial N} \right)_{T,V} = -kT \ln \left(\frac{\zeta}{N} \right) \qquad \text{(independent identical particles, no overlap).}$$

D   The partition function for a gas

We are now able to calculate the partition function for a system that contains identical subsystems in terms of the partition function of any one of them. We will illustrate how to do this for the case of a gas that has N identical diatomic molecules. Partition functions for other common (but simpler) systems can be calculated in the homework problems.

D.1   The various degrees of freedom

The energy ε of a single molecule might be stored in translational (ε_t), rotational (ε_r), and vibrational (ε_v) degrees of freedom. It is also conceivable that some energy could be stored in the excitation of electrons to higher energy levels (ε_e) or in excited nuclear states (ε_n), and perhaps in other ways as well:

$$\varepsilon = \varepsilon_t + \varepsilon_r + \varepsilon_v + \varepsilon_e + \varepsilon_n + \cdots.$$

In calculating the partition function for a single molecule, we would have to sum over all these various kinds of states:

$$\begin{aligned}
\zeta &= \sum_{t,r,v,e,n,\ldots} e^{-\beta(\varepsilon_t + \varepsilon_r + \varepsilon_v + \varepsilon_e + \varepsilon_n + \cdots)} \\
&= \sum_t e^{-\beta\varepsilon_t} \sum_r e^{-\beta\varepsilon_r} \sum_v e^{-\beta\varepsilon_v} \cdots \\
&= \zeta_t \zeta_r \zeta_v \zeta_e \zeta_n \cdots.
\end{aligned} \tag{18.14}$$

That is, the partition function for the molecule can be separated into a product of the partition functions for the different ways in which energy can be stored.

Some of these energy terms can be ignored. If we measure energies relative to the ground state,[4] the partition function for any degree of freedom becomes

$$\zeta = \sum_i e^{-\beta\varepsilon_i} = e^{-0} + e^{-\beta\varepsilon_1} + e^{-\beta\varepsilon_2} + \cdots.$$

If the energy of the first excited state is large compared to kT, then $e^{-\beta\varepsilon}$ is negligible for all excited states and so we can ignore this degree of freedom:[5]

$$\zeta \approx 1 + 0 + 0 + \cdots = 1 \qquad \text{(if } \varepsilon_1 \gg kT\text{)}. \tag{18.15}$$

In the homework problems it can be shown that the temperature would have to be thousands of degrees Kelvin for typical electronic excitations and billions of degrees for nuclear excitations. Consequently, we can ignore these and expect only

---

[4] Although convenient, it is not necessary to measure energies relative to the ground state. You may choose any reference level you wish.

[5] If the ground state is n times degenerate, the first term here will be n instead of 1. But the partition function can always be changed by a constant factor without affecting any result obtained from it (homework), so we make this constant factor equal to unity.

translational, rotational, and possibly vibrational degrees of freedom to contribute to the partition function:

$$\zeta = \zeta_t \zeta_r \zeta_v \tag{18.16}$$

D.2 The translational part

The translational part of the partition function, ζ_t, can be calculated by converting the sum over translational quantum states to an integral, according to equation 1.7:

$$\sum_t \longrightarrow \int \frac{d^3r\,d^3p}{h^3}.$$

Doing this gives (see Appendix E)

$$\zeta_t = \sum_t e^{-\beta p^2/2m} = \int \frac{d^3r\,d^3p}{h^3}\,e^{-\beta p^2/2m}$$

$$= \frac{V}{h^3} \int d^3p\,e^{-\beta p^2/2m} = V\left(\frac{2\pi m kT}{h^2}\right)^{3/2}. \tag{18.17}$$

D.3 The rotational part

For a diatomic molecule with rotational inertia I (see subsections 1B.6 or 4C.1),

$$\varepsilon_r = \frac{L^2}{2I}, \qquad \text{where } L^2 = l(l+1)\hbar^2,\ l = 0, 1, 2, \ldots \tag{18.18}$$

The state with angular momentum quantum number l is $2l + 1$ times degenerate, since there are $2l + 1$ possible orientations of this angular momentum ($l_z = 0, \pm 1, \pm 2, \ldots, \pm l$). Therefore our sum over the rotational states r is given by

$$\zeta_r = \sum_r e^{-\beta \varepsilon_r} = \sum_l (2l+1)e^{-\beta \hbar^2 l(l+1)/2I}.$$

If kT is small compared with the energy needed for excitation, or equivalently if $T \ll T_e$,[6] then the system is confined to the ground state and $\zeta_r = 1$. But in the other extreme, $T \gg T_e$, the sum can be converted to an integral. Since the angular momentum quantum number changes in units of 1 ($\Delta l = 1$), we can write

$$\zeta_r = \sum_l \Delta l (2l+1)e^{-\beta \hbar^2 l(l+1)/2I} \approx \int\limits_0^\infty dl\,(2l+1)e^{-\beta \hbar^2 l(l+1)/2I}.$$

[6] As we saw in Chapter 4, excitation temperatures for molecular rotations are typically around 5 to 10 K. Even when confined to the rotational ground state, the translational motion of molecules guarantees many more quantum states than particles and so we don't have to worry about possible overlap.

This integral is simplified by using the substitution of variables

$$x = l(l+1) = l^2 + l \qquad \text{with} \qquad dx = (2l+1)dl,$$

so that we have

$$\zeta_r = \int_0^\infty dx \, e^{-\beta \hbar^2 x/2I} = \frac{2IkT}{\hbar^2}.$$

Thus the rotational contribution to a molecule's partition function is as follows:

$$\zeta_r = \begin{cases} \sum_l (2l+1)e^{-\beta \hbar^2 l(l+1)/2I} & \text{always,} \\ 1 & \text{if } T \ll T_e, \\ \dfrac{2IkT}{\hbar^2} & \text{if } T \gg T_e. \end{cases} \qquad (18.19)$$

D.4 The vibrational part

We now examine the vibrations of atoms bound within a molecule. As we learned in subsection 4B.1, for small displacements from equilibrium the potential energy of any bound particle is that of a harmonic oscillator. In quantum mechanics we learn that the energy of the nth level is given by (equation 1.18)

$$\varepsilon_n - \varepsilon_0 = n\hbar\omega, \qquad (18.20)$$

where energies are measured relative to the ground state energy ε_0 and where the angular frequency ω depends on the masses and binding strengths. With this expression for the energies, the vibrational part of a molecule's partition function is

$$\zeta_v = \sum_v e^{-\beta \varepsilon_v} = \sum_n e^{-n\beta \hbar \omega}.$$

This is a series of the form

$$\sum_n a^n = 1 + a + a^2 + a^3 + \cdots, \qquad \text{where} \quad a = e^{-\beta \hbar \omega},$$

which gives

$$1 + a + a^2 + a^3 + \cdots = \frac{1}{1-a} \qquad (a < 1), \qquad (18.21)$$

as can be easily demonstrated by multiplying both sides by $1 - a$. Therefore, the partition function for molecular vibrations is given by

$$\zeta_v = \frac{1}{1 - e^{-\beta \hbar \omega}}. \qquad (18.22)$$

In the limit $kT \ll \hbar\omega$ (or $T \ll T_e$) this gives $\zeta_v = 1$, as expected, and in the other limit, $kT \gg \hbar\omega$ (or $T \gg T_e$), this gives $\zeta_v = kT/\hbar\omega$ (homework). That is,

$$\zeta_v = \begin{cases} \dfrac{1}{1 - e^{-\beta \hbar \omega}} & \text{always,} \\ 1 & \text{if } T \ll T_e, \\ kT/\hbar\omega & \text{if } T \gg T_e. \end{cases} \qquad (18.23)$$

Table 18.1. *The partition function for a molecule of a diatomic gas*

Degrees of freedom			Contribution to ζ	
Kind	Number	T_e	General	For $T \gg T_e$
translational	3	≈ 0 K	$\int \dfrac{d^3 r \, d^3 p}{h^3} e^{-\beta p^2/2m}$	$V \left(\dfrac{2\pi m k T}{h^2} \right)^{3/2} = \text{constant} \times V T^{3/2}$
rotational	2	≈ 10 K	$\sum\limits_{l=0}^{\infty} (2l+1) e^{-\beta \hbar^2 l(l+1)/2I}$	$\left(\dfrac{2I k T}{\hbar^2} \right)^{2/2} = \text{constant} \times T^{2/2}$
vibrational	2	≈ 1000 K	$\sum\limits_{n=0}^{\infty} e^{-n\beta \hbar \omega} = \dfrac{1}{1 - e^{-\beta \hbar \omega}}$	$\left(\dfrac{k T}{\hbar \omega} \right)^{2/2} = \text{constant} \times T^{2/2}$

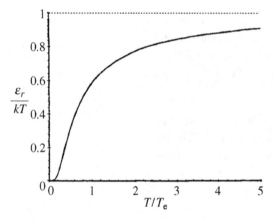
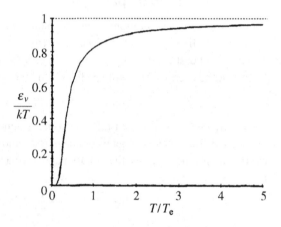

Figure 18.2 Plots of the average rotational (left) and vibrational (right) energies per molecule as a function of the temperature for a diatomic gas. The energy is in units of kT and the temperature is in units of the excitation temperature T_e, which is typically 5–10 K for rotations and 1000–2000 K for vibrations.

For a typical diatomic molecule such as N_2, the characteristic frequency ω is about 2×10^{14} Hz, giving T_e a value of about 1500 K.

D.5   The final result

Combining the results 18.11a, b and 18.16 we can write the partition function for a gas of N identical diatomic molecules as

$$Z = \left(\frac{e \zeta_t \zeta_r \zeta_v}{N} \right)^N, \tag{18.24}$$

where our results for ζ_t, ζ_r, and ζ_v are summarized in Table 18.1. You can see that, as long as $T \gg T_e$, each degree of freedom, contributes a factor proportional to $T^{1/2}$ to a molecule's partition function. Consequently, for particles with ν degrees of freedom,

$$\zeta = \zeta_t \zeta_r \zeta_v = \text{constant} \times V T^{\nu/2}$$

and the partition function for the entire gas is given by

$$Z = C \left(\frac{V T^{v/2}}{N} \right)^N \qquad (C \text{ is a constant}). \qquad (18.25)$$

A monatomic gas would have no rotational or vibrational degrees of freedom.

When the temperature is not large compared with T_e, we use the "general" form of the partition functions, given in the fourth column of Table 18.1. Once we have the partition function, the formulas of Section B will reveal any property we wish. Figure 18.2 displays the results of this calculation for the average rotational and vibrational energies of a diatomic molecule.

Summary of Section D

The partition function for a single molecule can be written as the product of partition functions for the translational, vibrational, rotational, electronic, nuclear, etc. degrees of freedom (equation 18.14):

$$\zeta = \zeta_t \zeta_r \zeta_v \zeta_e \zeta_n \cdots .$$

Whenever kT is small compared with the energy of the first excited state, or equivalently whenever $T \ll T_e$, we can ignore contributions from that degree of freedom (equation 18.15):

$$\zeta = \sum_i e^{-\beta \varepsilon_i} \approx 1 + 0 + 0 + \cdots = 1 \qquad (\text{if } kT \ll \varepsilon_1).$$

Normally the excited electronic and nuclear states are inaccessible, so we need only consider the translational, rotational, and possibly vibrational degrees of freedom. Their contributions are listed in Table 18.1 for a diatomic gas whose molecules have mass m, rotational inertia I, and fundamental frequency ω.

The partition function for a system of N identical molecules is (equation 18.24)

$$Z = \left(\frac{e \zeta_t \zeta_r \zeta_v}{N} \right)^N .$$

Every degree of freedom makes a contribution to the partition function that is proportional to $T^{1/2}$. If each molecule has v degrees of freedom then the partition function for a gas of N such molecules is given by (equation 18.25)

$$Z = C \left(\frac{V T^{v/2}}{N} \right)^N \qquad (C = \text{constant}).$$

D.6 Examples

Now that we know the partition function for a gas, it becomes a useful tool for calculating various properties. We illustrate with two examples.

Example 18.1   If each molecule has v degrees of freedom, what is the internal energy of the gas?

Using 18.4 for $\overline{E}$, 18.25 for Z, and the fact that $\beta = 1/kT$, we find

$$\overline{E} = -\frac{\partial}{\partial \beta} \ln Z = kT^2 \frac{\partial}{\partial T} \ln \left[C \left(\frac{T^{v/2}}{\rho} \right)^N \right] = \frac{Nv}{2} kT.$$

This result shows that with each of the Nv degrees of freedom of the gas is associated an average energy of $(1/2)kT$, in agreement with the equipartition theorem.

Example 18.2   Suppose that, for a certain gas, each gas molecule has three translational and two rotational degrees of freedom only. What is the chemical potential for this gas?

Using equation 18.13 for $\overline{\mu}$ and Table 18.1 for $\zeta = \zeta_t \zeta_r$ gives

$$\overline{\mu} = -kT \ln \left(\frac{\zeta_t \zeta_r}{N} \right) = -kT \ln \left(\frac{CVT^{5/2}}{N} \right) = -kT \ln \left(\frac{CT^{5/2}}{\rho} \right),$$

where $\rho = N/V$ is the density of the molecules and the constant C is given by (Table 18.1)

$$C = \left(\frac{2\pi mk}{h^2} \right)^{3/2} \left(\frac{2Ik}{\hbar^2} \right).$$

You might compare this with the more generic form 14.12, $\mu = -kT \ln[f(T)/\rho]$ where $f(T)$ is a function of T.

Problems

Section A

1. A certain system has only two accessible states, the ground state, with energy zero, and an excited state, with energy 1 eV. What is the value of the partition function for this system at a temperature of (a) 300 K, (b) 30 000 K?

2. Consider a system at 290 K for which the energy of the nth state is $E_n = n(0.01 \text{ eV})$. Using $1 + a + a^2 + a^3 + \cdots = 1/(1 - a)$ for $a < 1$ where necessary, find
   (a) the value of the partition function,
   (b) the constant C in the expression $P_s = Ce^{-\beta E_s}$,
   (c) the probability that the system is in the $n = 0$ state,
   (d) the average energy of the system, $\overline{E}$,
   (e) the value of the partition function at a temperature of 100 K.

3. When interacting with a reservoir, the probability for a small system to be in state s is $P_s = Ce^{-\beta \varepsilon_s}$. See whether you can derive this result from the first and second laws, without referring to Chapter 15.

4. Show that the average energy of a system depends on the temperature, even though the energies of the accessible states do not, by considering a system with two accessible states having energies 0 and 0.1 eV, respectively.
    (a) What is the probability for the system to be in the excited state at 300 K? At 600 K?
    (b) What is the average internal energy of the system at each of these two temperatures?
    (c) Does the average energy increase with T, even though the energies of the individual states do not?

Section B

5. Show that $\overline{E^2} = (1/Z)(\partial^2 Z/\partial \beta^2)_{V,N}$.

6. Show that the square of the standard deviation for the energy of a system interacting with a reservoir, $\sigma^2 = \overline{E^2} - \overline{E}^2$, is given by $\sigma^2 = (\partial^2 \ln Z/\partial \beta^2)_{V,N}$.

7. Using the chain rule, find how the derivative with respect to $\beta = 1/kT$ converts into a derivative with respect to T.

8. Using equation 18.4, prove that $\overline{F} = -kT \ln Z$ is a solution to the differential equation 18.8.

9. You are going to first show that any function of the form $\overline{F} = -kT \ln Z + f(V, N)T$ is a solution to equation 18.8. Then using equation 18.7a ($\overline{S} = -(\partial \overline{F}/\partial T)_{V,N}$) you will show that the third law ($S \to 0$ as $T \to 0$) demands that $f(V, N) = 0$. To do this, show that:
    (a) Any function $\overline{F}$ of the above form is a solution of equation 18.8.
    (b) $\overline{S} = -f(V, N) + k \ln Z + kT(\partial \ln Z/\partial T)_{V,N}$.
    (c) As $T \to 0$, $Z \to 1$ and therefore $k \ln Z \to 0$. (Measure all energies relative to the ground state. You may wish to write out the first few terms in Z to see what happens as $T \to 0$.)
    (d) As $T \to 0$, $kT(\partial \ln Z/\partial T)_{V,N} \to 0$. (Hint: Write $\partial \ln Z/\partial T = (1/Z) \partial Z/\partial T$ and express Z explicitly as a sum. Then use l'Hospital's rule to find the limit of terms of the form $e^{-x}x$ as $x \to \infty$.)
    Combining the answers to parts (b), (c), and (d), show that $f(V, N) = 0$ and therefore that $F = -kT \ln Z$.

10. For a certain system there is only one accessible state and it has energy $E_s = -CVT^2$, where C is a constant and V is the volume. Find the following in terms of T, V, N: (a) the partition function, (b) the average pressure, using the second of equations 18.7 and equation 18.9.
    (In real systems, the energies of the states do not usually depend on the temperature of the reservoir, but we ignore this in these examples.)

11. For a certain system, there is only one accessible state and it has energy $E = -NkT \ln(V/V_0)$, where V_0 is a constant. Find the following in terms

of T, V, N (use equations 18.7 and 18.9): (a) the partition function, (b) the average pressure, (c) the average chemical potential.

12. Repeat the above problem for a system with many states. The energy of state s is given by $E_s = f_s(T) - NkT \ln(V/V_0)$, where V is the volume, V_0 is a constant, and $f_s(T)$ is some function of the temperature only that depends on the state s.

13. For a certain system, the energy of each state s is given by

$$E_s = kT[(C_s + (3/2)N \ln(\beta/\beta_0) - N \ln(V/V_0)],$$

where β_0 and V_0 are fixed constants and C_s is a constant whose value depends on the state. Calculate the following for this system in terms of T, V, N: (a) the partition function, (b) the average internal energy, (c) the average pressure, (d) the average chemical potential, (e) the standard deviation σ for fluctuations in internal energy, (f) the average entropy.

14. The partition function for a certain system is given by $Z = (\beta/\beta_0)^{-3N} e^{N(V/V_0)}$, where β_0, V_0 are constants. Find the following as a function of T, V, N: (a) the average internal energy, (b) the standard deviation of the energy fluctuations, (c) the Helmholtz free energy, (d) the entropy, (e) the pressure, (f) the chemical potential.

Section C
15. System A contains one particle, x. System B contains two particles, y and z. The three particles are distinguishable. Each of these systems has the same two accessible single-particle states. How many different states are accessible to (a) system A, (b) system B, (c) the combined system?

16. In how many different ways can you arrange three different particles among three different boxes, with no more than one per box, if the particles are (a) distinguishable, (b) identical?

17. In how many different ways can you arrange three particles among four different boxes, with no more than one per box if the particles are: (a) distinguishable, (b) identical?

18. The results of Section C in the main text are not correct if there is a significant probability that two systems will occupy the same state simultaneously. Show this for the following simple case. Suppose that there are two subsystems (i.e., two particles) and the same two accessible states for each (i.e., two "boxes"). If both subsystems may occupy the same state, how many different arrangements of the system are there if the two subsystems are (a) distinguishable, (b) identical? (c) Do these two answers differ by a factor 2!.

19. Repeat the above problem for (a) two particles and three states, (b) two particles and 100 states. (c) For which of these two cases do the two answers differ by a factor that is closer to 2!?

20. Consider a system, at temperature 400 K, consisting of 1000 molecules, each having only two accessible states, with energies $\varepsilon_0 = 0$ and $\varepsilon_1 = 0.1$ eV, respectively. Find
    (a) the probability for a molecule to be in the excited state,
    (b) the average energy of the system.

21. Consider a system of N identical molecules, each having only two accessible states, with energies $\varepsilon_0 = 0$ and $\varepsilon_1 = 0.05$ eV, respectively. Give the following as a function of (T, V, N): (a) the partition function, (b) the Helmholtz free energy, (c) the chemical potential.

22. Derive $\mu = -kT \ln(\zeta/N)$ from equations 18.7, 18.9, and 18.11.

Section D

23. Find the excitation temperature for
    (a) electronic excitation if the first excited state is 0.3 eV above the ground state,
    (b) nuclear excitation if the first excited state lies 0.2 MeV above the ground state.

24. Do the integrals in equation 18.17 to see whether you get the stated results. For integration over the momenta, you may wish to use spherical coordinates and the tables of integrals in Appendix E.

25. Electrostatic interactions with neighbors cause atoms in solids to be held in place as if bound together by tiny springs. The excitation temperature for vibrations in many metals is around 200 K.
    (a) If the mass of a metal atom is 2.0×10^{-25} kg, what is a typical value for the force constant κ ($F_{spring} = -\kappa x$) with which the atoms are bound in place?
    (b) Show that the series $1 + a + a^2 + a^3 + \cdots$ converges to $1/(1 - a)$ for $|a| < 1$.
    (c) For a harmonic oscillator in three dimensions, the energies of the various excitation levels, measured relative to the ground state, are given by $\varepsilon - \varepsilon_0 = (n_x + n_y + n_z)\hbar\omega$, where n_x, n_y, n_z are integers that indicate the level of excitation for oscillations in each of the three dimensions. In terms of $\beta\hbar\omega$, what is the partition function for such an oscillator?
    (d) What is the average energy of an oscillator in terms of ω and T?

26. Using the results 18.4, 18.5, 18.7, 18.9, and 18.25, find the average values of the following properties of a monatomic gas as functions of T, V, N:
    (a) E, (b) S, (c) p, (d) μ, (e) $\overline{E^2} - \overline{E}^2$.

27. The oxygen (O_2) molecule is composed of two oxygen atoms, each with mass of 2.7×10^{-26} kg and a center-to-center separation 1.24×10^{-10} m. For this molecule, find

(a) the rotational inertia I about the perpendicular bisector of the line joining the two atoms,

(b) the energy of the first excited ($l = 1$) rotational state,

(c) the rotational excitation temperature.

28. The first line of equation 18.19 expresses the partition function for the rotational degrees of freedom of a diatomic gas molecule, for all temperatures, in terms of a sum. Using this, find the general expression for the average energy per molecule stored in the rotational degrees of freedom as a function of I and T.

29. The vibrations of an oxygen molecule have a characteristic frequency ω of about 2.98×10^{14} Hz. For this molecule, what is

(a) the energy of the first excited vibrational state,

(b) the vibrational excitation temperature?

30. The partition function for the vibrational degrees of freedom at all temperatures for a single molecule is given by equation 18.22. Using equation 18.4, find the average energy per molecule stored in the vibrational degrees of freedom as a function of ω and T, for all values of T.

31. From equation 18.22 show that $\zeta_v = kT/\hbar\omega$ for $T \gg T_e$.

32. Using equation 18.4 and the values of ζ_t, ζ_r, and ζ_v listed in Table 18.1, calculate the values of the average energy per molecule for each of these modes in the high-temperature limit. How does each of these compare with $(\nu/2)kT$, where ν is the number of degrees of freedom in that mode?

33. Using $E_r = L^2/2I$, derive the result 18.19 for the case $T \gg T_e$.

34. Consider the monatomic gas argon, whose atoms each have mass 6.68×10^{-26} kg and which is at standard temperature and pressure (273 K, 1.013×10^5 Pa). With the help of equations 18.13 and 18.17, calculate the value of the chemical potential in units of eV. (Hint: For V/N, use the ideal gas law $V/N = kT/p$.)

35. Consider the ideal monatomic gas of the above problem. Using equations 18.7, 18.12, and 18.17:

(a) show that $S = Nk \ln(\zeta/N) + (5/2)Nk$.

(b) with the result (a) and equation 18.13, show that for this (ideal) gas,
$$E = TS - pV + N\mu.$$

(c) What is the molar entropy of this gas in units of J/(K mole)?

Part VII
Quantum statistics

Chapter 19
Introduction to quantum statistics

We now shift our attention from classical to quantum statistics. As a brief review, recall that in both approaches we are considering a small system interacting with a large reservoir (Figure 15.1). The probability P for any particular configuration is proportional to the number of accessible states, Ω, which is related to the entropy through the definition $S \equiv k \ln \Omega$. Thus

$$P \propto \Omega = e^{S/k}.$$

We calculate the probability for the small system to be in any particular state by the effect that this would have on the entropy of the reservoir. If it would remove ΔE, ΔV, ΔN from the reservoir then the probability for it to be in this state is related to the change in the reservoir's entropy by (equation 15.2)

$$P = Ce^{\Delta S/k} = Ce^{-\beta(\Delta E + p\Delta V - \mu \Delta N)},$$

where C is a constant of proportionality and $\beta = 1/kT$. The two most common ways of applying this result depend on our "small system" (Figure 15.2).

- In *classical statistics* the small system is a single particle (or group of particles) that could be in any of various different states.

- In *quantum statistics* the small system is a single state that could be occupied by various numbers of particles.

The properties of systems depend on two things. In classical statistics they are:

- the spectrum of accessible states;
- The probability P_s for the small system to be state s.

In quantum statistics, they are:

- the spectrum of accessible states;
- the average number of particles in each (the "occupation number," $\bar{n}$).

In both approaches the spectrum of accessible states is the more elusive of the two ingredients. It depends on the system and the kinds of particle that are in it.

However, the other ingredient in both approaches is simple and universal. We have already derived it for the case of classical statistics (equation 15.5):

$$P_s = Ce^{-\beta \varepsilon_s},$$

where ε_s is the energy of state s. In the preceding chapters we have shown how to use this result in the study of many common systems.

Now, in this chapter, we turn our attention to the quantum approach. We will first derive the expression for the occupation number of a quantum state and then show how to use it to study the properties of various common systems.

A   The occupation number

A.1   General

According to the result 15.6, the probability that a certain state contains n particles is[1]

$$P_n = Ce^{-n\beta(\varepsilon - \mu)}, \qquad \text{where } C = \frac{1}{\sum_n e^{-n\beta(\varepsilon - \mu)}}$$

and where ε is the energy of a particle in that state. It is more convenient to write this as

$$P_n = Ce^{-nx}, \qquad \text{where } x = \beta(\varepsilon - \mu) \text{ and } C = \frac{1}{\sum_n e^{-nx}}.$$

The average number of particles occupying the state, the occupation number, is given by

$$\bar{n} = \sum_n n P_n = C \sum_n n e^{-nx}. \tag{19.1}$$

---

[1] This is equation 15.2 with $\Delta V = 0$, because the volume of a quantum state is fixed.

We now use the trick that

$$ne^{-nx} = -\frac{\partial}{\partial x} e^{-nx}$$

to write

$$\bar{n} = \frac{1}{\sum_n e^{-nx}} \left(\sum_n n e^{-nx} \right) = -\frac{1}{\sum_n e^{-nx}} \left(\frac{\partial}{\partial x} \sum_n e^{-nx} \right)$$

$$= -\frac{\partial}{\partial x} \ln \left(\sum_n e^{-nx} \right). \qquad (19.2)$$

A.2 Fermions and bosons

The problem is now reduced to evaluating the sum $\sum_n e^{-nx}$ in equation 19.2. For Nature's most fundamental particles there are only two possibilities.

- For fermions, no more than one particle may occupy a given state, so the sum becomes

$$\sum_{n=0}^{1} e^{-nx} = 1 + e^{-x},$$

and the occupation number of a state for fermions is given by

$$\bar{n}_{\text{fermions}} = -\frac{\partial}{\partial x} \ln(1 + e^{-x}) = \frac{1}{e^x + 1}, \qquad \text{where } x = \beta(\varepsilon - \mu). \qquad (19.3\text{a})$$

- For bosons any number may occupy a state, so the sum over n goes from 0 to ∞. As long as x is positive this series converges, according to equation 18.21:

$$\sum_{n=0}^{\infty} e^{-nx} = 1 + e^{-x} + e^{-2x} + \cdots = \frac{1}{1 - e^{-x}}.$$

So equation 19.2 gives the following result for the occupation number of a boson state:

$$\bar{n}_{\text{bosons}} = -\frac{\partial}{\partial x} \ln \left(\frac{1}{1 - e^{-x}} \right) = \frac{1}{e^x - 1}, \qquad \text{where} \quad x = \beta(\varepsilon - \mu). \quad (19.3\text{b})$$

As you can see from Equations 19.3a, b, the occupation numbers for fermion and boson states differ only by the sign within the denominator. This similarity sometimes makes it convenient to work problems for both types of particles simultaneously, using

$$\bar{n} = \frac{1}{e^{\beta(\varepsilon-\mu)} \pm 1} \qquad \text{with} \qquad \begin{cases} + & \text{for fermions,} \\ - & \text{for bosons.} \end{cases} \qquad (19.3)$$

The two forms reflect the two major subdivisions of quantum statistics, called Fermi–Dirac and Bose–Einstein statistics, respectively. In neither case does the occupation number tell us anything about the accessible states. It simply tells us the average number of particles that would be in a state with energy ε, if such a

$$\bar{n}(\varepsilon) = \frac{1}{e^{\beta(\varepsilon-\mu)} + 1}$$

$$\bar{n}(\varepsilon) = \frac{1}{e^{\beta(\varepsilon-\mu)} - 1}$$

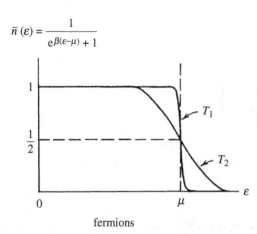

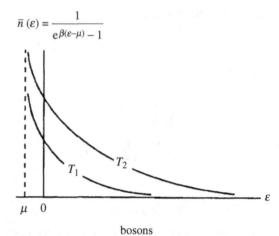

fermions   bosons

Figure 19.1 Plot of occupation number vs. energy at a low temperature T_1 and a high temperature T_2, for fermions and bosons.

state exists. Figure 19.1 shows how these occupation numbers vary with energy at high and low temperatures.

Although the two subdivisions of quantum statistics discussed above are both very fundamental and very important, equation 19.2 can be applied to other kinds of systems as well. For example, suppose that a certain type of large molecule may have between 0 and m identical atomic groups attached, each with binding energy ε (e.g., oxygen to hemoglobin, nitrates to an enzyme, or hydroxides to a catalyst). In such cases the sum would go from 0 to m, and equation 19.2 would tell us the average number of atomic groups attached to each large molecule.

A.3 Fluctuations

In addition to the average number of particles occupying a quantum state, we may also be interested in a measure of the fluctuations: this is provided by the standard deviation

$$\sigma^2 = \overline{\Delta n^2} = \overline{n^2} - \bar{n}^2.$$

Using the notation of the previous section,

$$\overline{n^2} = \sum_n n^2 P_n = C \sum_n n^2 e^{-nx} = \frac{1}{\sum_n e^{-nx}} \left(-\frac{\partial}{\partial x} \right)^2 \sum_n e^{-nx}$$

and, with the help of equation 19.2, this can be turned into (homework)

$$\overline{\Delta n^2} = \overline{n^2} - \bar{n}^2 = -\frac{\partial}{\partial x}\bar{n} = \bar{n} \mp \bar{n}^2,$$

where the minus sign applies to fermions and the plus sign to bosons. That is, the fluctuation in the occupation number of a quantum state is given by

$$\overline{\Delta n^2} = \begin{cases} \bar{n}(1 - \bar{n}) & \text{for fermions,} \\ \bar{n}(1 + \bar{n}) & \text{for bosons.} \end{cases} \tag{19.4}$$

You can see that fermions would only experience significant fluctuations near the transition around $\varepsilon \approx \mu$, where the occupation number $\bar{n} \approx 1/2$

(Figure 19.1). Below and above this region $\overline{n} \approx 1$ and $\overline{n} \approx 0$, respectively, so the fluctuations would be small. In contrast, you can see that the fluctuation for bosons, $\sqrt{\overline{n}(1 + \overline{n})}$, is larger than the occupation number itself, so large fluctuations should be expected.

Summary of Section A

If the energy per particle in a given quantum state is ε then the average number of particles occupying that state, the "occupation number" of that state, is given by (equation 19.2)

$$\overline{n} = \sum_n n P_n = \frac{1}{\sum_n e^{-nx}} \sum_n n e^{-nx} = -\frac{\partial}{\partial x} \ln \left(\sum_n e^{-nx} \right),$$

where $x = \beta(\varepsilon - \mu)$. To evaluate the sum $\sum_n e^{-nx}$, we need to know the maximum number of particles that could occupy the state. For the important case of fundamental particles, Nature provides only two possibilities, as follows.

   Fermions have half-integer spin quantum numbers and obey Fermi–Dirac statistics. No two identical fermions may occupy the same quantum state, so the sum has just two terms, $n = 0$ and $n = 1$. Bosons have integer spin quantum numbers and obey Bose–Einstein statistics. Any number of identical bosons may occupy the same quantum state, so the sum goes from $n = 0$ to $n = \infty$. Performing these sums and taking the indicated derivative (19.2) gives the following result (equation 19.3) for the respective occupation numbers:

$$\overline{n} = \frac{1}{e^{\beta(\varepsilon - \mu)} \pm 1} \quad \text{with} \quad \begin{cases} + & \text{for fermions,} \\ - & \text{for bosons.} \end{cases}$$

A measure of the fluctuation in the number of particles occupying a quantum state is (equation 19.4)

$$\overline{\Delta n^2} = \overline{n^2} - \overline{n}^2 = \begin{cases} \overline{n}(1 - \overline{n}) & \text{for fermions,} \\ \overline{n}(1 + \overline{n}) & \text{for bosons.} \end{cases}$$

B Comparison with classical statistics

B.1 How they differ

You might think that quantum statistics could be easily derived from classical statistics. Shouldn't the average number of particles in any quantum state s be the product of the number of particles N times the probability P_s for a particle to be in that state? That is, shouldn't we have

$$\overline{n}_{\text{classical}} = N P_s = N C e^{-\beta \varepsilon_s}, \qquad \text{with } C = \frac{1}{\sum_s e^{-\beta \varepsilon_s}}? \tag{19.5}$$

This reasoning is indeed correct, but only if the particle densities in phase space are small.

Figure 19.2 Occupation number $\bar{n}$ vs. energy for boson, classical, and fermion systems, for a given temperature and chemical potential. At high energies, where the occupation numbers are small, all three are nearly the same. But they differ at low energies where occupation numbers are larger.

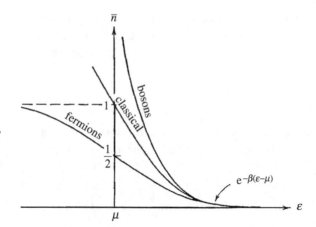

To show this, we will look at quantum statistics for bosons and fermions in the region of high energies and small occupation numbers, where $e^{\beta(\varepsilon-\mu)} \gg 1$. Then, for each state,

$$\bar{n}_{\text{quantum}} = \frac{1}{e^{\beta(\varepsilon-\mu)} \pm 1} \approx e^{-\beta(\varepsilon-\mu)} = e^{\beta\mu}e^{-\beta\varepsilon}. \qquad (19.6)$$

Because the total number of particles is the sum of the number in each state, we can write

$$N = \sum_s \bar{n}_s \approx \sum_s e^{-\beta(\varepsilon_s-\mu)} = e^{\beta\mu} \sum_s e^{-\beta\varepsilon_s}$$

$$\Rightarrow \qquad e^{\beta\mu} = \frac{N}{\sum_s e^{-\beta\varepsilon_s}}.$$

Putting this expression for $e^{\beta\mu}$ into equation 19.6 shows that it is the same as Equation 19.5. So, indeed, in the region of relatively high energies and low occupation numbers both classical and quantum approaches give the same result (Figure 19.2):

$$\bar{n}_{\text{classical}} \approx \bar{n}_{\text{bosons}} \approx \bar{n}_{\text{fermions}} \approx e^{-\beta(\varepsilon-\mu)}, \qquad \text{for} \quad \varepsilon - \mu \gg kT.$$

However, for states of lower energy and higher occupation number, we cannot ignore the 1 in the denominator of 19.6 in comparison with $e^{\beta(\varepsilon-\mu)}$. For these states, the classical and quantum results all differ:

$$\frac{1}{e^{\beta(\varepsilon-\mu)} - 1} > e^{-\beta(\varepsilon-\mu)} > \frac{1}{e^{\beta(\varepsilon-\mu)} + 1}$$

$$\Rightarrow \qquad \bar{n}_{\text{bosons}} > \bar{n}_{\text{classical}} > \bar{n}_{\text{fermions}}. \qquad (19.7)$$

B.2 Why they differ

To understand physically why the classical and quantum results differ for states with higher occupation numbers, consider the distribution of two particles between two states. As illustrated on the left in Figure 19.3, there are four such configurations for *distinguishable* particles, two with only one particle per state (single occupancy) and two where one state has both particles (double occupancy).

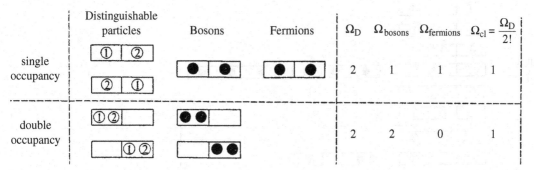

Table 19.1. *Summary of the number of different configurations for two particles in two states*

	Distinguishable particles	Identical particles		
		classical	bosons	fermions
Single occupancy	2	1	1	1
Double occupancy	2	1	2	0
Total configurations	4	2	3	1

Figure 19.3 The possible arrangements of two particles between any two states (empty states are not shown) for the cases of distinguishable particles (Ω_D), identical bosons (Ω_{bosons}), and identical fermions ($\Omega_{fermions}$). In the first three columns on the right the numbers of different states are given for each case. The predictions of the corrected classical formula $\Omega_{cl} = \Omega_D/2!$ are also given, in the final column on the right. For identical particles with single occupancy, the number of states in all three cases is correctly predicted by the corrected classical formula. But for double occupancy, there are two configurations for bosons, one for the corrected classical prediction, and none for the fermions. This is a simple example of the more general statement that for single occupancy all three approaches agree, but for multiple occupancy $\Omega_{bosons} > \Omega_{cl} > \Omega_{fermions} = 0$.

But if the particles are *identical*, we must correct for the duplicate counting of identical configurations, and also in the case of fermions, we must ensure that no two particles occupy the same state. For duplicate counting, classical statistics uses the correction factor $1/N!$ (subsection 6D.2, Figure 6.4). In the present example, with two particles, the correction factor $1/2!$ reduces the number of distinguishable states from four to two, one with single occupancy and one with double occupancy. As illustrated in Figure 19.3 and Table 19.1, the classical prediction is correct for the configurations of single occupancy, but not for those of double occupancy, being too small for bosons and too large for fermions.

Figure 19.4 displays the corresponding analysis for the distribution of three particles between any set of three states. In the homework problems this can be done for other cases (two particles in three states, etc.). In each case, we find that the classical approach agrees with the quantum approach as long as there is no more than one particle per state. But for states of multiple occupancy, the classical prediction is too small for bosons and too large for fermions.

Of course, real systems might have more like 10^{24} identical particles and perhaps $10^{10^{24}}$ states. But the same ideas that apply to small systems carry over into these larger systems. In the lower-energy states, with higher occupation numbers, the three approaches differ. Compared with the classical prediction, identical bosons have more states with multiple occupancy, so the average number of particles per state is larger. Conversely, fermions have zero states with multiple occupancy, so the average number of particles per state is correspondingly lower.

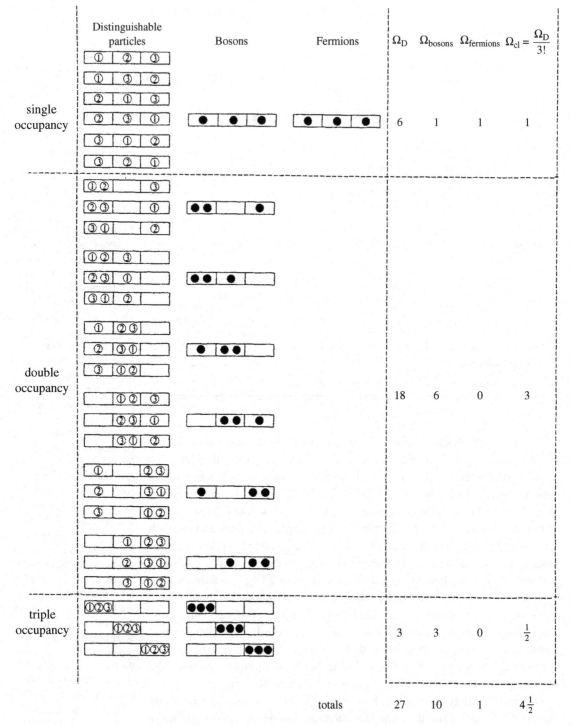

Figure 19.4 Illustration of the possible arrangements of three particles between any three states (empty states are not shown), for the cases of distinguishable particles (Ω_D), identical bosons (Ω_{bosons}), and identical fermions ($\Omega_{fermions}$). In the first three columns on the right, the numbers of states are given for each case. The predictions of the corrected classical formula $\Omega_{cl} = \Omega_D/3!$ are also given, in the final column on the right. For identical particles with single occupancy, all three approaches agree, but for the cases of multiple occupancy, $\Omega_{bosons} > \Omega_{cl} > \Omega_{fermions} = 0$.

C The limits of classical statistics

In the preceding section we learned that the classical approach is valid only if there is little likelihood of two or more identical particles trying to occupy the same state. The condition for the validity of classical statistics, then, is low particle densities in six-dimensional phase space (Figure 19.5):

$$\text{number of quantum states} \gg \text{number of particles}$$

$$\Rightarrow \quad \frac{V_r V_p}{h^3} \gg N. \tag{19.8}$$

If the characteristic separation between particles is r and the characteristic range in momentum is p then

$$V_r \approx N r^3 \quad \text{and} \quad V_p \approx p^3,$$

so inequality 19.8 becomes

$$\frac{r^3 p^3}{h^3} \gg 1.$$

By taking the cube root of both sides and using the equipartition theorem to express the momentum in terms of temperature ($p^2/2m \approx (3/2)kT$), this criterion for the validity of the classical approach becomes

$$\frac{rp}{h} \gg 1 \quad \text{with } r \approx \left(\frac{V}{N}\right)^{1/3} \text{ and } p \approx \sqrt{3mkT}. \tag{19.9}$$

Summary of Sections B and C

For states with low occupation numbers, where there is little likelihood that two identical particles will try to occupy the same state, the classical, Bose–Einstein, and Fermi–Dirac statistics all give the same result for the occupation number of a state:

$$\overline{n}_{\text{classical}} \approx \overline{n}_{\text{bosons}} \approx \overline{n}_{\text{fermions}} \approx e^{-\beta(\varepsilon-\mu)}, \quad \text{for } \varepsilon - \mu \gg kT.$$

But in states of lower energy and higher occupation number, the three approaches differ (equation 19.7):

$$\overline{n}_{\text{bosons}} > \overline{n}_{\text{classical}} > \overline{n}_{\text{fermions}}.$$

Because the classical approach works for low particle densities, the criterion for the validity of the classical approach is (equation 19.8)

$$\frac{V_r V_p}{h^3} \gg N.$$

If r is the characteristic separation between particles and p the characteristic range in momentum for each, then this criterion for the validity of the classical approach becomes (equation 19.9)

$$\frac{rp}{h} \gg 1 \quad \text{with } r \approx \left(\frac{V}{N}\right)^{1/3} \text{ and } p \approx \sqrt{3mkT}.$$

Figure 19.5 The classical approach works when particle separations in phase space are large compared with the dimensions of the individual quantum states, so that there is little chance that more than one particle will occupy the same state.

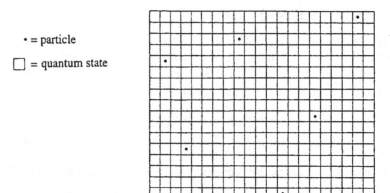

• = particle

☐ = quantum state

D   The spectra of accessible states

D.1   Overview

Of the two things that are needed to understand the behaviors of quantum systems, we now know one, the occupation number $\bar{n}$ (equations 19.3a, b). Consequently, the success or failure of our endeavors hinges on our understanding of the other ingredient, the spectrum of accessible states. This varies from one system to the next and is the object of intensive experimental and theoretical research.

The two ingredients usually combine in such a way that the experimental data determine the type of theoretical model that is most appropriate and the theoretical model, in turn, helps us to extrapolate our knowledge into areas beyond the data.

Many very interesting systems can be treated as quantum gases using the tools of the next chapter. These include not only ordinary gases and plasmas but also photons in an oven, vibrations in a solid, conduction electrons and holes, some types of condensed matter, and stellar interiors.

Other important cases include systems of particles that are tightly bound. For these the neighboring states may be widely separated, so that the occupation number changes considerably from one state to the next. Some systems are hybrids, having both discrete and continuous components. For example, electrons in stellar atmospheres have discrete states when bound to atoms, but a continuum of states when free. We will also study systems whose states come in bands separated by gaps.

Whether occupation numbers change smoothly or abruptly between neighboring states depends on how the spacing between states compares with kT. There are some materials whose properties vary continuously at normal temperatures but show discrete changes, or "quantum effects," at very low temperatures where kT is small. Examples include the transition to superconductivity displayed by some conductors and the transition to superfluidity displayed by liquid helium.

D.2 Summing over states

If we know the distribution of quantum states as a function of their energies ε we can determine important features of the system directly. For example, the total number of particles in a system would be the sum of the particles in each state, and the total internal energy of the system would be the sum of the energies of the particles in each state. Calculating the mean value of a property f using classical statistics would require summing over states, weighting each value by the probability of being in that state. Thus we have

$$N = \sum_s \bar{n}_s, \qquad E = \sum_s \bar{n}_s \varepsilon_s, \qquad \bar{f} = \sum_s P_s f_s. \qquad (19.10)$$

In order to do these sums, however, we must know the spectrum of states to be summed over.

For very small systems of one or a few particles in bound states, the states are discrete and can be summed individually. But for larger systems, or systems whose particles are not bound, the various energy levels are highly degenerate and extremely closely spaced. For these, the sum over states must be replaced by continuous integration:

$$\sum_s \longrightarrow \int g(\varepsilon) d\varepsilon,$$

where $g(\varepsilon)$ is the density of states (subsection 1B.5). For example, the equations in 19.10 for the total number of particles and the total internal energy of the system would become

$$N = \int g(\varepsilon) \bar{n}(\varepsilon) d\varepsilon, \qquad E = \int g(\varepsilon) \bar{n}(\varepsilon) \varepsilon d\varepsilon. \qquad (19.11)$$

The differential form of 19.11 tells us how many particles or how much energy is stored in the states that lie in energy increment $d\varepsilon$:

$$dN = g(\varepsilon) \bar{n}(\varepsilon) d\varepsilon, \qquad dE = g(\varepsilon) \bar{n}(\varepsilon) \varepsilon d\varepsilon. \qquad (19.12)$$

The corresponding distributions (per unit energy increment) are

$$\frac{dN}{d\varepsilon} = g(\varepsilon) \bar{n}(\varepsilon), \qquad \frac{dE}{d\varepsilon} = g(\varepsilon) \bar{n}(\varepsilon) \varepsilon. \qquad (19.12')$$

E   The chemical potential

E.1   General

All calculations in quantum statistics involve the chemical potential μ, which appears in expressions 19.3a, b for the occupation number. Its value is determined by what we know about the system. For example, equation 14.2 expresses μ in terms of energy, volume, and entropy per particle. Alternatively, we could find μ

Figure 19.6 Plots of occupation number $\bar{n}$ vs. the energy ε of a state for fermions (left) and bosons (right), each for two different values of the chemical potential μ. A system with more particles would have a higher chemical potential in order to accommodate the additional particles. Hence, one way of determining the chemical potential is to insist that it should give the correct number of particles for the system.

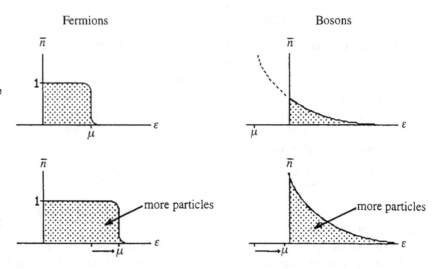

by insisting that the total number of particles is the sum of those in the individual quantum states (Figure 19.6):

$$N = \sum_s \bar{n}_s = \sum_s \frac{1}{e^{\beta(\varepsilon_s - \mu)} \pm 1}. \tag{19.13}$$

This is an implicit equation that can be solved for μ in terms of N and T. As we will soon see, the solution is rather easy for systems in the limits of low and high particle densities. But in general we have to be creative in determining μ, choosing methods appropriate to the particular system we study.

E.2   Low- and high-density limits

We will now solve equation 19.13 in the limits of low and high particle densities (in six-dimensional phase space). In subsection B.1 of this chapter we saw that in the "classical" limit of low particle densities, the exponential dominates. The chemical potential is then found from

$$e^{\beta\mu} = \frac{N}{\sum_s e^{-\beta\varepsilon_s}} \qquad \Rightarrow \qquad \mu_{\text{low densities}} = kT \ln\left(\frac{N}{\sum_s e^{-\beta\varepsilon_s}}\right). \tag{19.14}$$

Notice that we need to know the spectrum of accessible states in order to do the sum.

The opposite limit of high particle densities is even simpler. In fact, nearly *all* properties of a degenerate system, i.e., one confined to the configuration of lowest energy (subsection 9H.2 and Figure 9.8), are easy to study. For bosons we don't even need to know the spectrum of accessible states. All are in the very lowest state, so none of the other states matter.

Because degenerate bosons are all in the one lowest state, the occupation number of that state would be N,

$$N = \bar{n}_0 = \frac{1}{e^{\beta(0-\mu)} - 1} \qquad \text{(degenerate bosons)}.$$

We can solve this for the chemical potential, getting

$$\mu_{\text{degenerate bosons}} = -kT \ln\left(1 + \frac{1}{N}\right) \approx -\frac{kT}{N}. \qquad (19.15)$$

Fermions require a little more insight, because in a system of N identical fermions[2] the lowest N states are occupied and all those above are empty. So a knowledge of the spectrum of accessible states – at least the low-energy ones which are occupied – is helpful in understanding the system.

For degenerate fermions, all states are filled up to the Fermi surface, ε_f. As seen in Figures 19.1 or 9.8, this is equal to the chemical potential at $T = 0$:[3]

$$\mu_{T=0} = \varepsilon_f = \text{Fermi-surface energy}. \qquad (19.16)$$

Each quantum state has a volume h^3 in phase space, so a system of N degenerate fermions occupies a volume Nh^3. We can use this fact to estimate the momentum and kinetic energy of the particles at the Fermi surface via[4]

$$Nh^3 = V_r V_p \approx V p_f^3.$$

This gives

$$p_f \approx h\left(\frac{N}{V}\right)^{1/3},$$

so that

$$\mu_{\text{degenerate fermions}} = \varepsilon_f = \frac{p_f^2}{2m} \approx \frac{h^2}{2m}\left(\frac{N}{V}\right)^{2/3}. \qquad (19.17)$$

One thing you should notice about this result is that the chemical potential depends on the particle density N/V in position space. The greater the density, the higher the Fermi surface. A system of degenerate fermions occupies a fixed volume Nh^3 in six-dimensional phase space. So the more we constrain them in position space (smaller V_r), the more room they need in momentum space (larger V_p).

Many common fermion systems remain nearly degenerate at surprisingly high temperatures. As we have seen, the width of the tail in the fermion occupation

---

[2] Being "identical" includes having the same spin orientation.

[3] For nonzero temperatures, the chemical potential in fermion systems is called the "Fermi level."

[4] We have expressed the volume in momentum space as $V_p \approx p_f^3$, which is roughly correct, but the exact relationship depends on the spectrum of accessible states. For example, in the next chapter we will find that for unconstrained momenta, as in a gas, the volume would be that of a sphere, $V_p = (4/3)\pi p_f^3$. Here we use the p=0 state as our zero energy reference level.

number (i.e., the region where it goes from nearly 1 to nearly 0) is very roughly kT. So a system of fermions will remain mostly in the lowest possible states until the temperature reaches the point where thermal energies are comparable with ε_f:

$$\text{fermions are degenerate if } \frac{3}{2}kT \ll \varepsilon_f. \qquad (19.18)$$

In the homework problems, you will demonstrate that the Fermi surface for conduction electrons in a typical metal may correspond to a thermal temperature of over 100 000 K. And for electrons in collapsed stars or nucleons in a nucleus the Fermi surfaces correspond to more than a billion kelvins!

Summary of Sections D and E

The spectrum of accessible states is the subject of great experimental and theoretical interest. When combined with the occupation number, it determines the properties of a system.

The states of some systems are those of a quantum gas. Others come in bands. Some are discrete and widely separated, so that occupation numbers very greatly from one state to the next. Some are hybrids of the above. Whether the occupation number varies smoothly or abruptly between neighboring states depends on whether kT is large or small compared with the spacing of the states. The properties of some systems vary smoothly at high temperatures but display abruptly changing quantum effects at low temperatures. A "degenerate" system is one confined to the configuration of lowest possible energy.

The total number of particles in a system is the sum of the particles in each state, and the total internal energy of the system is the sum of the energies of the particles in each state. Calculating the mean value of some property f also requires summing over states, weighting each value by the probability of being in that state. Thus we have (equations 19.10)

$$N = \sum_s \bar{n}_s, \qquad E = \sum_s \bar{n}_s \varepsilon_s, \qquad \bar{f} = \sum_s P_s f_s.$$

For large numbers of accessible states, we often describe their distribution in energy by the density of states $g(\varepsilon)$. The distribution of particles is the product of the density of states times the average number of particles in each, and the distribution of energy is the product of the density of states times the average energy in each state (equation 19.12′):

$$\frac{dN}{d\varepsilon} = g(\varepsilon)\bar{n}(\varepsilon), \qquad \frac{dE}{d\varepsilon} = g(\varepsilon)\bar{n}(\varepsilon)\varepsilon.$$

Although the occupation number $\bar{n}(\varepsilon)$ has the same universal form for all systems (equations 19.3), the density of states $g(\varepsilon)$ varies from one system to the next.

All calculations involve the chemical potential, which appears in the occupation number $\bar{n}$. How we determine its value depends on what we know about the system. One method is to ensure that the total number of particles is the sum of those in the individual quantum states (equation 19.13):

$$N = \sum_s \bar{n}_s = \sum_s \frac{1}{e^{\beta(\varepsilon_s - \mu)} \pm 1}.$$

This is an implicit equation that can be solved for μ in terms of N and T.

For degenerate identical bosons all N particles are in the single lowest state, so the occupation number of that state would be N. From this we find that the chemical potential for a system of N degenerate bosons at temperature T is given by (equation 19.15)

$$\mu_{\text{degenerate bosons}} = -kT \ln\left(1 + \frac{1}{N}\right) \approx -\frac{kT}{N}.$$

For a degenerate system of identical fermions, the lowest N states are filled up to the Fermi surface, which is equal to the chemical potential at absolute zero (equation 19.16):

$$\mu_{T=0} = \varepsilon_f = \text{Fermi-surface energy.}$$

Because the N identical fermions occupy a fixed volume Nh^3 in phase space, the volumes they occupy in position and momentum space are related. The smaller the one, the larger the other. With small variations depending on the system, the energy and momentum of the Fermi surface are related to the particle density in position space through (equation 19.17)

$$p_f \approx h\left(\frac{N}{V}\right)^{1/3} \quad \text{or} \quad \varepsilon_f \approx \frac{h^2}{2m}\left(\frac{N}{V}\right)^{2/3}.$$

Fermions remain fairly degenerate until thermal energies are comparable to the energy of the Fermi surface.

Problems

Section A

1. At room temperature (295 K), find the occupation number for a fermion state that is (a) 0.01 eV below μ, (b) 0.01 eV above μ, (c) 0.1 eV below μ, (d) 0.1 eV above μ.

2. At room temperature (295 K), find the occupation number for a boson state that is (a) 0.01 eV above μ, (b) 0.1 eV above μ, (c) 1.0 eV above μ.

3. Is it possible for a boson to have the z-component of its angular momentum equal to $(1/2)\hbar$?

4. A certain state lies 0.004 eV above the chemical potential in a system at 17 °C. What is the occupation number for this state if the system is composed of (a) fermions, (b) bosons?

5. For a temperature of 300 K, compute the ratio $\bar{n}_{bosons}/\bar{n}_{fermions}$ for $\varepsilon - \mu$ equal to (a) 0.002 eV, (b) 0.02 eV, (c) 0.2 eV, (d) 2.0 eV.

6. A certain quantum state lies 0.1 eV above the chemical potential. At what temperature would the average number of bosons in this state be (a) 5, (b) 0.5, (c) 0.05?

7. The hydrogen atom is made of two spin-1/2 particles, a proton and an electron.
   (a) When the electron is in an s-orbital (i.e., an orbital with $l = 0$), what are the possible values of the total angular momentum quantum number j for the atom? (Hint: $\mathbf{J} = \mathbf{L} + \mathbf{S}$. Think about the z-component.)
   (b) When the electron is in an orbital of angular momentum $L = \sqrt{2}\,\hbar$, what are the possible values of the total angular momentum quantum number j for the atom?
   (c) The atmosphere of a certain star is atomic hydrogen gas. Would this be a boson or a fermion gas?
   (d) Deeper down toward the star's violent interior, the hydrogen atoms are stripped of their electrons. Would the leftover protons be a gas of bosons or fermions? Would the stripped electrons be a gas of bosons or fermions?

8. (a) Draw a plot of $\bar{n}$ versus ε, showing how it would look in the limits of extremely high and extremely low temperatures. Assume that the chemical potential doesn't vary with temperature. (In a later problem, it will be seen that this is a bad assumption.) Do this for (a) bosons, (b) fermions.

9. Suppose that there is a third class of fundamental particles, called "goofions," for which 0, 1, or 2 particles may occupy any state. Derive the occupation number $\bar{n}$ as a function of ε, μ, and β for these goofions. (Needless to say, goofions aren't found in our part of the Universe. But composite particles that behave like this are found.)

10. Repeat the above problem for "daffyons," which may have only 0, 2, or 5 particles per state.

11. The ground state of a boson system at 290 K has energy $\varepsilon = 0$. If the occupation number of this state is 0.001, what is the chemical potential?

12. The chemical potential of photons is zero.
    (a) What is the occupation number for the ($\varepsilon = 0$) ground state of a photon gas?

(b) Why isn't this a problem?

(c) The total energy held by the photons in any state is the product of the number of photons times the energy of each. Find the energy carried by all the photons in the ground state by writing this product for finite ε and taking the limit as $\varepsilon \to 0$. (Answer in terms of kT.)

13. Fill in the missing steps in the development leading to equation 19.4.

Section B

14. What is the occupation number $\bar{n}$ of a state with energy $\varepsilon = \mu$ according to (19.3, 19.6) (a) Fermi–Dirac statistics, (b) classical statistics, (c) Bose–Einstein statistics?

15. Regarding the occupation number $\bar{n}$:
    (a) For what value of $(\varepsilon - \mu)/kT$ is the classical prediction half that for boson systems?
    (b) For this value of $(\varepsilon - \mu)/kT$, what is the ratio of the classical prediction and that for fermions?

16. If a system is at 1000 K, what is the occupation number of a state of energy 0.1 eV above the chemical potential according to (a) Fermi–Dirac statistics, (b) classical statistics, (c) Bose–Einstein statistics. (d) Repeat these calculations for a temperature of 300 K.

17. Consider a system of three flipped coins. According to classical statistics, how many different arrangements are available to this system:
    (a) if the coins are distinguishable,
    (b) if the coins are identical?
    (c) For what fraction of the arrangements in (a) are all three heads or all three tails?
    (d) What is the true number of different heads/tails configurations available to a system of three identical flipped coins?
    (e) For what fraction of these arrangements are all three heads or all three tails?

18. Consider a system of two rolled dice, each having six possible states available to it (six different numbers of dots showing upward). According to classical statistics, how many different arrangements are available to this system
    (a) if the dice are distinguishable,
    (b) if the dice are identical.
    (c) According to classical statistics, in what fraction of the total number of different configurations do the two dice show the same number of dots?
    (d) What is the true number of distinguishable configurations available to two identical dice?

(e) For what fraction of the distinguishable configurations in (a) do the two dice show the same number of dots?

19. Consider the possible arrangements of two particles between four quantum states. Some arrangements will have no more than one particle per state (single occupancy) and some will have both particles in the same state (double occupancy). Give the numbers s and d of different arrangements having single and double occupancy according to
    (a) classical statistics for distinguishable particles,
    (b) classical statistics for identical particles,
    (c) Bose–Einstein statistics,
    (d) Fermi–Dirac statistics.

20. Consider the distribution of two particles between three states. List all possible arrangements of the system if the particles are (a) distinguishable, (b) identical bosons, (c) identical fermions.

21. In the above problem, what is the probability of finding the two particles in the same state according to (a) classical statistics, (b) Bose–Einstein statistics, (c) Fermi–Dirac statistics?

22. Consider the distribution of three particles among two states. List the possible arrangements of the system if the particles are (a) distinguishable, (b) identical bosons, (c) identical fermions.

23. In the problem above, what is the probability of finding the system at any instant in an arrangement where all three particles are in the same state according to
    (a) classical statistics for distinguishable particles,
    (b) classical statistics for identical particles,
    (c) Bose–Einstein statistics,
    (d) Fermi–Dirac statistics?

Section C

24. Consider a system of nonrelativistic electrons in a white dwarf star at a temperature of 10^9 K. Very roughly, what would be their density if the system is degenerate? How does this compare with typical electron densities in ordinary matter of about 10^{30} electrons/m^3?

25. Consider nitrogen (N_2) gas at room temperature and atmospheric pressure (1.013×10^5 Pa). The mass of each molecule is 4.67×10^{-26} kg.
    (a) Determine whether classical statistics would be appropriate for its study. (Hint: $N/V = p/kT$.)
    (b) If the particle density remains unchanged, at what temperature would the number of accessible quantum states and the number of N_2 molecules become about equal?

26. Consider the conduction electrons in copper. Each copper atom contributes one electron to the conduction electrons. Copper has atomic weight 64 and density 8.9 g/cm³.
    (a) What is the volume per conduction electron?
    (b) What is the characteristic separation of conduction electrons, r?
    (c) Can these conduction electrons be treated using classical statistics at room temperature?
    (d) Could these conduction electrons be treated using classical statistics at 10 000 K?
    (e) Can the copper *atoms* be treated using classical statistics at room temperature?

27. The characteristic separation of molecules of liquid water is about 0.3 μm. Could water molecules at room temperature be studied using classical statistics?

28. Thermal energies for nucleons in large nuclei are comparable with their binding energies of about 6 MeV.
    (a) To what temperature does this correspond?
    (b) Within nuclear matter, identical nucleons are separated by about 2.6×10^{-15} m. What is roughly the minimum temperature needed for them not to be degenerate (i.e., for the number of accessible states to be much larger than the number of particles.)?

Section D

29. Iron atoms have mass 9.3×10^{-26} kg and are bound in place by electrostatic forces with an effective force constant $\kappa = 2$ N/m.
    (a) What is the fundamental frequency of vibration for these atoms?
    (b) What is the excitation temperature?
    (c) At a temperature of 20 K, is the separation of neighboring states large or small compared with kT?

30. Consider a single conduction electron in a cube of metal measuring 1 cm on a side.
    (a) How much energy separates the ground state from the first excited state for this electron? (Hint: Each dimension must be an integer number of half wavelengths. So in the ground state the wave number in each of the three dimensions would be $(2\pi/2)$ cm^{-1}.)
    (b) What is the excitation temperature?
    (c) At a temperature of 20 K, is the separation of neighboring states large or small compared with kT?

31. Consider the electrons (spin 1/2) in a conduction band, which are confined to volume V and for which the energy and momentum are related through $\varepsilon = \varepsilon_0 + p^2/2m$, where ε_0 is some constant reference level. Suppose that this system can be treated as a gas.

(a) Start with $d^3r d^3p/h^3$ and integrate over all volume and all angles of the momentum to find an expression for the number of accessible states as a function of the magnitudes of p, m, and V. (Hint: Write $d^3p = p^2 dp \sin\theta\, d\theta d\phi$ and integrate over all solid angles.)

(b) Convert p into ε to find an expression for the density of states in terms of m, V, ε, and ε_0.

(c) What is the distribution of particles, $dN/d\varepsilon$, for these fermions in terms of m, V, ε, ε_0, μ, and T?

32. (Computational problem. It would help to have a spreadsheet or a programmable calculator.) Consider a system of bosons at 290 K with $\mu = -0.01$ eV, for which the energies of the various states are given by $\varepsilon_m = m0.01$ eV, where $m = 0, 1, 2, 3, \ldots$

(a) What is the total number of particles in this system, on average?

(b) What is the total energy of this system, on average?

(c) Repeat parts (a) and (b) for the case where the mth state is $m+1$ times degenerate.

33. Repeat the above problem for a system of fermions at 290 K for which $\mu = +0.03$eV.

34. Suppose that the density of states for some system is given by $g(\varepsilon) = C\varepsilon^\alpha$, where C and α are constants. Set up the integrals for the number of particles and for the total energy of this system, if the particles are (a) fermions, (b) bosons.

Section E

35. For a certain boson system, the density of states is constant $(g(\varepsilon) = C)$, and the occupation numbers of all states are sufficiently small that we can write $\bar{n} \approx e^{-\beta(\varepsilon-\mu)}$. Show that the chemical potential, μ, depends on the temperature and the total number of particles in the system according to $\mu = -kT \ln(CkT/N)$.

36. Consider a system of fermions and a system of bosons that each have the same spectrum of states. In order to have the same number of particles altogether, which chemical potential must be larger? (Hint: See Figure 19.2.)

Section F

For problems 37–39, assume that the fermions are truly identical, including having the same spin orientations.

37. Using equation 19.17, give a rough estimate of the energy of the Fermi surface in eV and the corresponding thermal temperatures, for each of the following systems:

(a) conduction electrons in copper metal, for which there are about 8×10^{28} conduction electrons per cubic meter (comparable with the density of copper atoms);

(b) conduction electrons in aluminum. Its density is 2.7 g/cm^3, its atomic mass number is 27, and each atom gives one electron to the conduction band.

38. Using equation 19.17, give a rough estimate of the energy of the Fermi surface in eV and the corresponding thermal temperatures, for each of the following systems:

(a) the electrons in a white dwarf star, whose electron density is about $10^{35}/m^3$;

(b) the protons in a white dwarf star, whose proton density is comparable with that of the electrons;

(c) the neutrons in an iron nucleus, which is a sphere of radius 5.0×10^{-15} m and which contains 30 neutrons.

39. White dwarf stars are essentially plasmas of free electrons and free protons (hydrogen atoms that have been stripped of their electrons). Their densities are typically 5×10^9 kg/m^3 (i.e., 10^6 times the average density of Earth). Their further collapse is prevented by the fact that the electrons are highly degenerate, that is, all the low-lying states are filled and no two identical electrons can be forced into the same state.

(a) Estimate the temperature of this system. (Assume nonrelativistic electrons and that $p_f^2/2m = (3/2)kT$.)

(b) Now suppose that the gravity is so strong that the electrons are forced to combine with the protons, forming neutrons (with the release of a neutrino). If the temperature remains the same, what would be the density of this degenerate neutron star?

Chapter 20

Quantum gases

Although the occupation number has the same form for all fermion or all boson systems, the spectrum of accessible states does not. We begin our study of this second and more elusive of the two ingredients in the study of such systems by looking into quantum gases.

A The density of states

The number of quantum states within the six-dimensional element of phase space $d^3r d^3p$ is given by equation 1.5:

$$\text{number of states} = \frac{d^3r d^3p}{h^3}.$$

Converting this to the form $g(\varepsilon)d\varepsilon$, where $g(\varepsilon)$ is the density of states, could be difficult, because for many systems, interactions may severely constrain the accessible regions of phase space and make the energy a complicated function of the position and momentum.

For a gas, however, the particle momenta and positions are unrestricted. As we showed in subsection 1B.5, we can integrate over all volume ($\int d^3r = V$) and all angles for the momentum,

$$\int p^2 dp \sin\theta \, d\theta d\phi \rightarrow 4\pi \int p^2 dp,$$

and then convert momentum to energy according to[1]

$$\varepsilon = \frac{p^2}{2m} \quad \text{(nonrelativistic)}, \qquad \varepsilon = pc \quad \text{(relativistic)},$$

to get (equation 1.9)

$$g(\varepsilon) = \frac{2\pi V(2m)^{3/2}}{h^3}\sqrt{\varepsilon} \qquad \text{(nonreleativistic)},$$

$$g(\varepsilon) = \frac{4\pi V}{h^3 c^3}\varepsilon^2 \qquad \text{(relativistic)}.$$

Each system of otherwise identical particles is actually composed of subsystems that are distinguished from each other by the orientations of the spins. A massive spin-s particle (subsection 1B.6) has $2s + 1$ possible spin orientations. (However, if it is massless and moving at the speed of light, its spin has only $2s$ orientations.) If we include this spin degeneracy in our density of states then equation 1.9 becomes

$$g(\varepsilon) = C\varepsilon^\alpha, \qquad (20.1)$$

where

$$C = \begin{cases} (2s + 1)\dfrac{2\pi V(2m)^{3/2}}{h^3}, & \alpha = \dfrac{1}{2} & \text{(nonrelativistic particles)}, \\[2ex] 2s\dfrac{4\pi V}{h^3 c^3}, & \alpha = 2 & \text{(relativistic particles)}. \end{cases}$$

B Distributions and mean values

Using the density of states and the occupation numbers, we can now calculate various properties. For example, the density of particles is the product of the density of states and the average number of particles in each, and the distribution of energy is the product of the density of states and the average energy in each (equations 19.3 and 19.12′):

$$\frac{dN}{d\varepsilon} = g(\varepsilon)\overline{n}(\varepsilon) = \frac{C\varepsilon^\alpha}{e^{\beta(\varepsilon-\mu)} \pm 1},$$

$$\frac{dE}{d\varepsilon} = g(\varepsilon)\overline{n}(\varepsilon)\varepsilon = \frac{C\varepsilon^{\alpha+1}}{e^{\beta(\varepsilon-\mu)} \pm 1}. \qquad (20.2)$$

More generally, the mean value of any property f, averaged over all particles, is given by

$$\overline{f} = \frac{1}{N}\int f\,dN = \frac{1}{N}\int f(\varepsilon)\underbrace{g(\varepsilon)\overline{n}(\varepsilon)d\varepsilon}_{dN}. \qquad (20.3)$$

The particle distributions $dN/d\varepsilon$ for quantum gases are shown on the right in Figures 20.1 through 20.3 for the following four cases, each at both high- and

---

[1] In the homework problems this conversion can be carried out for the general case of all speeds, with $\varepsilon = \sqrt{p^2 c^2 + m^2 c^4} - mc^2$.

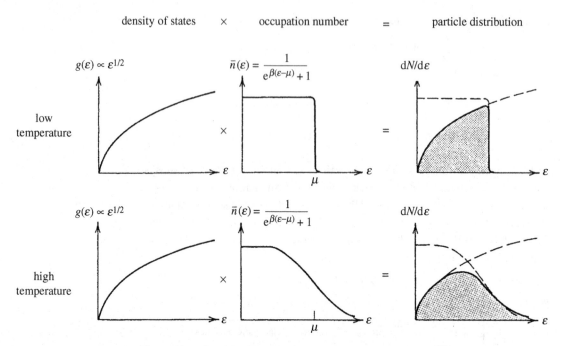

Figure 20.1 Particle distribution in a nonrelativistic ($g(\varepsilon) \propto \varepsilon^{1/2}$) fermion gas at low and high temperatures.

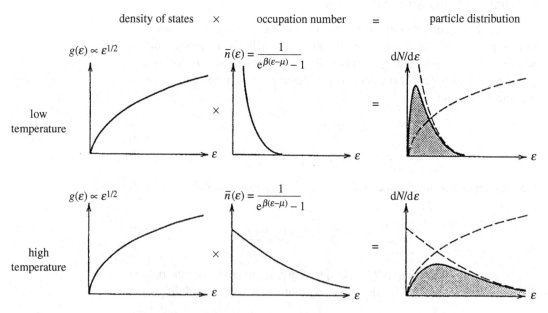

Figure 20.2 Particle distribution in a nonrelativistic ($g(\varepsilon) \propto \varepsilon^{1/2}$) boson gas at low and high temperatures.

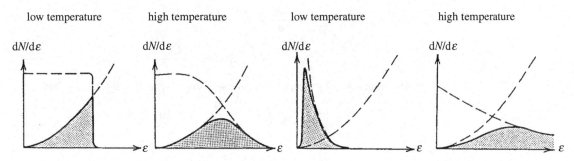

Figure 20.3 Particle distributions in relativistic ($g(\varepsilon) \propto \varepsilon^2$) quantum gases at low and high temperatures.

low-temperature extremes: nonrelativistic fermions, nonrelativistic bosons, relativistic fermions, relativistic bosons. You can see that in the low-energy limit the exponential is of little consequence and in the high-energy limit it dominates. So for the particle distributions in these two limits we have[2]

$$\frac{\mathrm{d}N}{\mathrm{d}\varepsilon} \approx \begin{cases} \varepsilon^{\alpha} & \text{at low energies,} \\ e^{-\beta\varepsilon} & \text{at high energies.} \end{cases}$$

Summary of Sections A and B

The number of quantum states within the six-dimensional element of phase space $\mathrm{d}^3 r \mathrm{d}^3 p$ is given by equation 1.5.

$$\text{number of states} = \frac{\mathrm{d}^3 r \mathrm{d}^3 p}{h^3}.$$

For the important case of gases, we integrate over all volume and momentum directions, convert momentum to energy, and sum over all possible spin orientations of the spin-s particles to get the following expression for the density of states (equation 20.1):

$$g(\varepsilon) = C\varepsilon^{\alpha},$$

where

$$C = \begin{cases} (2s+1)\dfrac{2\pi V(2m)^{3/2}}{h^3}, & \alpha = \dfrac{1}{2} \quad \text{(nonrelativistic particles),} \\ 2s\dfrac{4\pi V}{h^3 c^3}, & \alpha = 2 \qquad \text{(relativistic particles).} \end{cases}$$

Combining the expression for the density of states $g(\varepsilon)$ with the occupation numbers of equation 19.3, we can now use equation 19.12' to find the distribution of

---

[2] For the case of bosons with $\mu = 0$, the behavior will be slightly different, as can be shown in the homework problems.

particles and energy in a quantum gas (equation 20.2).

$$\frac{dN}{d\varepsilon} = g(\varepsilon)\bar{n}(\varepsilon) = \frac{C\varepsilon^{\alpha}}{e^{\beta(\varepsilon-\mu)} \pm 1},$$

$$\frac{dE}{d\varepsilon} = g(\varepsilon)\bar{n}(\varepsilon)\varepsilon = \frac{C\varepsilon^{\alpha+1}}{e^{\beta(\varepsilon-\mu)} \pm 1}.$$

The mean value of any property f, averaged over all particles, is (equation 20.3)

$$\bar{f} = \frac{1}{N}\int f dN = \frac{1}{N}\int f(\varepsilon)g(\varepsilon)\bar{n}(\varepsilon)d\varepsilon.$$

C   Internal energy and the gas laws

In Section 15H, we used classical statistics to show that a gas of N particles should have a total translational kinetic energy of

$$E_{\text{translational}} = \tfrac{3}{2}NkT \qquad \text{(classical gas)}.$$

But now we know that classical statistics is not quite correct for systems whose particles are sufficiently dense that more than one particle may attempt to occupy the same state.

In the case of identical bosons, the low-lying states have larger occupation numbers than the classical prediction (See Section 19B and Figures 19.2 and 20.4), so the total internal energy of a system of identical bosons should be correspondingly lower:

$$E_{\text{translational}} < \tfrac{3}{2}NkT \qquad \text{(boson gas)}.$$

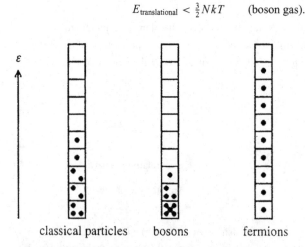

Figure 20.4   Compared with classical particles, bosons have more particles in the low-lying states of multiple occupancy. Therefore, a boson gas has a smaller internal energy than the classical prediction. For identical fermions, however, no two may occupy the same state. This exclusion forces them into higher levels and therefore a fermion gas has a higher internal energy than the classical prediction.

In the case of identical fermions, the low-lying states have smaller occupation numbers than the classical prediction. The exclusion of particles from states already occupied means that many fermions are forced into higher levels (Section 19B or Figure 20.4). Consequently, the total internal energy of a system of identical fermions is greater than the classical result:

$$E_{\text{translational}} > \tfrac{3}{2} NkT \qquad \text{(fermion gas)}.$$

In the homework problems these predictions can be verified by carrying out the steps in the following derivation. We examine the integrals of equations 20.2 for the total number of particles and the total energy of the system:

$$N = \int dN = C \int_0^\infty \varepsilon^\alpha \, \bar{n}(\varepsilon) d\varepsilon, \qquad (20.4a)$$

$$E = \int dE = C \int_0^\infty \varepsilon^{\alpha+1} \bar{n}(\varepsilon) d\varepsilon. \qquad (20.4b)$$

The similarity of these two expressions allows us to integrate one by parts to express it in terms of the other. Integrating the first by parts gives

$$N = 0 - \frac{C}{\alpha+1} \int_0^\infty \varepsilon^{\alpha+1} d\bar{n}.$$

Into this integral we substitute the following expression, which is obtained by differentiating equation 19.3 for $\bar{n}$:

$$d\bar{n} = -\beta \left(\bar{n} \mp \bar{n}^2 \right) d\varepsilon.$$

The result is[3]

$$N = \frac{C}{(\alpha+1)kT} \int_0^\infty \varepsilon^{\alpha+1} \left(\bar{n} \mp \bar{n}^2 \right) d\varepsilon.$$

Multiplying both sides by $(\alpha + 1)kT$ and breaking the integral into two parts gives

$$(\alpha+1)NkT = C \int_0^\infty \varepsilon^{\alpha+1} \bar{n} d\varepsilon \mp C \int_0^\infty \varepsilon^{\alpha+1} \bar{n}^2 d\varepsilon.$$

The first of the two integrals is the internal energy of the system as given in equation 20.4b above. Shifting this to one side of the equation and the other two terms to the other gives (refer to equation 20.1 for the values of α)

$$E = (\alpha+1)NkT \pm \delta, \qquad (20.5)$$

[3] We always use the convention that the upper sign is for fermions and the lower sign for bosons.

where

$$\delta = C \int_0^\infty \varepsilon^{\alpha+1} \overline{n}^2 d\varepsilon,$$

$$\alpha + 1 = \begin{cases} \frac{3}{2} & \text{(nonrelativistic gas)}, \\ 3 & \text{(relativistic gas)}. \end{cases}$$

This result shows that the kinetic energy of a quantum gas differs from the classical prediction $E = (\alpha + 1)NkT$, the fermion energy being larger $(+\delta)$ and the boson energy being smaller $(-\delta)$. Furthermore, because of the factor $\overline{n}^2$ in the integral for δ, the difference between the quantum and classical results is only significant when the occupation numbers are relatively large. This is exactly what we expected.

This difference in kinetic energy carries directly over into the pressure, giving the corresponding modification of the ideal gas laws. It is fairly easy to show (homework) that the pressure of a gas is given by

$$pV = \tfrac{2}{3} E_{\text{translational}} \qquad \left(pV = \tfrac{1}{3} E_{\text{translational}}, \text{ if relativistic} \right). \tag{20.6}$$

Using result 20.5, this gives the following ideal gas law for nonrelativistic quantum gases:

$$pV = NkT \pm \tfrac{2}{3}\delta. \tag{20.7}$$

So, compared with the classical result for any given temperature, the pressure in a fermion gas is larger and that in a boson gas is smaller.

D   Internal energy and the chemical potential

In the preceding chapter (subsection 19E.1, equation 19.13) we saw that we can determine the chemical potential implicitly from the requirement that the total number of particles equals the sum of those in the individual quantum states:[4]

$$N = \sum_s \overline{n}_s = \sum_s \frac{1}{e^{\beta(\varepsilon_s - \mu)} \pm 1}. \tag{20.8}$$

Once μ is determined, we then know the occupation numbers $\overline{n}_s$ for all states. This enables us to determine other properties. For example, the internal energy is the sum of the energies of the particles in each state:

$$E = \sum_s \overline{n}_s \varepsilon_s = \sum_s \frac{\varepsilon_s}{e^{\beta(\varepsilon_s - \mu)} \pm 1}. \tag{20.9}$$

D.1   Degenerate gases

As we saw in the previous chapter, it is particularly easy to solve these equations in the low-temperature "degenerate" limit, where the particles are in the lowest

---

[4] The dependence of μ on N and T (through $\beta = 1/kT$) is evident in this implicit equation. The dependence on V is more subtle, as it enters via the spectrum of states being summed over.

states possible. For degenerate bosons, all N particles are in the one state of very lowest energy ($\varepsilon = 0$), making equations 20.8 and 20.9 particularly easy to solve for μ and E (equation 19.15):

$$\mu_{\text{degenerate bosons}} = -kT \ln\left(1 + \frac{1}{N}\right) \approx -\frac{kT}{N},$$

$$E_{\text{degenerate bosons}} = 0. \tag{20.10}$$

This result does not depend on the spectrum of states, since only the lowest-energy state is occupied. Consequently, it is correct for all degenerate boson systems, gases or otherwise.

For fermions, however, the distribution of states *does* matter. Because of the exclusion principle, as many different states must be occupied as there are identical fermions. For summing over such a large number of states, it is easiest to replace the discrete sums in equations 20.8 and 20.9 by continuous integrals as in equations 20.4, using the density of states for gases, $g(\varepsilon) = C\varepsilon^{\alpha}$, from equation 20.1:[5]

$$N = \sum_s \bar{n}_s = \int g(\varepsilon)\bar{n}(\varepsilon)d\varepsilon \xrightarrow[\text{gases}]{} C \int \varepsilon^{\alpha}\bar{n}(\varepsilon)d\varepsilon, \tag{20.11a}$$

$$E = \sum_s \bar{n}_s\varepsilon_s = \int g(\varepsilon)\bar{n}(\varepsilon)\varepsilon d\varepsilon \xrightarrow[\text{gases}]{} C \int \varepsilon^{\alpha+1}\bar{n}(\varepsilon)d\varepsilon. \tag{20.11b}$$

For degenerate fermions, the lowest N states up to the Fermi surface ($\varepsilon_f = \mu_{T=0}$) are occupied and those above it are empty. This fact makes it easy to evaluate the integrals of equations 20.11, because the occupation number is a step function (Figure 19.1):

$$\bar{n}(\varepsilon) = \begin{cases} 1 & \text{for } \varepsilon < \varepsilon_f, \\ 0 & \text{for } \varepsilon > \varepsilon_f. \end{cases}$$

Thus we obtain

$$N = C \int_0^{\varepsilon_f} \varepsilon^{\alpha}d\varepsilon = \frac{C}{\alpha+1}\varepsilon_f^{\alpha+1}, \tag{20.12a}$$

$$E = C \int_0^{\varepsilon_f} \varepsilon^{\alpha+1}d\varepsilon = \frac{C}{\alpha+2}\varepsilon_f^{\alpha+2}. \tag{20.12b}$$

Solving the first of these for the chemical potential ($\mu = \varepsilon_f$) gives

$$\mu_{\text{degenerate fermions}} = \varepsilon_f = \left[\frac{(\alpha+1)N}{C}\right]^{1/(\alpha+1)}, \tag{20.13}$$

---

[5] Notice that, in replacing the discrete sum by an integral, the ground state is excluded because $g(0) = 0$. Although generally not significant for fermions, this exclusion would be significant for low-temperature boson systems, where large numbers of particles might be in this excluded state.

Table 20.1. *The chemical potential $\mu(V, N, T)$ for quantum gases (λ is the number of spin orientations. $\alpha = 1/2$ if the gas is nonrelativistic and $\alpha = 2$ if it is relativistic)*

All temperatures, bosons and fermions		
$N =$	$\lambda C(kT)^{\alpha+1} f(y)$, where $f(y) = \int \dfrac{x^\alpha dx}{e^{x-y} \pm 1} \left(y = \dfrac{\mu}{kT}\right)$	
Degenerate, high-occupancy, $T \to 0$ limit	nonrelativistic ($\alpha = 1/2$)	relativistic ($\alpha = 2$)
$\mu_{\text{bosons}} =$	$-\dfrac{kT}{N}$	$-\dfrac{kT}{N}$
$\mu_{\text{fermions}} =$	$\dfrac{h^2}{2m}\left(\dfrac{3N}{\lambda 4\pi V}\right)^{2/3}$	$hc\left(\dfrac{3N}{\lambda 4\pi V}\right)^{1/3}$
classical, low-occupancy, $T \to \infty$ limit		
$\mu_{\text{both bosons and fermions}} =$	$-kT \ln\left[\dfrac{\lambda V}{Nh^3}(2\pi m kT)^{3/2}\right]$	$-kT \ln\left[\dfrac{\lambda 8\pi V}{Nh^3 c^3}(kT)^3\right]$
nearly degenerate spin-1/2 fermions		
$\mu \approx$	$\varepsilon_f\left[1 - \dfrac{\pi^2}{12}\left(\dfrac{kT}{\varepsilon_f}\right)^2\right]$	

where the values of α and C are given by equation 20.1. And the ratio of the two results 20.12a, b gives the average energy per particle in a degenerate fermion gas:

$$\bar{\varepsilon} = \frac{E}{N} = \frac{\alpha + 1}{\alpha + 2}\varepsilon_f. \tag{20.14}$$

Inserting the values of α and C from equation 20.1 into equations 20.13 and 20.14 gives the following expressions for the chemical potential and average energy per particle for a system of degenerate spin-1/2 fermions:

in the nonrelativistic case,

$$\mu_{\text{degenerate fermions}} = \varepsilon_f = \frac{h^2}{2m}\left(\frac{3N}{8\pi V}\right)^{2/3}, \qquad \bar{\varepsilon} = \frac{3}{5}\varepsilon_f, \tag{20.15a}$$

and in the relativistic case,

$$\mu_{\text{degenerate fermions}} = \varepsilon_f = hc\left(\frac{3N}{4\pi V}\right)^{1/3}, \qquad \bar{\varepsilon} = \frac{3}{4}\varepsilon_f. \tag{20.15b}$$

The results for N and μ are collected together in Table 20.1. Important degenerate (or nearly degenerate) spin-1/2 fermion gases include the following:

nonrelativistic case
- conduction electrons in metals (free electron model)
- inner electrons in large atoms (Thomas–Fermi model)
- nucleons in large nuclei and neutron stars

relativistic case
- electrons in collapsed stars
- quarks, electrons, neutrinos in the early Universe

D.2 The classical limit

In the "classical limit" of high temperatures, the occupation numbers are small[6] and can be approximated as follows:

$$\overline{n}(\varepsilon) = \frac{1}{e^{\beta(\varepsilon-\mu)} \pm 1} \approx e^{-\beta(\varepsilon-\mu)}.$$

Our equations 20.8 and 20.9 then become

$$N = Ce^{\beta\mu} \int e^{-\beta\varepsilon} \varepsilon^\alpha d\varepsilon, \tag{20.16a}$$

$$E = Ce^{\beta\mu} \int e^{-\beta\varepsilon} \varepsilon^{\alpha+1} d\varepsilon. \tag{20.16b}$$

The integrals are now easy. Solving the first equation for μ gives (homework)

$$\mu_{cl} = \begin{cases} -kT \ln\left[\dfrac{(2s+1)V}{Nh^3} (2\pi mkT)^{3/2}\right] & \text{(nonrelativistic, } \alpha = 1/2), \\[3mm] -kT \ln\left[\dfrac{2s\,8\pi V}{Nh^3c^3} (kT)^3\right] & \text{(relativistic, } \alpha = 2) \end{cases}$$

$$\tag{20.17}$$

where we have inserted the values of C and α from equation 20.1. Knowing μ we could now use equation 20.16b to calculate the system's energy E (homework). But this is not necessary, because we can integrate equation 20.16a for N by parts to get equation 20.16b for E, showing (as we did in the preceding section) that

$$E_{cl} = (\alpha + 1)NkT.$$

D.3 Intermediate temperatures

We now have expressions for the chemical potential and energy for boson and fermion gases in the degenerate ($T \to 0$, high-occupancy) and classical ($T \to \infty$, low-occupancy) limits. But between these two limits, equations 20.11 are difficult to solve. Making the substitutions $x = \varepsilon/kT$, $y = \mu/kT$ they can be turned into the forms (problem 16)

$$N = C(kT)^{\alpha+1} f(y), \qquad \text{where } f(y) = \int \frac{x^\alpha dx}{e^{x-y} \pm 1}, \tag{20.18a}$$

and

$$E = C(kT)^{\alpha+2} g(y), \qquad \text{where } g(y) = \int \frac{x^{\alpha+1} dx}{e^{x-y} \pm 1}. \tag{20.18b}$$

The integrals on the right-hand sides of equations 20.18 can be done numerically, so that we can make plots of $f(y)$ and $g(y)$ vs. y (Figure 20.5). The

---

[6] It is not obvious from looking at the fermion occupation number that it should be small in the high-temperature (small-β) limit. But it is. The physical reason is that the increased accessible volume in momentum space means that the same number of particles is spread out over a larger number of states. The mathematical reason is that μ becomes large and negative.

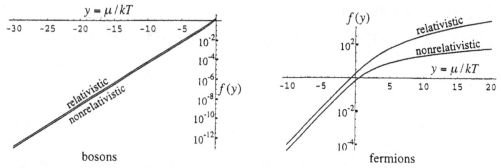

bosons   fermions

Figure 20.5 Plots of $f(y)$ vs. $y = \mu/kT$ for quantum gases, which are used in determining the chemical potential via equations 20.18. Why do we only have $y < 0$ on the boson plot? (Hint: When y and hence μ reach 0, how many bosons would be in the ground ($\varepsilon = 0$) state alone?)

correct value of $y = \mu/kT$ is the one that gives the correct number of particles in equation 20.18a.

Therefore, in order to determine the correct value of y (and hence μ) for a system, we can do the following with equation 20.18a:

- calculate the value of the constant C for the system using equation 20.1;
- solve equation 20.18a for $f(y)$ (using $f(y) = N/C(kT)^{\alpha+1}$);
- go to the plot of $f(y)$ vs. y to find the value of y.

Having determined the correct value of y (and hence μ) in this manner, we can now use it in the occupation number $\bar{n}$ to calculate any property of the system. For example, we could now use equation 20.18b to find the system's internal energy.

As you can see, this technique for determining the chemical potential is tedious. Fortunately, we seldom need to use it. Many systems are in either the degenerate or classical regimes, for which the chemical potentials are given by equations 20.10, 20.15, and 20.17. And in many cases, we won't need to know the chemical potential at all. Even in the intermediate regions we can sometimes determine the chemical potential cleanly and simply from other considerations.

But one important system for which we cannot always avoid the integrals of equations 20.18 is that of nearly (but not completely) degenerate nonrelativistic fermion gases, such as conduction electrons in metals. For these, we integrate by parts ($\int u\,dv = uv|_0^\infty - \int v\,du$), noting that we can do both integrals, $f(y)$ and $g(y)$, in the following manner:

$$I(y) = \int_0^\infty \underbrace{\left(\frac{1}{e^{x-y}+1}\right)}_{u} \underbrace{(x^\kappa dx)}_{dv} = \underbrace{\left(\frac{1}{e^{x-y}+1}\right)}_{u} \underbrace{\left(\frac{x^{\kappa+1}}{\kappa+1}\right)}_{v} \Bigg|_0^\infty$$

$$-\int_0^\infty \underbrace{\left(\frac{x^{\kappa+1}}{\kappa+1}\right)}_{v} \underbrace{\left(\frac{-e^{x-y}dx}{(e^{x-y}+1)^2}\right)}_{du},$$

and where $\kappa = 1/2$ for $f(y)$ and $\kappa = 3/2$ for $g(y)$. The first term on the right ($uv|_0^\infty$) is zero at both end points, and the second term ($-\int v\,du$) is evaluated by noting that the fermion occupation number varies significantly only near $\varepsilon \approx \mu$

(Figure 19.1). Hence, $du(= d\overline{n})$ is only non-zero near the point $x = y$, and it is convenient to expand our expression for v in a Taylor series in $z = (x - y)$ around that point. We carry this out in Appendix F and obtain the following results for the chemical potential and average energy per particle in a nearly degenerate nonrelativistic fermion gas (ε_f is given in equation 20.15):

$$\mu \approx \varepsilon_f\left[1 - \frac{\pi^2}{12}\left(\frac{kT}{\varepsilon_f}\right)^2\right],$$

(20.19)

$$\overline{\varepsilon} \approx \frac{3}{5}\varepsilon_f\left[1 + \frac{5\pi^2}{12}\left(\frac{kT}{\varepsilon_f}\right)^2\right].$$

Summary of Sections C and D

The translational kinetic energy of a quantum gas differs from the classical prediction. A fermion system has more energy, because some fermions are forced into higher levels if the lower levels are already occupied. In the case of bosons, the low-lying states have larger occupation numbers than the classical prediction, so that the total internal energy of a boson system is correspondingly lower. The calculations give (equation 20.5)

$$E = (\alpha+1)NkT \pm \delta, \qquad \text{where} \quad \delta = C\int_0^\infty \varepsilon^{\alpha+1}\overline{n}^2 d\varepsilon,$$

$$\alpha + 1 = \begin{cases} \frac{3}{2} & \text{(nonrelativistic gas)}, \\ 3 & \text{(relativistic gas)}. \end{cases}$$

and where the constant C is given in equation 20.1; δ is a positive function of μ and T and represents the difference between the classical and quantum predictions. This difference becomes larger for higher particle densities.

This difference in translational energies carries directly over into the pressure exerted by these gases. The ideal gas law becomes (equation 20.7)

$$pV = NkT \pm \tfrac{2}{3}\delta \qquad \text{(nonrelativistic gas)}.$$

The chemical potential appears in the expression for the occupation numbers, and so it is needed when performing a complete analysis of a system. It can be calculated by ensuring that the sum of the particles in all quantum states is equal to the total number of particles in the system (equation 20.11a):

$$N = \sum_s \overline{n}_s = \int g(\varepsilon)\overline{n}(\varepsilon)d\varepsilon \xrightarrow[\text{gases}]{} C\int \varepsilon^\alpha\overline{n}(\varepsilon)d\varepsilon.$$

This is an implicit equation that can be solved for μ in terms of N, V, and T. The internal energy is given by a similar expression (equation 20.11b):

$$E = \sum_s \overline{n}_s\varepsilon_s = \int g(\varepsilon)\overline{n}(\varepsilon)\varepsilon d\varepsilon \xrightarrow[\text{gases}]{} C\int \varepsilon^{\alpha+1}\overline{n}(\varepsilon)d\varepsilon.$$

Fortunately, these expressions become simple in the degenerate ($T \to 0$) and classical ($T \to \infty$) limits, so the integrals are easily done. The results in the degenerate limit (high occupancy, $T \to 0$) are (equations 20.10, 20.15):

$$\text{bosons,} \qquad \mu \approx -\frac{kT}{N}, \qquad E = 0;$$

$$\text{nonrelativistic fermions,} \qquad \mu = \varepsilon_\text{f} = \frac{h^2}{2m}\left(\frac{3N}{8\pi V}\right)^{\frac{2}{3}},$$

$$\text{relativistic fermions,} \qquad \mu = \varepsilon_\text{f} = hc\left(\frac{3N}{4\pi V}\right)^{\frac{1}{3}}.$$

In the classical limit (low occupancy, $T \to \infty$), for both bosons and fermions we have (equation 20.17)

$$\mu = \begin{cases} -kT\ln\left[\dfrac{(2s+1)V}{Nh^3}(2\pi mkT)^{3/2}\right] & \text{nonrelativistic,} \\[3ex] -kT\ln\left[\dfrac{2s8\pi V}{Nh^3c^3}(kT)^3\right] & \text{relativistic.} \end{cases}$$

At intermediate temperatures, things are more difficult. With the substitution of variables $x = \varepsilon/kT$, $y = \mu/kT$, expressions 20.11a,b can be put into the forms (20.18a, b)

$$N = C(kT)^{\alpha+1}f(y), \qquad \text{where} \quad f(y) = \int \frac{x^\alpha dx}{e^{x-y} \pm 1},$$

$$E = C(kT)^{\alpha+2}g(y), \qquad \text{where} \quad g(y) = \int \frac{x^{\alpha+1}dx}{e^{x-y} \pm 1}.$$

The first of these can be solved for μ by finding the value of $y\,(=\mu/kT)$ that gives the correct value for the number of particles, N. This is fairly tedious but fortunately, we seldom need to do it. Many systems we study are in either the degenerate or classical regimes, for which the chemical potentials are given by equations 20.10, 20.15, and 20.17. And in many cases, we won't need to know the chemical potential at all. Even in the intermediate region we can sometimes determine the chemical potential cleanly and simply from other considerations.

For nearly degenerate nonrelativistic fermions, however, such as conduction electrons in metals, we must do the integrals. Using a Taylor series expansion, we obtain for nearly degenerate fermions (equation 20.19)

$$\mu \approx \varepsilon_\text{f}\left[1 - \frac{\pi^2}{12}\left(\frac{kT}{\varepsilon_f}\right)^2\right]$$

$$\overline{\varepsilon} \approx \frac{3}{5}\varepsilon_\text{f}\left[1 + \frac{5\pi^2}{12}\left(\frac{kT}{\varepsilon_f}\right)^2\right].$$

Problems

Section A

1. The correct relativistic formula that relates momentum to energy at all speeds is $\varepsilon^2 = p^2c^2 + m^2c^4$. Using this, find the expression corresponding to the density of states of equation 1.9 or 20.1 that is correct at all speeds. (Equations 1.9 and 20.1 only give the answer for kinetic energy, $\varepsilon_{kin} = \varepsilon - mc^2$, in the high- and low-speed extremes.)

2. Consider the neutrino to be a massless spin-1/2 fermion. How many different orientations of its intrinsic spin angular momentum can a neutrino have? How about a photon, with spin 1? How about a massive vector boson with spin 1?

3. Using powers of 10, estimate the number of quantum states accessible to an outer electron on an atom, by estimating the size of an atom and knowing that typical binding energies are a few eV. (That is, if the kinetic energy is greater than a few eV then the electron is no longer on the atom.) The number of states $= V_r V_p / h^3$.)

Section B

4. Using equation 20.1, estimate the density of states at energy $\varepsilon = kT$ and at $T = 295$ K for:
   (a) air molecules in a room of volume 30 m³ (assume no intrinsic spin angular momentum and take the mass of an air molecule to be 4.8×10^{-26} kg);
   (b) electrons in a metal of volume 10^{-5} m³.

5. Using the densities of states from the previous problem, find the distribution of particles, $dN/d\varepsilon$, in particles per eV, around the energy kT for:
   (a) air molecules, which are bosons with $\mu = -0.01$ eV;
   (b) electrons, which are fermions with $\mu = +0.08$ eV;

6. Sketch diagrams similar to Figures 20.1 and 20.2 for fermions and bosons in the high- and low-temperature limits, if the density of states increases linearly with energy ε.

7. Suppose that the density of states for a certain system is a constant, $g(\varepsilon) = C_1$. In addition, suppose that the occupation number is given as $\bar{n}(\varepsilon) = C_2 e^{-\beta\varepsilon}$.
   (a) What is the average energy per particle in this system in terms of C_1, C_2, and kT?
   (b) Repeat for the case where the density of states is $g(\varepsilon) = C_1\varepsilon$.

8. Suppose that in a certain system, particle densities are small so that you can use the approximation $\bar{n}(\varepsilon) = e^{-\beta(\varepsilon-\mu)}$.
   (a) Using $\bar{p} = (1/N)\int p\,dN$ show that the average value of the magnitude of the momenta for particles in a nonrelativistic quantum gas is given by $(2s + 1)(2\pi V/Nh^3)e^{\mu/kT}(2mkT)^2$.

(b) In Chapter 16 we found that the average value of the magnitude of the momenta is given by $\sqrt{8mkT/\pi}$. Use this and your answer to part (a) to find an expression for the dependence of chemical potential on temperature in a classical gas.

9. Show that, for a boson system with $\mu = 0$, the distribution of particles $dN/d\varepsilon$ is proportional to $\varepsilon^{\alpha-1}$ for very low energies $\varepsilon \ll kT$.

Section C

10. Show that the internal energy of a quantum gas is given by equation 20.5, by starting with equations 20.4 and filling in the intermediate steps.

11. In this problem, we prove that $pV = (2/3)E_{trans}$ for a nonrelativistic gas, and $pV = (1/3)E_{trans}$ for a relativistic gas.
    (a) Suppose that there are N particles of mass m in a container of volume V, all moving in the x dimension with nonrelativistic speed v_x. Half are going to the right and half to the left. What is the flux of particles moving in the $+x$ direction in terms of N, V, m, and v_x?
    (b) When one of these collides elastically with a wall in the yz-plane, what is the momentum transfer?
    (c) What is the rate of momentum transfer (with units force per unit area)? (Flux $= \rho_+ v_x$, where ρ_+ is the density of particles moving in the $+x$ direction.)
    (d) Now suppose that they are moving in all directions with a variety of speeds, so you have to average over the velocities, replacing v_x^2 with its average $\overline{v_x^2}$. Show that $pV = (2/3)E_{trans}$.
    (e) Repeat the above for a relativistic gas, where the speed in the x direction is $c\cos\theta$ and the x-component of momentum is $p\cos\theta$, where θ is the angle relative to the x-axis. At the end you will have to average $\cos^2\theta$ over the forward hemisphere. For a relativistic particle the translational kinetic energy is $\varepsilon = pc$.

Section D

12. In the next chapter we will see that photons are massless bosons for which the chemical potential is zero. What does this say about the number of photons in the $\varepsilon = 0$ state?

13. Starting with the integral of equation 20.11a, fill in the steps, and see whether you can arrive at the chemical potential of equation 20.13 for degenerate fermions.

14. According to equation 20.14 the average energy per fermion is *not* halfway between 0 and ε_f. Why is this?

15. Evaluate the integral for N of equation 20.16a and solve for μ. Then put in the values of C from equation 20.1 to see whether you get the answers indicated. (You'll need to know that $\int e^{-x} x^{1/2} dx = \pi^{1/2}/2$.)

16. Show that equations 20.18 are the same as equations 20.11.

17. Using equation 20.17, estimate the chemical potential for air molecules in a room ($N/V = 2.6 \times 10^{25}/\text{m}^3$, $m = 4.8 \times 10^{-26}$ kg, $T = 290$ K, $s = 0$).

18. Use the value of the chemical potential of equation 20.17 in equation 20.16b for the internal energy. Then evaluate the integral to see whether you get the classical result for the internal energy.

19. Using equations 20.15, find the Fermi level in eV, the average energy per particle, and using $\varepsilon_f \approx (3/2)kT$ estimate the temperature above which the system begins becoming non-degenerate, for each of the following systems:
    (a) conduction electrons in copper metal, of which there are about 8×10^{28} per cubic meter (which is comparable with the density of copper atoms);
    (b) conduction electrons in aluminum, with density 2.7 g/cm^3 and atomic mass number 27, assuming that each atom gives one electron to the conduction band.
    (c) For each system, by how much would the chemical potential change in going from $T = 0$ to a room temperature of 290 K? (See equation 20.19.)

20. Using Equation 20.15, find the Fermi level in eV and using $\varepsilon_f \approx (3/2)kT$ estimate the temperature above which the system begins becoming non-degenerate, for each of the following systems (assume that each is nonrelativistic):
    (a) the electrons in a white dwarf star, whose electron density is about $10^{35}/\text{m}^3$;
    (b) the protons in a white dwarf star, whose proton density is comparable with that of the electrons;
    (c) the 30 neutrons in an iron nucleus, which is a sphere of radius 5.0×10^{-15} m.

21. Consider a degenerate gas of electrons at a density of $10^{29}/\text{m}^3$ (about the density of conduction electrons in metals) and at a temperature of 0 K.
    (a) What would be the Fermi level?
    (b) What would be the average energy per electron?
    (c) If you let this gas expand freely until the electrons were sufficiently far apart that they could be treated as a classical gas, what then would be the temperature of the gas?

Chapter 21
Blackbody radiation

We now examine the very important and elegant application of quantum statistics that initiated the quantum revolution at the beginning of the twentieth century – the study of electromagnetic radiation.

A Photons in an oven

For reasons that we still do not fundamentally understand, the energy in electromagnetic waves is quantized in discrete packets, called "photons." The energy of each photon (symbol γ), depends only on its frequency and nothing else:

$$\varepsilon_\gamma = h\nu = \hbar\omega.$$

Photons are massless spin-1 particles that travel at the speed of light and have two possible spin orientations.[1]

Consider a gas of these photons that is held within some oven. According to equation 20.1, the density of states for a gas of relativistic photons with two spin orientations is

$$g(\varepsilon) = \left(\frac{8\pi V}{h^3 c^3}\right)\varepsilon^2,\tag{21.1}$$

---

[1] They may be labeled right-handed or left-handed, according to whether their spin orientation is forwards or backwards along their direction of motion. The two independent transverse linear polarizations may be written as appropriate combinations of two circular polarizations, and vice versa.

and, according to equation 19.3b, the occupation number is

$$\bar{n}_\gamma = \frac{1}{e^{\beta\varepsilon} - 1}.$$ (21.2)

Notice that the photon's chemical potential μ is zero. To understand the reason for this, consider a photon gas inside a rigid oven that is insulated from the rest of the Universe, so that the energy and volume of the combined system are constant ($dE = dV = 0$). The number of photons can vary as they are created or absorbed by the oven's walls ($dN_\gamma \neq 0$). According to the first law (equation 8.4), the entropy of the *combined* system (photons plus oven) may change according to

$$T dS = dE + p dV - \mu_\gamma dN_\gamma \xrightarrow[dE=dV=0]{} -\mu_\gamma dN_\gamma.$$

When in equilibrium the entropy of the combined system is a maximum (second law). Hence, its derivatives must be zero. In particular,[2]

$$-\mu_\gamma = T \left(\frac{\partial S}{\partial N_\gamma}\right)_{E,V} = 0.$$ (21.3)

Equations 21.1 and 21.2 give us the following energy distribution (see equations 20.2):

$$dE = g(\varepsilon)\bar{n}(\varepsilon)\varepsilon d\varepsilon = \left(\frac{8\pi V}{h^3 c^3}\right) \frac{\varepsilon^3 d\varepsilon}{e^{\beta\varepsilon} - 1}.$$

Dividing by the volume gives the energy density in the range $d\varepsilon$:

$$du = \frac{dE}{V} = \left(\frac{8\pi}{h^3 c^3}\right) \frac{\varepsilon^3 d\varepsilon}{e^{\beta\varepsilon} - 1},$$ (21.4)

which is also frequently expressed in photon frequencies or wavelengths, via

$$\varepsilon = \hbar\omega = \frac{hc}{\lambda},$$ (21.5)

as

$$du = \left(\frac{\hbar}{\pi^2 c^3}\right) \frac{\omega^3 d\omega}{e^{\beta\hbar\omega} - 1} = 8\pi hc \frac{\lambda^{-5} d\lambda}{e^{\beta hc/\lambda} - 1}.$$ (21.6)

Figure 21.1 displays these distributions as $du/d\varepsilon$ versus ε and $du/d\lambda$ versus λ. In the homework problems it can be shown that the respective peaks are at[3]

$$\varepsilon_{max} = 2.82kT \quad \text{and} \quad \lambda_{max} = \frac{2.90\,\text{mm K}}{T}.$$ (21.7)

So, at higher temperatures, the spectrum peaks at higher energies and shorter wavelengths.

[2] This proof applies to any type of particle whose total number can vary (e.g., phonons, discussed in Chapter 22).

[3] The peak or maximum is where the derivative is zero. But the derivative with respect to ε is not the same as the derivative with respect to λ. So the two distributions do not peak at the same place.

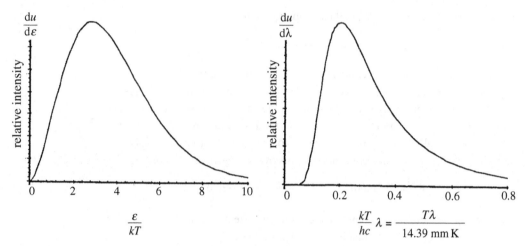

relative intensity

0   2   4   6   8   10

$$\frac{\varepsilon}{kT}$$

$\dfrac{du}{d\lambda}$

relative intensity

0   0.2   0.4   0.6   0.8

$$\frac{kT}{hc}\lambda = \frac{T\lambda}{14.39 \text{ mm K}}$$

Figure 21.1 Plots of photon distribution in energy and in wavelength for a photon gas in equilibrium with an oven. The distributions peak at $\varepsilon_{max} = 2.82kT$ and $\lambda_{max} = 2.90\text{mm } K/T$, respectively.

We can integrate equation 21.4 over all energies to find the total energy density in a photon gas, using the substitution $x = \beta\varepsilon = \varepsilon/kT$:

$$u = \frac{8\pi}{h^3 c^3}\int_0^\infty \frac{\varepsilon^3 d\varepsilon}{e^{\beta\varepsilon} - 1} \qquad \longrightarrow \qquad \frac{8\pi(kT)^4}{h^3 c^3}\int_0^\infty \frac{x^3 dx}{e^x - 1}$$

The expression on the right includes a standard integral that has the value $\pi^4/15$, so the total energy density u becomes

$$u = aT^4, \qquad \text{where} \qquad a = \frac{8\pi^5 k^4}{15 h^3 c^3} = 7.56 \times 10^{-16}\frac{\text{J}}{\text{m}^3 \text{ K}^4}. \tag{21.8}$$

B Principle of detailed balance

According to the result 21.4, the energy spectrum of a photon gas depends only on the temperature (through $\beta = 1/kT$) and not at all on the nature of the oven. For example, it doesn't depend on the oven's shape, or whether the oven's walls are smooth or rough, shiny or black, made of marble or wood, or whether they are red, green, or purple. Because it has no effect on the photon spectrum, we conclude that the wall must put back into the photon gas exactly what it takes out. That is, the absorbed and emitted intensities must be identical at all photon energies and wavelengths. This is called the "principle of detailed balance."

> **Principle of detailed balance**
>
> When in thermal equilibrium with a photon gas, the intensity of radiation emitted by an object must be equal to the intensity absorbed at each energy and wavelength.

For example, green oven walls would reflect green, absorbing less green and more of the other colors. Therefore, they must also emit less green and more of the other colors in order to have no net effect on the photon energy distribution

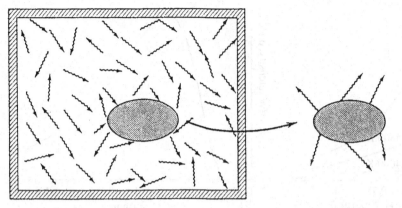

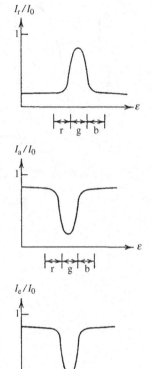

Figure 21.3 For an object in equilibrium with the photons in an oven, the spectrum of photons it emits must be identical to the spectrum of photons it absorbs. When in an oven, a perfectly black body absorbs all incident radiation and it must emit the same spectrum that it absorbs. So the spectrum of photons emitted by a black body must be identical to the spectrum of photons inside an oven of that temperature.

(Figure 21.2). Colors that are absorbed more strongly must also be emitted more strongly.

Consider an object that is perfectly black, absorbing all incident radiation perfectly. Imagine that this object is in thermal equilibrium inside an oven (Figure 21.3). Because it has no effect on the spectrum of photons, it must emit photons with an energy spectrum that is exactly the same as the spectrum that it absorbs – namely the spectrum of photons that are inside the oven. When removed from the oven and kept at the same temperature the thermal motions of its atoms and molecules remain the same, so it continues to radiate the same spectrum. We conclude that the spectrum of energy radiated from perfectly black bodies must be exactly the same as that within an oven with the same temperature. For this reason, we use the term "blackbody radiation" for either the radiation spectrum within an oven, or equivalently, the radiation emitted by perfectly black bodies.

In Figure 21.4 you can see that our Sun radiates approximately as a blackbody at temperature 5800 K. You can also see that the cosmic background radiation left over from the Big Bang origin of our Universe, which cooled as the Universe expanded, now emulates the radiation of a blackbody at a temperature of 2.7 K.

Figure 21.2 Plots of the fractions of photons reflected (I_r/I_0), absorbed (I_a/I_0), and emitted (I_e/I_0) as functions of the photon energies, for photons incident on a green object (i.e., one that reflects green). The red, green, and blue portions of the spectrum are indicated. All incident radiation is either reflected or absorbed; $I_0 = I_r + I_a$. In thermal equilibrium, the object does not affect the photon distribution, which means that photons of each energy must be emitted in the same proportions as they are absorbed: $I_e = I_a$.

C Energy flux

For photons in an oven, we now examine the rate of flow of energy in one direction, call it the z direction. At any instant, half the photons are moving in the $+z$ direction, and half in the $-z$ direction. The energy density u_{+z} for the former group will therefore be only half the total ($u_{+z} = u/2$). Averaging the velocity

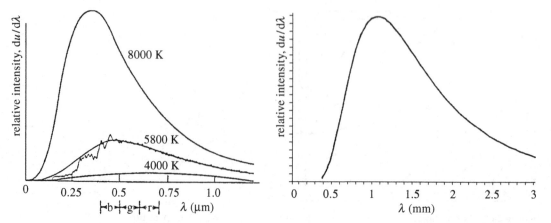

Figure 21.4 (Left) Plots of relative intensities for radiation emitted by blackbodies of various temperatures as a function of photon wavelength. The red (r), green (g), and blue (b) portions of the visible spectrum are indicated. The wiggly line is the Sun's spectrum, which approximates a blackbody at 5800K. (Right) The variation of intensity vs. wavelength for the cosmic background radiation, which is the remnant of the primordial "Big Bang." The best fit to the data points is a blackbody curve for a temperature of 2.735 K. At the low-wavelength end, the blackbody curve goes as $e^{-\beta hc/\lambda}$; at the high-wavelength end it goes as λ^{-4}. The error bars are smaller than the width of the line. (COBE satellite data.)

component $v_z = c\cos\theta$ (in spherical coordinates) over the positive $z-$ hemisphere gives their average velocity in this direction as $c/2$ (homework).

The energy flux J_z is the product of the energy density times the average z-velocity (Figure 16.3):

$$J_z = u_{+z}\bar{v}_{+z} = \left(\frac{u}{2}\right)\left(\frac{c}{2}\right) = \frac{uc}{4}. \tag{21.9}$$

Combining this with our expression 21.8 for the energy density u gives

$$J = \sigma T^4, \qquad \text{where} \quad \sigma = \frac{ac}{4} = 5.67 \times 10^{-8}\,\frac{\text{W}}{\text{m}^2\,\text{K}^4}. \tag{21.10}$$

The constant σ is called the Stefan–Boltzmann constant.

Most things are not perfect absorbers and emitters, so we define the "emissivity" (symbol e) of a surface to be the ratio of the energy flux that it emits to the flux that would come from a blackbody at the same temperature. Its value ranges from 0 (emits nothing) to 1 (perfect blackbody), and it depends on the nature of the material, its temperature and the photon energy:

$$e = e(T, \varepsilon), \qquad 0 \le e \le 1.$$

Using this and equation 21.9, we can write the emitted flux carried by photons whose energies are in the range $d\varepsilon$ as

$$dJ_{\text{actual}} = e(T, \varepsilon)dJ_{\text{blackbody}} = e(T, \varepsilon)\frac{c}{4}du, \tag{21.11}$$

where du is given by equation 21.4. Integration over all photon energies gives

$$J_{total} = e\sigma T^4, \tag{21.12}$$

where e is the emissivity averaged over all emitted energies.

As an object radiates energy into its environment, it also receives energy from its environment. If the object's and the environment's temperatures are T_0, T_e), respectively, then the *net* energy radiated is[4]

$$J_{net} = e\sigma\left(T_0^4 - T_e^4\right). \tag{21.13}$$

Because an object in equilibrium with a photon gas must absorb exactly as it emits at all energies and because that which is not absorbed is reflected, we have the following measures of absorptivity and reflectivity:

$$\text{absorptivity} = \text{emissivity} = e(T, \varepsilon),$$
$$\text{reflectivity} = 1 - \text{emissivity} = 1 - e(T, \varepsilon). \tag{21.14}$$

Figure 21.4 shows that the Sun is a good blackbody for the green and longer wavelengths with $e(T, \varepsilon) \approx 1$, but its emissivity falls well below that of a blackbody for the shorter wavelengths. You can also see that the relic radiation left over from the Big Bang birth of our Universe is extremely close to that of a perfect blackbody.

Example 21.1 The surface of the Sun acts like a blackbody of temperature 5800 K. What is the rate at which energy leaves each square meter of the Sun's surface?

Putting $T = 5800$ K into equation 21.10, the energy flux from the Sun's surface is

$$J_{Sun} = \sigma T^4 = 5.67 \times 10^{-8} \frac{W}{m^2\,K^4} \times (5800\,K)^4 = 6.4 \times 10^7 \frac{W}{m^2}.$$

In the homework problems you will show that by the time this energy reaches the Earth, it has spread out to the point where the flux is only 1.4×10^3 W/m^2.

Example 21.2 The radiation inside the Sun can be approximated as a photon gas at $T = 10^7$ K. The volume of the Sun is 1.4×10^{27} m^3. How much energy is contained in this photon gas?

The total energy is the product of the energy density times the volume. With a temperature of 10^7K, equation 21.8 gives the energy density as

$$u = 7.56 \times 10^{-16} \frac{J}{m^3\,K^4} \times (10^7\,K)^4 = 7.6 \times 10^{12} \frac{J}{m^3}.$$

Multiplying this by the Sun's volume gives the total energy in the Sun's photon gas:

$$E = uV = (7.6 \times 10^{12}\,J/m^3)(1.4 \times 10^{27}\,m^3) = 1.1 \times 10^{40}\,J.$$

---

[4] Although the principle of detailed balance guarantees that the emissivity and absorptivity must be identical at each wavelength, different temperatures involve different wavelength spectra. So the average emissivity at temperature T_0 might differ from that at T_e. But we ignore that here

Even if its thermonuclear fusion stopped, the Sun could continue shining at its present rate for nearly 8 million years before using up all this energy that is stored inside it.

Summary of Sections A–C

The electromagnetic radiation within an oven can be treated as a photon gas. Photons are spin-1 massless relativistic particles having two possible spin orientations and $\mu = 0$. Their equilibrium distribution is the product of the density of states times the occupation number, which gives the energy density in a photon gas as (equation 21.4)

$$du = \left(\frac{8\pi}{h^3 c^3}\right) \frac{\varepsilon^3 d\varepsilon}{e^{\beta\varepsilon} - 1}.$$

If expressed as a distribution in photon frequencies or wavelengths, it becomes (equation 21.6)

$$du = \left(\frac{\hbar}{\pi^2 c^3}\right) \frac{\omega^3 d\omega}{e^{\beta\hbar\omega} - 1} = 8\pi hc \frac{\lambda^{-5} d\lambda}{e^{\beta hc/\lambda} - 1}.$$

The respective peaks in these two distributions are at (equation 21.7)

$$\varepsilon_{max} = 2.82\, kT \quad \text{and} \quad \lambda_{max} = \frac{2.90\,\text{mm K}}{T},$$

and the total energy density is (equation 21.9)

$$u = aT^4, \quad \text{where} \quad a = \frac{8\pi^5 k^4}{15 h^3 c^3} = 7.56 \times 10^{-16} \frac{\text{J}}{\text{m}^3\,\text{K}^4}$$

Because the equilibrium distribution of photons in an oven depends only on the oven's temperature, neither the oven walls nor any other objects within the oven can alter this distribution. Those wavelengths that are absorbed more strongly must also be emitted more strongly. This is the principle of detailed balance. Perfect absorbers are called blackbodies, and the spectrum of photons emitted from a blackbody at temperature T must be exactly the spectrum of the photons in a photon gas at that temperature.

The flux of energy emitted from a blackbody at temperature T is given by (equation 21.10)

$$J = \sigma T^4, \quad \text{where} \quad \sigma = 5.67 \times 10^{-8} \frac{\text{W}}{\text{m}^2\,\text{K}^4}.$$

The emissivity e is the ratio of the intensity of radiation emitted by real objects to that of a blackbody of the same temperature. The flux emitted in energy increment $d\varepsilon$ is (equation 21.11)

$$dJ = e(T, \varepsilon)\frac{c}{4} du,$$

and, integrated over all photon energies, this gives for the total flux of radiation emitted (equation 21.12)

$$J_{\text{total}} = e\sigma T^4,$$

where e is the emissivity averaged over all emitted energies.

An object (temperature T_0) both radiates energy into its environment (temperature T_e) and receives energy from its environment, so the net flux of energy from a body is given by (equation 21.13)

$$J_{\text{net}} = e\sigma(T_0^4 - T_e^4).$$

The principle of detailed balance relates emissivity, absorptivity and reflectivity (equation 21.14):

$$\text{absorptivity} = \text{emissivity} = e(T, \varepsilon)$$
$$\text{reflectivity} = 1 - \text{emissivity} = 1 - e(T, \varepsilon)$$

D Heat shields

D.1 Layered foils

To begin our study of heat shields, we now consider the radiative energy transfer between two regions that are separated by n layers of foil, as illustrated in Figure 21.5. For simplicity, we assume that all surfaces have the same emissivity, and we use foils in order to concentrate entirely on radiative transfer and not worry about impedance due to low thermal conductivities. We assume that the foils' temperatures are $T_1, T_2, \ldots, T_n$, respectively, and that the two regions they separate are represented by surfaces held at temperatures T_0 and T_{n+1}.

The first surface emits a flux $e\sigma T_0^4$, but then a flux $e\sigma T_1^4$ comes back towards it from the neighboring foil, foil 1. So the net energy flux through the first layer (from the first surface to foil 1) is $J_{01} = e\sigma(T_0^4 - T_1^4)$, that through the second

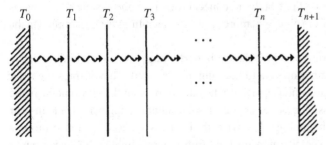

Figure 21.5 Consider the flow of radiated energy from one surface at temperature T_0 to another at temperature T_{n+1}, where the two surfaces are separated by n layers of foil whose temperatures are $T_1, T_2, \ldots, T_n$, respectively. In the steady state, the rate of flow through all layers is the same.

layer is $J_{12} = e\sigma(T_1^4 - T_2^4)$, etc. In the steady state heat enters, leaves, and flows through all intermediate points of our layered system at the same rate (like water through a hose):

$$\text{rate in} = \text{rate out} = \text{rate through each intermediate layer.}$$

Hence

$$\underset{\substack{\text{total through} \\ n \text{ layers of foil}}}{J} = \underbrace{e\sigma\left(T_0^4 - T_1^4\right)}_{\text{first layer}} = \underbrace{e\sigma\left(T_1^4 - T_2^4\right)}_{\text{second layer}} = \cdots. \tag{21.15}$$

If there were no foil layers, the net flux from left (where the temperature is T_0) to right (where the temperature is T_{n+1}) would be

$$J_{\text{no foil layers}} = e\sigma\left(T_0^4 - T_{n+1}^4\right).$$

By subtracting and adding T_1^4, T_2^4 etc., we can write this as

$$J_{\text{no foil layers}} = e\sigma\left(T_0^4 - T_1^4\right) + e\sigma\left(T_1^4 - T_2^4\right) + \cdots + e\sigma\left(T_n^4 - T_{n+1}^4\right).$$

Each of the $n + 1$ terms on the right is equal to the total flux when there are n layers of foil (equation 21.15). Hence

$$\underset{\substack{\text{total through} \\ n \text{ layers of foil}}}{J} = \frac{1}{n + 1} J_{\text{no foil layers}}. \tag{21.16}$$

So, by introducing n layers of foil the radiative heat transfer is reduced by a factor $1/(n + 1)$. One familiar consequence is that layered clothing keeps you warmer.

D.2   The greenhouse effect

Radiation from hotter objects is concentrated at shorter wavelengths. Radiation from the Sun or light bulbs is concentrated near the visible wavelengths whereas radiation from objects at more earthly temperatures is concentrated in the infrared.

The glass covering of a greenhouse is fairly transparent to incoming visible wavelengths but rather opaque to the outgoing infrared. Thus it traps the solar heating inside the greenhouse, making the interior warmer than it would be in the absence of the glass. The same happens in solar water heating panels, and in your car when it sits in the sunlight. Likewise, the interiors of glass-covered terrariums and incubators are kept warm by the light from external light bulbs. The trapping layer need not be glass. The Earth's atmosphere is also rather transparent to incoming solar energy and opaque to outgoing infrared, so it acts like the glass covering of a greenhouse by trapping solar heat and keeping us warm.

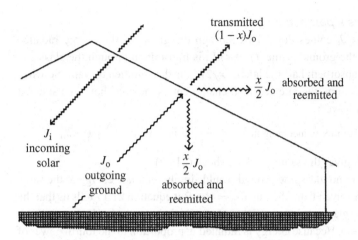

transmitted
$(1-x)J_\mathrm{o}$

$\frac{x}{2}J_\mathrm{o}$ absorbed and reemitted

J_i
incoming solar

J_o
outgoing ground

$\frac{x}{2}J_\mathrm{o}$
absorbed and reemitted

Figure 21.6 A greenhouse. J_i is the incoming solar radiation and J_0 is the radiation going outward from the ground, of which a fraction x is absorbed by the glass. That is, xJ_0 is absorbed and the remaining $(1-x)J_0$ is transmitted. Of the absorbed radiation, half, $(x/2)J_0$, is radiated outward and the other half inward.

We now develop a quantitative treatment for the greenhouse effect, using an actual greenhouse as our prototype (Figure 21.6). We assume that the temperatures of the ground and glass have stabilized so that, for each, the rates of inward and outward energy flow are the same. We use the symbols J_i for the incoming solar energy flux, which is mostly in the visible part of the spectrum, and J_o for the radiation flowing outward from the Earth. Some of this is reflected sunlight but most is absorbed and then emitted in the infrared wavelengths. Finally, we use the symbol x for the fraction of the outward-going radiation that is absorbed by the glass. This is then reradiated from the glass: one half, $(x/2)J_\mathrm{o}$, goes back downwards towards the ground and the other half, $(x/2)J_\mathrm{o}$, goes upwards and out.

The quantitative equation is obtained by noting that when equilibrium is established, the rate of heat coming in must equal the rate going out. This can be done from the perspective of either the ground or the glass.

1 The ground's perspective

Of the energy received by the ground, some comes from the Sun (J_i) and some from the glass – that part of the outward-flowing J_o that was absorbed by the glass and then reradiated back down to the ground, $(x/2)J_\mathrm{o}$. When the ground is in equilibrium,

$$\text{outgoing flux} = \text{incoming flux} \quad \Rightarrow \quad J_\mathrm{o} = J_\mathrm{i} + \left(\frac{x}{2}\right) J_o.$$

Solving for J_o gives

$$J_o = \left(\frac{2}{2-x}\right) J_\mathrm{i}. \qquad (21.17)$$

2   The glass's perspective

The solar flux J_i comes downward through the glass. Of the energy radiated upward from the ground, some, $(1 - x)J_o$, is transmitted through the glass and some, xJ_o, is absorbed; half the latter, $(x/2)J_o$, is then emitted upward. So when all is in equilibrium, the rate at which energy enters the greenhouse must equal the rate that it leaves:

$$\text{outgoing flux} = \text{incoming flux} \quad \Rightarrow \quad (1 - x)J_o + \left(\frac{x}{2}\right)J_o = J_i.$$

Solving for J_o gives the same result as above (21.17).

If there were no glass, the ground would simply reradiate energy at the same rate that it is received from the Sun ($J_o = J_i$). But equation 21.17 tells us that the presence of the glass has increased the rate at which the ground radiates energy by a factor $2/(2-x)$. Because energy is radiated in proportion to the fourth power of the temperature (equation 21.12), the ground's temperature must have increased according to

$$\frac{T_0'^4}{T_0^4} = \frac{2}{2 - x} \quad \Rightarrow \quad \frac{T_0'}{T_0} = \left(\frac{2}{2 - x}\right)^{1/4}. \tag{21.18}$$

Example 21.3   Averaged over all latitudes and all times of day and night, the flux of solar energy reaching the Earth's surface is about 175 watts/m^2. Of this, about 90% is absorbed and 10% reflected. Assuming that the Earth's emissivity in the infrared e is 0.9, what would be the average value of the Earth's surface temperature if it had no atmosphere?

The surface absorbs 90% of the incoming solar energy for a rate of 158 watts/m^2. In equilibrium it must reradiate at the same rate. Hence

$$e\sigma T_0^4 = 158 \text{ watts/m}^2.$$

Solving for the temperature T_0, using $e = 0.9$ and $\sigma = 5.67 \times 10^{-8}$ gives an average surface temperature

$$T_0 = 236 \text{ K } (-37\,°\text{C})$$

Example 21.4   Repeat the above example but include the effects of the Earth's atmosphere, which effectively absorbs nearly 100% ($x = 1$) of the outgoing infrared radiation.

According to equation 21.18, the Earth's surface temperature is increased by

$$\frac{T_0'}{T_0} = \left(\frac{2}{2 - 1}\right)^{1/4} = 1.19.$$

With T_0 from the previous example, we find that the atmosphere makes the average Earth surface temperature rise to

$$T_0' = 1.19 T_0 = 281 \text{ K } (8\,°\text{C}).$$

E   Entropy and adiabatic processes

We now examine the adiabatic expansion or compression of photon gases, which occurs as the Universe expands and also as a dying star collapses. According to result 21.8, the energy of a photon gas depends only on its temperature and volume:

$$u = aT^4 \quad \Rightarrow \quad E = uV = aVT^4.$$

Because the chemical potential is zero, $\mu_\gamma \, dN_\gamma = 0$, and the first law becomes

$$dE = dQ - p \, dV.$$

From this, the heat capacity is easy to calculate:

$$C_V = \left(\frac{\partial Q}{\partial T}\right)_V = \left(\frac{\partial E}{\partial T}\right)_V = 4aVT^3,$$

and the entropy of the photon gas is then

$$S(T, V) = \int_0^T \frac{dQ}{T'} = \int_0^T \frac{C_V \, dT'}{T'} = \frac{4a}{3} VT^3. \tag{21.19}$$

For (quasistatic) adiabatic expansions, the entropy is constant. Hence

$$VT^3 = \text{constant} \quad \text{(adiabatic expansion)}. \tag{21.20}$$

(In the homework problems this same result can be derived in a different way.) While expanding adiabatically, a photon gas loses energy by doing work on the walls of the container.

In a free expansion,[5] however, no work is done so the total energy of a photon gas remains unchanged. Since the energy density is proportional to T^4, we have

$$VT^4 = \text{constant} \quad \text{(free expansion)} \tag{21.21}$$

Shortly after its Big Bang birth our Universe was opaque, meaning that radiation created in this explosion could not go very far without colliding with a charged particle (e.g., an electron or quark). But now our Universe is transparent (otherwise our telescopes would be of little use), so a photon from this remnant "cosmic background radiation" could probably cross the Universe without hitting a thing. Here are questions for you. Was the early expansion of this radiation adiabatic or free? Has this changed?

---

[5] Review: free expansion occurs when a gas expands into a preexisting void. No boundary is moved and so no work is done. It is not isentropic even if no heat is added, because entropy increases (more available states) due to the increased volume in position space. Change in entropy and heat addition are only related (heat added $= T \, dS$) for systems in equilibrium, which does *not* apply to free expansion.

F   Thermal noise and the Nyquist theorem

A phenomenon related to blackbody radiation is thermal "noise," which is generated in all systems. It is often noticed as fluctuations in the voltage across resistive
elements in sensitive electrical circuits. We can think of the circuit element as a
one-dimensional system of length L and carrying electromagnetic waves of two
possible polarizations.[6] The sum over quantum states is given by (using energy
$= hf = pc$)

$$2 \int \frac{\mathrm{d}x \, \mathrm{d}p_x}{h} = \frac{2L}{h} \int \mathrm{d}p_x = \frac{2L}{c} \int \mathrm{d}f,$$

and the average energy in each state is the product of the occupation number
$1/(e^{hf/kT} - 1)$ and the average energy per wave, hf. Hence, the total energy carried
in these waves is

$$E_{\text{thermal}} = \underbrace{\frac{2L}{c} \int \mathrm{d}f}_{\substack{\text{sum over} \\ \text{states}}} \underbrace{\frac{hf}{e^{hf/kT} - 1}}_{\substack{\text{average energy} \\ \text{per state}}}.$$

Suppose that we are interested in frequencies in the range Δf, such as the
bandwidth of the circuit or the frequency range of our measurements, for which
$hf \ll kT$ (i.e., frequencies below about 10^{12} Hz at room temperature). Then we
can expand the exponential e^x as $1 + x$, so that the integrand becomes

$$\frac{hf}{e^{hf/kT} - 1} \approx kT,$$

and integrate over the range Δf to get

$$E_{\text{thermal}} = \frac{2L}{c} kT \int \mathrm{d}f = \frac{2L}{c} kT \Delta f.$$

These thermally generated waves appear throughout the element. So to reach
either end, the average wave would travel half the length of the element, taking
time $t = L/2c$. Consequently, the rate at which we detect the thermal noise exiting
the circuit element would be

$$\text{power} = \frac{\text{energy}}{\text{time}} = \frac{2LkT\Delta f/c}{L/2c} = 4kT\Delta f \tag{21.22}$$

Writing this in terms of the mean square voltage fluctuations over a resistance R
gives the Nyquist theorem for thermal noise in electrical circuits:[7]

$$\text{power} = \frac{\langle V^2 \rangle}{R} \qquad \Rightarrow \qquad \langle V^2 \rangle = 4RkT\Delta f. \tag{21.23}$$

---

[6] To keep things simple, we assume that that all signals reaching the ends leave (with no reflection),
and that the leakage out the ends has little effect on the thermal distribution of noise within the
circuit element.

[7] H. Nyquist, *Phys. Rev.* **32**, 110 (1928).

Summary of Sections D–F

When n layers of foil are inserted between regions at two different temperatures, the radiative heat transfer between the two regions is reduced by a factor of $1/(n + 1)$ (equation 21.16):

$$J_{\substack{\text{total through} \\ n \text{ layers of foil}}} = \frac{1}{n + 1} J_{\text{no foil}}.$$

Many materials that are fairly transparent to visible wavelengths are rather opaque to the infrared. Examples include the glass of greenhouses, automobile windshields, solar water panels, terrariums and incubators, as well as the Earth's atmosphere. Thus, they tend to allow the visible radiation (from the Sun, light bulbs, etc.) to enter but block the infrared radiation (from the ground, car seats, etc.) from leaving, trapping the heat inside. If the semitransparent layer absorbs a fraction x of the outgoing infrared radiation, then the outgoing flux J_0 of infrared radiation emitted by the interior surfaces exceeds the incoming visible radiation J_i by a factor $2/(2 - x)$ (equation 21.17):

$$J_o = \left(\frac{2}{2 - x} \right) J_i$$

The interior temperature is raised by factor (equation 21.18)

$$\frac{T_0'}{T_0} = \left(\frac{2}{2 - x} \right)^{1/4}.$$

Because the energy content of a photon gas depends only on its volume and temperature, it is easy to calculate its heat capacity and entropy (equation 21.19):

$$C_V = \left(\frac{\partial Q}{\partial T} \right)_V = \left(\frac{\partial E}{\partial T} \right)_V = 4aVT^3,$$

$$S(T, V) = \int_0^T \frac{\mathrm{d}Q}{T'} = \int_0^T \frac{C_V \mathrm{d}T'}{T'} = \frac{4a}{3} VT^3.$$

Consequently, for (quasistatic) adiabatic expansions (equation 21.20),

$$VT^3 = \text{constant} \qquad \text{(adiabatic expansion)}.$$

In a free expansion, the energy of the photon gas remains constant, so (equation 21.21)

$$VT^4 = \text{constant} \qquad \text{(free expansion)}.$$

Thermal noise is generated in all systems, and is particularly noticeable in resistive elements of sensitive electrical circuits. We can calculate the energy carried in these one-dimensional systems by thermally generated electromagnetic waves, and we find that the rate at which this thermal noise energy arrives at the ends of the elements is given by (equation 21.22)

$$\text{power} = \frac{\text{energy}}{\text{time}} = 4kT\Delta f,$$

where Δf is the bandwidth of frequencies in which we are interested. Expressing this in terms of thermal voltage fluctuations, we find that (equation 21.23)

$$\langle V^2 \rangle = 4RkT\Delta f.$$

Problems

Section A

1. Starting from Equation 21.4, show that the photon energy distribution $du/d\varepsilon$ is proportional to: (a) ε^2, for $\varepsilon \ll kT$, (b) $e^{-\beta\varepsilon}$, for $\varepsilon \gg kT$.
   (With L'Hospital's rule, you can show that exponentials dominate over polynomials when the variable is large.)

2. Starting from equation 21.6, show that $du/d\lambda$ is proportional to (a) λ^{-4}, for $\lambda \gg \beta hc$, (b) $e^{-\beta hc/\lambda}$, for $\lambda \ll \beta hc$.

3. Show that the plot of $du/d\varepsilon$ for the energy distribution in a photon gas peaks at $\varepsilon_{max} = 2.82\,kT$. (Hint: Start with expression 21.4 giving $du/d\varepsilon$. At the maximum, the derivative of this with respect to ε is zero. The result is an implicit equation for ε_{max}/kT, which you might best solve by trial and error.)

4. Show that the plot of $du/d\lambda$ for the energy distribution in a photon gas peaks at $\lambda_{max} = 0.201\beta hc$. If written in the form $\lambda_{max} = \text{constant}/T$, what is the value of the constant in units of mm K? (That is, check out the value given in equation 21.7.)

5. From the answers to problems 3 and 4 or from equation 21.7, does $\varepsilon_{max} = hc/\lambda_{max}$? If not, why not?

6. Given that the distribution $du/d\varepsilon$ peaks at $\varepsilon_{max} = 2.82kT$, we can write an expression in the form $\lambda_{max} = C/T$, where λ_{max} is now the wavelength at which the distribution $du/d\varepsilon$ peaks and C is a constant.
   (a) What is the value of the constant C in units of mm K? (Note: The peak in $du/d\varepsilon$ will be at a different place from the peak in $du/d\lambda$, as you showed in problem 5.)
   (b) At what wavelength does the energy distribution $du/d\varepsilon$ peak for the photons of the 2.735 K blackbody radiation left over from the primordial Big Bang that initiated our present Universe?
   (c) At what wavelength does the energy distribution $du/d\varepsilon$ from our Sun peak, if the visible surface acts like a blackbody of temperature 5800 K?

(d) What are the answers to parts (b) and (c) for the distribution in wave-lengths, $du/d\lambda$?

7. Assume that a neutrino is a massless particle that travels at the speed of light (which is almost true). Unlike a photon, it has spin 1/2 instead of spin 1. If neutrinos had zero chemical potential, what would be the expression for the distribution of energies, $du/d\varepsilon$, in a neutrino gas in an appropriate oven?

Section B

8. A helium–neon laser emits only orange photons. Could these be in equilib-rium with anything? Explain.

9. A ball painted perfectly black has a bright green light shining on it that keeps it heated at temperature T. Describe or sketch the distribution of energies absorbed and emitted by the ball. (Hint: The green light is not in equilibrium with the ball – or with anything, for that matter – because it doesn't have the equilibrium distribution in frequencies.)

10. What is the energy density of the energy held in the 2.735 K blackbody radiation left over from the Big Bang? How does this compare with the energy density (mc^2) of particulate matter in the Universe which amounts to an average of 0.2 protons (mostly in hydrogen atoms) per cubic meter?

11. Calculate the total energy contained in the 2.7 K blackbody background radiation left over from the Big Bang. The Universe has a radius of about 14 billion light years, and a light year is equal to about 9.5×10^{15} m. Estimate the temperature of the Universe at a time just after the original explosion when it had a volume of 1 m³. To do this, assume that all the energy of the present background radiation was in the Universe at that time and ignore any coupling to matter.

Section C

12. The Sun's surface radiates like a blackbody at temperature 5800 K. Its radius is 7×10^8 m.
    (a) How many joules per second does the Sun radiate altogether?
    (b) Inside, the Sun is much hotter. In fact, averaging T^4 throughout the volume of the Sun gives $\overline{T^4} = (10^7 \text{ K})^4$. What is the total energy of the photon gas stored inside the Sun?
    (c) If the thermonuclear fusion in the Sun's core stops tomorrow, how many more years could the Sun radiate energy at the present rate, before exhausting all this energy? (1 year $= 3.16 \times 10^7$ s.)

13. If you double the temperature of an oven, by what factor do the following things increase?
    (a) The total energy of the photon gas.

(b) The position of the peak in the energy distribution, ε_{max}.

(c) The rate at which energy strikes a wall of the oven.

14. (a) Sketch a plot of $\bar{n}(\varepsilon)$ versus ε for bosons with $\mu = 0$ for two different temperatures.

(b) Set up the integral for the total number of photons in a system, and show that it increases as T^3. (Hint: See equation 20.2, and make the substitution $x = \beta\varepsilon = \varepsilon/kT$.)

(c) Set up the integral for the total energy in the system, and show that it increases as T^4.

(d) Does the average energy *per photon* increase, decrease, or remain the same as temperature increases?

15. An object is in equilibrium with a photon gas at a temperature of several thousand kelvins, and I_0 is the intensity of radiation incident on its surface. Make qualitative plots of the fractions of reflected, absorbed, and emitted radiation (I_r/I_0, I_a/I_0, and I_e/I_0, respectively) as a function of photon energy ε, if the object looks red when illuminated by sunlight.

16. The radius of the Sun is 7×10^8 m and the radius of the Earth's orbit is 1.5×10^{11} m. The energy flux leaving the Sun's surface is 6.4×10^7 W/m^2. What is the flux of solar energy at the distance of the Earth?

17. You have a half cup of very hot coffee, a half cup of room-temperature water, and only two minutes in which to get the coffee as cool as possible. Should you add the cool water first and then wait two minutes, or first wait two minutes and then add the water? Why?

18. Suppose that the Earth acts as a blackbody of radius 6.4×10^6 m and effective temperature 240 K. (This is not the same as our surface temperature, due to our atmosphere's intervention. Most outgoing infrared ultimately leaves from the upper atmosphere.)

(a) What is the flux of energy leaving the Earth's surface for outer space?

(b) What is the total rate of emission of energy by the entire Earth?

(c) The flux of sunlight sweeping through space at our position is 1.4 kW/m^2. Earth's temperature remains constant on the average. What fraction of the incident solar radiation is absorbed by the Earth? (Caution: The area intercepting the sunlight is not the same as the total area of the Earth that emits.)

19. In a bad dream, you are naked in outer space. You only emit radiation, as there is little or none to absorb from your environment. Your skin acts like a blackbody at the infrared wavelengths that it radiates, and your body's surface area is about 2 m^2. Estimate the following:

(a) the temperature and area of your skin,

(b) the rate at which energy is emitted by your body,

(c) the kilocalories of energy (1 kcal=1 food Calorie) emitted by your skin per day,

(d) the number of milkshakes (each containing about 400 kcal of energy) that you would have to drink per day in order to compensate for the energy lost from your skin.

(e) Why do you suppose it is biologically advantageous for the blood vessels in your skin to constrict when your environment is cold?

20. Your skin temperature is about 300 K, and that of your clothing and immediate environment is normally about 290 K. Assume that your skin has a total area of 2 m^2 and acts like a blackbody at infrared wavelengths.

(a) What is the *net* rate at which your body radiates energy into the environment?

(b) One "food Calorie" is actually a kilocalorie $= 4.2 \times 10^3$ J. How much food energy do you have to consume per day to replace the energy lost into your clothing and environment?

21. Newton's law of cooling states that the rate at which an object cools is proportional to the temperature difference, ΔT, between it and its environment. You are going to derive this.

(a) Consider a blackbody of area A and temperature $T + \Delta T$ in an environment of temperature T. Calculate the difference between the rates of energy emission and absorption, keeping only terms to first order in $\Delta T/T$. The answer should be in terms of σ, A, T, and ΔT.

(b) Suppose that the object has heat capacity C. Show that the rate of cooling is given by $d(\Delta T)/dt = -\text{constant} \times \Delta T$, where the value of the constant is given in terms of C, σ, A, and T.

(c) Show that the object cools off in such a way that the difference between its temperature and that of the environment, ΔT, decreases exponentially in time.

22. Why is a teapot shiny silver, rather than black? Why is your car radiator black?

23. Suppose that you were to direct photons of various wavelengths at the Sun and wish to know whether they would be reflected or absorbed when they got there. Using Figure 21.4, plot (qualitatively) the fraction absorbed vs. wavelength for photons incident on the Sun. On the same graph, plot the fraction reflected vs. wavelength.

24. Show that when averaged over the $+z$ hemisphere the average value of $\cos\theta$ is $1/2$, where θ is the angle with the $+z$ axis. How is this relevant to the calculation of photon flux?

Section D

25. Why do building codes in some cold climates permit twice the window area for double-paned glass?

26. Suppose that your body temperature is 300 K and that of your environment is 273 K (freezing point). Your body's emissivity is 1 and its srface area is $2\ m^2$. Using radiative losses only, estimate the net rate of heat loss from your body if you are (a) naked, (b) wearing one layer of clothing, (c) wearing five layers of clothing.

27. If three identical foils separate regions held at 100 K and 300 K, what are the temperatures of the three foils? (Hint: See equation 21.15.)

Section E

28. Derive the result 21.20 by using equation 20.6 ($pV = E/3$) to write p in terms of E and V and then integrating $dE = -pdV$ for adiabatic processes ($dQ = 0$). Then use equation 21.8 to write E in terms of T and V.

29. Comparing results 21.20 and 21.21, you can see that entropy (hence, the number of accessible states) is not conserved in the free expansion of a photon gas. Why not?

30. The Universe presently has a radius of about 14 billion light years, and the temperature of the cosmic background radiation is 2.74 K. Ignoring any coupling to matter, estimate its temperature when the Universe's radius was only 1 light year, or when it was only 1 m. 1 light year $= 9.5 \times 10^{15}$ m. (Think about whether the expansion is free or not.)

31. A photon gas expands adiabatically from initial temperature and volume T_i, V_i to a final volume with temperature T_f. Show that the work done is given by $W = \left(8\pi^5 k^4/15h^3c^3\right) V_i T_i^3 \left(T_i - T_f\right)$.

Section F

32. What would be the root mean square voltage fluctuation caused by thermal noise across a circuit element having 5000 ohms resistance if the circuit is at room temperature (293 K) and you are measuring frequencies in the range $3 \times 10^{11} - 4 \times 10^{11}$ Hz?

33. How would the result 21.23 change if, when the thermally generated noise reaches the end of the circuit element, half exits and the other half is reflected?

Chapter 22
The thermal properties of solids

A Overview

As we learned in Sections 4B and 10E, the atoms in solids are anchored in place by electromagnetic interactions with neighboring atoms, making each a harmonic oscillator in three dimensions. The solid may have conduction electrons as well (Figure 22.1). The "classical" counting of degrees of freedom would give each atom six (three kinetic and three potential) and each conduction electron three (translational only). Because an average thermal energy $(1/2)kT$ is associated with each degree of freedom, the classical prediction for the heat capacity of a solid with N_a atoms and N_e conduction electrons would be

$$C_V = \left(\frac{\partial E}{\partial T} \right)_V = 3N_a k + \frac{3}{2} N_e k \quad \text{(classical prediction)}. \quad (22.1)$$

Experimental measurements of the heat capacities of solids show this to be wrong. For sufficiently low temperatures, the contribution from the lattice of atoms always drops well below the classical prediction (Figure 15.5), and the contribution from the conduction electrons is almost nonexistent.

In this chapter, we learn that quantum effects are responsible for these failures of the classical approach. Because the atomic vibrations are quantized (subsection 1B.8 and Section 10E), many are confined to the ground state as the temperature

Figure 22.1 We can view a solid as a system of atomic oscillators, and it might also include a gas of conduction electrons that are confined within its boundaries.

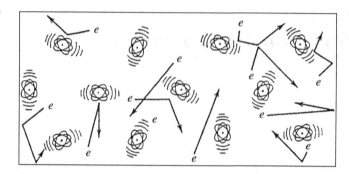

Figure 22.2 Only electrons near the Fermi level ($\varepsilon_f \approx \mu$) have any freedom at all, because only these electrons might be able to find vacant neighboring states into which they can move in response to thermal agitation or other stimuli.

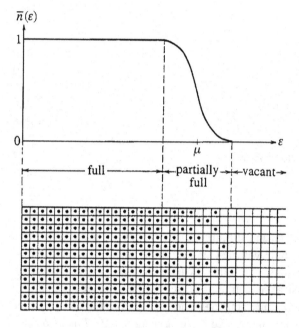

is lowered. This quantum confinement reduces the number of degrees of freedom and causes a corresponding reduction in the solid's thermal energy.

Quantum effects are even more restrictive for the conduction electrons, because at normal temperatures they are nearly degenerate. Nearly all electron states up to the Fermi level ($\varepsilon_f = \mu$) are occupied, and nearly all those above this level are vacant (Figures 19.1 and 22.2). Because the low-lying states are full, the electrons in these states can find no vacant neighboring states into which to move. This means that they cannot change their state in response to thermal agitation, applied electric fields, or any other stimulus. Only those near the Fermi level have any freedom. But these electrons are a small minority, so they can make only a very small contribution to the thermal properties of a solid.

In this chapter we use quantum statistics to learn how these quantum effects influence the thermal properties of solids, and we separate the latter into two parts, those due to the lattice of atoms and those due to the conduction electrons.

B Lattice vibrations

B.1 Background

In a 1907 paper, Albert Einstein proposed a very simple model to explain lattice vibrations in solids and the failure of the classical approach at low temperatures. Its predictions were qualitatively correct (homework). According to this model, for each of the $3N_a$ atomic simple harmonic oscillators there is a single quantum state of energy $\hbar\omega_0$.[1] The occupation number of this state determines the level of excitation of that oscillator. The significance of this paper was that it demonstrated the importance of using quantum statistics (via the occupation number) to explain the low-temperature behavior.

B.2 The Debye model

This early success led in 1912 to an important refinement by Peter Debye, who treated the vibrations in solids as a phonon gas. As we learned in subsection 10E.1, phonons are quantized vibrations. They are massless, and their energy is given by

$$\varepsilon = hf = \frac{hc_s}{\lambda}, \tag{22.2}$$

where f is the frequency, λ is the wavelength, and c_s is their speed through the solid.

This speed may be different for different polarizations. The longitudinal waves normally travel faster than do transverse waves of either polarization. Since we must sum over the three polarizations in determining the total number of phonon states, we find it convenient to define an average according to

$$\frac{1}{c_1^3} + \frac{1}{c_2^3} + \frac{1}{c_3^3} \equiv \frac{3}{c_s^3} \tag{22.3}$$

where c_1, c_2, c_3 are the speeds for the three polarizations. We use this prescription in the density of states of equation 20.1 for a gas of massless particles:

$$g(\varepsilon) = C\varepsilon^2, \quad \text{with} \quad C = \frac{12\pi V}{h^3 c_s^3}. \tag{22.4}$$

Since phonons have three polarizations, they can be treated as spin-1 bosons, and their chemical potential is zero for the same reason as for photons in the

---

[1] ω_0 is adjusted to fit the data. Each atom can oscillate in three dimensions, making a total of $3N_a$ harmonic oscillators. Energies are measured relative to the ground state energy $\varepsilon_0 = (1/2)\hbar\omega_0$ (equation 1.18).

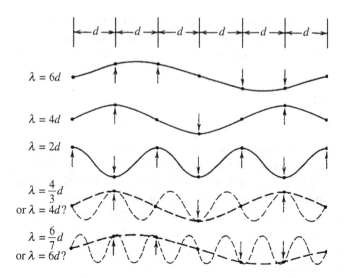

Figure 22.3 Elastic waves in solids must have wavelengths greater than twice the spacing between atoms ($\lambda \geq 2d$). As illustrated by the bottom two lines, wavelengths shorter than this are equivalent to longer ones.

previous chapter (equation 21.3). That is, the number of lattice vibrations increases and decreases in response to changes in the solid's temperature and thermal energy. Their number adjusts to maximize the entropy in compliance with the second law. When the entropy is a maximum, its derivative is zero,

$$\mu = -\left(T\frac{\partial S}{\partial N}\right)_{E,V} = 0.$$

The distribution of particles is the product of the density of states and the number of particles in each (equation 19.12). Using equation 22.4 for the density of states and equation 19.3b for the occupation number, the distribution of phonons is

$$dN = g(\varepsilon)\bar{n}(\varepsilon)d\varepsilon = C\frac{\varepsilon^2 d\varepsilon}{e^{\beta\varepsilon} - 1}, \qquad \text{where} \quad C = \frac{12\pi V}{h^3 c_s^3}. \tag{22.5}$$

B.3 The Debye cutoff

One important way in which phonons must differ from other gases is that there is an upper limit to their energy, which is called the Debye cutoff energy and is given the symbol ε_D. As illustrated in Figure 22.3, the phonon wavelength cannot be shorter than twice the atomic spacing. This lower limit on wavelength places an upper limit on phonon energy. Since the volume per atom is V/N_a, the average distance between atoms is $(V/N_a)^{1/3}$ and so the maximum phonon energy is (see equation 22.2)

$$\varepsilon_D = \frac{hc_s}{\lambda_{\min}} \approx \frac{hc_s}{2(V/N_a)^{1/3}} = 0.5hc_s\left(\frac{N_a}{V}\right)^{1/3}. \tag{22.6}$$

Table 22.1. *Debye cutoff energies* (ε_D) *and Debye temperatures* $\Theta_D(\varepsilon_D = k\Theta_D)$ *for the lattice vibrations, and Fermi energies* ($\varepsilon_f \approx \mu$) *for the conduction electrons for various solids*

Material	ε_D (10^{-2} eV)	Θ_D (K)	μ (eV)	Material	ε_D(10^{-2} eV)	Θ_D (K)	μ (eV)
sodium	1.36	158	3.23	copper	2.96	343	7.00
magnesium	3.45	400	7.13	silver	1.94	225	5.48
aluminum	3.69	428	11.63	gold	1.42	165	5.51
zinc	2.82	327	9.39	lead	0.90	105	9.37
potassium	0.78	91	2.12	silicon	5.56	645	–
calcium	1.98	230	4.68	diamond	27.8	2300	–
iron	4.05	470					

Debye suggested another way to determine the maximum phonon energy, and it gives nearly the same answer. We know that the motion of one simple harmonic oscillator can be described by one characteristic frequency, that of two coupled oscillators can be described by two characteristic frequencies, and so on. So a solid of $3N_a$ coupled oscillators has $3N_a$ states, each with its own characteristic frequency. Therefore, the sum over all states must give $3N_a$. This places an upper limit (ε_D) on the oscillator energy:

$$3N_a = \int_0^{\varepsilon_D} g(\varepsilon)d\varepsilon = \frac{12\pi V}{h^3 c_s^3}\int_0^{\varepsilon_D} \varepsilon^2 d\varepsilon = \frac{4\pi V}{h^3 c_s^3}\varepsilon_D^3. \tag{22.7}$$

Solving this for ε_D gives

$$\varepsilon_D = hc_s\left(\frac{3N_a}{4\pi V}\right)^{1/3} = 0.62hc_s\left(\frac{N_a}{V}\right)^{1/3}, \tag{22.8}$$

where we have written $(3/4\pi)^{1/3} = 0.62$ in order to see how close the result 22.8 is to the result 22.6, which we obtained by requiring the minimum phonon wavelength to be twice the atomic spacing.

From this result, you can see that the Debye cutoff energy is a simple function of the density of atoms N_a/V and the phonon speed c_s. It is sometimes expressed in terms of the Debye frequency ω_D, or the Debye temperature Θ_D which are defined by

$$\varepsilon_D \equiv \hbar\omega_D \equiv k\Theta_D. \tag{22.9}$$

Values of ε_D and Θ_D for several common solids are listed in Table 22.1.

The result 22.8 allows us to write the constant factor in the phonon density of states (equation 22.4) as:

$$C = \frac{12\pi V}{h^3 c_s^3} = \frac{9N_a}{\varepsilon_D^3},$$

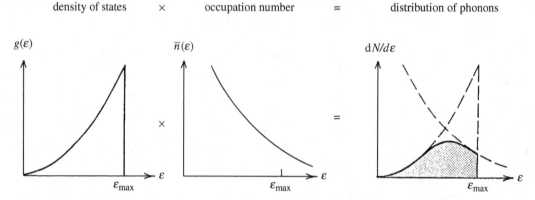

density of states   ×   occupation number   =   distribution of phonons

$g(\varepsilon)$ $\bar{n}(\varepsilon)$ $dN/d\varepsilon$

Figure 22.4 The distribution of phonons in a solid is the product of the density of phonon states $g(\varepsilon)$ times the occupation number $\bar{n}(\varepsilon)$.

so the distribution of phonons 22.5 is often written in the following way (Figure 22.4):

$$dN = \begin{cases} \left(\dfrac{9N_{\mathrm{a}}}{\varepsilon_{\mathrm{D}}^{3}}\right)\dfrac{\varepsilon^{2}d\varepsilon}{e^{\beta\varepsilon}-1} & \text{for } 0 \leq \varepsilon \leq \varepsilon_{\mathrm{D}}, \\ 0 & \text{for } \varepsilon > \varepsilon_{\mathrm{D}}. \end{cases} \tag{22.10}$$

The total energy of the system is the sum of the energies of the individual phonons:

$$E = \int \varepsilon\, dN = \frac{9N_{\mathrm{a}}}{\varepsilon_{\mathrm{D}}^{3}} \int_{0}^{\varepsilon_{\mathrm{D}}} \frac{\varepsilon^{3}d\varepsilon}{e^{\beta\varepsilon}-1}$$

$$= \frac{9N_{\mathrm{a}}}{\varepsilon_{\mathrm{D}}^{3}}(kT)^{4} \int_{0}^{\varepsilon_{\mathrm{D}}/kT} \frac{x^{3}dx}{e^{x}-1},$$

where the last line follows by substitution of $x = \beta\varepsilon = \varepsilon/kT$. This result for the thermal energy of the lattice is often expressed as the product of the classical value $(3N_{\mathrm{a}}kT)$ times the Debye function $D(T)$:

$$E_{\ell} = 3N_{\mathrm{a}}kTD(T), \quad \text{where} \quad D(T) = 3\left(\frac{kT}{\varepsilon_{\mathrm{D}}}\right)^{3} \int_{0}^{\varepsilon_{\mathrm{D}}/kT} \frac{x^{3}dx}{e^{x}-1}. \tag{22.12}$$

In general, evaluating the Debye function $D(T)$ is difficult and best done numerically. It increases from 0 in the low-temperature limit to 1 in the high-temperature limit, where the classical result $E = 3N_{\mathrm{a}}kT$ is valid (Figure 22.5). For these two extremes, it can be shown (homework) that the thermal energy of lattice vibrations has the following values:

$$E_{1} = \begin{cases} \dfrac{3N_{\mathrm{a}}\pi^{4}k^{4}}{5\varepsilon_{\mathrm{D}}^{3}}T^{4} & \text{(low temperature limit, } kT \ll \varepsilon_{\mathrm{D}}), \\ 3N_{\mathrm{a}}kT & \text{(high temperature limit, } kT \gg \varepsilon_{\mathrm{D}}). \end{cases} \tag{22.13}$$

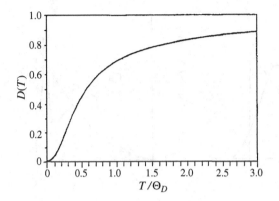

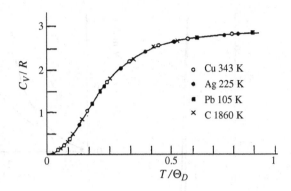

B.4 Comparison with experiment

The experimental determination of thermal energy is accomplished through measurements of heat capacities, which later in this chapter we investigate further. For now, we simply note that the Debye model gives remarkably accurate results, as is illustrated for some representative solids in Figure 22.5. As you can see in this figure, all solids fit the same curve if we measure the temperature for each in terms of its Debye temperature (T/Θ_D). The same would be true for Figure 15.5.

In real solids, there may be differences in atomic configurations, binding strengths, and spacing in different directions. The wave speed may depend on the direction of travel and the phonon polarization. Therefore, in real solids the density of states is a superposition of many components. The Debye model, in contrast, assumes just one average atomic spacing, one average speed of sound, and one average cutoff value. The Debye-model density of states corresponds to an average of all the individual components of the densities of states for a real solid (Figure 22.6).

You may wonder how the Debye-model results can be so accurate, if real densities of states are so different from that of the Debye model. The reason is that all measured thermal properties require a summation over states, giving us an integrated average. The actual details of the densities of states, such as local maxima and minima, are therefore much less important than the overall features. The Debye model does indeed represent an appropriate summed average for the individual components of the densities of states, and that is why it works so well.

Figure 22.5 (Left) The Debye function $D(T)$ vs. temperature measured in terms of the Debye temperature $\Theta_D = \varepsilon_D/k$. ($\Theta_D$ values for various common solids are listed in Table 22.1.) (Right) Molar heat capacities for various solids as a function of temperature. The solid line is the Debye model prediction. (After M. A. Omar, *Elementary Solid State Physics*, Addison-Wesley, 1975.)

B.5 Low-temperature fluctuations

From earlier work, we know that the relative fluctuation for N events is proportional to $N^{-1/2}$. So at high temperatures, where all $3N_a$ phonon states are filled, fluctuations in the lattice vibrations are far too small to be detected. But at low temperatures, where only a small fraction of the phonon states are filled, the

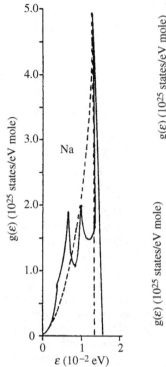

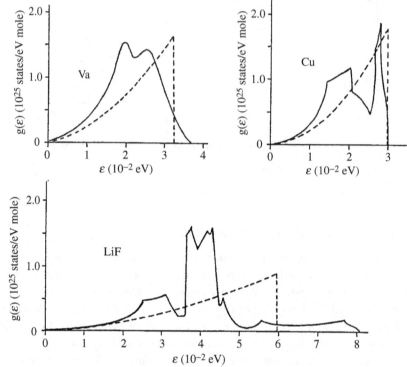

Figure 22.6 Plots of actual densities of states (solid lines) and those from the Debye model (broken lines), all drawn to the same scale, for sodium, vanadium, copper, and lithium fluoride.

situation is different. Equation 18.6' ($\sigma_E^2 = kT^2 C_V$) and the result 22.13 can be used to show that the relative fluctuation at low temperatures is given by

$$\frac{\sigma_E^2}{E^2} = \frac{20}{3\pi^4 N_a} \left(\frac{\varepsilon_D}{kT}\right)^3 = \frac{0.0684}{N_a} \left(\frac{\Theta_D}{T}\right)^3 . \tag{22.14}$$

Thus a 1 cm^3 sample of a typical material would need to be below about 100 μK in order to experience relative fluctuations above the 1% level (homework).

Summary of Sections A and B

The classical prediction for the heat capacity of a solid with N_a atoms and N_e conduction electrons is wrong. Quantum effects cause considerable modification. The electron contribution is almost nonexistent, and the lattice contribution falls well below $3N_a kT$ at low temperatures. Einstein proposed a very simple model for the lattice vibrations, which produced the correct qualitative behavior and demonstrated the importance of quantum statistics.

The Debye model envisions lattice vibrations as a phonon gas, where the phonons are discrete quanta of vibrational energy that travel through the solid with speed c_s. They are massless bosons with zero chemical potential. In order to include the three polarizations in the sum over states, we find it convenient to define an average according to (equation 22.3)

$$\frac{1}{c_1^3} + \frac{1}{c_2^3} + \frac{1}{c_3^3} \equiv \frac{3}{c_s^3}.$$

With this, the distribution of phonons in a solid according to the Debye model is (equation 22.5)

$$dN = C \frac{\varepsilon^2 d\varepsilon}{e^{\beta\varepsilon} - 1}, \qquad \text{where} \quad C = \frac{12\pi V}{h^3 c_s^3}.$$

There is an upper limit to phonon energies. A solid of N_a atoms can be regarded as $3N_a$ coupled harmonic oscillators, so there are $3N_a$ characteristic frequencies. Therefore the sum over all states must equal $3N_a$. This provides the upper limit on the energy of a state (equation 22.7):

$$3N_a = \int_0^{\varepsilon_D} g(\varepsilon)d\varepsilon = \frac{12\pi V}{h^3 c_s^3} \int_0^{\varepsilon_D} \varepsilon^2 d\varepsilon = \frac{4\pi V}{h^3 c_s^3} \varepsilon_D^3.$$

Solving for ε_D gives (equation 22.8)

$$\varepsilon_D = hc_s \left(\frac{3N_a}{4\pi V} \right)^{1/3} = 0.62 hc_s \left(\frac{N_a}{V} \right)^{1/3},$$

which is close to the answer we get by insisting that the wavelength can be no shorter than twice the atomic spacing. This upper limit is called the Debye cutoff energy, and is frequently expressed as the corresponding Debye frequency or Debye temperature (equation 22.9):

$$\varepsilon_D \equiv \hbar\omega_D \equiv k\Theta_D.$$

The resulting distribution of phonons in a solid is given by (equation 22.10)

$$dN = \begin{cases} \left(\dfrac{9N_a}{\varepsilon_D^3} \right) \dfrac{\varepsilon^2 d\varepsilon}{e^{\beta\varepsilon} - 1} & \text{for } 0 \leq \varepsilon \leq \varepsilon_D, \\ 0 & \text{for } \varepsilon > \varepsilon_D. \end{cases}$$

The total energy of the system is the sum of the energies of the individual phonons. With the substitution of variables $x = \beta\varepsilon$ this can be written in the form (equation 22.12)

$$E_\ell = 3N_a kT D(T), \qquad \text{where} \quad D(T) = 3 \left(\frac{kT}{\varepsilon_D} \right)^3 \int_0^{\varepsilon_D/kT} \frac{x^3 dx}{e^x - 1}.$$

The Debye function $D(T)$ ranges in value from 0 to 1 in the low- and high-temperature limits, respectively, and reproduces the experimental data well.

The Debye model is surprisingly accurate in view of the fact that the real density of phonon states is a superposition of many different components. The reason for this is that model predictions require us to sum over all states, and the Debye density of states represents a good integrated average of the many different components that are present in real solids.

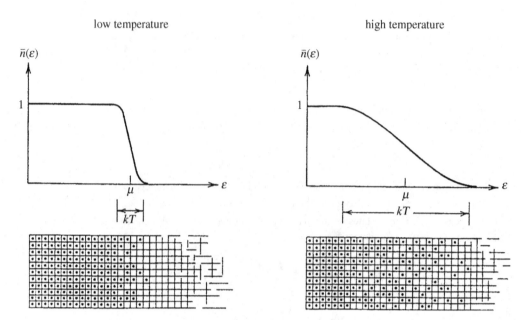

low temperature   high temperature

Figure 22.7 The characteristic width of the tail of the Fermi distribution is equal to kT: the tail widens with increased temperature. Only electrons in the tail have any freedom, as only they have neighboring states into which they can move. (For illustrative purposes, the above tail widths have been exaggerated.)

C Conduction electrons

C.1 Overall properties

Most electrons remain on their atoms and vibrate with them. But conduction electrons are not bound to individual atoms and make up a separate system. They are spin-1/2 fermions, and in most conductors they are highly degenerate. The occupation number has a narrow tail near the Fermi level (Figure 22.7). The width of the tail, coming from the factor $e^{(\varepsilon - \mu)/kT}$, is roughly equal to kT and is small compared with the depth of the Fermi sea μ, so the fraction of electrons in the tail is typically between 10^{-2} and 10^{-3}:

$$\text{fraction of electrons in tail} \approx \frac{kT}{\mu} < 10^{-2}.$$

Only electrons in the tail have vacant neighboring states into which they can move. So only those in the tail have the freedom of movement that allows them to contribute to the thermal properties of the system.

As the temperature increases, both the number of electrons in the tail and the average energy of each electron increase in proportion to kT. Therefore, their contribution to the system's thermal energy is proportional to T^2:

$$E_e = \text{number of free electrons} \times \text{thermal energy of each} \propto T^2.$$

Thus the internal energy of the conduction electrons is that of the degenerate Fermi sea plus a small "tail" contribution proportional to T^2:

$$E_e = E_0 + bT^2 \qquad (b \text{ is a positive constant}). \tag{22.15}$$

C.2 A nearly degenerate fermion gas

The behavior 22.15 is also what we expect if we treat this nearly degenerate fermion gas with the tools of subsection 20D.3. From equations 20.19 and 20.15, the thermal energy of the electrons is given by

$$E_e \approx N_e \frac{3}{5} \varepsilon_f \left[1 + \frac{5\pi^2}{12} \left(\frac{kT}{\varepsilon_f} \right)^2 \right] \qquad (22.16)$$

with

$$\varepsilon_f = \frac{h^2}{2m} \left(\frac{3 N_e}{8\pi V} \right)^{2/3}. \qquad (22.17)$$

Several things are noteworthy about this result.

- First, we can see that the thermal energy of a degenerate electron gas does indeed increase quadratically with the temperature as we expected (22.15). Furthermore, we now know the values of the constants, so we can make quantitative as well as qualitative predictions.
- Second, if we put into equation 22.17 the electron density N_e / V for typical metals, we find that the Fermi level is around 4 to 8 eV (homework, or see Table 22.1). If we compare these values with kT, which is about 0.025 eV at room temperature, we can see that typical thermal energies are extremely small compared with the Fermi level. As expected, the gas is highly degenerate, and thermal agitation has relatively little effect on it.
- Third, if we wish to write the Fermi energy in the form $\varepsilon_f = p^2/2m$, where $p = h/\lambda$ (equation 1.2), then we find from equation 22.17 (homework) that

$$\lambda = \left(\frac{8\pi}{3} \right)^{1/3} \left(\frac{V}{N_e} \right)^{1/3} = 2.03 \left(\frac{V}{N_e} \right)^{1/3}.$$

So the wavelength of these highest-energy electrons is roughly twice the average electron spacing. Notice the similarity with lattice vibrations, for which the wavelength of the highest-energy phonons is also roughly twice the atomic spacing.

Summary of Section C

In a typical conductor, more than 99% of the conduction electrons are trapped in the Fermi sea, and less than 1% are in the tail. Only electrons in the tail can contribute to the thermal properties. The product of the number of relatively free electrons ($\propto kT$) times the thermal energy of each ($\propto kT$) is quadratic in temperature. Thus the internal energy of the conduction electrons is that at absolute zero plus a small increase proportional to T^2 (equation 22.15):

$$E_e = E_0 + bT^2.$$

We can treat the conduction electrons as a nearly degenerate fermion gas, for which the internal energy is given by (equation 22.16)

$$E_e \approx N_e \frac{3}{5} \varepsilon_f \left[1 + \frac{5\pi^2}{12} \left(\frac{kT}{\varepsilon_f} \right)^2 \right],$$

with (equation 22.17)

$$\varepsilon_f = \frac{h^2}{2m} \left(\frac{3N_e}{8\pi V} \right)^{2/3}.$$

Interesting features of this include the following.

- It is what we anticipated from 22.15, and it also gives the numerical values of the constants in 22.15.
- It confirms our expectation that the conduction electrons should be highly degenerate.
- It reveals that the wavelengths associated with the highest-energy electrons are about twice the electron spacing.

D Heat capacities

We now examine the heat capacity of a solid.[2] At normal temperatures the lattice contribution dominates. Taking the derivative of the Debye-model prediction for the lattice's thermal energy, 22.12, we find that the model is in excellent agreement with the experimental data at all temperatures (Figure 22.5). Although the result is complicated for intermediate temperatures, it becomes simple in the high- and low-temperature limits (see equation 22.13):

$$C_{V,\text{Debye}} = \begin{cases} 3N_a k, & \text{high-temperature limit,} \\ K_1 T^3 & \text{low-temperature limit,} \end{cases}$$

where

$$K_1 = \frac{12 N_a \pi^4 k^4}{5 \varepsilon_D^3}. \tag{22.18}$$

In the homework problems you will show that the Einstein model gives the same high-temperature result but differs at low temperatures.

Regarding the conduction electrons, their contribution to the heat capacity is obtained by taking the derivative of their thermal energy from equation 22.16, which yields

$$C_{V,e} = K_e T, \qquad \text{where} \quad K_e = \frac{N_e \pi^2 k^2}{2\varepsilon_f}. \tag{22.19}$$

---

[2] For constant volume $dV = 0$, so the first law reads $dE = dQ$. (Ignore μdN because $\mu_{\text{phonon}} = 0$.) Therefore the heat capacity at constant volume is $C_V = (\partial Q/\partial T)_V = (\partial E/\partial T)_V$. As we saw at the end of subsection 10F.1, C_V and C_p are nearly the same for solids, so we normally don't make a distinction between the two.

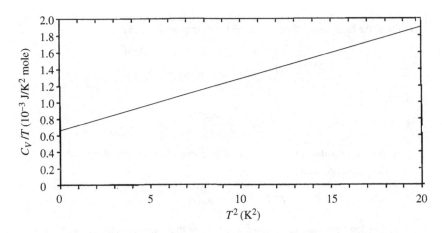

Figure 22.8 Plot of molar heat capacity C_V/T vs. T^2 for copper at very low temperatures. The intercept at $T^2 = 0$ represents the contribution from the electron gas, and the fact that the plot is linear in T^2 demonstrates that the Debye model describes the lattice's contribution correctly at these low temperatures. (From Corak, Garfunkel, Satterthwaite, and Wexler, *Phys. Rev.* **98**, 1699, 1955.)

Since the number of electrons in the tail of the Fermi distribution is so small, they have a negligible effect on the heat capacity of metals at normal temperatures, where the $6N_a$ degrees of freedom of the lattice dominate. However, at low temperatures, the heat capacity of the electron gas goes to zero linearly with T whereas that of the lattice goes to zero more rapidly as T^3. This means that, at sufficiently low temperatures, the small electron contribution dominates. From equations 22.18 and 22.19, we see that the combined heat capacity from the lattice and conduction electrons at low temperatures is

$$C_V = K_e T + K_l T^3 \qquad \text{(low temperatures)} \qquad (22.20)$$

or, equivalently,

$$\frac{C_V}{T} = K_e + K_l T^2. \qquad (22.21)$$

So we expect that the plot of C_V/T vs. T^2 should be linear in T^2, the intercept at $T = 0$ representing the free electron contribution K_e. As is seen in Figure 22.8, this is indeed what happens. Furthermore, the experimental values of both the slope K_l and the intercept K_e generally agree with the values calculated from equations 22.18 and 22.19 (homework). This gives us confidence that our models represent Nature fairly well. Treating the lattice vibrations as a phonon gas and the conduction electrons as a degenerate fermion gas seems to work.

Summary of Section D

Because the number of electrons in the tail of the Fermi distribution is so small, the thermal energy of the lattice dominates that of the conduction electrons at ordinary temperatures. The Debye-model prediction for the solid's heat capacity fits the experimental data well over a wide range of temperatures, giving the following results in the high- and low-temperature limits (equation 22.18):

$$C_{V,\text{Debye}} = \begin{cases} 3N_a k & \text{high temperature limit,} \\ K_1 T^3 & \text{low temperature limit,} \end{cases}$$

where

$$K_1 = \frac{12 N_a \pi^4 k^4}{5 \varepsilon_D^3}.$$

Treating the conduction electrons as a degenerate fermion gas, their heat capacity is (equation 22.19)

$$C_{V,e} = K_e T, \qquad \text{where} \quad K_e = \frac{N_e \pi^2 k^2}{2 \varepsilon_f}.$$

At very low temperatures, the heat capacity of the lattice goes to zero as T^3, whereas that of the electron gas goes to zero linearly with T. So, at sufficiently low temperatures, the lattice contribution no longer dominates and the total heat capacity has the form (equation 22.20)

$$C_V = K_e T + K_1 T^3 \qquad \text{(low temperatures)}.$$

If we plot C_V / T versus T^2, we get the following form, which agrees with measurements (equation 22.21):

$$\frac{C_V}{T} = K_e + K_1 T^2.$$

Problems

Section A

1. Make a rough estimate of the force constant κ that holds the atoms of a typical solid in place. Use the facts that the average energy per degree of freedom is $(1/2)kT$ and that a typical root mean square value for the amplitude of oscillation at room temperature is 10^{-11} m.

2. We are going to investigate the feasibility of treating atoms as harmonic oscillators, even if their potential wells are not parabolic. Consider a particle in one dimension that has potential energy $V(x) = -V_0 e^{-x^2}$.
   (a) What is the equilibrium position of the particle? (I.e., where is $V(x)$ a minimum?)
   (b) Write out the Taylor series expansion for $V(x)$ about the equilibrium position, keeping only the zeroth, first, and second order terms in x. (Note that this makes it a harmonic oscillator.)
   (c) By what percentage does this expansion differ from the real value of $V(x)$ at $x = 0.1$? At $x = 0.5$?

3. If the atoms in conductors behaved as classical harmonic oscillators and the conduction electrons as a classical ideal gas, what would be the molar heat

capacity ($C_V = dE/dT$) of a metal if each atom contributes two electrons to the conduction band?

4. In quantum mechanics we learn that the average energy of a simple harmonic oscillator is given by $\varepsilon = (n + 1/2)\hbar\omega$, with $n = 0, 1, 2, \ldots$ Classically, the energy is given by $\varepsilon = (1/2m)p_x^2 + (1/2)\kappa x^2$.
   (a) What would be the classical value of the heat capacity for N of these oscillators? (Use equipartition.)
   (b) If the heat capacity fell well below the classical value for temperatures below 190 K, what would you estimate the frequency ω to be?
   (c) In real solids having N_a atoms, we might expect $6N_a$ degrees of freedom, each with an average energy $(1/2)kT$, giving a total thermal energy of $3N_a kT$. But at temperatures below the Debye temperature, the internal energy is lower than this. Does this violate the equipartition theorem? Explain.

Section B

5. According to the Einstein model, all $3N_a$ harmonic oscillators have the same fundamental frequency, given by $\varepsilon_0 = \hbar\omega_0$, and the level of excitation of each is given by the occupation number $\bar{n} = 1/(e^{\varepsilon_0/kT} - 1)$. The value of ε_0 (or ω_0) is chosen to fit the data.
   (a) Show that the lattice's thermal energy is $3N_a\varepsilon_0/(e^{\varepsilon_0/kT} - 1)$ according to this model.
   (b) We find that, for lead, $E \approx 3N_a kT$ for temperatures above about 100 K. Use this information to estimate very roughly the value of ε_0 or ω_0 for lead.
   (c) Using the variable $x = \varepsilon_0/kT$, write down an expression for the ratio $E/3N_a kT$ as a function of x.
   (d) Make a qualitative plot of $E/3N_a kT$ vs. the variable x. Is $x = 0$ the low- or high-temperature limit?

6. (a) Show that the Einstein model (problem 5) predicts the internal energy of a solid to be $3N_a\varepsilon_0 e^{-\varepsilon_0/kT}$ in the low-temperature limit and $3N_a kT$ in the high-temperature limit.
   (b) What would be the heat capacity, C_V, in each of these limits?

7. Using the Debye model and the parameter $x = \varepsilon_D/kT$, express the ratio $E/3N_a kT$ as a function of x.

8. We find that $E \approx 3N_a kT$ at temperatures above 100 K for lead and above 2300 K for diamond. The mass of a lead atom is 3.4×10^{-25} kg, and the mass of a carbon atom is 2.0×10^{-26} kg. For each, give a rough estimate of the value of the force constant κ holding an atom in place. Use $\omega_D \approx (\kappa/m)^{1/2}$.

9. The oscillations for a phonon traveling in the positive x direction can be approximated by a plane wave, $y = A \cos(kx - \omega t)$, where A is the amplitude, $k = 2\pi/\lambda$ is the wave number, and $\omega = 2\pi/T$ is the angular frequency.
   (a) What is the speed of this phonon, c_s, in terms of ω and k?
   (b) If the momentum and energy are given by $p = \hbar k$ and $\varepsilon = \hbar\omega$, how are the momentum, energy, and wave speed interrelated?

10. (a) Make a sketch similar to Figure 22.3 for the displacement of atoms when the phonon's wavelength is exactly $2/5$ of the atomic separation d.
    (b) Is this distinguishable from a phonon with wavelength $(2/3)d$?
    (c) Both these would be indistinguishable from a phonon whose wavelength λ is longer than the atomic separation: $\lambda = xd$ with $x \geq 2$. What is the value of x?

11. In a certain solid the atomic separation is about 0.2 nm and the speed of sound (i.e., the speed of travel of the phonon vibrations) is 1000 m/s. What is the maximum phonon energy according to the Debye model?

12. The density of lead is 13.6 g/cm^3 and the mass of an atom is 3.4×10^{-25} kg. Calculate the following:
    (a) $(N_a/V)^{1/3}$,
    (b) the speed of sound c_s. (Use the value of ε_D for lead from Table 22.1.)

13. For a certain solid, the Debye temperature is 290 K. What are the Debye cutoff energy ε_D and Debye frequency ω_D for this material?

14. Two different solids consist of the same number of atoms, and vibrations travel with the same speed in both. But the atomic spacing in solid A is twice as large as the atomic spacing in solid B.
    (a) For which solid will the cutoff energy ε_D be the larger? By how many times?
    (b) At low temperatures, which solid will have the largest internal energy? By how many times?

15. Show that the Debye function $D(T)$ in equation 22.12 and Figure 22.5 has the following values in the given limits:

    $$D(T) \to (\pi^4/5)(kT/\varepsilon_D)^3 \text{ for } kT \ll \varepsilon_D \text{ (low-temperature limit)};$$
    $$D(T) \to 1 \text{ for } kT \gg \varepsilon_D \text{ (high-temperature limit)}.$$

    You may need to use $\int_0^\infty x^3 dx/(e^x - 1) = \pi^4/15$, and the expansion $e^x \approx 1 + x$, for $x \ll 1$. Use this to confirm the result 22.13 for the thermal energy in the low- and high-temperature limits.

16. Consider the vibrations in a two-dimensional lattice of N_a atoms (i.e., a thin film) to be represented by a two-dimensional phonon gas. Assume that longitudinal and transverse phonon modes both travel with the same speed,

c_s. The total area of this lattice is A. Find expressions for the following in terms of c_s, N_a, and A:

(a) the density of phonon states $g(\varepsilon)$,

(b) the maximum phonon energy.

17. Use equations 22.13 and 18.6′ to show that the relative fluctuation in the energy of the lattice vibrations for a solid in contact with a cold reservoir is given by $(\sigma_E^2/E^2) = (0.0684/N_a)(\theta_D/T)^3$. (For constant volume we have $dV = 0$, so the first law reads $dE = dQ$. Ignore μdN because $\mu_{phonon} = 0$. Therefore $C_V = (\partial Q/\partial T)_V = (\partial E/\partial T)_V$.) Consider a typical solid with atomic density $\approx 10^{29}/m^3$ and $\theta_D \approx 300\,K$. If a sample measures 1 mm on a side, at what temperature would the relative fluctuation reach 1%?

18. Let's see how the phonon model fits energy excitations in liquid helium at very low temperatures. The speed of sound in liquid helium is 238 m/s and only longitudinal waves can propagate through a liquid. Its density is 145 kg/m³ and its molar mass is 6.64×10^{-3} kg. Calculate the Debye temperature and the specific heat capacity for cold liquid helium. (Beware: From equation 22.3 you can see that you will have to make the replacement $3/c_s^3 \rightarrow 1/c_s^3$ in our formulas since there is only one wave polarization.) Compare your answer with the experimental value, $c_V = (20.4\,J/kg\,K^4)T^3$.

Section C
Use the free electron model where appropriate.

19. The spacing between identical (i.e., same spin direction) protons or neutrons in a nucleus is about 3×10^{-15} m, and their mass is 1.67×10^{-27} kg.

(a) What is the characteristic kinetic energy of a nucleon in a nucleus? (Hint: Use the uncertainty principle.)

(b) At what temperature would such a system of fermions have to be, in order not to be degenerate?

20. What would the characteristic spacing of conduction electrons have to be in order for them to be reasonably nondegenerate at 295 K? (See condition 19.9.)

21. (a) Write down an expression for the characteristic kinetic energy $p^2/2m$ for conduction electrons, in terms of m, h, and N_e/V. (Note that because the electrons may be spin up or spin down, the density of *identical* fermions is $N_e/2V$. Use the uncertainty principle.)

(b) How does the result (a) compare with the Fermi energy of a system of conduction electrons as given in equation 22.17?

(c) Show that the Fermi energy of equation 22.17 corresponds to an electron wavelength of about twice the average electron spacing.

22. In a certain system of degenerate conduction electrons, only 10^{15} are "free" at 200 K.
    (a) Roughly what is the total thermal energy of these electrons?
    (b) At 400 K, how many electrons are free and what is their thermal energy?

23. In a certain solid at 40 K ($\ll \Theta_D$), the lattice contribution to the thermal energy is 10^4 times greater than the contribution from the conduction electrons.
    (a) At what temperature would the thermal energy of both systems be the same?
    (b) At what temperature would the electrons have 100 times greater thermal energy than the lattice?

24. According to equation 22.16, we can write the internal energy of a gas of degenerate electrons as $E = (3/5)N_e\varepsilon_f(1 + bT^2)$. Given that typical values of ε_f for conduction electrons in metals are around 7 eV, roughly what is the value of the constant b?

25. The radius of a uranium nucleus is 7×10^{-15} m. It has atomic number 92 and mass number 238. Protons and neutrons are each spin-1/2 particles. Find
    (a) the densities of the neutrons and of the protons,
    (b) the Fermi temperature ε_f/k for each of these.

26. The density of matter in the center of the Sun is about 150 g/cm^3, and the temperature is about 1.2×10^7 K. Nearly all the mass is made up of individual protons, and there is a roughly equal number of free electrons. Are the protons nearly degenerate? How about the electrons?

27. The ^{3}He isotope is a fermion. It has two protons, one neutron, two electrons, and its net spin is 1/2. Hence, a gas of ^{3}He atoms is a fermion gas. At standard temperature and pressure (1 atm, 0°C), a mole occupies 22.4 liters. What is the Fermi temperature, ε_f/k? Is the gas degenerate?

28. What is the ratio kT/ε_f for the conduction electrons in a metal with $\varepsilon_f = 7$ eV at room temperature?

29. The number of electrons within a small range of energies $\Delta\varepsilon$ is given by the product of the number of states $g(\varepsilon)\Delta\varepsilon$ and the occupation number of each, $\bar{n}(\varepsilon)$. For electrons in the "tail" of the Fermi distribution, we have $\varepsilon \approx \varepsilon_f$, $\Delta\varepsilon \approx kT$, and $\bar{n}(\varepsilon)$ averaging 1/2 (it goes from 1 to 0 in this region). Using equation 22.17 to write the number of electrons N_e in terms of the Fermi energy ε_f and equation 20.1 for the density of states, show that the fraction of the conduction electrons in the "tail" of the Fermi distribution at room temperature would be about $(3/4)(kT/\varepsilon_f)$. What is this fraction at room temperature (295 K) for a typical value of ε_f, 7 eV?

30. In a certain metal there are 6.4×10^{28} conduction electrons per cubic meter. Assuming that the fermion gas model is correct, what is the energy of the Fermi surface, in eV?

31. The density of gold is $19.3 \times 10^3 \, \text{kg/m}^3$, and its atomic mass number is 197. Each gold atom gives one electron to the conduction band. For these electrons, what is (a) the density, N_e/V, (b) the chemical potential, μ? (Hint: According to equation 20.19 the chemical potential is equal to the Fermi energy to within a factor $(kT/\varepsilon_f)^2$.)

32. The density of copper is $8.9 \times 10^3 \, \text{kg/m}^3$ and its atomic mass number is 64. Each copper atom gives one electron to the conduction band. What is the Fermi energy for these conduction electrons?

33. Consider a system of conduction electrons with $\varepsilon_f = 8 \, \text{eV}$. How much larger is their internal energy at room temperature (295 K) than at absolute zero?

Section D

34. In a mole of a certain solid, $\varepsilon_D = 0.02 \, \text{eV}$, $\varepsilon_f = 10 \, \text{eV}$, and $N_a = N_e = N_A$.
    (a) Using this information, evaluate the constants K_1 and K_e that appear in equation 22.20 for the molar heat capacity of a solid at low temperatures. (See equations 22.18 and 22.19.)
    (b) At what temperature would the electron and lattice contributions to the molar heat capacity be equal?
    (c) At a temperature one tenth of that calculated in part (b), which contribution to the heat capacity will be larger and by how many times?

35. From the slope and intercept of the data displayed in Figure 22.8, calculate the Debye cutoff energy ε_D for the lattice and the Fermi level ε_f for the conduction electrons for copper.

36. The speed of sound in copper is $c_s = 2.60 \, \text{km/s}$, its density is $8.9 \, \text{g/cm}^3$, its atomic mass number is 64, each atom contributes one electron to the conduction band, and the Fermi energy is 7.00 eV. From this information, calculate the following:
    (a) the densities N_a/V and N_e/V,
    (b) the values of ε_D, ω_D, Θ_D,
    (c) the molar heat capacity of the copper lattice at $T \gg \Theta_D$,
    (d) the value of the two constants K_1 and K_e for a mole of copper (equations 22.18 and 22.19).
    (e) Compare the two values in (d) with those you found from the graph in Figure 22.8 (problem 35).
    (f) From the results of part (d), beneath what temperature would you expect the heat capacity of conduction electrons to dominate that of the lattice?

37. The density of gold is $19.3 \times 10^3 \, \text{kg/m}^3$, its atomic mass number is 197, and the Debye energy is 0.0142 eV. The Fermi level for its conduction electrons is at $\varepsilon_f = 5.51 \, \text{eV}$. At a temperature of 2 K, find (a) the lattice's contribution to the molar heat capacity, (b) the conduction electrons' contribution to the molar heat capacity.

38. Show that the Debye-model prediction for the lattice heat capacity is $C = 3N_a k[4D(T) - (3\varepsilon_D/kT)/(e^{\varepsilon/kT} - 1)]$. You may need to use a fundamental theorem of calculus, $(d/dx)\int_a^x f(t)dt = f(x)$, and the chain rule.

39. Consider a "supermetal," which will not melt at any temperature. Carefully make a qualitative plot (not to scale) of the heat capacity vs. temperature for this material, which takes into consideration the contributions of both the lattice and the conduction electrons for all ranges of temperature. Let $N = N_a = N_e$ and label the C-axis in units of Nk. Be sure that your plot considers the following effects.

    (a) For very low T we have $T^3 \ll T$, so only the conduction electrons contribute.
    (b) For medium T we have $T^3 > T$, so the lattice term dominates.
    (c) For high T ($kT > \varepsilon_D$), the lattice term becomes constant but then the electron term has a slow linear increase in T.
    (d) For very high T the conduction electrons are no longer degenerate, and the classical value of the metal's heat capacity is reached.

Chapter 23
The electrical properties of materials

In this chapter, we use quantum statistics to help us understand the distinctive electrical properties of conductors, semiconductors, and insulators. We have previously learned that although the occupation number has the same form for all systems, the spectrum of accessible states varies from one to the next. For this reason, we begin this chapter with a brief and simplified overview of band structure.

A Band structure

A.1 The splitting of levels

As atoms are brought close together, the overlapping of their electron clouds allows electrons to move from one atom to another. These interactions with their

Figure 23.1 (Top) A schematic diagram, showing that when as identical atoms are brought together, their outer electrons may be shared. The identical electron states of the isolated atoms split into slightly different states for mutually shared electrons. (Bottom) When *N* atoms come together, each electron state splits into a band of *N* different states. The larger the overlap of atoms, the greater the spreading. So in general the bands are wider when the atoms are closer together and the bands for outer states are wider than those for inner states.

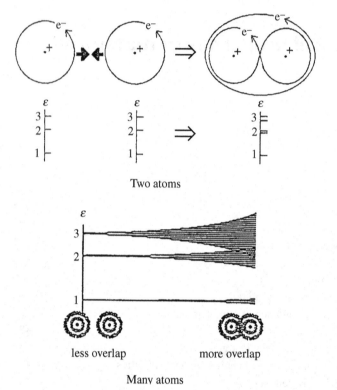

Two atoms

Many atoms

neighbors cause shifts in the allowed electron energies. What were initially states of identical energies in isolated identical atoms turn into "bands" of very closely spaced states for the shared electrons in groups of atoms.

A.2 Band widths and structure

In general, the outer states of higher energy experience greater overlap, which usually results in greater splitting and wider bands (Figure 23.1). Within any band, the density of states usually is largest near the middle and falls off near the edges (Figure 23.2). Electrons preferentially fill the lowest energy states, so at low temperatures the lower bands are full and the higher bands are empty.

The highest completely filled band in the $T \to 0$ limit is called the "valence band." Because it is full, there are no empty states into which these valence electrons can move. (Although pairs could trade places, they are identical particles, so it is the same as staying put.) So we can think of these valence electrons as confined to their parent atoms. They cannot contribute to the electrical or thermal properties of the material.

The next higher band is called the "conduction band." It is either empty or partially filled, so it contains a myriad of vacant states through which electrons

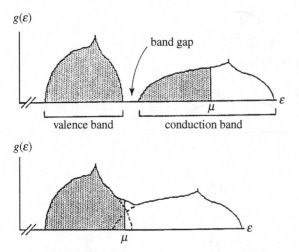

Figure 23.2 Typical shapes of the density of states as a function of energy. The lower-energy bands tend to be narrower and therefore more compressed. A good conductor must have an unfilled outer band, because the electrons must have empty states into which they can move. (top) This would happen if the atoms had unpaired outer electrons, thus giving the outer band fewer electrons than states. (bottom) It could also happen if the last filled band overlaps with the next higher empty band.

can move in response to external stimuli such as electrical fields. Conductors have large numbers of electrons in this band, whereas semiconductors and insulators have few or none.

B   Conductors

B.1   Unfilled bands

Good conductors must have both large numbers of mobile electrons in the conduction band and large numbers of vacant states into which they can move. There are two common ways in which these requirements can be met.

First, the original atoms may have unpaired outer electrons or partially filled levels, thereby contributing fewer electrons to the band than there are quantum states (Figure 23.2(top)). For example, silver has 47 electrons and there is one unpaired electron in its outermost state ($5s^1$). When N silver atoms are together in a metal these outermost states split up into a band that can accommodate $2N$ electrons, N with spin up, and N with spin down. That is twice as many states as there are electrons.

Second, there may be overlapping bands (Figure 23.2(bottom)). Even if the outer electrons could fill one band completely, overlap with the next higher empty band would guarantee that there will be plenty of vacant neighboring states for some of these electrons. Overlapping bands are especially prominent in transition

Figure 23.3 (a) In light divalent metals, the last filled band barely overlaps with the next higher band. So the Fermi level is at a minimum in the density of states. Fewer available states mean a lower electrical conductivitiy. (b) In transition metals the s-band is tall, narrow, and overlaps with the lower and broader d-band. If the s-band is only half full, the density of states at the Fermi level is very large, so these metals are excellent conductors. (c) When the Fermi level lies somewhere in the d-band, the density of states at the Fermi level is smaller, so the conductivities are correspondingly lower.

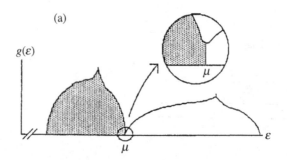

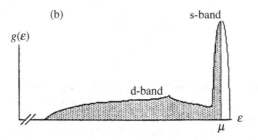

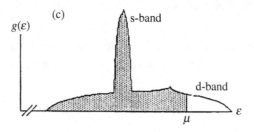

and heavy metals, where the outer electron states are so closely spaced that even modest band spreading guarantees overlap. Overlap is also important for divalent metals, such as magnesium and calcium, as otherwise the two outer electrons would completely fill the outer band.

B.2 The electrons in the Fermi tail

Conduction electrons in metals are often modeled as a nearly degenerate fermion gas (subsection 20D.3, section 22C). The conduction band is essentially filled to the Fermi level and empty above that. Electrons deep within the Fermi sea are surrounded by filled states. Only those in the "tail" region of the Fermi distribution have access to nearby empty states (Figure 22.7), which allows them to respond to external stimuli. Therefore, the density of states near the Fermi level is crucial in the study of electrical conduction by metals.

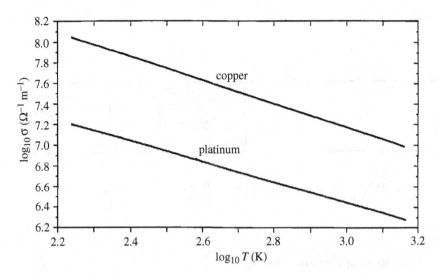

Figure 23.4 Log–log plot of electrical conductivity vs. temperature for platinum and copper, with temperatures ranging from 170 K to 1300 K. For both the slope is close to -1, so the conductivity decreases with temperature roughly as $\sigma \approx 1/T$.

In some light divalent metals, the last filled band barely overlaps with the next higher band. Therefore the Fermi level falls near a local minimum in the density of states (Figure 23.3a). This restricts the number of electrons and vacant states in the tail and gives these metals relatively lower electrical conductivities.

In the transition metals, the s-bands overlap with the d-bands[1]. The s-bands tend to be narrow and dense, whereas the d-bands are broader and have lower state densities (Figures 23.3b, c). Consequently, for those metals with half-filled s-bands the Fermi level lies where the density of states is very large (Figure 23.3b). The electrons in the Fermi tail have access to large numbers of states, which makes these metals excellent conductors. Examples are copper (half-filled 4s band), silver (half-filled 5s band), and gold (half-filled 6s band). Metals neighboring copper, silver, and gold on the periodic table have somewhat lower electrical conductivities, however, often because the Fermi level lies in the d-band, where the density of states is smaller (Figure 23.3c).

B.3 Temperature dependence of conductivity

Because the width of the Fermi tail is proportional to the temperature, higher temperatures mean more "free" electrons in this tail region. But the lattice vibrations also increase with temperature, and they impede the electron flow. The rate of collision with lattice vibrations increases faster than does the width of the Fermi tail, so that electrical conductivity in metals actually decreases with increased temperature (Figure 23.4).

---

[1] The letters identify the orbital angular momentum in the isolated atoms: s means $l = 0$ and d means $l = 2$.

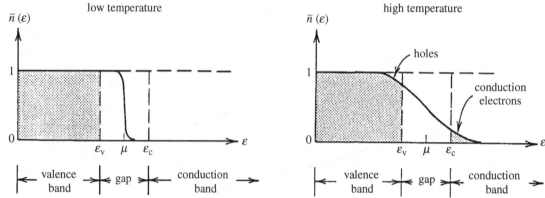

Figure 23.5 Plot of occupation number vs. energy for semiconductors or insulators. (Left) At low temperatures the valence band is completely full and the conduction band completely empty. (Right) At higher temperatures, the Fermi tail may reach across the band gap. These excitations put some electrons in the conduction band and leave some holes in the valence band. (We have exaggerated the width of the Fermi tail. At room temperature the width is about 0.025 eV, a very small fraction of the band gap.)

Summary of Sections A and B

As atoms are brought together and their electron clouds overlap, the electron states become bands of closely spaced states. Higher-lying states generally experience greater overlap, greater splitting, and wider bands. The last completely filled band in the $T \to 0$ limit is called the valence band, and the next one above it is called the conduction band.

For a material to be a good conductor, the conduction band must contain both large numbers of electrons and large numbers of vacant states into which they can move in response to external fields. This happens if the atoms have unpaired outer electrons or partially filled bands or if outer bands overlap. Only electrons in the Fermi tail have neighboring empty states, so the conductivity depends on the width of the tail and the density of states near the Fermi surface.

As the temperature rises the width of the tail in the Fermi distribution broadens, providing increased numbers of mobile electrons. However, collisions with lattice vibrations increases even faster, so that the electrical conduction in metals normally decreases with increasing temperature.

C   Semiconductors

For semiconductors and insulators, the conduction band is empty in the $T \to 0$ limit (Figure 23.5, Table 23.1). With no mobile charge carriers, the material is an insulator. But as the temperature increases, excitation of valence electrons across the band gap becomes increasingly likely.

If the material contains no impurities and no crystalline imperfections, these two bands are the only states available to the outer electrons. Such perfectly pure materials are called "intrinsic" semiconductors. If impurities are present, we have a "doped" semiconductor. Impurities provide states with energies intermediate

Table 23.1. *Size of the band gap in various semiconductors at 0 K and 300 K (after C. Kittel, Introduction to Solid State Physics, fourth edition, John Wiley and Sons, 1971)*

Material	$\Delta\varepsilon_{gap}$ (eV)		Material	$\Delta\varepsilon_{gap}$ (eV)	
	0 K	300 K		0 K	300 K
Si	1.17	1.14	PbSe	0.17	0.27
Ge	0.74	0.67	PbTe	0.19	0.30
InSb	0.23	0.18	CdS	2.58	2.42
InAs	0.36	0.35	CdSe	1.84	1.74
InP	1.29	1.35	CdTe	1.61	1.45
GaP	2.32	2.36	ZnO	3.44	3.2
GaAs	1.52	1.43	ZnS	3.91	3.6
GaSb	0.81	0.78	ZnSb	0.56	0.56
SnTe	0.13	0.18			

between the two bands. Those impurities that accept electrons from the valence band are called "acceptors," and those that donate electrons to the conduction band are called "donors." We will examine intrinsic semiconductors first.

C.1 Thermal excitation of the charge carriers

Although a typical material has about 10^{29} electrons per cubic meter (or $10^{23}/cm^3$) in the valence band, only a very small fraction of these would reach the conduction band. But any that do are surrounded by vacant states. In addition, they leave behind vacancies, or "holes," in the valence band (Figure 23.5), which act like positively charged particles (Figure 10.6). Thus, electrical current in intrinsic semiconductors is carried both by electrons in the conduction band and by holes in the valence band, which are created in pairs and are collectively referred to as "charge carriers."

C.2 Electrical conductivity and charge carrier mobilities

In Section 16C (Figure 16.3) we learned that a current density is the product of density times velocity, $\mathbf{J} = \rho\mathbf{v}$. The density of electrical charge is the product of the density of charge carriers times the charge of each: $\rho_e = ne$. Hence, the electric current density is

$$\mathbf{J} = \rho_e\mathbf{v} = ne\mathbf{v}, \tag{23.1}$$

Table 23.2. *Charge carrier mobilities at room temperature for various semiconductors*

Mobility (m²/(V s))			Mobility (m²/(V s))		
Material	Electrons	Holes	Material	Electrons	Holes
diamond	0.18	0.16	CdS	0.034	0.002
Si	0.135	0.048	CdTe	0.03	0.007
Ge	0.39	0.19	PbS	0.055	0.060
GaAs	0.85	0.04	PbSe	0.102	0.093
GaSb	0.40	0.14	PbTe	0.162	0.075
InAs	3.30	0.046	ZnS	0.012	0.0005
InP	0.46	0.015	ZnSe	0.053	0.002
InSb	8.0	0.075			

where $\mathbf{v}$ is the average drift velocity.[2] The electrical current density is proportional to the applied field, and the constant of proportionality σ is called the "electrical conductivity:"

$$\mathbf{J} = \sigma \mathbf{E}.$$

Equating these two gives the following expression for the electrical conductivity:

$$\sigma = ne\mu, \qquad \text{where} \quad \mu = |\mathbf{v}|/E. \tag{23.2}$$

The quantity μ is called the "mobility".[3] As its name implies, it measures how responsive the charge carriers are to an applied electrical field. Because the charge carriers may include both electrons in the conduction band and holes in the valence band, we can write

$$\sigma = e(n_e \mu_e + n_h \mu_h). \tag{23.3}$$

The charge carrier mobility depends on the overlap in electron clouds between neighboring atoms. Outer orbits have greater overlap and therefore provide greater ease of movement between atoms. This is one reason why electrons in the conduction band generally have greater mobility than do holes in the valence band (Table 23.2).

---

[2] Don't confuse the density of charge carriers, n, with occupation number, $\bar{n}$. In this chapter we use the following symbols for densities (number/m³) of charge carriers, atoms, and quantum states: "n" for charge carrier densities (n_e and n_h for conduction electrons and valence holes, respectively) "N" for impurity densities (N_d and N_a for donors and acceptors, respectively) "$\mathcal{N}$" for densities of quantum states ($\mathcal{N}_d$, $\mathcal{N}_a$, $\mathcal{N}_c$ and $\mathcal{N}_v$ for donor states, acceptor states, conduction band edge equivalent states, and valence band edge equivalent states, respectively)

[3] Not to be confused with chemical potential or magnetic moment, even though it has the same symbol.

Electrical conductivity (equation 23.3) clearly depends on the densities n_e and n_h of the charge carriers that are created by excitation of valence electrons across the band gap and into the conduction band. The probability for such excitations increases exponentially with the temperature. This exponential increase in charge carrier density (i.e., in n_h and n_e) produces a corresponding exponential increase in electrical conductivity and dominates over the reduction in mobility due to collisions.

C.3 Band-edge equivalent states

The Fermi level μ lies in the band gap, and the occupation number falls off rapidly in energy. Therefore, only states very near the edge of the band gap have any charge carriers in them (Figure 23.5). This encourages us to replace these states by an appropriate number of "band-edge equivalent states" all having the same energy and occupation number. Here is how we do it.

Because the valence band is nearly full and the conduction band is nearly empty, the distance of the Fermi level from either band is large compared with kT. So we can write the occupation numbers for electron states in the conduction band and for holes (i.e., electron vacancies) in the valence band as follows:

$$\overline{n}_e = \frac{1}{e^{\beta(\varepsilon-\mu)} + 1} \approx e^{-\beta(\varepsilon-\mu)}, \qquad \varepsilon - \mu \gg kT, \tag{23.4a}$$

$$\overline{n}_h = 1 - \overline{n}_e = 1 - \frac{1}{e^{\beta(\varepsilon-\mu)} + 1} \approx e^{\beta(\varepsilon-\mu)}, \qquad \mu - \varepsilon \gg kT. \tag{23.4b}$$

(Note that $\varepsilon - \mu$ is positive for the conduction band and negative for the valence band.)

For the conduction band we write the total density of electrons as the integral of the occupation number times the density of states (equation 19.11):

$$n_e = \int_{\varepsilon_c} e^{-\beta(\varepsilon-\mu)} g(\varepsilon)\mathrm{d}\varepsilon.$$

We need not worry about the integral's upper limit because the exponential falls off quickly and cuts out the higher states. We break the exponential's argument into two parts by adding and subtracting ε_c, which marks the edge of the conduction band, and then we pull outside the integral the part that does not depend on ε. First we write

$$\varepsilon - \mu = (\varepsilon - \varepsilon_c) + (\varepsilon_c - \mu),$$

so that

$$n_e = e^{-\beta(\varepsilon_c-\mu)} \overbrace{\int_{\varepsilon_c} e^{-\beta(\varepsilon-\varepsilon_c)} g(\varepsilon)\mathrm{d}\varepsilon}^{\mathcal{N}_c} \equiv e^{-\beta(\varepsilon_c-\mu)} \mathcal{N}_c, \tag{23.5}$$

Upon identifying $\mathcal{N}_c$ we have accomplished our objective. We have replaced a summation over a large number of states with varying energies and occupation numbers by an appropriate density $\mathcal{N}_c$ of band-edge equivalent states, all with the *same* energy ε_c and the *same* occupation number $\bar{n}(\varepsilon_c) \approx e^{-\beta(\varepsilon_c - \mu)}$. Likewise, we can write the density of holes in the valence band as the product of the occupation number at the top of the valence band times the density of band-edge equivalent states for holes (homework):

$$n_e = e^{-\beta(\varepsilon_c - \mu)} \mathcal{N}_c, \qquad n_h = e^{\beta(\varepsilon_v - \mu)} \mathcal{N}_v. \qquad (23.6)$$

Notice that the charge carrier density, and hence the electrical conductivity, varies exponentially in the temperature and the energy gap, $\varepsilon_{c \text{ or } v} - \mu$. This exceptional sensitivity is what makes semiconductors so important in modern electronics.

C.4 Fermi-gas model

The density $\mathcal{N}_c$ or $\mathcal{N}_v$ of band-edge equivalent states can be determined either experimentally by measuring the density of charge carriers (n_e or n_h) or theoretically by using a model for the density of states that would allow us to evaluate the integral in equation 23.5. For example, we frequently model the electrons as a Fermi gas, for which the density of states can be found from equation 20.1:

$$g(\varepsilon) = \frac{4\pi (2m^*)^{3/2}}{h^3} (\varepsilon - \varepsilon_c)^{1/2} \qquad (23.7)$$

Notice three modifications.

- *Volume V* We have divided out the volume from expression 20.1, so that our final answer for n_e or n_h will be in units of charge carriers per cubic meter.
- *Kinetic energy,* $\varepsilon - \varepsilon_c$ Electrons with zero drift velocity occupy the lowest states in the conduction band. Hence, momenta and therefore kinetic energies are measured relative to the band edge: $\varepsilon_{\text{kinetic}} = \varepsilon - \varepsilon_c$.
- *Effective mass,* m^{*4} The effective mass m^* can be larger or smaller than the electron's actual mass, depending on whether states within the band are denser (more compressed) or less dense (expanded) compared with those of a free gas. It also reflects the inertia of the charge carriers. The two are related, because narrower, denser, bands reflect a smaller overlap of neighboring electron clouds and hence greater difficulty for electrons to travel from one atom to the next. So electrons (or holes) in narrower, denser, bands also have greater inertia.

Using the density of states 23.7 the integral in equation 23.5 is easily evaluated, giving

$$\mathcal{N}_c = \frac{2 (2\pi m^* kT)^{3/2}}{h^3}. \qquad (23.8)$$

We get the same answer for holes in the valence band (homework). Because the value of m^* is usually near that of the actual electron mass and the temperature

---

[4] Beware! This is just one of many different model-dependent definitions of effective mass that you might encounter in various fields.

is usually near 300 K, we can use these numbers in the above equation to get

$$\mathcal{N}_{c\,\text{or}\,v} = (2.51 \times 10^{25}/\text{m}^3)\left(\frac{m^*}{m}\frac{T}{300\,\text{K}}\right)^{3/2} \tag{23.9}$$

This is the Fermi-gas-model result for replacing the actual density of states with the band-edge equivalent density of states. Although there are small variations due to effective masses and temperatures (as indicated), you can see that this density of states is typically given by

$$\mathcal{N}_c \approx \mathcal{N}_v \approx 2.5 \times 10^{25}/\text{m}^3. \tag{23.9'}$$

You might wonder how this could give the correct carrier density, as it is considerably smaller than the actual density of states in a band, which is typically around $10^{29}/\text{m}^3$:

$$\frac{\text{band-edge equivalent density of states}}{\text{actual density of states}} \approx 2.5 \times 10^{-4}. \tag{23.10}$$

The reason that the model works is that occupation numbers fall off quickly as you go into higher energies within a band. Therefore, smaller numbers of states at the band edge give the same number of charge carriers as larger numbers of states further into the band.

C.5 Law of mass action

One very interesting consequence of result 23.6 is revealed by multiplying the densities of the two types of charge carriers (n_e and n_h) together. We obtain the "Law of mass action,"

$$n_e n_h = \mathcal{N}_c \mathcal{N}_v e^{-\beta(\varepsilon_c - \varepsilon_v)}. \tag{23.11}$$

Notice that for any given band gap $\varepsilon_c - \varepsilon_v$ and temperature T, this product is a constant. If we increase the conduction electrons (e.g., by adding donor impurities), there must be a corresponding decrease in valence holes. And if we increase the valence holes (e.g., by adding acceptor impurities), there must be a corresponding decrease in conduction electrons. (More foxes mean fewer rabbits.) The product of the two remains unchanged.

In intrinsic semiconductors, every electron in the conduction band comes from the valence band, so the number of conduction electrons is equal to the number of valence holes. Consequently, equation 23.11 is sometimes written as

$$n_e n_h = n_{\text{intrinsic}}^2. \tag{23.12}$$

If we combine this with equations 23.11 and 23.9' and take the square root, we get

$$n_{\text{intrinsic}} \approx (2.5 \times 10^{25}/\,\text{m}^3)e^{-\beta(\varepsilon_c - \varepsilon_v)/2}. \tag{23.13}$$

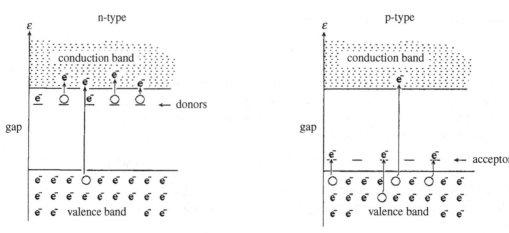

Figure 23.6 Doped
semiconductors. (Left)
n-type: although there are
far fewer electrons in
donor levels than in the
valence band, the smaller
excitation energy makes
the donors dominate.
(Right) p-type: although
there are far fewer
vacancies in acceptor
levels than in the
conduction band, the
smaller excitation energy
makes the acceptors
dominate. As the
temperature rises,
excitations across the
band gap increase and
eventually dominate
those from impurity
levels. The semiconductor
then becomes effectively
intrinsic.

C.6 Doped semiconductors

A typical donor or acceptor impurity has one more or one less outer electron than
is needed to bind to the neighboring atoms in its lattice position, so readily donates
or accepts an extra electron. For example, it might come from the column in the
periodic table that is either to the right or to the left of the parent semiconductor
material. Each donor state lies typically only a few hundredths of an eV below the
conduction band, reflecting the rather small amount of thermal energy required to
shake a surplus electron loose. Likewise, acceptor states lie only a few hundredths
of an eV above the valence band, reflecting the relative ease of exciting a valence
electron into one of these levels. Each electron that moves into a vacant acceptor
level leaves behind a vacancy at its former residence. So we think of acceptors as
exchanging holes with the valence band in the same way that donors exchange
electrons with the conduction band.

Because probabilities fall exponentially in energy, excitation of electrons to or
from these intermediate impurity levels is much more probable than excitation all
the way across the band gap (Figure 23.6). Consequently, a small concentration of
impurities may have a large effect. A material is "n-type" or "p-type" depending
on whether the electrical properties are dominated by negatively charged electrons
from donors or positively charged holes from acceptors.

Normally, each impurity atom can give or take one electron, but that electron
could be in either of two states, spin up or spin down. Therefore, the density
of impurity states ($\mathcal{N}_d$ or $\mathcal{N}_a$) is *twice* the density of the corresponding impurity
atoms (N_d or N_a) and also twice the density of the electrons they may donate or
accept:

$$\text{donor atom density} = N_d = \mathcal{N}_d/2,$$
$$\text{acceptor atom density} = N_a = \mathcal{N}_a/2. \tag{23.14}$$

The same reasoning that applies to these impurity states also applies to the conduction and valence bands. Any electron that jumps out of the valence band or

into the conduction band can be in either of two states: spin up or spin down. But our use of band-edge equivalent states means that $\mathcal{N}_c$ and $\mathcal{N}_v$ are reduced from the true density of states by a factor of about 2.5×10^{-4} (equation 23.10). So the conversion from the density of band edge equivalent states to the density of atoms in the host material would be

$$\mathcal{N}_c/2 = \mathcal{N}_v/2 \approx 2.5 \times 10^{-4} N_{\text{atoms}}. \tag{23.14$'$}$$

C.7 The Fermi level

For intrinsic semiconductors, each electron that reaches the conduction band leaves a hole in the valence band. This means that the Fermi level ($\approx \mu$) lies very near the center of the band gap (Figure 23.5), as can be seen by equating n_e and n_h in equations 23.6 and solving for μ:

$$\mu = \frac{\varepsilon_c + \varepsilon_v}{2} + \frac{kT}{2} \ln\left(\frac{\mathcal{N}_v}{\mathcal{N}_c}\right) \quad \text{(intrinsic)} \tag{23.15}$$

The term on the right is very small, typically less than 0.01 eV (homework).

In doped materials, however, impurities cause the Fermi level to move. It lies closer to the conduction band in n-type materials because there are more conduction electrons than valence holes ($n_e > n_h$), and closer to the valence band in p-type materials because there are more valence holes than conduction electrons ($n_h > n_e$). We can calculate its position if we know the dopant concentration, as we now illustrate for *n*-type materials (Figure 23.7). (The analysis for *p*-type materials would be the same, except that we consider the exchange of holes between acceptors and the valence band rather than electrons between donors and the conduction band.) The results are summarized in Table 23.3.

At absolute zero, there are no excitations at all. All $\mathcal{N}_d/2$ donor electrons[5] remain on the donor atoms. Since *half* the $\mathcal{N}_d$ donor states are occupied[5], $\bar{n}(\varepsilon_d) = 1/2$ and so the Fermi level lies precisely on the donor level:

$$\mu_{T=0} = \varepsilon_d \quad \text{(n-type).} \tag{23.16}$$

As the temperature rises, however, excitations become possible. Although the probability of excitation from a donor level to any one state in the conduction band may be small, the conduction band has a huge number of such states into which it may go. The sum of many small probabilities is large, so at normal temperatures, nearly all the $\mathcal{N}_d/2$ donor electrons are in the conduction band. Equating this number to that given by equation 23.5, we have

$$N_d = \frac{\mathcal{N}_d}{2} \approx n_e = \mathcal{N}_c e^{-\beta(\varepsilon_c - \mu)}.$$

Solving for μ gives

---

[5] Remember, the donor electron could be either spin up or spin down, so there are twice as many states as electrons. When the energy of the state is equal to μ, the occupation number is $\bar{n} = 1/(e^0 + 1) = 1/2$.

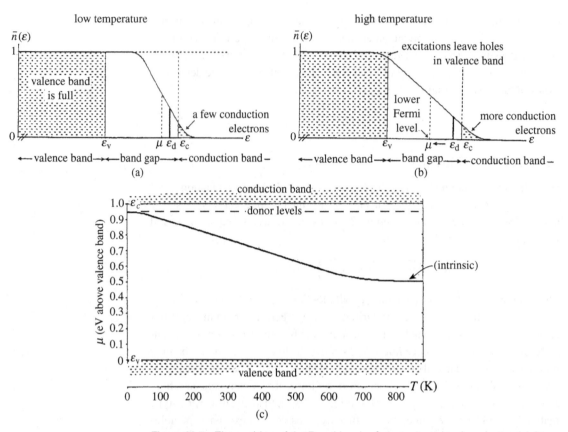

Figure 23.7 The position of the Fermi level μ for an n-type semiconductor. (a) At normal temperatures it lies below the donor levels but above the middle of the band gap. All conduction electrons come from donor levels, leaving most donor levels empty. (b) As temperature increases, the Fermi tail broadens and the Fermi level moves downward toward the middle of the band gap. Excitations across the gap increase, eventually dominating as the material becomes intrinsic. (c) Plot of the position of the Fermi level as a function of the temperature for a typical n-type semiconductor with donor levels 0.04 eV below the conduction band, a band gap of 1.0 eV, and with $N_d = \mathcal{N}_d/2 = 2.5 \times 10^{21}/m^3$ (roughly 0.025 ppm).

$$\mu \approx \varepsilon_c - kT \ln\left(\frac{2\mathcal{N}_c}{\mathcal{N}_d}\right) \qquad (\textit{n}\text{-type}). \qquad (23.17)$$

This shows that the Fermi level moves downward as the temperature increases (Figure 23.7). Both the broadening of the tail and the movement of μ downward increase the number of holes in the valence band. We should expect this, of course, because as temperature increases, excitations across the band gap become increasingly likely.

Table 23.3. *Position of the Fermi level ($\approx \mu$) as a function of the temperature and densities of states for intrinsic, n-type, and p-type semiconductors (ε_c, conduction band edge; ε_v, valence band edge; ε_d, donor level; ε_a, acceptor level; $\mathcal{N}_c$, density of conduction band-edge equivalent states; $\mathcal{N}_v$, density of valence band-edge equivalent states; $\mathcal{N}_d$, density of donor states; $\mathcal{N}_a$, density of acceptor states)*

intrinsic	μ(all temperatures) $= \dfrac{\varepsilon_c + \varepsilon_v}{2} + kT \ln\left(\dfrac{\mathcal{N}_v}{\mathcal{N}_c}\right) \approx \dfrac{\varepsilon_c + \varepsilon_v}{2}$ (midgap)		

		$\mu(T = 0)$	μ(normal temperatures)	temperature of transition to intrinsic
doped:	n-type	ε_d	$\varepsilon_c - kT \ln\left(\dfrac{2\mathcal{N}_c}{\mathcal{N}_d}\right)$	$T \approx \dfrac{\varepsilon_c - \varepsilon_v}{2k \ln\left(2\mathcal{N}_v/\mathcal{N}_d\right)}$
	p-type	ε_a	$\varepsilon_v + kT \ln\left(\dfrac{2\mathcal{N}_v}{\mathcal{N}_a}\right)$	$T \approx \dfrac{\varepsilon_c - \varepsilon_v}{2k \ln\left(2\mathcal{N}_c/\mathcal{N}_a\right)}$

Conversion of dopant concentrations from density of states to density of atoms:
$$\frac{N_d}{N_{\text{atoms}}} \approx 2.5 \times 10^{-4} \frac{\mathcal{N}_d}{\mathcal{N}_c}, \qquad \frac{N_a}{N_{\text{atoms}}} \approx 2.5 \times 10^{-4} \frac{\mathcal{N}_a}{\mathcal{N}_v}$$

C.8   Transition to intrinsic behavior

Because excitations across the band gap require much more energy than excitations to or from impurity levels, a band gap excitation is much less probable. However, there are far more electrons in the valence band than in donor levels (Figure 23.6). So although the probability for any particular band gap excitation is smaller, there are many more possibilities, and the sum of many small probabilities may be large. As the temperature rises, the probability for excitations from the valence to the conduction band increases exponentially and eventually dominates over excitations from impurity levels. The material becomes intrinsic.

The temperature for this transition from doped to intrinsic behavior can be estimated by finding the temperature at which the number of excitations from the impurity levels equals that from excitations across the band gap (Figure 23.8). Using the law of mass action 23.11 for the intrinsic excitations, we can write this criterion as

$$(n_{\text{donor excitations}})^2 = (n_{\text{intrinsic excitations}})^2 \quad \Rightarrow \quad \left(\frac{\mathcal{N}_d}{2}\right)^2 = \mathcal{N}_v \mathcal{N}_c e^{-\beta(\varepsilon_c - \varepsilon_v)}.$$

Solving for T, assuming that $\mathcal{N}_v \approx \mathcal{N}_c$, gives (homework):

$$T_{\text{intrinsic}} \approx \frac{\varepsilon_c - \varepsilon_v}{k \ln\left(4 N_v \mathcal{N}_c / \mathcal{N}_d^2\right)} \longrightarrow \frac{\varepsilon_c - \varepsilon_v}{2k \ln\left(2\mathcal{N}_c / \mathcal{N}_d\right)} \quad (n\text{-type}). \qquad (23.18)$$

This result confirms that the transition temperature is higher if the valence electrons have a larger gap to overcome or if they have a larger number of donor states with which to compete.

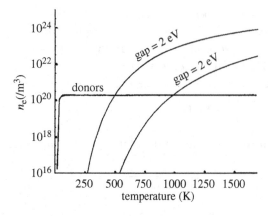
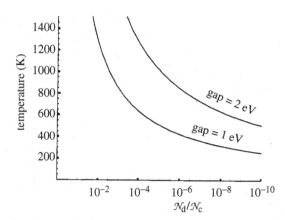

Figure 23.8 (a) The density of conduction electrons as a function of temperature for an *n*-type semiconductor, indicating those electrons coming from donor levels (using $N_d = N_a/2 = 2.5 \times 10^{20}/m^3$) and those from excitation across the band gap, for widths of 1 eV and 2 eV. (b) The temperature for transition to intrinsic vs. donor level concentration N_d/N_c for band gaps of 1 eV and 2 eV. (N_d/N_c is very roughly 4000 times the dopant concentration by number of atoms.)

Example 22.1 What is the minimum donor concentration needed to keep a semiconductor *n*-type up to 600 K, if the band gap is 1 eV?

Putting $\varepsilon_c - \varepsilon_v = 1.00$ eV and $T = 600$ K into equation 23.18 yields

$$600\,\text{K} = \frac{1.00\,\text{eV}}{2k \ln\left(2N_c/N_d\right)}.$$

Solving for $N_d/2$ gives

$$\frac{N_d}{2} = 6.3 \times 10^{-5} N_c,$$

where (equation 23.9)

$$N_c \approx (2.5 \times 10^{25}/\,\text{m}^3)\left(\frac{600\,\text{K}}{300\,\text{K}}\right)^{3/2} = 7.1 \times 10^{25}/\text{m}^3.$$

So the impurity concentration is around $4.5 \times 10^{21}/\text{m}^3$. (In terms of relative number of atoms this is around 0.05 ppm.)

Summary of Sections C and D

For semiconductors and insulators, the filled valence band is separated from the empty conduction band by a band gap. At low temperatures, there is insufficient thermal energy for electrons to jump this gap, but as the temperature increases, these transitions become increasingly likely. Therefore, whether these materials are semiconductors or insulators depends on the size of the gap and the temperature.

When an electron leaves the valence band it leaves behind a positively charged hole. Both electrons in the conduction band and holes in the valence band are referred to as charge carriers. The mobility μ of the charge carriers measures how well they move in response to external fields. The outer, higher-energy, orbits generally have greater overlap, providing the electrons with greater ease of movement. Collisions with lattice vibrations increase with increasing temperature. The electrical conductivity of a material depends both on the density of charge carriers and their mobilities (equations 23.2, 23.3):

$$\sigma = e(n_e\mu_e + n_h\mu_h), \qquad \text{where} \quad \mu = |\mathbf{v}|/E.$$

We can replace the states in a band with a reduced number of band edge equivalent states, all having the same energy and the same occupation number. For example, we can write the density of electrons in the conduction band as (equation 23.5)

$$n_e = \int_{\varepsilon_c} e^{-\beta(\varepsilon-\mu)} g(\varepsilon)d\varepsilon \; = e^{-\beta(\varepsilon_c-\mu)} \int_{\varepsilon_c} e^{-\beta(\varepsilon-\varepsilon_c)} g(\varepsilon)d\varepsilon$$
$$= e^{-\beta(\varepsilon_c-\mu)} \mathcal{N}_c.$$

The same procedure for holes in the valence band gives a corresponding result, so for electrons and holes we have (equation 23.6)

$$n_e = e^{-\beta(\varepsilon_c-\mu)} \mathcal{N}_c \qquad n_h = e^{\beta(\varepsilon_v-\mu)} \mathcal{N}_v$$

The density $\mathcal{N}_c$ or $\mathcal{N}_v$ of band-edge equivalent states can be determined either experimentally by measuring the density of charge carriers (n_e or n_h), or theoretically by using a model for the density of states. Using the Fermi gas model gives (equation 23.9)

$$\mathcal{N}_{c\,or\,v} = (2.51 \times 10^{25}/\,\mathrm{m}^3)\left(\frac{m^*}{m}\frac{T}{300\,\mathrm{K}}\right)^{3/2},$$

where the effective mass m^* reflects the density of states within a band and also the effective inertia of the charge carriers.

If we multiply the density of electrons in the conduction band by the density of holes in the valence band we get the law of mass action (equations 23.11, 23.12):

$$n_e n_h = \mathcal{N}_c\mathcal{N}_v e^{-\beta(\varepsilon_c-\varepsilon_v)} \quad \text{or} \quad n_e n_h = n_{\mathrm{intrinsic}}^2.$$

At any given temperature, the product of the two types of charge carrier is constant.

Donor states lie close to the conduction band, reflecting the small energy required to excite the loosely held electrons into the conduction band. Acceptor states lie close to the valence band, reflecting the small energy required to excite valence electrons into these states. Each impurity atom can donate or accept electrons in either spin-up or spin-down states. Therefore, there are twice as many impurity states as there are impurity atoms (equation 23.14):

$$\text{density of donor atoms} = N_d = \mathcal{N}_d/2,$$
$$\text{density of acceptor atoms} = Na = \mathcal{N}_a/2.$$

Because of the smaller energies involved, there is a much higher probability for any given electron to be excited to or from these intermediate states than to jump the entire band gap. However, there are far more electrons in the valence band than in the donor states, and far more empty states in the conduction band than in acceptor states. So as the temperature increases and transitions across the band gap become increasingly probable, the band gap transitions eventually dominate. The material becomes intrinsic.

Under normal conditions in n-type materials, most donor atoms have given up their electron to the conduction band, and the Fermi level lies somewhat below them. As the temperature rises, transitions across the band gap become increasingly probable. The tail in the Fermi distribution broadens and the Fermi level moves downward towards the center of the band gap. The corresponding pattern is followed in p-type materials. Formulas for the location of the Fermi level in both intrinsic and doped materials and the temperature at which a doped semiconductor becomes intrinsic are listed in Table 23.3.

D p–n junctions

D.1 Diffusion across the junction

We learned in Chapters 9 and 14 that particles diffuse towards regions of lower chemical potential and that the chemical potential depends on (a) the potential energy and (b) the particle density.

An increase in one of these can be offset by a sufficient decrease in the other. When diffusive equilibrium is reached the chemical potentials in the regions are equal. The region of higher potential energy has the lower particle density.

Doped semiconductors that are p-type have few if any electrons in the conduction band, and n-type materials have few if any holes in the valence band. Therefore, when they are brought together to form a "p–n junction," conduction electrons from the n-side diffuse across to the p-side, where conduction electrons are scarce. Likewise, valence holes from the p-side diffuse to the n-side where holes are scarce (Figure 23.9a).

The diffusion of charged particles between the two sides creates a buildup of negative charge on the p-side of the junction and positive charge on the n-side (Figure 23.9b). This causes a corresponding shift in the potential energies of the charge carriers on the two sides. When this potential barrier becomes large enough, it will oppose further diffusion. That is, when the increase in potential energy offsets the decrease in particle concentrations, the chemical potentials in the two regions are equal. At this point, diffusive equilibrium is reached, and there is no more net diffusion.

In equilibrium, then, there is a skewed charge distribution across a p-n junction, causing the energy of conduction electrons to increase as they enter the p-side, and the energy of the valence holes to increase as they enter the n-side. In both cases the increase in potential energy is offset by a decrease in particle concentration.

D.2 Drift and diffusion currents

We now examine the behavior of the electrons and holes near a p-n junction. The electrons diffuse in the opposite direction to the holes and carry the opposite electrical charge. Therefore, the directions of the resulting electrical currents are

(a)

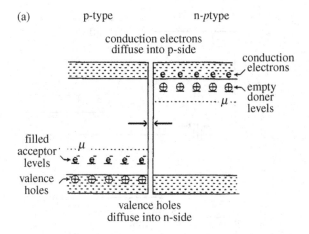

Figure 23.9 (a) When p-type and n-type materials are first brought together, charge carriers of each type diffuse from where there are many of that type to where there are few: conduction electrons diffuse towards the p-side and valence holes (encircled plus signs) towards the n-side. (b) The resulting separation of electrical charge creates a shift in potential energies across the junction. The diffusion continues until this shift is sufficient for the chemical potential (colloquially, the Fermi level) on both sides to be the same. (c) The shift in potential energies due to this charge separation is similar to that for electrons across a charged capacitor. Electron energies increase upwards, so hole energies increase downwards.

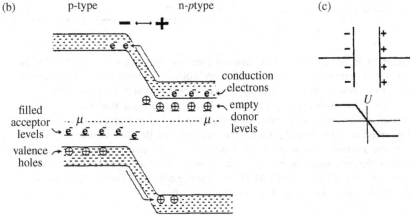

the same. The diffusion of electrons toward the p-side and holes toward the n-side is referred to as the "diffusion current." It is countered by an opposing current, called the "drift current," which is due to the charge carriers flowing "downhill" towards the region of lower electrical potential energy.[6] In diffusive equilibrium, the drift and diffusion current densities are equal and opposite and so the net flow is zero (Figure 23.10a):

$$j_{net} = j_{diffusion} - j_{drift} = 0 \quad \text{(in equilibrium).}$$

All electrons approaching the junction from the p-side continue on, because they are going "downhill." The same is true for the holes approaching from the n-side. The tendency to flow downhill is independent of the height of the hill so the resulting drift current is constant (at any given temperature):

$$j_{drift} = j_0 = \text{constant.}$$

---

[6] Beware! Holes have the opposite electrical charge to electrons, so on electron energy diagrams the potential energy of holes increases in the *downward* direction.

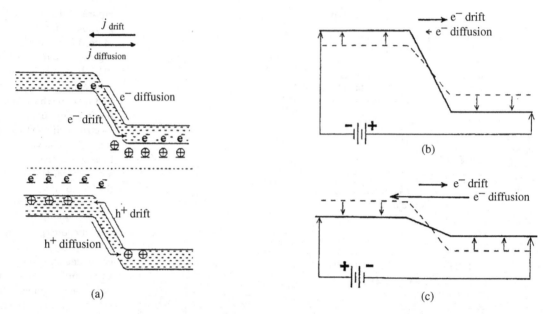

Figure 23.10 The movement of electrons (or holes) across a p–n junction can be separated into two components. The "drift" current is constant and results simply from the tendency of the charge carriers to flow downhill to lower potential energies. The "diffusion current" results from the opposing tendency to flow to regions of lower concentration, albeit higher potential energy, and varies exponentially with the height of the potential barrier. (b) With a "reverse bias" the potential barrier is increased and the diffusion current goes exponentially to zero. (c) With a "forward bias" the barrier is decreased and the diffusion current increases exponentially. (Caution: Because of the negative charge of electrons, the current direction is opposite to the direction of electron flow.)

The situation is quite different for the diffusion current. There is a huge number of conduction electrons on the n-side of the junction and a huge number of holes on the p-side. But as they approach the junction they have a potential energy barrier to overcome. Because the occupation number decreases exponentially with energy, the fraction of electrons or holes that can overcome the barrier decreases exponentially with its height,

$$j_{\text{diffusion}} \propto e^{-\beta \varepsilon_{\text{barrier}}}.$$

We now combine these expressions for the drift and diffusion currents and impose the constraint that when in diffusive equilibrium the two currents are equal and opposite. The result for the net current across the junction is

$$j_{\text{net}} = j_{\text{diffusion}} - j_{\text{drift}} = j_0 \left(e^{-\beta \Delta \varepsilon} - 1 \right), \tag{23.19}$$

where $\Delta \varepsilon$ represents any change in the height of the barrier from its equilibrium value.

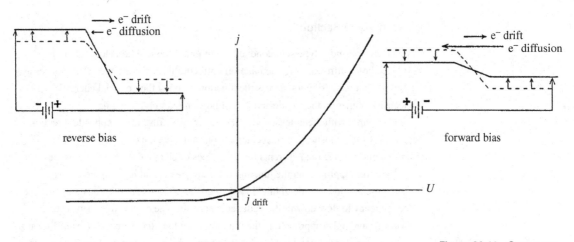

Figure 23.11 Current vs. voltage across a p–n junction, showing the exponential variation in the diffusion current.

D.3 Applied voltage

The important feature of the result 23.19 is that the diffusion current changes exponentially with changes in the height of the energy barrier, whereas the drift current is constant (Figure 23.11). Because kT is only a few hundredths of an electron volt, a very small change in the height of the barrier causes a huge change in the diffusion current.

In a "forward bias," we connect the negative terminal to the n-side and the positive terminal to the p-side. The reversed situation is called a "reverse bias." With a forward bias of U volts, the height of the barrier is reduced by eU for both the electrons and holes:

$$\Delta\varepsilon = -eU.$$

Putting this barrier height into equation 23.19, we have

$$j_{\text{net}} = j_0\,(e^{\beta eU} - 1). \tag{23.20}$$

Notice that the diffusion current increases exponentially in the applied voltage and dominates the drift current going the other direction.

$$j_{\text{net}} \approx j_0 e^{\beta eU} \qquad \text{(forward bias, } eU \gg kT\text{).} \tag{23.21}$$

For a reverse bias (negative U), the diffusion current goes equally quickly to zero, and the net current is dominated by the relatively small drift current in the reverse direction:

$$j_{\text{net}} \approx -j_0 \qquad \text{(reverse bias).} \tag{23.22}$$

This behavior is illustrated in Figure 23.11. This very important property of the p-n junction makes it useful in a wide variety of electronic devices.

Summary of Section D

When p-type and n-type semiconductors are joined, conduction electrons and valence holes diffuse across the junction until equilibrium is attained. This causes a skewed charge distribution across the junction, making the potential energy of electrons higher on the p-side and that of holes higher on the n-side.

Compared with the holes, conduction electrons diffuse in the opposite direction and carry the opposite electrical charge. Consequently, the directions of the corresponding electrical currents for both types of charge carriers are the same.

Electrons approaching the junction from the p-side and holes approaching from the n-side are flowing "downhill" in a drift current. This current is constant, because the tendency to flow downhill is independent of the size of the "hill." However, electrons and holes approaching the junction from the other direction (the diffusion current) encounter a potential barrier. Their probability of surmounting this barrier depends exponentially on the height of the barrier. When in equilibrium, these two currents are equal and opposite, so there is no net current across the junction. But we can change the height of the barrier and drastically alter the diffusion current by applying a voltage across the junction. If we apply a bias of U volts, then the net current across the junction is given by (equation 23.20)

$$j_{\text{net}} = j_{\text{diffusion}} - j_{\text{drift}} = j_0 \left(e^{\beta e U} - 1 \right),$$

where j_0 is the drift current.

With a forward bias, the diffusion current increases exponentially in the applied voltage and dominates the drift current. And with a reverse bias the diffusion current goes to zero, which leaves only the relatively small drift current in the reverse direction (equations 23.21, 23.22):

$$j_{\text{net}} \approx j_0 e^{\beta e U} \qquad \text{(forward bias, } eU \gg kT\text{)},$$
$$j_{\text{net}} \approx -j_0 \qquad \text{(reverse bias)}.$$

Problems

Sections A and B

1. (a) Why can outer electrons migrate between neighboring atoms more easily than the inner electrons?
   (b) Why are the outer bands generally wider than the inner bands?

2. Suppose that we model the temperature dependence of electrical conductivity in metals by $\sigma = CT^{\alpha}$, where C and α are constants. From Figure 23.4, estimate the values of C and α for platinum and copper.

3. Suppose that Figure 23.2a is a sketch of the band structure for aluminum. Sketch the corresponding figure for silicon. Aluminum has 13 electrons, with one in the outermost level, and silicon has 14 electrons.

4. Comparing Figure 23.3a and 23.3b, why do you suppose transition metals such as copper are generally better electrical conductors than divalent metals, such as calcium?

Section C

(Assume $m = m^*$ unless otherwise indicated. But don't forget that N_c and N_v depend on T.)

5. The density of conduction electrons in most metals is about $10^{29}/m^3$.
   (a) With this, equation 23.2, and the conductivities from Figure 23.4, estimate the average mobility of the conduction electrons in copper and platinum at room temperature.
   (b) Comparing these with those for semiconductors given in Table 23.2, why do you suppose the average electron mobilities in conductors are so much lower?

6. Show that: (a) equations 23.4a, b are true for $|\varepsilon - \mu| \gg kT$; (b) equation 23.7 follows from 20.1. (c) Starting with equations 23.5 and 23.7 derive expressions 23.8 and 23.9 for the number density of electrons in the conduction band. Do the same for holes in the valence band.

7. The density of silicon is 2.42×10^3 kg/m³ and that of germanium is 5.46×10^3 kg/m³. Their respective atomic masses are 28.1 and 72.6. Estimate the actual number of states (states/m³) in the conduction band for each. Assume that state degeneracies make $g(\varepsilon)$ four times denser than in equation 23.7. There we assumed two electron states (spin up and spin down) per atom. How do these numbers compare with the number of band-edge equivalent states? Take $T \approx 300$ K.

8. Suppose that the Fermi level is 0.50 eV below the conduction band in a certain semiconductor. Estimate the density of conduction electrons in the conduction band for temperatures of 50 K, 200 K, and 600 K.

9. (a) Starting with equations 23.6, derive equation 23.15 for the location of μ in an intrinsic material. What conditions are needed?
   (b) Consider an intrinsic semiconductor at 300 K whose valence holes have twice the effective mass of the conduction electrons. How far above the midpoint of the band gap would the Fermi level lie?

10. Consider excitations from level $\varepsilon_i = -4$ eV with 10^{25} states/m³ to level $\varepsilon_j = -2.5$ eV with 10^{24} states/m³ in a semiconductor at 295 K. (Treat the states as band-edge equivalent states if you wish.)
    (a) How many excitations are there per cubic meter?
    (b) What is the energy of the Fermi level?
    (c) Repeat for the same case except that level i now has 10^{26} states/m³ and the temperature is 500 K.

11. Estimate the density of charge carriers for an intrinsic semiconductor with gap and temperature as follows: (a) 1.0 eV, 290 K, (b) 2.0 eV, 290 K, (3) 1.0 eV, 580 K.

12. Consider an intrinsic semiconductor at 290 K with $\Delta\varepsilon_{gap} = 1.1$ eV.
    (a) How far above or below the center of the band gap will the Fermi surface be if $m_v^*/m_c^* = 2.1$?
    (b) Repeat for temperature 700 K.

13. Consider a semiconductor at 295 K with $\Delta\varepsilon_{gap} = 1.1$ eV. It is intrinsic, so that excitations involve the valence and conduction bands only.
    (a) Roughly what is the density of electrons in the conduction band?
    (b) If we add an impurity that increases the density of electrons (the number per cubic meter) in the conduction band tenfold, what will be the density of holes in the valence band?

14. Using room-temperature values for the band gap and carrier mobilities from Tables 23.1 and 23.2, estimate the electrical conductivities of the following intrinsic semiconductors (a) germanium at 290 K, (b) silicon at 290 K, (c) silicon at 670 K.

15. Consider each of the following intrinsic semiconductors at 300 K: Si, Ge, GaAs, InSb.
    (a) From the band gaps of Table 23.1, estimate the density of conduction electrons and valence holes.
    (b) Combine these results with the mobilities in Table 23.2 to estimate their electrical conductivities.

16. For an intrinsic semiconductor with a band gap of 2.0 eV, estimate the temperature at which there would be only one conduction electron per cubic meter of material.

17. The electrical conductivity of an intrinsic semiconductor at 295 K is 1.0×10^{-3} A/(V m). If its charge carrier mobilities are $\mu_e = 0.15$ m^2/(V s) and $\mu_h = 0.09$ m^2/(V s), roughly what is the band gap?

18. Consider an intrinsic material having a band gap of 2.5 eV. At what temperature would the density of conduction electrons be (a) 1/m^3, (b) 10 000/m^3. (c) What percentage change in temperature causes this 10 000-fold increase in electrical conductivity?

19. Consider an intrinsic material with a band gap of 1.2 eV at 300 K. By how many degrees must its temperature be increased if the electrical conductivity is to be doubled?

20. Consider an intrinsic material at 300 K. Under pressure, the band gap is reduced from 1.6 to 1.5 eV. By what factor does the electrical conductivity of this material increase?

21. Suppose you wish an intrinsic material to remain an electrical insulator at 800 K, having less than one conduction electron per cubic *centimeter* of material. What is the minimum band gap that is needed?

22. Consider an *n*-type semiconductor with a band gap of 0.8 eV, and with 10^{21} dopant atoms per m^3, giving 2×10^{21} donor levels per m^3, which are 0.04 eV below the conduction band.
    (a) At absolute zero, is the Fermi level above, below, or at the donor levels?
    (b) At what temperature is the Fermi level 0.08 eV below the donor level?
    (c) At 300 K, how many conduction electrons come from donor levels? From the valence band?
    (d) Above what temperature will this semiconductor become intrinsic?

23. Consider an *n*-type semiconductor at 300 K with a band gap of 1.1 eV, donor levels lying 0.03 eV below the conduction band, and $\mathcal{N}_d = 2.51 \times 10^{21}/\text{m}^3$.
    (a) How far below the conduction band is the Fermi level?
    (b) Roughly what is the density of conduction electrons?
    (c) Roughly what is the density of holes in the valence band?
    (d) Repeat parts (a), (b), and (c) for a temperature of 500 K.
    (e) Roughly what is the temperature above which it becomes intrinsic?

24. Repeat the above problem for $\mathcal{N}_d = 2.51 \times 10^{19}/\text{m}^3$.

25. Consider a *p*-type semiconductor at 300 K with band gap of 0.8 eV, acceptor levels lying 0.04 eV above the valence band, and $\mathcal{N}_a = 2.51 \times 10^{21}/\text{m}^3$.
    (a) How far above the valence band is the Fermi level?
    (b) Roughly what is the density of valence holes?
    (c) Roughly what is the density of electrons in the conduction band?
    (d) Repeat parts (a), (b), and (c) for a temperature of 500 K.
    (e) Roughly what is the temperature above which it becomes intrinsic?

26. Consider an *n*-type material at 300 K with donor levels 0.02 eV below the conduction band. The number of holes in the valence band is $10^{13}/\text{m}^3$, and that of electrons in the conduction band is $10^{21}/\text{m}^3$. What is (a) the band gap, (b) the value of $\mathcal{N}_d$, (c) the distance of μ below the conduction band?

27. Consider a *p*-type semiconductor at 300 K with $n_h = 10^{22}/\text{m}^3$ and $\mathcal{N}_a/\mathcal{N}_v = 10^{-6}$.
    (a) What is the band gap?
    (b) How far above the valence band is the Fermi level?
    (c) At what temperature does the semiconductor turn intrinsic?

28. A certain *p*-type semiconductor has a band gap of 0.8 eV, the acceptor levels being 0.03 eV above the valence band. What is the minimum density of acceptor atoms (in number per cubic meter) that would ensure that the semiconductor remains *p*-type up to a temperature of 400 K?

29. A certain semiconductor has 10^{21} donor atoms per m^3, and the donor levels are 0.03 eV below the conduction band. There are 10^{16} conduction electrons per m^3. What is the temperature?

Section D

30. The band displacement across a certain p-n junction is 0.4 eV. The temperature is 300 K and the density of conduction electrons on the n-side is $10^{22}/m^3$.
    (a) Show that the ratio of the densities of conduction electrons on the p-side and on the n-side is $n_{e,p}/n_{e,n} = e^{-\beta \Delta \varepsilon_c}$, where $\Delta \varepsilon_c$ is the displacement of the band across the junction.
    (b) What is the density of conduction electrons on the p-side?
    (c) How far below the conduction band edge does the Fermi level lie on the n-side? On the p-side?

31. Consider a p-n junction at 290 K with $\Delta \varepsilon_{gap} = 1.1$ eV. The Fermi level lies 0.2 eV from the valence and conduction band edges on the p- and n-sides, respectively.
    (a) What is the band displacement, $\Delta \varepsilon_c$ or $\Delta \varepsilon_v$, across the junction?
    (b) What are the charge carrier densities, n_h and n_e, on the p-side?
    (c) What are the charge carrier densities on the n-side?
    (d) What are the answers to part (b) at a temperature of 320 K?

32. The drift current can be estimated from the density of charge carriers and their average thermal speeds.
    (a) Why is it that only the charge density of electrons on the p-side and that of holes on the n-side need be considered?
    (b) Show that the drift current due to either charge carrier is given by $j_{drift} = en\bar{v}/4$, where n is the density of the appropriate charge carrier and $\bar{v}$ is its average thermal speed. (Hint: At any instant, what fraction of the n charge carriers per m^3 are moving toward the junction and what is the average component of their speed in that particular direction?)
    (c) What is $\bar{v}$ as a function of m^* and T? (See Chapter 16.)

33. In the above problem, we found that the drift current due to either charge carrier can be approximated by $j_{drift} = en\bar{v}/4$, where n is the density of the conduction electrons on the p-side, or of the holes on the n-side. We are now going to apply this to a typical p-n junction in a semiconductor at 300 K with a band gap of 1.1 eV and a band displacement across the junction of 0.6 eV. The Fermi level lies 0.25 eV below the conduction band on the n-side.
    (a) What is the density of conduction electrons on the p-side of the junction?
    (b) What is the density of valence holes on the n-side of the junction?
    (c) What is the average thermal speed of a conduction electron or hole (use $m^* = m$ for both)?

(d) From this information, estimate the drift current density, j_{drift}, across the junction.

34. Consider a p–n junction at 295 K with a drift current density of 6×10^{-4} A/m^2. What is the current density through this junction when we apply
    (a) a reverse bias of 0.01 volts? Of 0.2 volts?
    (b) a forward bias of 0.01 volts? Of 0.2 volts?

Chapter 24
Low temperatures and degenerate systems

In this chapter we return to the study of systems that are nearly degenerate. As illustrated in Figure 24.1, a degenerate system of N identical fermions fills the N lowest quantum states, one particle per state. For degenerate bosons, all are in the one single state of lowest energy.[1]

A degenerate system is confined to a small volume in phase space because of either:

- restricted volume in momentum space owing to small masses, or low temperatures,[2] or
- Restricted volume in coordinate space owing to high densities.

Important examples of each case will be studied in this chapter.

---

[1] A fermionic substructure may restrict the overlap between bosons in position space. The consequent larger volume in position space would then be offset by a correspondingly smaller volume in momentum space.

[2] Remember that the classical equipartition theorem ($p^2/2m = 3/2kT$) predicts a root mean square momentum of $p_{rms} = (3mkT)^{(1/2)}$. So volume in momentum space depends on both mass and temperature.

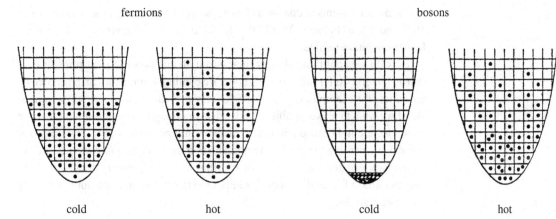

fermions   bosons

cold   hot   cold   hot

Figure 24.1 At high temperatures, particles are surrounded by vacant states, which gives them a great deal of freedom, and so the system's properties vary smoothly. At low temperatures, however, they have no vacant neighboring states. The only possible transitions are upward and often require an energy that is large compared with kT. Changes in physical properties may then be discrete rather than smooth.

The measures of high densities, small masses, and low temperatures are on a relative scale and are interdependent. At earthly densities, many systems become degenerate only if temperatures descend to a few kelvins or lower. Yet these same systems may be degenerate at several million kelvins when squashed together in extremely dense collapsed stars. The conduction electrons in metals have the same density and temperature as the atoms. Yet because of the difference in their masses, the electrons are degenerate and the atoms are not.

At high temperatures and/or low densities, the particles of a system are surrounded by vacant quantum states into which they can move (Figure 24.1). This freedom enables them to give rather smooth and continuous responses to varying environmental conditions. But in degenerate systems, quantum effects are more visible. Imprisoned particles are unable to move into neighboring states, which makes the system unresponsive to external stimuli. Excitations from the ground state may require discrete jumps in energy, and this may produce some very interesting properties.

A large number of important systems are degenerate or nearly degenerate in our normal earthly conditions, where either high densities (such as those of conduction electrons in metals or nucleons in a nucleus) or high excitation energies (such as those of electrons in inner atomic orbits or the vibrational excitations of some molecules) confine them to the lowest-energy arrangements. However, in this chapter we examine important systems that are *not* degenerate in our everyday environment but which become degenerate under more severe conditions. When degenerate, they display properties that are quite different from those with which we are familiar, so they intrigue us.

A   Low temperatures

The properties of systems depend on the ratio ε/kT, where ε is a state's energy and T the temperature. Because this ratio varies by the same factor either way, we might expect systems to display as large a range of interesting properties between 10^{-9} and 1 K as between 1 and 10^{+9} K. We begin this chapter by looking at the low-temperature regime.

How do we cool things below their earthly temperatures? Heat transfer (i.e., thermal interactions) can work for moderate cooling. Lemonade can be cooled by ice cubes, for example. Dry ice (195 K) may provide even cooler temperatures. But if we wish to make an object colder than anything in its environment then we must insulate it, because any heat transfer would be in the wrong direction. For this reason, cooling to very low temperatures must be done *adiabatically* (i.e., in a thermally isolated way), using either diffusive or mechanical interactions.[3] We now examine a few of the ways currently used to produce and measure extremely low temperatures.

A.1 Mechanical cooling through expansion

We are familiar with the adiabatic cooling of a gas through expansion. Rising air expands and cools, causing clouds and rainfall. By opening the valve on a tank of compressed CO_2, the escaping gas may cool to the point where dry ice snow is formed.

In the throttling process of subsection 12E.1, a gas expands as it passes through a constriction or resistive barrier. Cooling in this way depends on attractive interatomic forces and negative potential energies. Larger distances between molecules usually results in increased potential energy – hence decreased thermal energy. The throttling process may be repeated or done in stages (Figure 24.2). For example, throttled air does not cool from room temperature to the point where liquid nitrogen condenses (77 K) unless the process is repeated many times. The cooled air from one cycle is used to pre-cool the incoming air for the next cycle. Liquid nitrogen can then be used as the starting point for the throttling of hydrogen, and then liquid hydrogen for the throttling of helium.

Instead of expanding against itself as in throttling, a gas may also expand against a piston. This process cools the gas more efficiently, because the work is done on the receding piston rather than on the gas itself. That is, the internal energy of the gas is reduced by transfer to a different system (the piston) via the mechanical interaction. Challenges involve materials and machining. The large pressure changes require strong materials. Yet these must also have low heat capacities and thermal conductivities, so that they don't reheat the cooled gases. And the machining must be precise to reduce losses to leakage. Because of the inefficiency of throttling cycles, expansion against a piston is often preferred in spite of the higher equipment cost. Now much of our liquid helium is produced

---

[3] Remember, constant entropy means that the total six-dimensional volume in phase space $(V_r V_p)$ is constant. For quasistatic adiabatic expansion or diffusion, the increased volume in coordinate space produces a corresponding reduction in volume in momentum space (hence, lower temperature).

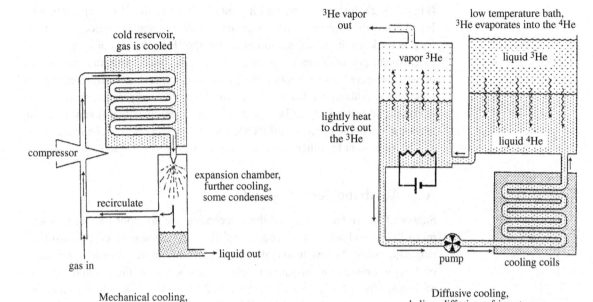

cold reservoir, gas is cooled

compressor

recirculate

gas in

→ liquid out

Mechanical cooling, throttling

expansion chamber, further cooling, some condenses

³He vapor out

low temperature bath, ³He evaporates into the ⁴He

vapor ³He

liquid ³He

lightly heat to drive out the ³He

liquid ⁴He

pump

cooling coils

Diffusive cooling, helium diffusion refrigerator

Figure 24.2 Adiabatic cooling through mechanical and diffusive interactions. (Left) Cooling and liquefying gases via expansion. (Right) The helium dilution refrigerator uses the evaporation of ³He into ⁴He, which cools in the same way that water evaporating from your skin cools you after you step out of a shower. Evaporation into ⁴He presents a smaller potential barrier than evaporation into the vacuum, and so it can continue to much lower temperatures.

this way. With a boiling point of 4.22 K at atmospheric pressure, liquid helium serves as the starting point for research at temperatures of a few kelvins and lower.

A.2 Diffusive cooling and helium dilution

Diffusive cooling is experienced whenever we step out of a shower. As moisture on our skin evaporates, the faster molecules leave and the slower, "cooler," molecules remain. The same happens with other liquids. If we put liquid nitrogen or helium into a partial vacuum, the evaporation of faster-moving molecules makes the remaining liquid cooler.

Unfortunately, there is a limit to the lowest temperature attainable by pumping a vacuum on liquid helium. The weak attraction between the helium atoms, which causes it to condense at very low temperatures, also puts the atoms of liquid helium into a shallow potential well. At temperatures below about 1 K, there is insufficient thermal energy for the helium atoms to surmount this potential barrier and the evaporation stops. More precisely, it falls to such a low rate that evaporative cooling no longer successfully competes with heat entering the system from other sources.

There is some improvement with the ³He isotope. By pumping a vacuum on liquid ³He, we can reach temperatures around 0.3 K before the vapor pressure goes to zero. And if we replace the vacuum by liquid ⁴He, we can go still lower. This process is called "helium dilution refrigeration," and it works as follows

(Figure 24.2). Instead of evaporating into a vacuum, the ^{3}He evaporates into liquid ^{4}He, which provides a considerably smaller potential barrier. In fact, as long as we keep the concentration of ^{3}He in the ^{4}He solution below about 8%, its chemical potential there is lower. The decrease in concentration more than offsets the increase in potential energy, so the evaporative cooling continues indefinitely. Although other processes can lower the temperature further, one important advantage of the helium dilution refrigerator is that it can maintain temperatures in the range of millikelvins for extended periods. It can also serve as a starting point for other processes, as follows.

A.3   Adiabatic demagnetization

Suppose that we have used the above techniques to cool our sample to a few millikelvins, and we wish to cool it still further. One way to do this involves magnetic fields. At low temperatures, considerably more entropy is available in the spin orientations of paramagnetic materials than in the thermal motions of atoms, which decrease as T^4 (equation 22.13). At 0.1 K, for example, the number of spin states in a typical solid is about 10^8 times larger than the number of vibrational degrees of freedom (homework).

"Adiabatic demagnetization" is the magnetic version of adiabatic expansion: $p\,\mathrm{d}V$ is replaced by $-B\,\mathrm{d}M$, where M is the component of the magnetic moment parallel to the magnetic field B. With this replacement, the first law $\mathrm{d}E = T\,\mathrm{d}S - p\,\mathrm{d}V$ becomes

$$\mathrm{d}E = T\,\mathrm{d}S + B\,\mathrm{d}M. \tag{24.1}$$

So you would think that the magnetic counterpart of cooling a gas by letting it expand ($\mathrm{d}V$ positive) under reduced pressure would be to let it demagnetize ($\mathrm{d}M$ negative) under a reduced magnetic field. This line of thinking is indeed correct, but the analogy with a gas is not so good at very low temperatures.

There is another and simpler way of understanding adiabatic demagnetization, which depends on the following two observations.

1 A larger magnetic moment means more ordering of the atomic spins and therefore smaller entropy, as illustrated in Table 24.1,

2 The magnetic moment of the sample depends on the ratio B/T (equation 17.9),

$$M = N\overline{\mu_z} = N\frac{\sum_{\mu_z}\mu_z e^{\mu_z B/kT}}{\sum_{\mu_z}e^{\mu_z B/kT}}.$$

According to the first observation, if we begin with a magnetized sample at a very cold temperature where all the entropy is in the magnetic degrees of freedom and reduce the magnetic field to zero quasistatically and adiabatically, so that the entropy remains constant, the magnetic moment cannot change (so the term

Table 24.1. *Some configurations of four spin-1/2 particles, each with magnetic moment μ. Ω is the number of states. The larger the magnetic moment of the system, the smaller its entropy*

Spin configurations	Ω	$S = k \ln \Omega$	Magnetic moment
↑↑↑↑	1	$k \ln 1$	4μ
↑↑↑↓, ↑↑↓↑, ↑↓↑↑, ↓↑↑↑	4	$k \ln 4$	2μ
↑↑↓↓, ↑↓↑↓, ↑↓↓↑, ↓↓↑↑, ↓↑↓↑, ↓↑↑↓	6	$k \ln 6$	0

"demagnetization" is misleading). But, according to the second observation, this means that the ratio B/T also remains constant:

$$\frac{B}{T} = \text{constant} \qquad (\text{adiabatic, M} = \text{constant}). \qquad (24.2)$$

Consequently, as we reduce the imposed external magnetic field adiabatically, the temperature falls proportionally.

The method of adiabatic demagnetization, then, is the following three-step process.

- Refrigerate a paramagnetic system in a strong magnetic field to the lowest temperature possible.
- Insulate the system from its refrigerated environment.
- Reduce the imposed magnetic field B to zero.

Equation 24.2 implies that we could reach absolute zero simply by making $B = 0$. Unfortunately, two things prevent us from quite getting there. First, as the spin temperature falls below that of the lattice, energy will flow from the lattice to the spin states, causing some "demagnetization" and preventing their reaching the absolute-zero goal. Although the lattice carries very little energy at these low temperatures, the energy is not zero.

Second, even though we can remove the external field completely there will still be a little residual magnetism internal to the material, as the internal magnetic moments interact among themselves. This turns out to be a larger problem than the heat from the lattice, so a great deal of effort goes into reducing this self-interaction impediment.

Sometimes nuclear magnetic moments are used instead, because they are much weaker and therefore have less interaction between neighbors. And sometimes, both are used in tandem. The adiabatic demagnetization of electrons in paramagnetic salts is used to get the sample environment down to the microkelvin range. Then nuclear magnetic moments are used in the next step to reduce the temperature yet further.

A.4 Optical methods

When an atom absorbs light from a laser, it recoils owing to the conservation of momentum. Of course the absorbed light is quickly reemitted, but in a random direction. So on average there is a net change in momentum in the direction of the incident laser beam.

One method of confining a gas and isolating it from its warmer surroundings uses the splitting of atomic energy levels in the presence of imposed nonuniform magnetic fields. Atoms leaving a central region encounter slightly different magnetic fields and therefore have slightly different line splittings compared with atoms that remain at the center. We can therefore tune our lasers carefully to the level splittings of atoms that have left the central region. These atoms will then absorb the laser photons and be pushed back, whereas those that have remained near the center will be unaffected.

Optical cooling uses the Doppler effect. Atoms approaching a light source "see" a slightly higher frequency than those at rest. Thermal motions tend to make a gas spread out, those with the fastest motions spreading fastest. So we bombard our gas from all directions with lasers whose light frequency will be absorbed by approaching atoms and not by those standing still. This pushes these faster-moving atoms back and reduces their motion. Temperatures in the microkelvin range have been achieved this way.

Further temperature reduction can be achieved by imposing the above optical methods and then letting the fastest of the remaining atoms go (like diffusive cooling). The slower-moving atoms remain near the center and have correspondingly lower temperatures. Temperatures in the nanokelvin range have been reached this way, and true Bose–Einstein condensation (see Section B) has been observed.

A.5 Measuring low temperatures

Once we have cooled the system, how can we tell what its temperature is? Sometimes, we can use the properties of the system itself. For example, in the case of optically cooled gases we can turn off the trapping mechanism and see how fast the gas expands. The temperature reflects the thermal motions of the particles, and therefore the rate of expansion. In the case of adiabatic demagnetization, we can use the definition of temperature, (equation 8.2)

$$\frac{1}{T} = \left(\frac{\Delta S}{\Delta E} \right)_{\Delta W = 0}.$$

We add a small amount of energy ΔE and determine the change in entropy ΔS by measuring the change in magnetic moment (as illustrated in Table 24.1). We ensure that no work is done, $\Delta W = B \Delta M = 0$, by ensuring that there is no magnetic field ($B = 0$).

In other cases, we include a different system in our cooled sample to serve as a thermometer. For example, we might use a material whose low-temperature electrical or magnetic properties are already known. We can then determine the temperature by measuring this property. Or, we might measure the relative populations of closely spaced nuclear or electronic levels in some material. These are related to the temperature through

$$\frac{N_i}{N_j} = \frac{P_i}{P_j} = \frac{Ce^{-\beta\varepsilon_i}}{Ce^{-\beta\varepsilon_j}} = e^{-\beta(\varepsilon_i - \varepsilon_j)}.$$

Taking the logarithm of both sides and solving for the temperature gives (homework)

$$T = \frac{\varepsilon_i - \varepsilon_j}{k \ln(N_j/N_i)}. \tag{24.4}$$

These are just some examples. In general, the measurement of low temperatures involves experimental ingenuity in choosing a thermometer that is appropriate for the system and has minimal impact on it. All measurements perturb the system. For example, the measurement of level populations or the rate of expansion involves the absorption or scattering of light, respectively. And because low-temperature heat capacities are so small, the addition of even very tiny amounts of energy cause huge increases in the system's temperature. This fact makes all forms of low-temperature studies (including temperature measurement) particularly delicate and challenging.

Example 24.1 Suppose that two levels are separated by 10^{-6} eV and that the population of the lower level is 10 times that of the upper level. What is the temperature?

According to equation 24.4, the answer is

$$T = \frac{10^{-6} \text{ eV}}{k \ln(10)} = 5 \text{ mK}.$$

Summary of Section A

A degenerate system is confined to a small volume in six-dimensional phase space. The cause of degeneracy may be a restricted volume in position space owing to high densities or a restricted volume in momentum space owing to small masses or low temperatures.

If we wish to make something colder than anything in its environment, then the cooling must be done adiabatically, because otherwise any heat flow would be in the wrong direction. These adiabatic cooling processes could be mechanical, such as the expansion of gases, or diffusive, such as evaporation.

Diffusive cooling includes the throttling process, in which the fluid does work on itself as it expands through a constriction. It also includes the more efficient expansion against a piston, although engineering challenges make the equipment more expensive.

The evaporation of liquid helium in a vacuum can produce temperatures down to around 1 K for ^{4}He and 0.3 K for ^{3}He. The helium dilution refrigerator works by evaporating liquid ^{3}He into ^{4}He, which reduces the potential barrier to the evaporating ^{3}He atoms. With this process, temperatures of millikelvins can be maintained.

Adiabatic demagnetization is the magnetic equivalent of volume expansion. At very low temperatures, virtually all the entropy is in the spin states, which determine the magnetic moment of the material. In quasistatic adiabatic processes the number of spin states remains constant, as does the magnetic moment, which depends on the ratio B/T. So if we reduce the external magnetic field adiabatically in this way, the temperature must decrease with it.

Optical methods use the fact that atoms recoil upon absorbing light. We tune the frequency of the laser light to match that of atoms whose absorption frequencies have been changed by imposed nonuniform magnetic fields or by their own motion. We can use these optical techniques for both confinement and cooling.

Methods for measuring low temperatures include measuring the properties of a material whose behavior is known, measuring expansion rates of gases, measuring the change in entropy with energy for a paramagnetic substance, and measuring the occupation of closely spaced levels. All such measurements perturb the system, so measurement is a particularly sensitive process at low temperatures where heat capacities are very small.

B   Degenerate boson systems

B.1   Bose–Einstein condensation

We now turn our attention to systems of noninteracting bosons whose total number N is fixed.[4] These systems display particularly interesting behaviors at low temperatures, as particles become trapped in the ground state – a process referred to as Bose–Einstein condensation.[5] This ground-state entrapment occurs at considerably higher temperatures than our "classical" intuition might suggest. There are two related reasons for this.

- First, the counting of states for identical bosons favors states of multiple occupancy. As we learned in Chapter 19 (see Figures 19.3 and 19.4), the probability of finding all N in the same state, such as the ground state, is roughly $N!$ times larger than the classical prediction.

---

[4] This is in contrast with the photons and phonons studied in Chapters 21 and 22, for which the number changes with temperature and the chemical potential is zero, as dictated by the second law.

[5] The 2001 Nobel Prize in Physics was awarded to Professors Cornell, Keterle, and Wieman for producing and studying Bose–Einstein condensation in dilute gases of alkali atoms.

- Second, the chemical potential for a system of degenerate bosons is extremely close to the ground state. So the occupation number falls off very rapidly, giving all excited states very low occupancies. To illustrate this, consider a small system of 10^{22} bosons at 0.1 K. According to equation 20.10 ($\mu = -kT/N$), the chemical potential would be only -10^{-27} eV. For comparison, the first excited state for a cubic centimeter of dense gas, (e.g. liquid helium), would be at about $+10^{-16}$ eV, therefore being 10^{11} times further from the ground state and having an occupation number that is 10^{11} times smaller. As can be shown in the homework problems, a single particle, or any number of *classical* particles, would have a nearly equal probability of occupying either the ground or the first excited state. So the effect of many identical particles is crucial here.

The Einstein model

For a more quantitative treatment of boson condensation, we sometimes use a model proposed by Einstein. In this model the distribution of excited states is that of a quantum gas, and the ground state is handled separately.[6] Also, because the chemical potential is so very close to the $\varepsilon = 0$ ground state (see the preceding paragraph), we set it equal to zero.

According to equation 20.1, the density of states in a nonrelativistic quantum gas is[7]

$$g(\varepsilon) = C\varepsilon^{1/2}, \qquad \text{with} \qquad C = \frac{2\pi V(2m)^{3/2}}{h^3}. \qquad (24.5)$$

The number of bosons in excited states is the product of the number of states times the occupation number of each. Using $x = \varepsilon/kT$ we have

$$N_{\text{excited}} = \int \bar{n}(\varepsilon)g(\varepsilon)d\varepsilon = C\int_0^\infty \frac{\varepsilon^{1/2}d\varepsilon}{e^{\beta\varepsilon} - 1} \quad \longrightarrow \quad C(kT)^{3/2}\int_0^\infty \frac{x^{1/2}dx}{e^x - 1}.$$

Numerical evaluation of the integral on the right gives the value 2.32, so our result is

$$N_{\text{excited}} = 2.32C(kT)^{3/2}. \qquad (24.6)$$

Something is obviously wrong with this result. It implies that the number of bosons in excited states increases indefinitely with temperature, whereas there can be no more than N. The source of this error is our assumption that the chemical potential remains equal to zero and does not change as the temperature is increased. But we learned in subsections 14A.1 and 20D.2 (equations 14.1 and 20.17) that for a system with a fixed number of bosons the chemical potential must decrease as the temperature increases (Figure 24.3).

---

[6] If we didn't treat the ground state separately, then when we switched from summation to integration over states, the density of states would cut out the ground state completely, because according to equations 20.1 or 24.5, $g(0) = 0$. This omission would be a disastrous omission at low temperatures.

[7] We will assume that the number of spin orientations for our bosons is 1.

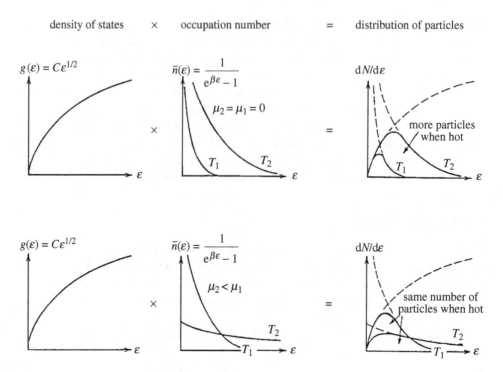

density of states   ×   occupation number   =   distribution of particles

Figure 24.3 The distribution of particles is the product of the density of states and the occupation number, which depends on the chemical potential. (Top) In the Einstein model, we assume that $\mu = 0$ at all temperatures below T_0. Consequently, the total number of particles in excited states (i.e., the area under the $dN/d\varepsilon$ curve) increases indefinitely with temperature. (Bottom) In reality, the chemical potential must decrease with increasing temperature, so that the number of particles in excited states does not grow past N.

In light of this problem, we make the crude approximation that the result 24.6 is valid up to the temperature T_0 at which all N bosons are in excited states. Above this temperature, the number in excited states remains at N, because there can be no more. To find this temperature T_0, we set the number of bosons in excited states equal to N and solve for T_0:

$$N = 2.32C(kT_0)^{3/2}, \tag{24.7}$$

so with C from equation 24.5,

$$kT_0 = \left(\frac{N}{2.32C}\right)^{2/3} = \frac{h^2}{2m}\left(\frac{N}{4.64\pi V}\right)^{2/3}. \tag{24.8}$$

There are two very interesting things about this result.

- If we write the momentum as $p = h/\lambda$ then this maximum energy $kT_0 = p^2/2m$ has a wavelength of about twice the interparticle spacing ($\lambda = (4.64\pi V/N)^{1/3} = 2.4(V/N)^{1/3}$). We have seen this same thing twice before in completely different systems (the Debye cutoff for phonons and the Fermi level for degenerate fermions).
- Using the atomic mass and density of liquid helium gives $T_0 \approx 3$ K (homework), which is not too much different from the observed temperature, 2.18 K, for the onset of condensation into the ground state. This gives us added confidence in the model.

For temperatures below T_0, the fraction of atoms in excited states, N/N_0, is found by dividing equation 24.6 by 24.7. Above T_0, all N atoms are in the excited

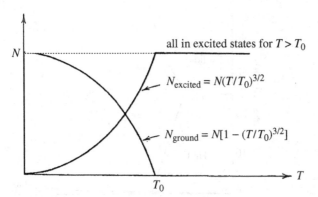

Figure 24.4 Plots of the number of bosons in the ground state and in excited states as a function of temperature for the Einstein model.

states. The number in the ground state is simply equal to the number that are not in excited states (Figure 24.4):

$$T \leq T_0, \qquad \frac{N_{\text{excited}}}{N} = \left(\frac{T}{T_0}\right)^{3/2}, \qquad \frac{N_{\text{ground}}}{N} = 1 - \left(\frac{T}{T_0}\right)^{3/2};$$

$$T > T_0, \qquad \frac{N_{\text{excited}}}{N} = 1, \qquad \frac{N_{\text{ground}}}{N} = 0. \qquad (24.9)$$

That is, for low temperatures the number of atoms in excited states increases as $T^{3/2}$ until all atoms are in excited states. Above temperature T_0, all N atoms are in excited states.

B.2   Superfluid liquid helium

Liquid helium takes on surprising properties. Although there are some very weak interatomic forces that cause it to liquefy at low temperatures, applying the preceding model of a gas of noninteracting bosons does give us some insight into its behavior.

The reason that liquid helium fails to freeze can be traced to the small atomic mass and weak interatomic forces. If we write the energy of the zero-point oscillations as

$$\varepsilon_{\text{zero point}} = \tfrac{1}{2}\hbar\omega = \tfrac{1}{2}\kappa A^2$$

and express the angular frequency in terms of the force constant and mass ($\omega = \sqrt{\kappa/m}$) the above equation tells us that the amplitude of these oscillations is given by

$$A = \frac{\hbar^{1/2}}{(\kappa m)^{1/4}}. \qquad (24.10)$$

For helium, the atomic mass and interaction strength (m and κ) are so small that the amplitude would be comparable with the atomic spacings (homework). Consequently, if helium were a solid then it would spontaneously melt from its own zero-point oscillations.

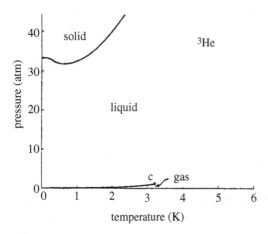

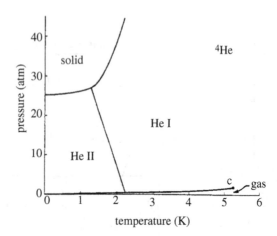

Figure 24.5 Phase diagrams for ^{3}He (left) and ^{4}He (right) at low temperatures.

The two isotopes of helium behave differently at low temperatures (Figure 24.5). One is a fermion and one a boson. For both isotopes the two electrons are in a state of zero total angular momentum, but the ^{3}He nucleus has a total spin of $1/2$ whereas that of the ^{4}He nucleus is zero. The ^{4}He isotope is 10^6 times more abundant, so unless ^{3}He is specifically mentioned, the name "liquid helium" refers to ^{4}He.

At atmospheric pressure, helium condenses into a normal liquid at 4.22 K. At 2.18 K it begins another phase transition to a "superfluid." This marks the onset of the condensation of ^{4}He bosons into the ground state. This temperature is called the "lambda point" because of the shape of the heat capacity curve (Figure 24.6). The phase above 2.18 K is called "helium I" and that below 2.18 K is "helium II." As the temperature is lowered still further, the superfluid component grows and the normal fluid component decreases.

The superfluid component gives the system the remarkable properties of seemingly zero viscosity and infinite thermal conductivity. The lack of viscosity means that the superfluid component can flow through the smallest crack without impediment, and if you put it into circular motion around the container, it keeps on going forever. You might wonder how bosons in the state of zero energy can flow. The explanation is that from their point of view they are standing still and the container is rotating.

Infinite thermal conductivity means that heat seems to be disbursed infinitely fast. Regardless of where you heat the liquid, it will evaporate from the top surface where the pressure and therefore the boiling point is lowest. This contrasts with normal fluids, in which the vapor bubbles tend to form and rise from the point where the heat is added. Therefore, the He I–He II phase transition is visually characterized by the disappearance of bubbles and boiling.

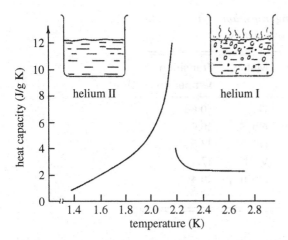

Figure 24.6 A plot of heat capacity vs. temperature for liquid helium. The phase transition between helium I and helium II is called the λ-point, because of the shape of this curve. Above the λ-point (right) liquid helium boils. But below the λ-point (left) the bubbles disappear. Owing to infinite thermal conductivity the temperature is uniform, so the helium II vaporizes from the top surface where the pressure is least.

B.3 Fermions?

It is particularly interesting that certain fermion systems also undergo Bose–Einstein condensation. At first thought this might seem impossible since identical fermions obey the exclusion principle, which prevents any two fermions from occupying the same quantum state. How then, could large numbers of them condense into the same ground state?

The answer to this riddle is that any group of an even number of fermions is itself a boson. Two spin-1/2 fermions, for example, have a combined spin of either 1 or 0 (in units of $\hbar$). Add to that any relative angular momentum, which is also an integer, and the total angular momentum of the system is an integer. Therefore, any pair of fermions is a boson. (Think of ^{4}He, for example, which consist of two protons, two neutrons, and two electrons. Together, these six fermions form a boson of zero total angular momentum.) The fermion substructure could restrict the overlap in position space. The resulting larger volumes in position space would require correspondingly smaller volumes in momentum space – and hence lower temperatures – in order for the substance to condense.

Particularly interesting are systems of ^{3}He, which undergoes a Bose–Einstein type of condensation into a superfluid at temperatures below 3 mK, and of conduction electrons, which may undergo a superconducting transition[8] at temperatures

---

[8] The electrical resistance of many ordinary metals drops suddenly to zero when cooled to a few kelvins. Once started, a current through a loop of one of these materials keeps going indefinitely.

Table 24.2. *Superconducting transition temperature in zero magnetic field for various metals*

Metal	Transition temperature (K)	Metal	Transition temperature (K)
Al	1.18	U	0.68
V	5.38	Nb_3Sn	18.05
Nb	9.20	Nb_3Al	17.5
Sn	3.72	V_3Si	17.1
Hg	4.15	La_3In	10.4
Pb	7.19		

Figure 24.7 Interactions between superconducting electrons are mediated by phonons. That is, an electron interacts with one lattice point, and the elastic wave that it produces travels through the solid until it transfers this energy to some other electron.

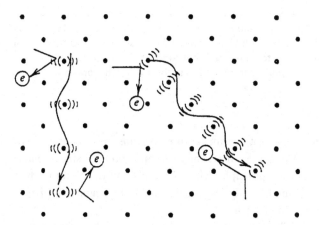

typically of a few kelvins (Table 24.2). These fermions form pairs having equal and opposite momenta (in the appropriate reference frame). This makes each pair a boson with zero total momentum and therefore in the zero-energy ground state. The individual members of each pair may be widely separated spatially, so the interactions that correlate their behaviors involve other intervening particles. In the case of superfluid ^{3}He the correlating interactions involve the nuclear magnetic moments, and in the case of a superconductor, the electron pairs interact via phonons traveling through the lattice (Figure 24.7).

Summary of Section B

We examine systems with a fixed number N of identical bosons at low temperatures. The entrapment of bosons in the ground state is called Bose–Einstein condensation. It occurs at higher temperatures than our classical intuition would suggest because of (1) the counting of states for identical particles and (2) the fact that the chemical potential is so close to the ground state that the occupation number for excited states is very small.

Einstein proposed a model for this in which the chemical potential is approximated as zero, the excited states are treated as those of a gas, and the ground state is treated separately. The number of bosons in excited states increases as $T^{3/2}$ until the temperature T_0 is reached where all N are in excited states (equation 24.9):

$$T \leq T_0, \qquad \frac{N_{excited}}{N} = \left(\frac{T}{T_0} \right)^{3/2} \qquad \frac{N_{ground}}{N} = 1 - \left(\frac{T}{T_0} \right)^{3/2};$$

$$T > T_0, \qquad \frac{N_{excited}}{N} = 1 \qquad \frac{N_{ground}}{N} = 0.$$

Here T_0 is given by (equation 24.8)

$$kT_0 = \left(\frac{N}{2.32C} \right)^{2/3} = \frac{h^2}{2m} \left(\frac{N}{4.64\pi V} \right)^{2/3}.$$

Although there are weak interatomic forces among helium atoms (which cause its liquefaction at low temperatures) it does undergo a condensation into the ground state analogous to a Bose–Einstein condensation. The two isotopes of helium behave differently, because one is a fermion and one a boson. The ^{4}He isotope is 10^6 times more abundant so, unless ^{3}He is specified, the name liquid helium refers to ^{4}He.

At atmospheric pressure, helium condenses into a normal liquid at 4.22 K. At the lambda point (2.18 K) it begins a phase transition to a superfluid, as atoms start becoming entrapped in the ground state. As the temperature is lowered still further, the superfluid component grows and the normal fluid component decreases. The liquid above 2.18 K is called helium I and below 2.18 K is called helium II. The He I–He II phase transition is visually characterized by the disappearance of bubbles and boiling.

Some fermion systems can undergo Bose–Einstein type condensations at low temperatures when the fermions pair up to form bosons. Interesting examples include ^{3}He, which undergoes a superfluid transition at 3 mK, and conduction electrons in metals, which undergo superconducting transitions at typically a few kelvins. Interactions that correlate the motions of the fermion pairs are carried through the intervening material by nuclear magnetic moments for ^{3}He and by lattice vibrations for conduction electrons.

C Stellar collapse

We now examine systems whose degeneracy is due to extremely high particle densities rather than to extremely low temperatures. That is, the scarcity of available quantum states is due to small volumes in position space rather than in momentum space. Stellar corpses are degenerate systems with huge masses and high temperatures. Some of these are produced in spectacular explosions, which release more energy in a few seconds than was released during the entire preceding lifetime of the star.

C.1   The death of a star

Our Sun is a typical star. Its diameter is over 100 times larger than Earth's, and it could hold more than a million Earths within it. But in spite of its extremely strong gravity and highly compressible materials, its average density is only slightly greater than that of water. Radiation pressure successfully opposes gravitational collapse. But some day the core's nuclear fuel will be exhausted, the thermonuclear reactor will shut down, and there will be no more radiation pressure to oppose the collapse. The remnant material will be compressed to an object about the size of the Earth and a million times denser.

Our Sun is typical. Gravitational collapse is the final destiny of all stars. These collapses happen quickly and, during one of these brief events, the outer layers may be blown out into space in a spectacular display. But the bulk of the material remains in a small, dense, collapsed object, whose final size is determined primarily by the fact that no two identical fermions may be forced into the same quantum state.

C.2   White dwarfs and neutron stars

From the uncertainty principle, we know that a particle confined to a region of length r has an associated momentum of roughly h/r, and therefore a kinetic energy of

$$\varepsilon_k = \frac{p^2}{2m} \approx \frac{(h/r)^2}{2m}.$$

Notice that particles with smaller masses have larger kinetic energies. Therefore less massive particles exert a greater pressure to oppose further collapse. This important feature is maintained in the relativistic treatment that follows.

Since electrons are much less massive than nucleons, the electrons normally acquire the bulk of the kinetic energy and limit the collapse. The resulting body is called a "white dwarf" which very slowly cools to "red" and then "brown" dwarf stages. If the star's mass is greater than 1.4 solar masses, however, the gravity is so strong that it forces electrons and protons into each other to form neutrons along with neutrinos which escape into space:

$$\text{e}^- + \text{p}^+ \rightarrow \text{n} + \nu. \tag{24.11}$$

The result is a "neutron star." If the Sun were the size of an average two-story house, a typical white dwarf would be the size of a grapefruit, and a neutron star would be the size of a grain of salt. Finally, if the collapsed star is more than about three times as massive as our Sun, it becomes an extremely small and dense "black hole."

We now look more closely at the electrons in a white dwarf or the neutrons in a neutron star. In a degenerate gas of N fermions, all the quantum states up to a certain energy are occupied, and all those above that energy are empty. We use this fact to determine the maximum momentum of the fermions, p_{max}, in the collapsed star. There are $N/2$ identical spin-up fermions and $N/2$ identical spin-down fermions, no two of which may occupy the same quantum state. Therefore, the number of states for either up or down fermions is $N/2$:

$$\frac{V_r V_p}{h^3} = \frac{N}{2} \quad \Rightarrow \quad \frac{\left(\frac{4}{3}\pi R^3\right)\left(\frac{4}{3}\pi p_{max}^3\right)}{h^3} = \frac{N}{2}, \tag{24.12}$$

where R is the radius of the collapsed star. Solving for p_{max} gives

$$p_{max} = \left(\frac{9}{32\pi^2}\right)^{1/3} N^{1/3} \frac{h}{R} \quad (R = \text{radius of collapsed star}).$$

It is easy to show (homework or equation 20.15) that in a degenerate gas the mean square momentum is $(3/5)\, p_{max}^2$, so we can write the root mean square momentum as[9]

$$p_{rms} = \sqrt{\frac{3}{5}} p_{max} = \eta N^{1/3} \frac{h}{R}, \tag{24.13}$$

where

$$\eta = \sqrt{\frac{3}{5}}\left(\frac{9}{32\pi^2}\right)^{1/3} = 0.237.$$

Notice that the momentum is inversely related to the star's radius R. The further the collapse, the larger the average momentum. (The smaller the volume in position space, the larger the volume in momentum space that is needed to accommodate the fermions.)

C.3 The size of a collapsed star

We now demonstrate that this simple result of quantum statistics determines the size of these stellar corpses. The gravitational energy released when a star of mass M collapses to radius R is proportional to GM^2/R. If the resulting object is of uniform density, and the final radius is small compared with the initial radius, then the constant of proportionality is $3/5$ (homework).[10] It is customary to write

$$\text{potential energy lost} = \Gamma\frac{M^2}{R}, \qquad \text{where} \quad \Gamma \approx \frac{3}{5}G.$$

Conservation of energy dictates that potential energy lost is equal to kinetic energy gained, and this (relativistic) kinetic energy is shared among the N degenerate fermions:

$$PE_{lost} = KE_{gained} \quad \Rightarrow \quad \Gamma\frac{M^2}{R} = N(\sqrt{p^2c^2 + m^2c^4} - mc^2). \tag{24.14}$$

---

[9] p_{rms} changes slightly from $(3/5)^{1/2}\, p_{max}$ to $(3/4)^{1/2}\, p_{max}$ in the extreme relativistic case. (So η would change from 0.237 to 0.265.) Although this must be considered in a more thorough treatment, we ignore it here as it has little effect on our results.

[10] In reality, the density is somewhat larger towards the center, so this constant is somewhat larger than $3/5$.

(We ignore the initial potential and kinetic energies of the particles and the energy radiated into space, because these are very small compared with the above terms.)

We replace p^2 in this expression by the result 24.13 for the root mean square momentum p_{rms}, and we solve equation 24.14 for the radius R of the collapsed star. The result is (homework)

$$R = R_0 \left(\frac{1}{x} - x \right), \tag{24.15a}$$

where

$$R_0 = \frac{\eta h N^{1/3}}{2mc}, \qquad x = \frac{\Gamma M^2}{\eta h c N^{4/3}}, \qquad \eta \approx 0.237, \qquad \Gamma \approx \frac{3}{5}G. \tag{24.15b}$$

This result can be greatly simplified by writing the number of fermions N in terms of the mass of the star M. We now do this. The result will be an expression for the radius of a collapsed star that depends only on its mass M.

The thermonuclear fusion that powers a typical star turns hydrogen into helium and some heavier nuclei. If the final stage is a white dwarf (as eventually our Sun will become), most of the remaining mass is in nuclei such as helium and carbon, for which the number of electrons is half the number of nucleons. If the final stage is a neutron star, there are no electrons left. All have combined with protons according to process 24.11. In either case, the number of nucleons doesn't change but the number of electrons does. For each proton that becomes a neutron, an electron disappears.

We use our Sun as a reference, for which the mass M_s and number of fermions N_s is given by

$$M_s = 1.99 \times 10^{30} \text{ kg} \tag{24.16}$$
$$N_s = 0.60 \times 10^{57} \text{ electrons when a white dwarf}$$
$$N_s = 1.19 \times 10^{57} \text{ nucleons}$$

(The number of electrons at present is 1.01×10^{57}.) We can then replace the factors M, N appearing in the expressions 24.15b by setting

$$M = M_s \left(\frac{M}{M_s} \right), \qquad N = N_s \left(\frac{N}{N_s} \right) = N_s \left(\frac{M}{M_s} \right).$$

In the last step on the right, we have used the fact that the fermion composition of all white dwarfs is about the same, and likewise that of all neutron stars. So the number of fermions (whether electrons or neutrons) is in proportion to the mass.

$$\frac{N}{N_s} = \frac{M}{M_s}.$$

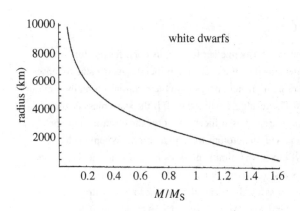

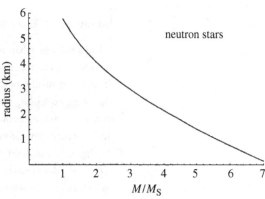

Figure 24.8 Plot of radius vs. mass for white dwarfs (left) and neutron stars (right). The mass is given in solar masses, and the radius in kilometers. The curves are sensitive to some approximations we have made, so the plots are only approximately correct.

With these substitutions, the expressions 24.15b for R_0 and x become

$$R_0 = \frac{\eta h N_s^{1/3}}{2mc}\left(\frac{M}{M_s}\right)^{1/3}, \qquad x = \frac{\Gamma M_s^2}{\eta h c N_s^{4/3}}\left(\frac{M}{M_s}\right)^{2/3}.$$

We now insert the numerical values of the constants η, h, m, c, G and use equations 24.16 for N_s and M_s and 24.15b for Γ to turn equation 24.15a into the following user-friendly form (homework):

$$R = R_0\left(\frac{1}{x} - x\right), \qquad (24.17a)$$

with, for white dwarf,

$$R_0 = (2430\,\text{km})\left(\frac{M}{M_s}\right)^{1/3}, \qquad x \approx (0.664)\left(\frac{M}{M_s}\right)^{2/3} \qquad (24.17b)$$

and for neutron stars,

$$R_0 = (1.66\,\text{km})\left(\frac{M}{M_s}\right)^{1/3}, \qquad x \approx (0.267)\left(\frac{M}{M_s}\right)^{2/3}. \qquad (24.17c)$$

The above results are displayed in Figure 24.8.

Take a moment to appreciate how simple these results are, and how they simply reflect the fact that no two identical fermions may be forced into the same quantum state. Also notice the following.

- The radii of neutron stars are only a few kilometers, and the radii of white dwarfs are comparable with that of the Earth (6370 km).
- Larger masses have smaller radii. When x reaches 1.0, the radius goes to zero. Notice that this corresponds to a star not much larger than our Sun (homework). So this line of reasoning predicts black holes, even though we have not considered the physics of nuclear forces or general relativity.

Summary of Section C

When a star has exhausted the thermonuclear fuels in its core, fusion stops and the star collapses to a size determined by the exclusion principle. As the star collapses, fermions of smaller masses gain greater energies, so the electrons normally limit the size of the collapsing star. The result is a white dwarf. If the star's mass is greater than 1.4 solar masses then its gravity is sufficient to force the electrons into the nuclei, where they combine with the protons to form neutrons. When this happens, the degenerate neutron system is what limits the star's final size, and a neutron star is formed. More massive stars collapse to smaller radii. If the corpse's mass is more than about three solar masses, the surface gravity becomes so strong that nothing – not even light – can leave. We call these massive and completely uncommunicative stars "black holes."

In a degenerate fermion gas, all accessible quantum states are occupied. We use this and the fact that in a degenerate gas $\langle p^2 \rangle = (3/5)p_{\max}^2$, to obtain (equation 24.13)

$$p_{\mathrm{rms}} = \eta N^{1/3} \frac{h}{R}, \qquad \text{where} \quad \eta \approx \sqrt{\frac{3}{5}} \left(\frac{9}{32\pi^2} \right)^{1/3} = 0.237.$$

The momentum is inversely related to the radius of the collapsed star, because the smaller the volume in coordinate space, the larger the volume in momentum space needed to accommodate all the fermions. Consequently, both the decrease in potential energy and the gain in kinetic energy are functions of the collapsed star's final size. We can solve the conservation of energy equation (equation 24.14)

$$PE_{\mathrm{lost}} = KE_{\mathrm{gained}} \quad \Rightarrow \quad \Gamma \frac{M^2}{R} = N \left(\sqrt{p^2c^2 + m^2c^4} - mc^2 \right)$$

for the radius of the collapsed star, obtaining (equations 24.17)

$$R = R_0 \left(\frac{1}{x} - x \right)$$

with for white dwarfs,

$$R_0 = (2430\,\mathrm{km}) \left(\frac{M}{M_\mathrm{s}} \right)^{1/3}, \qquad x \approx (0.664) \left(\frac{M}{M_\mathrm{s}} \right)^{2/3},$$

and for neutron stars,

$$R_0 = (1.66\,\mathrm{km}) \left(\frac{M}{M_\mathrm{s}} \right)^{1/3}, \qquad x \approx (0.267) \left(\frac{M}{M_\mathrm{s}} \right)^{2/3}.$$

Here M/M_s is the ratio of the star's mass to that of our Sun.

Problems

Section A

1. (a) Using classical statistics show that for a temperature $0.02T_\mathrm{e}$ ($kT_\mathrm{e} = \varepsilon_1 - \varepsilon_0$) there will be less than one particle out of 10^{21} in the first excited state.

(b) What is the excitation temperature for a system whose first excited state lies 0.1 eV above the ground state? 0.0001 eV above?

2. For a certain system, the first excited state lies 1 eV above the ground state. Is the system degenerate at room temperature? At 10^6 K?

3. We have said that, at room temperature, much less than 1% of the conduction electrons in a typical metal are in the tail of the Fermi distribution. Estimate what the real fraction is. The number of states in the tail is $g(\varepsilon)\Delta\varepsilon$, with $\varepsilon = \mu$, and the width of the tail is about $\Delta\varepsilon = kT$. Half these states have electrons in them, on average. Assume a fermion gas with a density of 10^{29} electrons per cubic meter. The density of states is $g(\varepsilon) = C\varepsilon^{1/2}$, with $C = 4\pi V (2m)^{3/2}/h^3$ and the Fermi level is given by equation 22.17.

4. In Chapter 22 we saw that the Debye temperature for a solid is typically several hundred kelvins. There are as many phonon states as there are atoms in the solid. Suppose that, for a certain solid at 100 K, the occupation number for a particular phonon state is $1/2$.
   (a) Roughly what is the occupation number of this state at 0.1 K?
   (b) Repeat the above for a state whose occupation number is $1/2$ at 5 K.

5. Consider a system of six distinguishable spin-$1/2$ particles each with magnetic moment μ. List the various possible magnetic moments for this system (in terms of μ) and give the entropy for each. (Remember: The number of different ways in which n of N elements can satisfy a criterion is $N!/n!(N-n)!$)

6. Consider a system of three distinguishable spin-1 particles each with magnetic moment μ. What is the entropy of the system if the total magnetic moment is (a) 3μ, (b) μ?

7. What is the root mean square spreading speed for a gas of rubidium atoms $(m = 1.42 \times 10^{-25}$ kg) at 10 μK?

8. Derive equation 24.4, which gives the temperature in terms of the relative populations of two energy levels. (Use classical statistics, because here we consider particles occupying various states, rather than vice versa.)

9. In the quasistatic adiabatic expansion of a gas, the entropy remains constant, although more volume in position space becomes available through the expansion. If the volume in position space increases, what must happen to the volume in momentum space and the temperature? In adiabatic demagnetization, if the magnetic moment of the spin system decreases as the external field is removed, what happens to the spin entropy of the system, the entropy in the kinetic and other degrees of freedom, and the temperature?

10. In Chapter 12 we found that for the throttling process the enthalpy $H = E + pV$ remains constant. For an ideal gas, $E = (\nu/2)nRT$ and

$pV = nkT$. For quasistatic adiabatic expansions we found that (equation 12.7) $TV^{\gamma-1} = \text{constant}$, where $\gamma = (\nu + 2)/\nu$. Consider an ideal gas, with $\nu = 5$ degrees of freedom per molecule, that begins at 290 K and expands to five times this volume. What is the final temperature if this expansion is done (a) via throttling, (b) via quasistatic adiabatic expansion?

11. In Chapter 21 we learned that the rate at which an object radiates energy is proportional to T^4. Liquid nitrogen at 77 K surrounds a Dewar (i.e., thermos bottle) holding liquid helium at 1 K.
    (a) By what factor is the heat energy radiated into the Dewar reduced compared with that which would be radiated if it were surrounded by materials at room temperature, 295 K?
    (b) Roughly how many times more energy is radiated from the liquid nitrogen toward the liquid helium than vice versa?

12. Give examples of cooling via mechanical interactions, or via diffusive interactions.

13. In earlier chapters we learned that, at high temperatures, a solid of N atoms has a total thermal energy $E = 3NkT$. In Chapter 22, however, we learned that at lower temperatures, some of these degrees of freedom are not available. Expression 22.13 tells us that at temperatures well below the Debye temperature ($kT \ll k\Theta_D = \varepsilon_D$) the internal energy is given by $E = 3NkT(\pi^4/5)(kT/\varepsilon_D)^3$. That is, the number of degrees of freedom is reduced from the high-temperature case by a factor $(\pi^4/5)(kT/\varepsilon_D)^3$.
    (a) Use this to estimate the average number of degrees of freedom per atom at 0.1 K for a solid whose Debye temperature is 100 K.
    (b) If each atom has one magnetic degree of freedom, how much more energy is carried in magnetic degrees of freedom than in vibrational degrees of freedom at this temperature?

14. Why is adiabatic demagnetization done adiabatically? (That is, why is the sample removed from the helium bath first?)

15. For the system of Table 24.1, when 10^{-7} eV of energy is added the magnetic moment changes from 4μ to 2μ. What is the temperature?

16. Consider a system that includes five spin-1/2 particles, each having magnetic moment μ. The external magnetic field has been turned off. Suppose that in addition to these five spin degrees of freedom, the system has three other degrees of freedom. When 10^{-5} eV of energy is added to the system, its total magnetic moment changes from $+3\mu$ to $+\mu$.
    (a) Assuming equipartition, how much added energy has gone into magnetic degrees of freedom?
    (b) By how much has the entropy of the magnetic moments increased?

(c) What is the temperature of the system, as determined from the magnetic properties of the system?

17. For N spin-1/2 particles, the number of different arrangements in which n of them are spin up and the remaining $N - n$ are spin down is given by $N!/[n!(N - n)!]$, where the factorials of large numbers can be calculated from Stirling's approximation $M! = (M/e)^M$. Consider a system of 10^{22} spin-1/2 particles, each having magnetic moment μ. Suppose that, in addition, the system has 0.1×10^{22} other degrees of freedom.
  (a) If the magnetic moment of the entire system is $1.1 \times 10^{21}\mu$, how many particles are spin up?
  (b) If the magnetic moment of the system were $1.0 \times 10^{21}\mu$ how many particles would be spin up?
  (c) What is the change in spin entropy if the magnetic moment goes from $1.1 \times 10^{21}\mu$ to $1.0 \times 10^{21}\mu$?
  (d) This change occurs when 2×10^{-6} J of energy is added to the system. What is the temperature?

18. Consider a system having 10^{24} spin-1/2 particles. The z-component of the magnetic moment of each is $\pm\mu_B$ ($\mu_B = 9.27 \times 10^{-24}$ J/T). They are sitting in a magnetic field, in the $+z$ direction, of 0.5 tesla.
  (a) If 70% of the particles are spin up, what is the temperature of the system? (Hint: Use classical statistics.)
  (b) If in state 1 70% of them are spin up, and in state 2 69.9% are spin up, find the change in entropy in going from one state to the other. (Hint: See the previous problem.)
  (c) What is the change in the magnetic interaction energy between the two states?
  (d) If you use $1/T = \Delta S/\Delta E$, do you get the same temperature as in part (a)?
  (e) If the system is insulated from its environment and the magnetic field is reduced to 0.000 001 tesla, the magnetic moment remaining unchanged, what is the new temperature of the system?

19. Two nuclear states have magnetic moments that differ by one nuclear magneton. When $B = 0.1$ tesla, the populations of the two levels are in the ratio $1 : 2$. What is the temperature?

20. In a certain type of molecule, electronic levels i and j are separated by 2×10^{-6} eV.
  (a) If the ratio of the numbers of molecules in level j and in level i is $1 : 20$, what is the temperature of the system?
  (b) If level k is 3×10^{-6} eV above level j, what is the ratio of the number of molecules in level k and the number in level i?

Section B

21. Using classical statistics, estimate the relative populations N_2/N_1 for two levels separated by 10^{-15} eV, if the temperature is 0.1 K. ($e^{-\varepsilon} \approx 1 - \varepsilon$ for $\varepsilon \ll 1$.)

22. (a) Estimate a typical momentum (in one dimension) of a helium atom in liquid helium, using the uncertainty principle along with the fact that the density of liquid helium is 0.15×10^3 kg/m^3.
    (b) A change in kinetic energy is given by $\Delta(p^2/2m) = (p/m)\Delta p$. Knowing that excitation from the ground state requires flow velocities in excess of 7 m/s for superfluid helium, estimate the change in momentum Δp to which this corresponds.
    (c) Combine the above two results to estimate the energy needed for excitations out of the ground state.

23. The zero-point oscillation energy of a simple harmonic oscillator is given by $\varepsilon = (1/2)\kappa A^2 = (1/2)\hbar\omega$, where $\omega = (\kappa/m)^{1/2}$ is the fundamental frequency of oscillation and A is the amplitude. If helium were a solid we would guess that the excitation temperature ($kT_e = \varepsilon_{\text{excitation}}$) would be about 0.1 K, because at 2.2 K all particles are excited and none are left in the ground state. The first excited state has energy $(3/2)\hbar\omega$.
    (a) Using this information, estimate the amplitude of zero-point oscillations if helium were a solid. (Hint: You should be able to find the value of ω from the excitation temperature.)
    (b) Compare your answer to part (a) with the average spacings of helium atoms. Liquid helium has a density of 0.15 g/cm^3.

24. If helium II has infinite thermal conductivity, explain why it vaporizes from the top surface rather than boiling uniformly throughout its volume.

25. Consider what happens after you stir helium II to get it to swirl around inside a circular container. If the normal-fluid component experiences friction with the walls of the container and the superfluid component does not, describe the subsequent motion of the two components.

26. The density of liquid helium is 0.15 g/cm^3. Use this to find the value of T_0 for the Einstein model.

27. At $T = T_0/2$, what fraction of the bosons are in the ground state, according to the Einstein model?

28. We are going to estimate what the temperature for transition to the superconducting state for a typical metal wire would be if there were no interactions at all among the electrons. That is, we will treat the electrons as if they were a gas of noninteracting fermions. In any one dimension (e.g., in the direction along the wire), the electrons can move both ways, so the wave function that describes them is a standing wave. This standing wave must have a node at

both ends, because the electrons cannot be found beyond the boundary of the metal. Therefore, the length of the metal must be equal to some number of half-wavelengths.

(a) Show that the momentum of an electron must be given by $p = nh/2L$, where n is an integer and L is the length of the metal.

(b) Show that the change in kinetic energy of the electrons between two neighboring states is given by

$$\Delta\varepsilon = (p/m)\Delta p = nh^2/4L^2m = 2\varepsilon/n.$$

(c) The density of electrons in a typical metal is $0.8 \times 10^{29}/m^3$. Using this and equation 22.17, estimate the energy of the Fermi level $\mu \approx \varepsilon_f = p_f^2/2m$ for the electrons in a typical metal. If the metal has length $L = 2$ cm, to what value of the integer n does this correspond?

(d) Substituting the values of L, μ, and n from part (c) into the answer to part (b), find the energy that separates two neighboring states for an electron near the Fermi surface in a typical metal. To what excitation temperature $(kT_e = \varepsilon_{excited})$ does this correspond?

(e) Can we conclude that this model (electrons as a noninteracting gas) is appropriate for explaining superconductivity?

Section C

29. Consider adding a shell of thickness dr to a spherical object of uniform mass density ρ and radius r. As the shell of material is brought in from infinity, its loss in gravitational potential energy is given by $dU = -GMdM/r$, where the mass of the object is $M = \rho(4/3)\pi r^3$ and that of the added shell is $\rho 4\pi r^2 dr$ (the volume equals the area of the shell times its thickness). Integrate this from $r = 0$ to $r = R$ to show that the gravitational potential energy lost in building up this object is given by $U = -(3/5)GM^2/R$.

30. The number of quantum states in a phase-space volume $d^3r d^3p$ is given by $d^3r d^3p/h^3$. If we integrate over all volume and all momentum angles, this becomes $(4\pi V_r/h^3)p^2 dp$. If the momenta of the particles range from $p = 0$ to $p = p_{max}$, show that the mean square momentum per particle is given by $(3/5)p_{max}^2$. Solve equation 24.12 for p_{max} and then use this with the preceding result to derive equation 24.13.

31. Starting with the conservation of energy (24.14) and the momentum of a fermion in a degenerate system (24.13), derive equations 24.15a, b for the radius of a collapsed star.

32. Derive Equations 24.17c from 24.15 and 25.16.

33. What is the radius of a collapsed star that has the same mass as our Sun? One that is three times as massive as our Sun?

34. According to our results 24.17a–c, what would be the upper mass limit (in solar masses) of a white dwarf? (Beyond this $R \to 0$, so it would be a neutron star.) At what mass would the radius of a neutron star go to zero?

35. Suppose that when a better model of stellar interiors is used, the value of Γ (see the equation before 24.14) becomes $(4/5)G$ rather than $(3/5)G$. In this case:
   (a) What would be the expressions for x in equations 24.17b, c?
   (b) What would be the answers to problem 33?
   (c) To question 34?

Appendices

Appendix A Magnetic moment and angular momentum

Consider a charge q moving in an elliptical orbit, as in Figure A.1. The area of the shaded triangle is

$$dA = \frac{1}{2} \text{ base} \times \text{height} = \frac{1}{2} r \, dr \sin\theta$$

or, in vector form,

$$d\mathbf{A} = \frac{1}{2} \mathbf{r} \times d\mathbf{r}.$$

We relate this to the particle's angular momentum by using $d\mathbf{r} = \mathbf{v} dt$ and $\mathbf{L} = \mathbf{r} \times m\mathbf{v}$, which gives

$$d\mathbf{A} = \frac{1}{2} \mathbf{r} \times \mathbf{v} dt = \frac{1}{2m} \mathbf{L} dt.$$

For constant angular momentum, integrating over one complete orbit gives a total area

$$\mathbf{A} = \frac{1}{2m} \mathbf{L} T, \tag{A.1}$$

where T is the time for the charge to orbit the loop once.

We now relate this to the magnetic moment $\boldsymbol{\mu} = i\mathbf{A}$. The current intensity, $i = q/T$, is the rate at which charge passes a given point on the loop. Combining this with the result A.1 for the area of the loop gives

$$\boldsymbol{\mu} = i\mathbf{A} = \left(\frac{q}{T}\right) \frac{1}{2m} \mathbf{L} T = \frac{q}{2m} \mathbf{L}. \tag{A.2}$$

Appendix B Taylor series expansion

Any analytic function of x may be written as a polynomial of the form

$$f(x) = c_0 + c_1(x-a) + c_2(x-a)^2 + c_3(x-a)^3 + \cdots$$
$$+ c_n(x-a)^n + \cdots \tag{B.1}$$

where a is a constant, as are the coefficients $c_0, c_1, c_2, \ldots$ If we take the derivatives of this expansion we get

$$f'(x) = 0 + c_1 + 2c_2(x-a) + 3c_3(x-a)^2 + \cdots + nc_n(x-a)^{n-1} + \cdots$$
$$f''(x) = 0 + 0 + 2c_2 + 6c_3(x-a) + \cdots + n(n-1)c_n(x-a)^{n-2} + \cdots$$

$$\vdots$$

$$f^{(n)}(x) = 0 + 0 + 0 + 0 + \cdots + n!c_n + \cdots$$

Figure A.1 An elliptical orbit. As indicated, the area of the shaded triangle is $(1/2)r dr \sin\theta$.

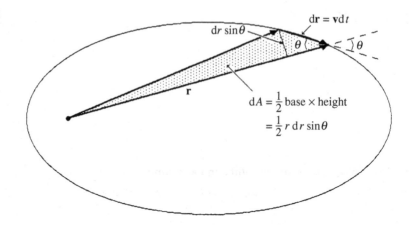

$$dA = \frac{1}{2} \text{base} \times \text{height}$$
$$= \frac{1}{2} r \, dr \, \sin\theta$$

By evaluating each of these above expressions at the point $x = a$, we see that

$$f(a) = c_0, \qquad f'(a) = c_1, \qquad f''(a) = 2!c_2, \ldots,$$
$$f^{(n)}(a) = n!c_n, \qquad \ldots$$

Hence, the coefficients are given by

$$c_n = \frac{1}{n!} f^{(n)}(a),$$

and so the expansion B.1 can be written as

$$f(x) = f(a) + f'(a)(x - a) + \frac{1}{2!} f''(a)(x - a)^2 + \cdots$$
$$+ \frac{1}{n!} f^{(n)}(a)(x - a)^n + \cdots$$
$$= \sum_{n=0}^{\infty} \frac{1}{n!} f^{(n)}(a)(x - a)^n \qquad (B.2)$$

This is the Taylor series expansion of the function $f(x)$ around the point $x = a$.

Appendix C   The states accessible to a system

We now do the calculations for the results of Table 6.2 with more mathematical rigor. Because the energy in the ith degree of freedom has the form $\varepsilon_i = b_i \xi_i^2$, e.g. $(1/2m)p_x^2$, we define a new set of variables $\alpha_i \equiv \sqrt{\varepsilon_i} = \sqrt{b_i}\xi_i$ and rewrite the integral in equation 6.9 as

$$\int d\xi_1 d\xi_2 d\xi_3 \cdots = (b_1 b_2 b_3 \cdots)^{-1/2} \int d\alpha_1 d\alpha_2 d\alpha_3 \cdots. \qquad (C.1)$$

The integral $\int d\alpha_1 d\alpha_2 d\alpha_3 \cdots$ over all $N\nu$ degrees of freedom is simplified by noting that the sum of the energies held in all the degrees of freedom must equal the total thermal energy of the system:

$$\varepsilon_1 + \varepsilon_2 + \cdots + \varepsilon_{N\nu} = \alpha_1^2 + \alpha_2^2 + \cdots + \alpha_{N\nu}^2 = E_{\text{therm}}$$

This equation defines the surface of an $N\nu$-dimensional sphere ($x_1^2 + x_2^2 + \cdots = R^2$) with radius $R^2 = E_{\text{therm}}$, so the integral $\int d\alpha_1 d\alpha_2 d\alpha_3 \cdots$ is the surface area of this sphere.

The area of an n-dimensional sphere is given by

$$\frac{(2\pi)^{n/2}}{(n-2)!!} R^{n-1} \qquad \text{if } n \text{ is even,} \qquad \sqrt{\frac{2}{\pi}} \frac{(2\pi)^{n/2}}{(n-2)!!} R^{n-1} \qquad \text{if } n \text{ is odd.}$$

At large n we can set $n - 2 \approx n - 1 \approx n$; we consider n to be even and, using Stirling's formula, we write

$$\frac{n!!}{2^{n/2}} = \left(\frac{n}{2}\right)\left(\frac{n-2}{2}\right) \cdots = \left(\frac{n}{2}\right)\left(\frac{n}{2} - 1\right) \cdots = \left(\frac{n}{2}\right)! \to \left(\frac{n}{2e}\right)^{n/2}. \tag{C.2}$$

So, with $n = N\nu$ and $R = \sqrt{E_{\text{therm}}}$, we get

$$\int d\alpha_1 d\alpha_2 \cdots d\alpha_{N\nu} \approx \frac{(2\pi)^{n/2}}{n!!} R^n = \left(\frac{2e\pi E_{\text{therm}}}{N\nu}\right)^{N\nu/2}. \tag{C.3}$$

Thus the integral in equation C.1 gives us a factor $(2e\pi E_{\text{therm}}/N\nu)^{1/2}$ for every degree of freedom. In addition, we must also include the following factors in our result:

- $b^{-1/2}$ for every degree of freedom (equation C.1);
- $1/h^{1/2}$ for every variable over which we integrate (equation 6.9);
- V from integration over the x, y, z coordinates for each particle when the energy is independent of the particle's position (e.g. in a gas);
- 2π for integration over the angle for each rotational degree of freedom;
- A factor $(e/N)^N$ when correcting for identical particles in a gas.

Altogether, then, the number of states $\Omega = \omega_c^N$ accessible to a system of $N\nu$ degrees of freedom is given by

$$\omega_c^N = \begin{cases} \left(\dfrac{eV}{Nh^{3/2}}\right)^N \dfrac{1}{h^{N\nu/2}} (b_1 b_2 \ldots b_{N\nu})^{-1/2} \left(\dfrac{2e\pi E_{\text{therm}}}{N\nu}\right)^{N\nu/2} & \text{gas,} \\[4mm] \dfrac{1}{h^{N\nu/2}} (b_1 b_2 \ldots b_{N\nu})^{-1/2} \left(\dfrac{2e\pi E_{\text{therm}}}{N\nu}\right)^{N\nu/2} & \text{solid,} \end{cases} \tag{C.4}$$

where the constant b has the following form:

$$b = \begin{cases} \dfrac{1}{2m} & \text{for each translational degree of freedom } \left(\varepsilon_i = \dfrac{1}{2m} p_i^2\right), \\[3mm] \dfrac{\kappa}{2} & \text{for each vibrational degree of freedom } \left(\varepsilon_i = \dfrac{\kappa}{2} x_i^2\right), \\[3mm] \dfrac{1}{2I} & \text{for each rotational degree of freedom } \left(\varepsilon_i = \dfrac{1}{2I} L_i^2\right). \end{cases}$$

There is also a factor 2π from the integration over the angle for each rotational degree of freedom. So, for the common examples used in this book, $\Omega = \omega_c^N$ where

$$\omega_c^N = \begin{cases} \dfrac{eV}{Nh^3} m^{3/2} \left(\dfrac{4e\pi E_{\text{therm}}}{3N}\right)^{3/2} & \text{monatomic gas } (\nu = 3), \\[4mm] \dfrac{eV}{Nh^5} 4\pi^2 I(m)^{3/2} \left(\dfrac{4e\pi E_{\text{therm}}}{5N}\right)^{5/2} & \text{diatomic gas } (\nu = 5), \\[4mm] \dfrac{1}{h^3} \left(\dfrac{m}{k}\right)^{3/2} \left(\dfrac{4e\pi E_{\text{therm}}}{6N}\right)^3 & \text{solid } (\nu = 6). \end{cases} \tag{C.5}$$

We can also write these in terms of the temperature using

$$\frac{E_{\text{therm}}}{N\nu} = \frac{1}{2} kT.$$

Appendix D   Energy distribution in interacting systems

We begin with equation 7.5 for the number of states accessible to two interacting systems with n_1 and n_2 degrees of freedom each (equation 7.5):

$$\Omega_0(E_1) = C E_1^{n_1/2} E_2^{n_2/2},$$

where $E_2 = E_0 - E_1$ and C is a constant of proportionality. Working with the logarithm of the number of states,

$$f(E_1) \equiv \ln \Omega_0 = \ln C + \frac{n_1}{2} \ln E_1 + \frac{n_2}{2} \ln(E_0 - E_1), \tag{D.1}$$

and expanding it in a Taylor series around $\overline{E}_1$ gives

$$f(E_1) = f(\overline{E}_1) + \left.\frac{\partial f}{\partial E_1}\right|_{\overline{E}_1} (E_1 - \overline{E}_1) + \frac{1}{2} \left.\frac{\partial^2 f}{\partial E_1^2}\right|_{\overline{E}_1} (E_1 - \overline{E}_1)^2 + \cdots. \tag{D.2}$$

We assume a distribution sharply peaked at $\overline{E}_1$. By setting the first derivative equal to zero, as must be true at a function's maximum, we find the position of the peak and get equipartition as a by-product (homework):

$$\left.\frac{\partial f}{\partial E_1}\right|_{\overline{E}_1} = 0 \quad \Rightarrow \quad \frac{\overline{E}_1}{\overline{E}_2} = \frac{n_1}{n_2}. \tag{D.3}$$

That is, the energy is distributed in proportion to the number of degrees of freedom. The second derivative is also easily evaluated from equation D.1 above, giving (homework)

$$\left.\frac{\partial^2 f}{\partial E_1^2}\right|_{\overline{E}_1} = -\frac{n_0}{2\overline{E}_1 \overline{E}_2}. \tag{D.4}$$

Now that we know the first and second derivatives, we can take the antilogarithm of the Taylor series D.2 to get the Gaussian form

$$\Omega_0(E_1) = \Omega_0(\overline{E}_1) e^{-(E_1 - \overline{E}_1)^2/2\sigma^2}, \quad \text{with } \sigma^2 = \frac{2\overline{E}_1 \overline{E}_2}{n_0} = \frac{n_1 n_2}{2 n_0}(kT)^2, \tag{D.5}$$

where we have used $\overline{E}_i = (n_i/2)kT$ in evaluating σ^2. Because the probability is proportional to the number of states ($P \propto \Omega_0$), and the sum over all possible distributions gives a total probability of 1, the distribution is the familiar Gaussian form of equation 3.7:

$$P(E_1) = \frac{1}{\sqrt{2\pi}\,\sigma} e^{-(E_1 - \overline{E}_1)^2/2\sigma^2}. \tag{D.6}$$

Appendix E   Standard integrals

$$\int_{-\infty}^{\infty} e^{-\alpha x^2} dx = 2 \int_{0}^{\infty} e^{-\alpha x^2} dx = \left(\frac{\pi}{\alpha}\right)^{1/2} \tag{E.1}$$

$$\int_{-\infty}^{\infty} e^{-\alpha x^2} x^{2n} dx = 2 \int_{0}^{\infty} e^{-\alpha x^2} x^{2n} dx$$

$$= \frac{1 \times 3 \times 5 \times \cdots \times (2n - 1)}{2^n \alpha^n} \left(\frac{\pi}{\alpha}\right)^{1/2} \tag{E.2}$$

$$\int_{-\infty}^{\infty} e^{-\alpha x^2} x^{2n+1} dx = 0 \tag{E.3}$$

$$\int_{0}^{\infty} e^{-\alpha x^2} x^{2n+1} dx = \frac{n!}{2\alpha^{n+1}} \tag{E.4}$$

Appendix F Nearly degenerate fermion gas

To evaluate the two integrals in equations 20.18 for nearly degenerate fermions, we integrate by parts. Both integrals have the form

$$
I(y) = \int_0^\infty \underbrace{\left(\frac{1}{e^{x-y}+1}\right)}_{u} \underbrace{(x^\kappa dx)}_{dv} = \underbrace{\left(\frac{1}{e^{x-y}+1}\right)}_{u} \underbrace{\left(\frac{x^{\kappa+1}}{\kappa+1}\right)}_{v} \Bigg|_0^\infty
$$

$$
- \int_0^\infty \underbrace{\left(\frac{x^{\kappa+1}}{\kappa+1}\right)}_{v} \underbrace{\left(\frac{-e^{x-y} dx}{(e^{x-y}+1)^2}\right)}_{du}
$$

where $\kappa = 1/2$ for $f(y)$ and $\kappa = 3/2$ for $g(y)$. The first term on the right (uv) is zero at both end points, and the second term ($-\int v du$) is evaluated by noting that the fermion occupation number varies significantly only near $\varepsilon \approx \mu$. Hence, du is only nonzero near the point $x = y$, and so it is convenient to expand v in a Taylor series around $x = y$: $v(x) = v|_{x=y} + v'|_{x=y}(x - y) + \cdots$. Substituting the expression for v we obtain.

$$
v = \frac{x^{\kappa+1}}{\kappa+1} = \frac{y^{\kappa+1}}{\kappa+1} + y^\kappa(x - y) + \frac{\kappa y^{\kappa-1}}{2}(x - y)^2 + \cdots
$$

We now change variables to $z = x - y$ and extend the lower limit from $x = 0$ to $x = -\infty$ (since $u = 1$, hence $du = 0$ in that region), so the above integral becomes

$$
I(y) = \frac{y^{\kappa+1}}{\kappa+1} \int_{-\infty}^\infty \frac{e^z dz}{(e^z+1)^2} + y^\kappa \int_{-\infty}^\infty \frac{e^z z dz}{(e^z+1)^2}
$$

$$
+ \frac{\kappa y^{\kappa-1}}{2} \int_{-\infty}^\infty \frac{e^z z^2 dz}{(e^z+1)^2} + \cdots
$$

Of these three integrals, the first gives unity, the second is zero (the integrand is odd),[1] and the third gives $\pi^2/3$. So our answer is

$$
I(y) = \frac{y^{\kappa+1}}{\kappa+1} + \frac{\kappa \pi^2 y^{\kappa-1}}{6} + \cdots
$$

Because $\kappa = 1/2$ for $f(y)$ and $\kappa = 3/2$ for $g(y)$, we have

$$
f(y) = \frac{2}{3} y^{3/2} \left(1 + \frac{\pi^2}{8y^2} + \cdots\right),
$$
$$
g(y) = \frac{2}{5} y^{5/2} \left(1 + \frac{5\pi^2}{8y^2} + \cdots\right).
$$
(F.1)

We put the result for $f(y)$ with $y = \mu/kT$ into equation 20.18a, divide both sides by $(2/3)C$, and use equation 20.13 for ε_f to write this as

$$
\frac{3N}{2C} = \varepsilon_f^{3/2} = \mu^{3/2}\left[1 + \frac{\pi^2}{8}\left(\frac{kT}{\mu}\right)^2 + \cdots\right]
$$

Now, for our approximation, we note that $kT/\mu \ll 1$ for degenerate fermions, so the correction term is small. So we can put $\mu \approx \varepsilon_f$ in the correction term and ignore all higher

---

[1] $e^z z/(e^z + 1)^2 = z/(e^{z/2} + e^{-z/2})^2$.

order terms to get

$$\varepsilon_{\mathrm{f}}^{3/2} \approx \mu^{3/2}\left[1 + \frac{\pi^2}{8}\left(\frac{kT}{\varepsilon_{\mathrm{f}}}\right)^2\right] \quad\Rightarrow\quad \mu \approx \varepsilon_{\mathrm{f}}\left[1 + \frac{\pi^2}{8}\left(\frac{kT}{\varepsilon_{\mathrm{f}}}\right)^2\right]^{-2/3}.$$

We next use $(1+\delta)^{-x} \approx 1 - x\delta$ for small δ to get

$$\mu \approx \varepsilon_{\mathrm{f}}\left[1 - \frac{\pi^2}{12}\left(\frac{kT}{\varepsilon_{\mathrm{f}}}\right)^2\right]. \tag{F.2}$$

Regarding the energy, we notice that the average energy per particle is given by

$$\bar{\varepsilon} = \frac{E}{N} = kT\frac{g(y)}{f(y)}.$$

Using $y = \mu/(kT)$, results F.1 for $g(y)$ and $f(y)$, and the above approximations we get

$$\bar{\varepsilon} = kT\frac{g(y)}{f(y)} \approx \frac{3}{5}\mu\left(1 + \frac{5\pi^2}{8y^2}\right)\left(1 + \frac{\pi^2}{8y^2}\right)^{-1} \approx \frac{3}{5}\mu\left(1 + \frac{\pi^2}{2y^2}\right).$$

With μ from equation F.2 this becomes

$$\bar{\varepsilon} \approx \frac{3}{5}\varepsilon_{\mathrm{f}}\left[1 - \frac{\pi^2}{12}\left(\frac{kT}{\varepsilon_{\mathrm{f}}}\right)^2\right]\left[1 + \frac{\pi^2}{2}\left(\frac{kT}{\varepsilon_{\mathrm{f}}}\right)^2\right] \approx \frac{3}{5}\varepsilon_{\mathrm{f}}\left[1 + \frac{5\pi^2}{12}\left(\frac{kT}{\varepsilon_{\mathrm{f}}}\right)^2\right]. \tag{F.3}$$

Further reading

Introductory, statistical approach

Terell Hill, *An Introduction to Statistical Thermodynamics*, Dover, 1986.
Charles Kittel, *Elementary Statistical Physics*, Krieger Publishing Co., 1988.
Franz Mandl, *Statistical Physics*, second edition, Wiley, 1988.
Frederick Reif, *Statistical Physics: Berkley Physics Course – Volume 5*, McGraw-Hill, 1967.
Daniel Schroeder, *An Introduction to Thermal Physics*, Addison Wesley, 2000.

Introductory, postulatory approach

Michael Abbot and Hendrick Van Ness, *Schaum's Outline of Thermodynamics with Applications*, McGraw-Hill, 1989.
Ralph Baierlein, *Thermal Physics*, Cambridge University Press, 1999.
Herbert Callen, *Thermodynamics and an Introduction to Thermostatistics*, second edition, Wiley, 1985.
Ashley Carter, *Classical and Statistical Thermodynamics*, Prentice Hall, 2001.
Enrico Fermi, *Thermodynamics*, Dover, 1937.
C. B. P. Finn, *Thermal Physics*, second edition, Routledge and Kegan Paul Books, 1986.
Charles Kittel and Herbert Kroemer, *Thermal Physics*, second edition, W. H. Freeman, 1980.
H. C. Van Ness, *Understanding Thermodynamics*, Dover, 1983.

More advanced

David Chandler, *Introduction to Modern Statistical Mechanics*, Oxford University Press, 1987.
Kerson Huang, *Introduction to Statistical Physics*, Taylor and Francis, 2001.
L. D. Landau and E. M. Lifshitz, *Statistical Physics*, third edition, Pergamon Press, 1980.
Michael Moran and Howard Shapiro, *Fundamentals of Engineering Thermodynamics*, Wiley, 2000.
R. K. Pathria, Statistical Mechanics, second edition, Butterworth-Heinemann, 1996.
L. E. Reichl, *A Modern Course in Statistical Physics*, second edition, Wiley, 1998 (very advanced).
Frederick Reif, *Fundamentals of Statistical and Thermal Physics*, McGraw Hill, 1965.
Richard Sonntag, Claus Borgnakke, and Gordon Van Wylen, *Fundamentals of Thermodynamics*, Wiley, 1997.

Problem solutions

Chapter 1

1. 6×10^{-6} **3.** The result of 20 flips is closer to 50–50 on average. The probability of exactly 10 heads is small (0.18). **7.** Foolish. The probability each time is $1/2$, regardless of past history. **9.** 12 **11.** (a) 0.2 nm (b) 3.3×10^{-24} kg m/s (c) Electron, 3.6×10^6 m/s; proton, 2.0×10^3 m/s (d) Electron, 38 eV; proton, 0.021 eV (e) 6200 eV **13.** $\Delta p_x = 6.6 \times 10^{-24}$ kg m/s, $\Delta v_x = 7.3 \times 10^6$ m/s **15.** 13 **17.** (a) 1.6×10^{-14} m (b) 3.2 MeV **19.** (a) $\times 2$ (b) $\times 2$ (c) $\times 8$ **21.** (a) 7.6×10^{54} states per joule or 1.2×10^{36} states per eV (b) 8.4 states per eV **23.** The proton and electron could be in a state with orbital angular momentum corresponding to $l = 1$, oriented opposite to the two parallel spins. **25.** $|\mu| = 2.44 \times 10^{-26}$ J/T, $\mu_z = \pm 1.41 \times 10^{-26}$ J/T, $U = \pm 1.41 \times 10^{-26}$ J **27.** $\pm 7.4 \times 10^{-24}$ J, $\pm 3.7 \times 10^{-24}$ J, 0 J **29.** If the kinetic energy were zero, the momentum would be zero, the wavelength would be infinite, and so the particle would not be contained within the potential well. **31.** $77 \, (2^{77} \approx 10^{23})$

Chapter 2

1. (a) $f + g = \sum P_s (f_s + g_s) = \sum P_s f_{ss} + \sum P_s g_s = \overline{f} + \overline{g}$ (b) $cf = \sum P_s c f_s = c \sum P_s f_s = \overline{cf}$ **3.** (a) $33\frac{1}{6}$ (b) $5\frac{1}{6}$ (c) $-\frac{4}{3}$ (d) $63\frac{1}{2}$ **5.** $7\frac{7}{8}$ **7.** (a) 3/8 (b) 6 (c) hhtt, htht, htth, thht, thth, tthh, yes **9.** 1/36 **11.** (a) 0.0042, 56 (b) 0.0046 **13.** (a) 0.313 (b) 10 **15.** 1820 **17.** 0.0252 **19.** (a) 6.7×10^{-6} (b) 4.5×10^{-2} (c) 2.1 $\times 10^{-35}$ **21.** 1.0×10^{29} **23.** (a) 0.062 (b) 0.062 **27.** 3/40 (there are 40 unseen cards and 3 of them are queens) **29.** 5.8 cents **31.** 12.25

Chapter 3

1. (a) 16.7 (b) 3.7 (c) 0.22 **3.** Justify each step in equation 3.3. **5.** (a) 10 (b) 2.6 (c) 0.26 **7.** (a) 50 (b) 5 (c) 0.080 (d) 0.067 **(9)** (a) (1) Definition of mean values with $n^2 = \sum n^2 P_n(n)$; (2) $n^2 p^n = \left(p \frac{\partial}{\partial p} \right)^2 p^n$; (3) binomial expansion, $\sum_n \frac{N!}{n!(N-n)!} p^n q^{N-n} = (p+q)^N$. (b) $\overline{n} = \left(p \frac{\partial}{\partial p} \right)(p+q)^n$ (c) Find the derivatives and evaluate them at $p + q = 1 \, (q = 1 - p$, etc.). **11.** 10^{116} **13.** We know with certainty (i.e., probability $= 1$) that a system must be in one of its possible configurations. Therefore, the sum over all configurations must give a total probability equal to unity. **15.** 0.607 **17.** (a) 60 (b) 7.07 (c) 0.0564 (d) 7.9×10^{-3} **19.** (a) 0.0133 (b) 5×10^{-198} **21.** (a) Gaussian, 0.44; binomial, 0.40 (b) Gaussian, 0.040; binomial, 0.054 (c) Gaussian, 0.24; binomial, 0.34 **23.** (a) Gaussian, 0.178; binomial, 0.176 (b) Gaussian, 0.120;

binomial, 0.120 (c) Gaussian, 0.015; binomial, 0.015, (d) Gaussian, 0.00030; binomial, 0.00018 **25.** $1 = A^2 \int e^{-B(x^2+y^2)} dx dy = A^2 \int e^{-Br^2} r \, dr \, d\theta = \pi A^2 / B$ **27.** (a) 0.20 m, 0.872 m (b) 80 m, 17.4 m (c) 4.4, 0.22 **29.** (a) 0 (b) 3.33×10^{-11} m^2 (c) 0 (d) 0.026 m (e) 1.08×10^5 s = 30 hours **31.** (a) 0, 1.73×10^{-10} m (b) 0, 5.5×10^{-4} m **33.** (a) 2.5×10^{-14} m (b) 5×10^{-10} m (c) 12.6 A **35.** 4.64 m, 36.2 m, 176 m (using the standard deviation as the characteristic spread) **37.** $\overline{S_N^2} - \overline{S_N}^2 = N\overline{s^2} + N(N-1)\bar{s}^2 - (N\bar{s})^2 = N(\overline{s^2} - \bar{s}^2) = N\sigma^2$

Chapter 4

3. (a) $f'(0) = 0$, $f''(0) = +2$ (b) $f(x) = -2 + 0x + x^2 + x^3 + 0x^4$ (c) $(-0.57, +0.54)$
5. $\sin x = \sum_{n=0}^{\infty} \frac{(-1)^n}{(2n+1)!} x^{2n+1}$, $\cos x = \sum_{n=0}^{\infty} \frac{(-1)^n}{(2n)!} x^{2n}$, $\ln(1+x) = \sum_{n=1}^{\infty} \frac{(-1)^{n+1}}{n} x^n$, $e^x = \sum_{n=0}^{\infty} \frac{1}{n!} x^n$ **7.** B, C, A **9.** (a) 1.42×10^{-46} kg m^2 (b) 4.9×10^{-4} eV (c) 5.7 K
11. (a) 4.58×10^{-48} kg m^2 (b) 1.5×10^{-2} eV (c) 176 K **13.** H$_2$O is not a linear molecule as is N$_2$. Therefore, H$_2$O has appreciable moments of inertia about all three rotational axes, and so rotations about all three axes can be excited. For N$_2$, however, the rotational inertia about one axis (the one that passes through both atomic nuclei) is extremely small, and so the energy of even the first excited rotational state about this axis is too high to be reached.
15. (a) 6 (b) 3 (c) no (d) decreased (e) The iron atoms are released from the potential wells in which they were bound when in the solid state, so their potential energy in the liquid state is higher (although still negative). **17.** 20.8 J/(mole K) **19.** (a) 0.0707 eV, 0.0707 eV (b) 0.0625 eV (c) $\Delta u_0 = +0.0625$ eV (d) It remains the same. **21.** (a) 35.3 J (1.74×10^{22} degrees of freedom) (b) 420 J (2.076×10^{23} degrees of freedom)

Chapter 5

1. (a) yes (b) Probably (c) Although the air and walls emit more radiation altogether, only a small fraction of that hits the rock. **3.** (a) It decreases, because the system does work, rather than having work done on it. (b) As the system contracts, the potential energy of each particle decreases. (That is why it contracts. The particles are seeking the configuration of lower potential energy.) Like a ball rolling downhill, each particle's average kinetic (hence thermal) energy increases as the potential well deepens. **5.** (a) 120 J (b) 796 K **7.** u_0 is negative, because thermal energy is released when the H$_2$SO$_4$ molecule enters the solution. The change in chemical potential is negative, because the molecules go into solution rather than out of solution. They go in the direction that lowers their chemical potential. **9.** (a) The temperature would fall because the same thermal energy would be distributed over more degrees of freedom, meaning less thermal energy per degree of freedom. (b) The temperature would fall because the gain in potential energy leaves less energy to be distributed among the thermal degrees of freedom. **11.** 15600J/mol **13.** 10700J/mol **15.** 1.6×10^9 K **17.** $\nu = 18.1$ **19.** (a) In liquid water the molecules are closer together. Their mutually attractive forces are stronger, and therefore the potential well is deeper at these closer distances. (b) In ice, the reduced thermal motion of the molecules allows them to orient and space themselves in a way that lowers their potential energy. (c) Compared with the molecules of oil, the water molecule is much more highly polarized, leading to a stronger electrostatic attraction between the charged parts of the water molecule and the charged salt ions. **21.** (a) 20.3 J (b) 4.9 cal (c) u_0 rises. (Less thermal energy but the same total

Problem solutions

internal energy.)   **23.** (a) 1.52×10^5 J   (b) $\Delta T = -22$ K   **25.** (a) exact   (b) exact
(c) inexact   (d) exact   (e) exact   **27.** (1) 47 (2) 47, yes   **29.** (a) $p_i(V_f - V_i)$
(b) $p_f(V_f - V_i)$ (c) yes   **31.** (a) 142   (b) 142   (c) 142   **33.** (a) $2w/y^3$ (b) $\frac{3}{2}\sqrt{yz}$
(c) $\frac{1}{2}\sqrt{z/y}$ (d) $2/3\sqrt[3]{wz}$ **35.** $y(z + 1)$ **37.** $s/2\sqrt{t} - se^{st}$

Chapter 6

1. (a) $10^{1.8 \times 10^{23}}$ (b) $10^{4.3 \times 10^{21}}$ (c) $10^{4.3 \times 10^{20}}$ **3.** (a) 10   (b) 10^{12} (c) 10^{900} (d) 10^{900}
(e) $10^{99 \times 10^{22}}$ **5.** (a) $+++, ++-, +-+, -++, +--, -+-, --+, ---$
(b) 3/8   (c) 2/8   **7.** (a) 1.3×10^{-24} (b) 0.30   (c) 0.48   **9.** (a) 2.9×10^{-25} eV   (b) 11.4
$\times 10^{-25}$ eV   (c) 8.6×10^{-25} eV   **11.** (a) hh, ht, th, tt   (b) hhh, hht, hth, thh, htt, tht, tth, ttt
(c) hhhh, hhht, hhth, hthh, thhh, hhtt, htht, htth, thht, thth, tthh, httt, thtt, ttht, ttth, tttt   (d) yes
13. (a) 64   (b) 7   **15.** (a) 1.1×10^9 (b) 3.5×10^9 (c) 3.7×10^{18} **17.** (a) 20, 10   (b)
9900, 4950   (c) 9.7×10^5, 1.6×10^5 (c) 1.0×10^9 1.7×10^8 **19.** (a) Macroscopic   (b)
4.6×10^{26} **21.** (a) 8.4×10^{24} (b) 1.26   (c) $10^{1.9 \times 10^{24}}$ **23.** (a) 3.0×10^7 J   (b)
4.64×10^{-67}(kg m/s)^3 (c) 2.0×10^{35} (d) 1.82×10^8 (e) $10^{2.48 \times 10^{28}}$
(f) $\omega_c(N_2) = 2.29 \times 10^8$, $\omega_c(O_2) = 8.63 \times 10^8$, $\Omega = 10^{2.54 \times 10^{28}}$ **25.** When it melts, the
volume in coordinate space accessible to each molecule increases immensely, because the
molecules become mobile and can move throughout the fluid. This volume is much larger and
affords access to many more states than did the much more restricted volume that was
represented in the three potential-energy degrees of freedom of the solid state.

Chapter 7

1. (a) Set $df/dE = 0$ for each, which gives $E = 5n/(n + m)$, where n is the first exponent
and m the second. For all three cases the ratio $n/(n + m)$ is 0.4, giving $E = 2$.
(b) $f(2)/f(1) = 1.7, 187, 10^{2.27 \times 10^{22}}$; $f(2)/f(3) = 1.5, 58, 10^{1.76 \times 10^{22}}$.
3. (a) $(E_1, E_2, \Omega_1, \Omega_2, \Omega_0) = (0, 4, 0, 16, 0), (1, 3, 1, 9, 9), (2, 2, 2.8, 4, 11.3), (3, 1, 5.2,$
$1, 5.2), (4, 0, 8, 0, 0)$, total $= 25.5$.   (b) 0.20   (c) (2, 2), 0.44 **5.** (a) $(E_1, E_2, \Omega_1, \Omega_2, \Omega_0)$=(0,
$6, 0, 216, 0), (1, 5, 1, 125, 125), (2, 4, 4, 64, 256), (3, 3, 9, 27, 243), (4, 2, 16, 8, 128), (5, 1,$
$25, 1, 25),(6, 0, 36, 0, 0)$, total=777 (b) (2, 4), 0.33   **7.** (a) $(E_1, E_2, E_3, \Omega_1, \Omega_2, \Omega_3, \Omega_0)$
$= (4, 0, 0, 16, 0, 0, 0), (3, 1, 0, 9, 1, 0, 0), (3, 0, 1, 9, 0, 1, 0), (2, 2, 0, 4, 5.7, 0, 0), (2, 1, 1,$
$4, 1, 1, 4), (2, 0, 2, 4, 0, 8, 0), (1, 3, 0, 1, 15.6, 0, 0), (1, 2, 1, 1, 5.7, 1, 5.7), (1, 1, 2, 1, 1, 8, 8),$
$(1, 0, 3, 1, 0, 27, 0), (0, 4, 0, 0, 32, 0, 0), (0, 3, 1, 0, 15.6, 1, 0), (0, 2, 2, 0, 5.7, 8, 0), (0, 1, 3,$
$0, 1, 27, 0), (0, 0, 4, 0, 0, 64, 0)$, total $= 17.7$ (b) (1, 1, 2), 0.45   **9.** All accessible states
have to be equally probable in order for the probability for any particular configuration to be
proportional to the number of states corresponding to that configuration.   **11.** (a) 10
(b) $10^{10^{23}}$ (c) $10^{99 \times 10^{23}}$ **13.** $10^{2.000\,000\,000\,000\,000\,000\,09 \times 10^{20}}$ **15.** (a) $(E_1, E_2, \Omega_1, \Omega_2, \Omega_0) =$
$(0, 5, 0, \exp(6.99 \times 10^{24}), 0), (1, 4, 1, \exp(6.02 \times 10^{24}), \exp(6.02 \times$
$10^{24})), (2, 3, \exp(3.61 \times 10^{24}), \exp(4.77 \times 10^{24}), \exp(8.38 \times 10^{24})), (3, 2, \exp(5.73 \times$
$10^{24}), \exp(3.01 \times 10^{24}), \exp(8.74 \times 10^{24})), (4, 1, \exp(7.22 \times 10^{24}), 1, \exp(7.22 \times$
$10^{24})), (5, 0, \exp(8.39 \times 10^{24}), 0, 0)$ **17.** $10^{-3.62 \times 10^{14}}$ **19.** $(E_1, E_2, \Omega_1, \Omega_2, \Omega_0) =$
$(0, 4, 0, \exp(3.61 \times 10^{24}), 0), (1, 3, 1, \exp(2.86 \times 10^{24}), \exp(2.86 \times 10^{24})),$
$(2, 2, \exp(0.60 \times 10^{24}), \exp(1.81 \times 10^{24}), \exp(2.41 \times 10^{24})), (3, 1, \exp(0.95 \times$
$10^{24}), 1, \exp(0.95 \times 10^{24})), (4, 0, \exp(1.20 \times 10^{24}), 0, 0)$, total $= \exp(2.86 \times 10^{24})$
(b) $10^{-4.5 \times 10^{23}}$, or one chance in $10^{4.5 \times 10^{23}}$ **21.** $10^{216\,976}$ **23.** (a) At the peak, the

distribution has a maximum, so its derivative is zero.   (b) Because each degree of freedom carries on average the same energy the total energy E_0 is apportioned in proportion to the number of degrees of freedom.   (c) The second derivative gives $-n_1/E_1{}^2 - n_2/E_2{}^2$, which you can manipulate using $n_1/E_1 = n_2/E_2 = n_0/E_0$.   (d) Substitute the results for the first and second derivatives into equation 7.5 and exponentiate it.   **27.** Heat flowing from cold to hot, fluids flowing from lower pressure to higher pressure, particles going from regions of low concentration to high concentration, friction speeding things up and cooling them off, etc. In short, for such processes energy is conserved but flows the wrong way.   **29.** 8   (b) 0.96 $\times 10^{-23}$ J/K, 1.91×10^{-23} J/K   (c) 2.87×10^{-23} J/K   **31.** (a) $10^{5 \times 10^{24}}$ (b) 63.6 J/K, 95.4 J/K   (c) 159.0 J/K   **33.** (a) $10^{1.30 \times 10^{22}}$ (b) $\Delta S = k \ln(\Omega_f/\Omega_i) = 0.41$ J/K   **35.** (a) 16 J   (b) 1151 J/K, 3499 J/K, 4650 J/K   **37.** (a) 6.21×10^5 J   (b) $10^{2.2 \times 10^{23}}$ (c) 6.9 J/K   **39.** There is a greater volume in momentum space. (For molecules with the same kinetic energy, the ones with the larger mass have the greater momentum.)

Chapter 8

1. (a) $3N/2$ (b) Write $S = k \ln \Omega$ with $\Omega = CE^{3N/2}$ and use this in equation 8.2 for $1/T$.   (c) Use $S = k \ln \Omega$ and equation 8.2 to get $E = \alpha N v k T$.   **3.** (a) 20, 16   (b) $E_1 = 5.56 \times 10^{-19}$ J, $E_2 = 4.44 \times 10^{-19}$ J   (c) 2.7×10^{-21} J/K   (d) 4020 K   **5.** 3.77×10^{-23} J/K, 137 K   **7.** (a) 0.079 J/K   (b) $10^{2.49 \times 10^{21}}$ **9.** (a) $+3.67 \times 10^{-3}$ J/K   (b) -3.53×10^{-3} J/K   (c) $+0.14 \times 10^{-3}$ J/K   **11.** (a) Use $p/T = (\partial S/\partial V)_{E,N}$, where $S = k \ln \Omega$, to get $pV = NkT$.   (b) $R = N_A k$ **13.** Use $1/T = (\partial S/\partial E)_{V,N}$, where $S = k \ln \Omega$, to get (a) $E = NkT$ (b) $E = 2NkT$ (c) $E = 3NkT$ **15.** (a) $\Omega_0 = CV_1^{N_1}(V_0 - V_1)^{N_2}$ (b) $dS_0 = 0 = dS_1 + dS_2 = (\partial S_1/\partial V_1 - \partial S_2/\partial V_2)dV_1$ (c) $p_1/T_1 = p_2/T_2$ **17.** 250 J/K   **19.** (a) About 10^7 to 10^8 J, depending on the room's size   (b) 660 J   (c) 60 J   (d) 2.8×10^{26} J   **21.** (a) 3.2×10^{-3} J/K   (b) 314 K   **23.** $1/T, 1/p$ **25.** (a) 3.3×10^{-3} J/K   (b) $10^{1.05 \times 10^{20}}$ **27.** (a) 3.57×10^{-4} J/K   (b) 3.57×10^{-4} J/K   (c) $10^{1.1 \times 10^{19}}$ for each   **29.** (a) 3.38×10^{-7} J/K   (b) $10^{1.06 \times 10^{16}}$ (c) 3.48×10^{-8} J/K, $10^{1.1 \times 10^{15}}$ **31.** 9.6×10^{-24} J/K   **33.** (a) 0   (b) Although there is greater room in momentum space, there is correspondingly less room in coordinate space, due to the smaller volume.   **35.** -0.25 eV   **37.** (a) 3.4 J/K   (b) 0.10 J/K   (c) 0.71 J/K   (d) 1.4 J/K   **39.** c_p would be larger. If the system is held at constant pressure and allowed to expand, the expansion would cause some cooling, resulting in a smaller net rise in temperature as heat is added. A smaller rise in temperature means a larger heat capacity.   **41.** (a) For a mole, $Nk = R$. Write the first law as $dE = dQ - pdV = (v/2)RdT$. Then look at dQ/dT for $dV = 0$ (c_V) and for $pdV = RdT$ (c_p).   **43.** (a) 7.26 J/K   (b) $10^{2.3 \times 10^{23}}$ **45.** (a) 13 700 J   (b) The heat released goes to some other system whose entropy increases as a result of the heat transfer. The total entropy of the combined system rises.   **47.** $+0.376$ eV   (b) 0.136 J/K   **49.** 8730 J/K

Chapter 9

1. (a) 1   (b) 1   **3.** Perhaps some manifestation of heat flowing from cold to hot, or something moving in the opposite direction to the net applied force, particles diffusing from low chemical potential to high chemical potential.   **5.** (a) (0, 0)   (b) 0   (c) 0   (d) negative   (e) negative   **7.** Carry out the derivation from equation 9.4 to equation 9.6, with the constraint that $dS_0 > 0$.   **9.** The condition 9.3a applies to systems interacting thermally only, which precludes particles entering or leaving (i.e. diffusive interactions).   **11.** $\mu_{crystal}$ < $\mu_{amorphous}$ **13.** Mix in some water and shake it. The sugar moves from the oil to the water, where its chemical potential is lower. Let it stand, so that the water separates, and then remove the water. If you want to recover the sugar, evaporate the water.   **17.** No, because the condition is for purely thermal interactions. The volume of melting ice is changing, so it is engaging in mechanical interactions as well. If the volume were held constant, the pressure would fall and the freezing point would rise ($\Delta T > 0$) as heat was added ($\Delta Q > 0$) to melt the ice. The condition would be met.   **19.** You must explain how to measure ΔT and ΔQ for thermal interactions, Δp and ΔV for mechanical interactions, or $\Delta \mu$ and ΔN for diffusive interactions. ($\Delta \mu$ will probably be the most difficult, unless you are using voltmeters for the diffusion of charged particles.)   **21.** (a) 2.04×10^{-19} J   (b) 41.7 K, 0.288×10^{-19} J (c) 0.141 for both   **23.** (a) 3780 J   (b) 3.78×10^{-9} J   (c) 1.0×10^{-12} **25.** 2.25 μm **27.** (a) 3.35×10^{10} (b) 3.2×10^{-6} **29.** (a) 6.1×10^{-3} (b) 6.1×10^{-12} **33.** (a) You could seal the oven and use the air pressure as a measure of its temperature. (b) Yes. You wouldn't be able to get food in or out without breaking the pressure seal.   **35.** (a) $46\,°C$ (b) $-281\,°C$ **39.** Whatever the values of TS and pV for the system, these values could be have been obtained in an infinite number of different ways, with different heat transfers and different amounts of work. For example, imagine starting with a system in a canister at fixed volume near absolute zero, where the pressure is nearly zero ($pV \approx 0$). Then simply add heat until the pressure rises to some arbitrary value. This value of pV was obtained without any work at all having been done. Similarly, you could get any value of TS through volume changes without any heat at all having been added.   **41.** Take the differential and subtract $dE = TdS - pdV + \mu dN$. **43.** In each case, write out the full derivative of the appropriate function in equation 9.14′ (e.g., $dF = -pdV - Vdp + \mu dN + Nd\mu$) and then substitute $-SdT + Vdp$ for $Nd\mu$.   **45.** Just carry out the steps for these proofs that are outlined in subsections F.2, F.3, and F.4.   **47.** $dH = TdS + Vdp + \mu dN$; $dH = TdS =$ heat added only if $dp = dN = 0$.   **49.** For the two interacting systems, $\Delta G_0 = \Delta G_1 + \Delta G_2 = (\mu_1 - \mu_2)\Delta N$, since $\Delta T = \Delta p = 0$. Make A_2 a reservoir, so that μ_2 is constant. Write $\overline{\mu}_1 = \mu_1 + \Delta\mu/2$. To first order $\Delta G_0 = 0$, because $\mu_1 = \mu_2$ in equilibrium. To second order $\Delta G_0 = \Delta\mu\Delta N/2$, and this term must be positive according to the result 9.3c. If first order terms are zero and second order terms are positive then the function is a minimum.   **51.** Each has the form $dw = fdx + gdy + hdz$, so you will get the relations corresponding to $(\partial f/\partial y)_{x,z} = (\partial g/\partial x)_{y,z}$, $(\partial f/\partial z)_{x,y} = (\partial h/\partial x)_{y,z}$, $(\partial g/\partial z)_{x,y} = (\partial h/\partial y)_{x,z}$.   **53.** (a) Add ΔN particles. To ensure that T and p retain their original values, two quantities must be adjusted, such as ΔV and ΔQ, until the two are at their original values. Then the net change in volume ΔV is recorded. (b) Add heat ($\Delta S = \Delta Q/T$) to a system of fixed number of particles adjust the volume (ΔV) so that the pressure returns to its original value. (c) Add heat to a system of a fixed number of particles kept at constant volume. Record both Δp and ΔT.   **55.** Solve $\omega_c = \frac{(eV/N)(4/3)\pi p^3}{h^3}$ for p, with $\omega_c = 1$ and the given V/N. KE $= p^2/2m$. (a) 6.4 eV   (b) 0.30 MeV

Chapter 10

1. Four **3.** (a) $1/T = CV^2/E$ or $E = CV^2T$ (b) $p/T = 2CV \ln E$ (c) $E = CV^2T$, $(\partial E/\partial V)_T = 2CVT = 2E/V$ (d) $p = (2x/V)e^x$, where $x = S/CV^2$; $\partial p/\partial V)_S = (-2x/V^2)(3 + 2x)e^x$ **5.** $E = (\nu/2)pV$, $(\partial E/\partial p)_{V,N} = (\nu/2)V$ **9.** (a) Two (b) $2N$ (c) 41.4 J (d) $\Omega = (CAE/N^2)^N$ (e) $S = Nk \ln(CAE/N^2)$ **11.** (a) $S = k \ln C + bkV^{4/5} + 2Nk \ln E$ (b) $E = 2NkT$ (c) $4N$ (d) $pV^{1/5} = (4/5)bkT$ **13.** (a) $N/2$

(b) Flux = density × velocity = number per area per second. So density × velocity × area = number per second = $(N/2V)v_x A$ (c) $2mv_x$ (d) Force = $\Delta p/\Delta t$ = number per second × impulse per particle = $(N/2V)v_x A(2mv_x)$ (e) Divide the force by the area: $p = (N/V)mv_x^2$ (f) Use the average value of $(1/2)mv_x^2 = (1/2)kT$ **15.** $\mu = kT[(\nu + 2)/2 - \ln\omega_c] = (E + pV - TS)/N$ **17.** Obtain $S = k\ln C + ka V^{1/2} + kbV \ln E$, and use then equations 8.10. (a) $T = E/(kbV)$ (b) $p = kT[(a/2V^{1/2}) + b\ln E]$ **19.** $H = E + pV = (6/2)RT + RT = 25\,700\,\text{J}$ **21.** 28.97 g, 28.64 g **23.** (a) $\omega_c = 3.8 \times 10^6$, $S = 126\,\text{J/(mole K)}$ (b) $\omega_c = 1.3 \times 10^{11}$, $S = 213\,\text{J/(mole K)}$ (c) $\omega_c = 250$, $S = 45.9\,\text{J/(mole K)}$ **25.** (a) $\Delta H = 7540\,\text{J/mole}$, $\Delta S = 23.5\,\text{J/(mole K)}$, $\Delta E = 7540\,\text{J/mole}$ (b) $\Delta H = 4.07 \times 10^4\,\text{J/mole}$, $\Delta S = 109\,\text{J/(mole K)}$, $\Delta E = 3.76 \times 10^4\,\text{J/mole}$ (c) $\Delta H = 3570\,\text{J/mole}$, $\Delta S = 8.5\,\text{J/(mole K)}$, $\Delta E = 2740\,\text{J/mole}$ **27.** (a) $3.1 \times 10^{-10}\,\text{m}$ (b) $33 \times 10^{-10}\,\text{m}$ (c)11 **29.** Since the repulsive potential energy would be proportional to $1/r \approx \nu^{-1/3}$, and work = $\int p dv$, we would expect an added pressure term $\approx \nu^{-4/3}$, giving $(p + a/\nu^2 - c/\nu^{4/3})(\nu - b) = RT$ **31.** water, $\nu = 0.0181 = 0.6\,b$; ethyl alcohol, $\nu = 0.0581 = 0.7b$ **33.** (a) From $p/T = (\partial S/\partial V)_{E,N}$ get $S = (AV/T)[1 + B(1 - V/2V_0)] + f(E, N)$ (or other equivalent forms) where f is any constant or function of E and N. (b) $\Omega = e^{S/k}$, where S is given in part (a). **35.** $x_i' = \sqrt{\kappa_i\kappa_1}\,x_i$ (b) $m_i = \sqrt{\kappa_1\kappa_i}\,m$ **37.** $-4.7\,\text{eV}$ **39.** (a) $2.04 \times 10^{-21}\,\text{J}$ (b) $k = 82\,\text{N/m}$ (c) $f = 6.8 \times 10^{12}\,\text{Hz}$ **41.** $C_p - C_v = p(\partial v/\partial T)_p + N_A[(\partial u_0/\partial T)_p - \partial(u_0/\partial T)_V]$ **43.** (a) Polyatomic (b) eight **45.** (a) $dV = (1/a)[-(2/p)dp + (1/3T)dT]$ (b) $\kappa = 2/apV$ (c) $\beta = 1/3aVT$ **47.** 0 (b) $-p/V$ (c) $-\gamma p/V$ **49.** You should first obtain the differential form, $V^2 dp + (2pV - aT)dV - (aV + b)dT = 0$, from which you can then read off the answers to parts (a) and (b): (a) $\beta = \frac{(a+b/V)}{2pV - aT}$, (b) $\kappa = \frac{V}{2pV - aT}$. **51.** (a) $p = 83.5\,\text{atm}$ (b) $\beta = 2.1 \times 10^{-3}/\text{K}$ **53.** Differential form is $Cdp + Ddv = RdT$, with $C = e^{Bv}v$, $D = e^{Bv}(p + pvB + AB)$. (a) $\beta = R/vD$ (b) $\kappa = C/vD$ (c) $C_p - C_V = pR/D$

Chapter 11

1. Definition of C_p **3.** $(\partial S/\partial p)_T = -(\partial V/\partial T)_p = -V\beta$ **5.** LHS = $(\partial S/\partial p)_T(\partial p/\partial V)_T = (\partial S/\partial p)_T/(\partial V/\partial p)_T = (-V\beta)/(-V\kappa) = \beta/\kappa$ **7.** LHS = $(\partial T/\partial p)_S\,(\partial V/\partial S)_p = (\partial V/\partial T)_p/(\partial S/\partial T)_p = V\beta/(C_p/T)$ **9.** LHS = $-(\partial p/\partial S)_V = -1/(\text{answer to problem 4})$ **11.** LHS = $-(\partial S/\partial V)_p/(\partial S/\partial p)_V = -(C_p/TV\beta)/(\kappa C_V/\beta T)$ **13.** $dV = (\partial V/\partial p)_T dp + (V/\partial T)_p dT = -V\kappa dp + V\beta dT$ **15.** $dE = T(\partial S/\partial T)_V dT + [T(\partial S/\partial V)_T - p]dV = C_V dT + (T\beta/\kappa - p)dV$ **17.** $dE = T(\partial S/\partial p)_V dp + [T(\partial S/\partial V)_p - p]dV = (C_V\kappa/\beta)dp + [(C_p/V\beta) - p]dV$ **19.** $dT = (\partial T/\partial p)_V dp + (\partial T/\partial V)_p dV = (\kappa/\beta)dp + (1/V\beta)dV$ **21.** $dE = [T - p(\partial V/\partial S)_p]dS - p(\partial V/\partial p)_s dp$, solve for dS: $dS = (1/A)dE - (pVC_V\kappa/C_p A)dp$, with $A = T(1 - pV\beta/C_p)$. **23.** (a) $dN = dV = 0$ (b) $dS = (\partial S/\partial T)_V dT$ (c) $dS = (C_V/T)dT$ **25.** (a) 1 (b) $dS = (\partial S/\partial N)_{T,p}dN$ (c) $dS = -(\partial\mu/\partial T)_{p,N}dN$ **27.** $\beta = 4.0 \times 10^{-5}/K$ (Ignore the smaller terms of order ΔT^2 and ΔT^3.) **29.** (a) Hold the gas in the cylinder at constant volume. Add heat ΔQ and measure the temperature change, $\Delta T\,C_V = (\Delta Q/\Delta T)_V$. (b) Put the liquid in the cylinder and let the volume change to keep the pressure constant as you add ΔQ and measure ΔT. $C_p = (\Delta Q/\Delta T)_p$. (c) Put a volume V of the liquid in the cylinder and measure the change in volume ΔV as you change the pressure by Δp. You will have to add or remove heat to keep the temperature constant as you do this; then $\kappa = (1/V)(\Delta V/\Delta p)_T$. (d) Immerse the solid in the liquid in the cylinder, and measure V_{total} and ΔV_{total} as above. To get V_s and ΔV_s for the solid alone, you have to subtract those of the liquid, determined as above. $(V_s = V_{total} - V_w, \Delta V_s = \Delta V_{total} - \Delta V_w) \kappa = (1/V)(\Delta V/\Delta p)_T$. **31.** $dE = TdS - pdV = T(\partial S/\partial T)_V dT + [T(\partial S/\partial V)_T - p]dV = C_V dT + (T\beta/\kappa - p)dV$. **33.** (a) 0.019 (b) 13 **35.** Write $\Delta S = (\partial S/\partial p)_V \Delta p + (\partial S/\partial V)_p \Delta V$ and then use Table 11.1 to convert the two partials into $C_p, C_V, \kappa, \beta, T, p, V$ as appropriate. **37.** Use $\partial^2 S/\partial p \partial T = \partial^2 S/\partial T \partial p$ and Maxwell's relation M10 for $\partial S/\partial p)_T$. **39.** (b) 4.9×10^6 Pa, 4.2×10^5 Pa **41.** (a) $V = NkT/p$, so $(\partial V/\partial T)_p = Nk/p$ and $(\partial^2 V/\partial T^2)_p = 0$. (b) From the van der Waals equation, $Avdp + Bpdv = RdT$, where $A = 1 - b/v$ and $B = 1 - a/pv^2 + 2ab/pv^3$, we get $(\partial v/\partial T)_p = R/pB$. The second derivative is $(\partial^2 v/\partial T^2)_p = -(R/pB^2)(\partial B/\partial T)_p$, where $(\partial B/\partial T)_p = (2a/pv^3)(1 - 3b/v)(\partial v/\partial T)_p$, and this last derivative is given above (R/pB). **43.** $\Delta E = T(\partial S/\partial T)_M \Delta T + [T(\partial S/\partial M)_T + B]\Delta M$

Chapter 12

1. (a) $dS = (C_p/TV\beta)dV$ (b) $dS = (C_p/T)dT$ **3.** $dV = V\beta dT$ **5.** (a) $dH = (C_p/V\beta)dV$ (b) $dF = -(S/V\beta + p)dV$ (c) $dG = -(S/V\beta)dV$ **7.** (a) $dE = (Cp/V\beta - p)dV + [T(\partial S/\partial N)_{p,V} + \mu]dN$ (b) $dE = [T(\partial S/\partial \mu)_{p,N} - p(\partial V/\partial \mu)_{p,N}]d\mu + [T(\partial S/\partial N)_{p,\mu} - p(\partial V/\partial N)_{p,\mu} + \mu]dN$ **9.** (a) $dE = (T\beta/\kappa - p)dV$ (b) $dE = (-TV\beta + pV\kappa)dp$ **11.** (a) $dH = (-TV\beta + V)dp$ (b) $dF = pV\kappa dp$ (c) $dG = V dp$ **13.** For an ideal gas, $\beta = 1/T, \kappa = 1/p$. **15.** (a) 530 K (b) 800 K (c) 61 atm **17.** (a) Use $pV = NkT$ to eliminate the volume V, and integrate. (b) 5.5 km **19.** Nine **21.** (a) Start with the rearranged first law, $dQ = dE + pdV$, with $E = (Nv/2)kT$ and use the definitions $C_V = (\partial Q/\partial T)_V$ and $C_p = (\partial Q/\partial T)_p$. (b) $\Delta(E + pV) = \Delta[(v/2)NkT + NkT] = [(v + 2)/2]Nk\Delta T = C_p \Delta T$ **23.** (a) 507 J, 410 J, 379 J (b) 1770 J, 410 J, 0 J (c) 1260 J, 0 J, −379 J (d) 1770 J, 0 J, −531 J **25.** (a) $dE = (pVC_V\kappa/C_p)dp$ (b) $dE = -pdV$ (c) $dE = (pC_V\kappa/T\beta)dT$ **27.** (a) 1.01×10^4 Pa (b) Use $\Delta T/T = (v\beta/C)\Delta p$ and get $\Delta T_{adiabatic} = 1.4 \times 10^{-4}$ K/m, which is greater than the lapse rate, so it is stable. **29.** As moist air rises and cools adiabatically, some of the moisture condenses, releasing latent heat. Consequently, rising moist air cools less rapidly with altitude than rising dry air. If the lapse rate falls between these two then the moist air will continue to rise whereas the dry air will not. **31.** $V = 0.226$ liters, $T = 531$ K, work $= 205$ J. **33.** Yes **35.** While moving, a part of the motions of all the molecules is coherent – they are all moving the same direction together. Friction turns this coherent motion into random thermal motion. To go back, this random thermal motion would have to turn back into synchronized coherent motion – all molecules going in the same direction together. This is very unlikely, i.e., it is a state of very low entropy, like the state that would occur if you flipped 10^{24} coins and they all landed heads. **37.** −0.32 °C **39.** (a) 324 K (b) 3.92×10^{-3}/K (c) 0.23 °C/atm **41.** $T_f - T_i = (2a/vR)(1/v_f - 1/v_i)$ **43.** (a) $\Delta E = \Delta Q = \Delta W = 0$ (b) $\Delta E = -\Delta W = -[NkT_i/(\gamma - 1)][1 - (V_i/V_f)^{\gamma - 1}]$, with $\gamma = (v + 2)/v$, $\Delta Q = 0$ (c) $\Delta E = 0, \Delta Q = \Delta W = NkT_i \ln(V_f/V_i)$ **45.** 256 (b) 25 (c) 9 **47.** Series: $\Delta T = \Delta T_1 + \Delta T_2 + \cdots = R_1 \dot{Q} + R_2 \dot{Q} + \cdots = (R_1 + R_2 + \cdots)\dot{Q} = R_{tot}\dot{Q}$. Parallel: $\dot{Q} = \dot{Q}_1 + \dot{Q}_2 + \cdots = \frac{1}{R_1}\Delta T + \frac{1}{R_2}\Delta T + \cdots = \left(\frac{1}{R_1} + \frac{1}{R_2} + \cdots\right)\Delta T = \frac{1}{R_{tot}}\Delta T$. **49.** (a) $R_w = 3.5 \times 10^{-2}$ K/W (b) $R_i = 9.5 \times 10^{-2}$ K/W (c) $R_s = 21.6 \times 10^{-2}$ K/W (d) $R_{total} = 2.30 \times 10^{-2}$ K/W (e) 6.9×10^9 J (f) Electricity, $286; gas, $95. **51.** Take the second derivative with respect to x, and compare it with $1/K$ times the first derivative with respect to t. Then use the fact that the Gaussian factor at $t = 0$ is an infinitely narrow spike beneath which the area is unity to show that $T(x, t = 0) = f(x)$.

Chapter 13

1. (a) $\Delta Q = [(v + 2)/2](nRT_i/V_i)(V_f - V_i)$, $\Delta W = (nRT_i/V_i)(V_f - V_i)$. (b) $\Delta Q = \Delta W = nRT_i \ln(V_f/V_i)$ (c) $\Delta Q = 0$, $\Delta W = (vnRT_i/2)[1 - (V_i/V_f)^{2/v}]$, $[(\gamma - 1) = 2/v]$ **3.** 1730 J
5. The diagram will have $-F$ on the vertical axis and L on the horizontal axis.
(1) Constant-length heat addition: straight up: (2) Adiabatic contraction: down and to the left. (3) Constant-length heat removal: straight down. (4) Adiabatic extension: up to the right. **7.** (a) p_2, V_1; $p_2(V_1/V_3)^\gamma$, V_3; $p_1(V_1/V_3)^\gamma$, V_3; p_1, V_1. (b) $\Delta Q = (p_2 - p_1)V_1/(\gamma - 1)$, $\Delta W = 0$; $\Delta Q = 0$, $\Delta W = [p_2V_1/(\gamma - 1)][1 - (V_1/V_3)^{\gamma-1}]$; $\Delta Q = (p_1 - p_2)(V_1/V_3)^\gamma V_3/(\gamma - 1)$, $\Delta W = 0$; $\Delta Q = 0$, $\Delta W = [p_1V_1/(\gamma - 1)]$
$[(V_1/V_3)^{\gamma-1} - 1]$. **9.** They will look like the diagrams on the right in Figure 13.5.
11. (a) Straight across, slope down, straight down, slope up to the left. (Adiabatic curve is steeper than isothermal curve, and both are concave upward.) (b) Slope up, straight across, slope down to the left, straight up. (Isochoric is steeper than isobaric, and both are concave upward.) (c) $(\Delta Q, \Delta W, \Delta E) = + + +(1)$, $+ + 0(2)$, $- 0 - (3)$, $0 - +(4)$. **13.** (a) Slope down, slope down, straight to the left, straight up. (Adiabatic is steeper than isothermal, and both are concave upward.) (b) Straight across, straight down, slope down to the left, slope up to the right. (Isochoric is steeper than isobaric, and both are concave upward.)
(c) $(\Delta Q, \Delta W, \Delta E) = + + 0 (1)$, $0 + - (2)$, $- - - (3)$, $+ 0 + (4)$. **15.** (a) 6 °C (b) 600 kg/s **17.** The gas turbine's hot reservoir is slightly hotter than the coil, because heat is flowing from the reservoir to the coil; vice versa for the refrigerator. **19.** (a) $2T/vR$
(b) $2T/(v + 2)R$ (c) 0 (d) The isochoric and isobaric cases curve upward, because they slope upward and the slope increases with increasing T. **21.** (a) $\Delta Q = \Delta W = \int p \, dV$, with $p = nRT/V$ (b) $\Delta S = \Delta Q/T$ (constant T) (c) Replace V_2/V_1 by p_1/p_2, because $V = \text{constant}/p$. **23.** (a) $\Delta W = nRT \ln(V_f/V_i) = p_i V_i \ln(V_f/V_i)$ (b) $p_f = p_i(V_i/V_f)$
25. (a) $\Delta W = p_1 V_1\{\ln(V_2/V_1) + (5/2)[1 - (V_2/V_3)^{2/5}]\}$ (b) $p_3 = (p_1 V_1/V_2)(V_2/V_3)^{7/5}$
27. 2.5 J (b) 21 J **29.** (a) 1020 J (b) 582 J (c) 873 J (d) 436 J (e) 0.43
31. (a) $\Delta Q = \Delta W = 161$ J (b) $\Delta Q = 1400$ J, $\Delta W = 400$ J (c) $\Delta Q = 0$, $\Delta W = 119$ J
33. (a) $\Delta Q = 0$, $\Delta W = (5/2)R(T_1 - T_2)$; $\Delta Q = \Delta W = (5/2)RT_2 \ln(T_2/T_1)$; $\Delta Q = (5/2)R(T_1 - T_2)$, $\Delta W = 0$ (b) $\Delta Q = 0$, $\Delta W = (5/2)R(T_1 - T_2)$; $\Delta Q = (7/2)RT_2[(T_2/T_1)^{5/2} - 1]$, $\Delta W = RT_2[(T_2/T_1)^{5/2} - 1]$; $\Delta Q = (5/2)RT_1[1 - (T_2/T_1)^{7/2}]$, $\Delta W = 0$ **35.** (a) In units of $p_0 V_0$, $(\Delta Q, \Delta W, \Delta E) = (7/2, 1, 5/2)$, $(\ln 2, \ln 2, 0)$, $(0, 5/2, -5/2)$. (b) In units of $p_0 V_0/nR$, $T = 2, 1, 0.76$. **37.** Start with $H = E + pV$ and then take the differential form with $dQ = 0$, so that $dH = V dp$. Then integrate using $pV^\gamma = \text{constant}$. **39.** (a) $(p(10^5 \text{ Pa}), V(10^{-3} \text{ m}^3), T(\text{K})) = (2, 2, 600)$, $(1.13, 3, 510)$, $(2.26, 1.5, 510)$, $(4, 1, 600)$ (b) $(\Delta Q, \Delta W, \Delta E)$ (in joules) $= (277, 277, 0)$, $(0, 150, -150)$, $(-236, -236, 0)$, $(0, -150, 150)$ (c) 0.15 **41.** If ΔQ were to flow from cold to hot then $\Delta S = \Delta Q/T_h - \Delta Q/T_c = [\Delta Q/(T_h T_c)](T_c - T_h) < 0$, violating the second law.
43. Suppose that on your p–V diagram you have isothermal expansion from V_1 to V_2, adiabatic expansion from V_2 to V_3, isothermal compression from V_3 to V_4, and adiabatic compression from V_4 to V_1. Then from the two isothermal parts you should be able to show that $Q_h/T_h = nR \ln(V_2/V_1)$ and $Q_c/T_c = nR \ln(V_3/V_4)$. From the two adiabatic lines you should get $T_h V_2^\gamma = T_c V_3^\gamma$ and $T_h V_1^\gamma = T_c V_4^\gamma$, from which $V_2/V_1 = V_3/V_4$; plug this into

your isothermal results.   **45.** (a) 0.48   (b) 0.44   **47.** Gasoline burns fast, so that the piston doesn't move much (and therefore, the volume doesn't change much) during the combustion. In turbine engines, the heat is added while the fluid is in an open tube in the heat exchanger, so the pressure is the same from one end of the tube to the other.

49. (a) p_1, V_2, $p_1 V_2/nR$; $p_1(V_1/V_2)^{7/2}$, $V_1(V_2/V_1)^{7/2}$, $p_1 V_1/nR$; p_1, V_1, $p_1 V_1/nR$.

(b) $\Delta Q = (7/2)p_1(V_2 - V_1)$, $\Delta W = p_1(V_2 - V_1)$, $\Delta E = (5/2)p_1(V_2 - V_1)$; $\Delta Q = 0$, $\Delta W = -\Delta E = (5/2)p_1(V_2 - V_1)$; $\Delta Q = \Delta W = (7/2)p_1 V_1 \ln(V_1/V_2)$, $\Delta E = 0$.

51. (a) 750 K, 2.5×10^6 Pa   (b) 1500 K   (c) 0.8   **53.** The line slopes steeply downward until it reaches the mixed phase, is horizontal across the mixed phase, and then slopes more gently downward in the gas phase.   **55.** (a) 2.9×10^{24} J   (b) 2.8×10^{20} J   (c) 0.01%   (d) 0.1%   (e) 7%   (f) 0.7%

Chapter 14

1. (a) 0.096 eV (b) −0.294 eV (c) −0.372 eV (d) −0.390 eV (e) −0.443 eV (f) 1.4 J goes into thermal energy, 2.8 J into u_0.   **3.** Liquid water's chemical potential falls faster, because above 0 °C the molecules prefer the liquid phase so it must have the lower chemical potential.   **5.** $(\Omega_1, \Omega_2, \Omega_2/\Omega_1) =$ (a) (2, 3, 1.5) (b) (3, 6, 2) (c) (3, 4, 1.33) (d) (6, 10, 1.67) (e) (10, 20, 2) (f) Yes (g) Yes   **7.** $S = k \ln \Omega = Nk \ln \omega_c$. So if you multiply your answers by k then you should find that $\partial S/\partial E = Nvk/2E = 1/T$, $\partial S/\partial V = Nk/V = p/T$, and $\partial S/\partial N = k[\ln \omega_c - (v + 2)/2]$. If you also multiply your answers by T, you should get $T\Delta S = \Delta E + p\Delta V - \mu \Delta N$, where $\mu = kT[(v + 2)/2 - \ln \omega_c] = \bar{\varepsilon} + p\bar{\varepsilon} - kT \ln \omega_c$, as in equation 14.2.   **9.** $V_p = 2.9 \times 10^{-68}$ (kg m/s)^3, $eV/N = 8.1 \times 10^{-29}$ m^3; $\omega_c = (e/N)(V_r V_p/h^3) = 8000$ **11.** (a) $\Delta E = +60.2$ eV, $\Delta W = -0.013$ eV, $\mu \Delta N = +400$ eV, $\Delta Q = \Delta E + \Delta W - \mu \Delta N = -339.8$ eV   (b) $\Delta E = -60.2$ eV, $\Delta W = +0.013$ eV, $\mu \Delta N = -720$ eV, $\Delta Q = \Delta E + \Delta W - \mu \Delta N = +659.8$ eV   (c) 219.8 eV   (d) Use $\mu = -kT \ln \omega_c + \bar{\varepsilon}$ and solve for ω_c; $\omega_{c,A} = 1470$ and $\omega_{c,B} = 40$.   **13.** 2.7g/m^3 **15.** (a) 1.66×10^{-29} m^3, 2.44×10^{-29} m^3, 9.67×10^{-29} m^3, 2.99×10^{-29} m^3 (b) 1.05×10^{-5} eV, 1.54×10^{-5} eV, 6.12×10^{-5} eV, 1.89×10^{-5} eV   (c) 2420, 1650, 416, 1350   **17.** (a) −0.505 eV   (b) 16 (from 14.5′).   **19.** 0.338 eV   **21.** 2.02×10^3 Pa, −40 Pa   **23.** 4.9×10^6 Pa (48 atm)   **25.** (a) 10^5 Pa, or about 1 atm   (b) Toward the hot side   (c) Somewhere around 200 000 to 1 000 000 K   **27.** No; $\mu_A + \mu_B < \mu_C$. So to minimize G, the system stays as $A + B$ rather than going to C.   **29.** To the right, because $2\mu_D + 3\mu_E < 3\mu_A + \mu_B + 4\mu_C$ **31.** (a) 189 J/(mole K) (b) 1.959×10^{-3} eV/K, 1.965×10^{-3} eV/K (c) $\Delta \mu (= -\sigma \Delta T) = -3.92 \times 10^{-3}$ eV   **33.** (a) $\rho(H^+) = 10^{-7}$mole/liter (b) 7 (c) 6.51   **35.** (a) Use equation 14.4, $\varepsilon = u_0 + (3/2)kT$, and $p\varepsilon = kT$.   (b) $[(2\pi m kT)^{3/2}/h^3]e^{-u_0/kT}$ **37.** The line separating the solid and liquid phases should be vertical. Pressure favors neither phase, so increased pressure has no effect on the temperature of the phase transition.   **39.** T_c, is higher for water. The self-attraction is stronger for the water molecules, so larger thermal energies are required to oppose the tendency for molecules to stick together and condense into a liquid.   **41.** (a) 3.6 atm   (b) 16 atm   (c) 97 atm   (d) 0.082 atm   (e) 0.0082 atm   (f) 36% error. We assumed L to be a constant, independent of the temperature.   **43.** (a) $v_{sol} = 7.154 \times 10^{-6}$ m^3, $v_{liq} = 7.904 \times 10^{-6}$ m^3 (b) $\Delta v = +0.750 \times 10^{-6}$ m^3 (c) The pressure would have to increase.   (d) 7.6×10^8 Pa (Integrate equation 14.15, $dp = (L/\Delta v)dT/T$.)   **45.** −5 °C   **47.** $pT^{-B/R}e^{A/RT} =$ constant.   **49.** (a) More volume in position space means more accessible states – hence increased entropy. During adiabatic expansion, the system cools, so that although the volume in position space increases, that in momentum space decreases, the net change in accessible states being zero.   **51.** (b) Show that the second derivative of $-T\Delta S_m$ is positive for all f.   **53.** Thermal energy is released when the particles fall into deeper potential wells. So the attraction between unlike particles must be stronger than that between like particles.

Chapter 15

1. (a) $S_R = S_0 - (\Delta E + p\Delta V - \mu\Delta N)/T$ (b) $e^{S_R/k}$, with S_R from part (a)   (c) $P = Ce^{-(\Delta E + p\Delta V - \mu\Delta N)/kT}$ **3.** 3.2×10^{10} Pa $= 3.2 \times 10^5$ atm   **5.** (a) $\Delta E = +10.2$eV, $p\Delta V = 2.5 \times 10^{-5}$ eV (b) 4.2×10^{10} Pa $= 4.1 \times 10^5$ atm   **7.** (a) 0.371 (b) 0.371 (c) 0.233   **9.** (a) 0.798 (b) 0.798 (c) 0.161   **11.** (a) 4.635 (b) 0.990 (c) 0.990 (d) 9.3×10^{-5} **13.** (a) 0.765 (b) 0.235   **15.** 10.2 times   **17.** (a) 1010 K (b) 2320 K   **19.** 11 600 K   **21.** 348 K   **23.** (a) 4.56×10^{-3} (b) 0.332, 0.333, 0.335 (c) 2.84×10^{-26} J/T (d) 0.017 J/T   **25.** (a) 0.368 (b) 0.288   **27.** (a) 0.08006 eV (b) 0.043   **29.** $dV = 0$, so $dQ/dT = dE/dT$ with $E = (N_A\nu/2)kT$.   **31.** (a) 1.16×10^{-48} kg m^2 (b) 0.0599 eV (c) 694 K   **33.** (a) 2320 K (b) 8.1 K (c) $3R$ **35.** 0.24 K **37.** $E/N = (\sum e^{-\varepsilon_i/kT}\varepsilon_i)/(\sum e^{-\varepsilon_i/kT})$.   **39.** (a) 0.772 (b) 0.076 (c) 0.228   **41.** 3×10^{-7} **43.** (a) $g_1(\varepsilon) = n_1/(b-a)$, $g_u(\varepsilon) = n_u/(d-c)$ (b) $[(b-a)/(d-c)](n_u/n_l)(e^{-\beta c} - e^{-\beta d})/ (e^{-\beta a} - e^{-\beta b})$ (c) $(n_u/n_l)e^{-\beta(c-a)}$ (d) $[(b-a)/(d-c)](n_u/n_l)e^{-\beta(c-a)}$ **45.** (a) 3 (b) 3.8×10^{-2} eV $= 6.1 \times 10^{-21}$ J (c) 508 m/s (d) 1920 m/s   **49.** 2.8×10^{-5}m/s **51.** (a) 4.07×10^{-21} J (b) 2.0×10^{-11} **radians or** 4.1×10^{-11} m

Chapter 16

1. 0.006   **3.** (a) 0.01 (b) 0.81   **5.** Use equation E.1 with $\alpha = \beta m/2$.   **7.** With $d^3v = v^2 dv \sin\theta \, d\theta \, d\phi$, integrate over the angles to get 4π and then use equation E.2 for the integral over v (from 0 to ∞) with $n = 1$, $\alpha = \beta m/2$.   **9.** (a) 0.39 (b) 0.11 (c) 7.4×10^{-3} **11.** (a) 0 (b) 470 m/s (c) 510 m/s (d) 416 m/s   **13.** 1.45   **15.** Use equation E.4 with $n = 0$.   **17.** (a) 3.11×10^{21} particles/s (b) 2.88×10^{-6}/s (c) $N = N_0 e^{-Ct}$, $N_0 = 1.08 \times 10^{27}$, $C = 2.88 \times 10^{-6}$/s (d) 67 hours   **19.** 122 s   **21.** (a) 20 m/s (b) 14 m/s (c) 14 m/s (d) 0   (e) Those moving south   **23.** Water vapor; 1.25 times   **25.** Put $v_1 = V + u/2$, $v_2 = V - u/2$ in 16.15.   **27.** (a) $+$ (b) $-x$ **29.** It would proceed more slowly, because fatter molecules don't go as far between collisions, and therefore transfer Q over smaller distances.   **31.** (a) $D = 4.8 \times 10^{-6}$ m^2/s (b) $K = 4.2 \times 10^{-3}$ W/(m K) (c) $\eta = 5.8 \times 10^{-6}$ kg/(m s^2)   **33.** 43 watts/m^2 **35.** (a) #/m^3 (b) #/m^6 (c) m^2/s (d) $3Da$

Chapter 17

1. (a) Out of the page (b) Radially inward, adds to (c) Increase, increase, oppose (d) Out of the page; radially outward, detracts from; decrease, decrease, oppose   **3.** The charge is overall neutral, but the negative charge is distributed farther out and the positive charge more centrally. (So thinking classically, the negative charge has a larger orbit as the particle spins, making the negative charge dominate the magnetic moment.)   **5.** (a) 45° (b) 35° (c) 30°

7. 1.88×10^{-28} kg **9.** 6.16×10^{-27} J/T **11.** Put $\mu_z = -(l_x + 2s_z)\mu_B$ and $x = \mu_B B/kT$ into equation 17.10. **13.** Show this $e^{nx} \gg e^{(n-1)x}$ for $x \gg 1$. This should justify keeping only the largest term in each sum in equation 17.10. **15.** (a) $kT = 2.5 \times 10^{-2}$ eV, $\mu_B B = 5.8 \times 10^{-5}$ eV; thermal energy dominates. (b) 3.7×10^{-4} K

17. (a) 1.88 J K/(T^2 · mole) (b) 3.75×10^{-3}J/T, 0.375 J/T (c) 6.7×10^{-5} K **19.** (a) 1.007 (b) 9.53 J/T **21.** (a) 1.11×10^{-6} J K/(T^2 mole) (b) 2.31×10^{-5} J K/(T^2 mole) (c) 7.83×10^{-8} J /(T mole) **23.** (a) 5.58 J/T (b) 22.3 J/T (c) 11.2 J/T **25.** (a) 9.38 J K/(T^2 mole) (b) 1.07 J/T (c) 91.7 J/T

Chapter 18

1. (a) $1 + 1.6 \times 10^{-17}$ (b) 1.68 **3.** Use $P \propto \Omega = e^{S/k}$, where the entropy of the small system in a given state is 1 and that of the reservoir is $S_0 - \Delta S$, ΔS being its loss in entropy owing to its supplying energy ε_s to the small system. **5.** With $Z = \Sigma_s e^{-\beta \varepsilon_s}$, show that the indicated operation gives $\Sigma_s P_s E_s^2$ with $P_s = (1/Z)e^{-\beta Es}$. **7.** Via the operation $-kT^2 \partial/\partial T$. **9.** (a) Use $\partial/\partial\beta = -kT^2 \partial/\partial T$. (c) $Z = 1 + e^{-\varepsilon_1/kT} + e^{-\varepsilon_2/kT} + \cdots \rightarrow 1 + e^{-\infty} + e^{-\infty} + \cdots = 1$ (d) The indicated operation should give $(kT/Z)[0 - (\varepsilon_1/kT^2)e^{-\varepsilon_1/kT} - (\varepsilon_2/kT^2)e^{-\varepsilon_2/kT} - \cdots]$, and so you need to show that terms of the form $e^{-ax}x^2 \rightarrow 0$ as $x \rightarrow \infty$. **11.** (a) $(V/V_0)^N$ (b) $p = NkT/V$ (c) $\mu = -kT\ln(V/V_0)$ **13.** (a) $Z = C(\beta/\beta_0)^{-3N/2}(V/V_0)^N$, where $C = \Sigma_s e^{-C_s}$ (b) $(3N/2)kT$ (c) NkT/V (d)$-kT[(3/2)\ln(\beta_0/\beta) + \ln(V/V_0)]$ (e) $\sqrt{3N/2}kT$ (f) $(3/2)Nk$ **15.** (a)2 (b)4 (c) 8 **17.** (a) 24 (b)418 **19.** (a) 9, 6, no (b)10 000, 5050, no (c) Case (b) **21.** (a) $[e(1 + e^{-580/T})/N]^N$ (b) $-NkT \ln[e(1 + e^{-580/T})/N]$ (c) $\mu = -kT\{\ln[e(1 + e^{-580/T})/N] - 1\}$ **23.** (a) 3480 K (b) 2.3×10^9 K **25.** (a) 137 N/m (b) Multiply both sides by $1 - a$. (c) $Z = (1 - e^{-\beta\hbar\omega})^{-3}$ (d) $3\hbar\omega/(e^{\beta\hbar\omega} - 1)$ **27.** (a) 2.08×10^{-46} kg m^2 (b) 3.35×10^{-4} eV (c) 3.9 K **29.** (a) 0.196 eV (b) 2280 K **31.** Use $e^{-x} \approx 1 - x$ for $x \ll 1$. **33.** Carry out the steps linking equations 18.18 and 18.19. **35.** (c) 153 J/(mole K)

Chapter 19

1. (a) 0.597 (b) 0.403 (c) 0.981 (d) 0.019 **3.** No **5.** (a) 25.9 (b) 2.72 (c) 1.001 (d) 1.000. **7.** (a) 1, 0 (b) 0, 1, 2 (c) Boson gas (d) Fermions, fermions **9.** $(e^x + 2)/(e^{2x} + e^x + 1)$, where $x = \beta(\varepsilon - \mu)$. **11.** -0.173 eV **15.** (a) 0.693 (b) 1.5 **17.** (a) 8 (b) 4/3 (c) 1/4 (d) 4 (e) 1/2 **19.** (a) $s = 12, d = 4$ (b) $s = 6, d = 2$ (c) $s = 6, d = 4$ (d) $s = 6, d = 0$ **21.** (a) 1/3 (b) 1/2 (c) 0 **23.** (a) 1/4 (b) 1/4 (c) 1/2 (d) 0 **25.** (a) Yes (b) 0.8 K **27.** Yes $(rp/h = 9)$ **29.** (a) 4.64×10^{12}/s (b) 35 K (c) Large **31.** (a) $(8\pi V/h^3)p^2dp$ (b) $g(\varepsilon) = [4\pi V(2m)^{3/2}/h^3](\varepsilon - \varepsilon_0)^{1/2}$ (c) $dN/d\varepsilon = [4\pi V(2m)^{3/2}/h^3](\varepsilon - \varepsilon_0)^{1/2}/(e^{\beta(\varepsilon-\mu)} + 1)$ **33.** (a) 4.05 (b) 0.130 eV (c) 17.0, 0.93 eV **35.** Write $N = \int dN = \int g(\varepsilon)n(\varepsilon)d\varepsilon$ and integrate. Solve for μ. **37.** (a) 28 eV, 2.2×10^5 K (b) 23 eV, 1.8×10^5 K **39.** (a) 2.4×10^{10} K (b) 3.9×10^{14} kg/m^3

Chapter 20

1. $g(\varepsilon) = (4\pi V/h^3 c^3)\sqrt{\varepsilon^2 - m^2 c^4}\,\varepsilon$ **3.** About 1 $(V \approx 0^{-29}$ m^3, $p^3 = 10^{-72}$ (kg m/s)^3, $h^3 = 10^{-100}$ (J s)^3) **5.** (a) 6.49×10^{34} particles/eV (b) 0.96×10^{22} particles/eV **7.** (a) kT (b) 2kT **9.** Use equation 19.14 and $e^x \approx 1 + x$ for $x \ll 1$.

11. (a) $f_x = N v_x / 2V$ (b) $2mv_x$ (c) pressure $= (N/V)mv_x^2$ (d) pressure $= (N/V)m\overline{v^2}/3 =$
$(2/3)(N/V)\varepsilon_{ave}$ (e) $f_x = Nc \cos\theta/2V$, $\Delta p_x = 2p \cos\theta$, pressure $= (N/V)pc \cos^2\theta$, then
average over $\cos^2\theta$ and p. **17.** -0.39 eV **19.** (a) 6.8 eV, 4.1 eV, 53 000 K (b) 5.6 eV, 3.4
eV, 44 000 K (c) -7.6×10^{-5}eV, -9.2×10^{-5} eV **21.** (a) 7.9 eV (b) 4.7 eV
(c) 61 000 K

Chapter 21

1. (a) Use $e^x \approx 1 + x$ for $x \ll 1$. (b) Use L'Hospital's rule on $Cx^3/(e^x - 1)$, where
$x = \beta\varepsilon$. **3.** Write the distribution as $Cx^3/(e^x - 1)$, where $x = \beta\varepsilon$. Set the derivative equal
to zero to get $x = 3(1 - e^{-x})$. Show that it is satisfied by $x = 2.82$. **5.** No. Using the chain
rule $d\varepsilon = (-hc/\lambda^2)d\lambda$, so the distribution in λ has an extra factor of $-hc/\lambda^2$ compared with
the distribution in ε. This extra factor shifts the position of the maximum.
7. $(4\pi/h^3c^3)\varepsilon^3/(e^{\beta\varepsilon} + 1)$ **9.** The energy absorbed is at green wavelengths and nothing else.
The energy emitted is a blackbody spectrum at temperature T. **11.** 4.0×10^{66} J, $1.5 \times$
10^{20} K **13.** (a) 16 (b) 2 (c)16 **15.** I_r/I_0 has a peak in the red. $I_a/I_0 = I_e/I_0 = 1 - I_r/I_0$
has a dip in the red. **17.** Wait two minutes before adding the water, because you want to
radiate away as much heat as possible, and the coffee radiates heat faster when hotter.
19. (a) About 300 K and 2 m^2 (b) 920 W (c) 1.9×10^4 kcal (d) 47 (e) A slightly cooler
skin temperature greatly reduces the rate of heat loss through radiation. **21.** (a) Net power
radiated $= 4A\sigma T^3\Delta T$ (b) $d(\Delta T)/dt = -C_1\Delta T$, with $C_1 = 4A\sigma T^3/C$ (c) Write this
equation as $d(\Delta T)/\Delta T = -C_1 dt$ and integrate, obtaining $\Delta T = C_2 e^{-C_1 t}$, where C_2 is the
temperature difference at the beginning. **23.** The absorptivity would wiggle between 0 and 1
(like the ratio of the Sun's curve to that of the 5800 K blackbody) for wavelengths shorter than
green, but would be about equal to 1 for green and longer wavelengths; reflectivity $= 1-$
absorptivity. **25.** Double glazing reduces the rate of heat loss **27.** 214 K, 253 K, 279
K **29.** Volume in coordinate space is gained without a corresponding loss in momentum
space (as would happen in an adiabatic process because of the reduction in momentum
resulting from collisions with receding walls). Free expansion is a nonequilibrium process,
and so entropy increases. **31.** $\Delta W = -\Delta E = E_i - E_f = a(V_i T_i^4 - V_f T_f^4)$, where VT^3 is
constant (constant entropy). **33.** The mean square voltage noise would be half as large.

Chapter 22

1. 81 N/m **3.** (a) $6R$ **5.** (b) 8.6×10^{-3} eV, 1.3×10^{13}/s (c) $x/(e^x - 1)$ (d) It starts at
unity and slopes downward (with positive curvature), approaching zero as $x \to \infty$ (the
high-temperature limit). **7.** $D(T) = (3/x^3)\int_0^x t^3 dt/(e^t - 1)$ **9.** (a) $c_s = \omega/k$ (b) $\varepsilon = pc_s$
11. 0.013 eV **13.** 0.025eV, 3.8×10^{13}/s **17.** 6×10^{-4} K **19.** (a) 90 MeV (b) 10^{12} K
21. (a) $\varepsilon = (h^2/2m)(N_e/2V)^{2/3}$ (b) It differs by a factor $(3/4\pi)^{2/3} \approx 0.38$. (c) $\varepsilon = p^2/2m$
where $p = h/\lambda$, so that $\lambda = (8\pi V/3N_e)^{1/3} = 2.03(V/N_e)^{1/3} = 2.03 \times$ spacing
23. (a) 0.4 K (b) 0.04 K **25.** (a) 1.02×10^{44}/m^3, 6.40×10^{43}/m^3 (b) 5.0×10^{11} K,
3.7×10^{11} K **27.** 0.07 K, no **29.** 0.0027 **31.** (a) 5.90×10^{28}/m^3 (b) $\mu \approx \varepsilon_f = 5.5$ eV
33. It is larger by 4.2×10^{-3}% **35.** A slope of 0.05×10^{-3}J/K^4 gives $\varepsilon_D = 0.029$ eV, and
an intercept of 0.68×10^{-3}J/K^2 gives $\varepsilon_f = 5.2$ eV. **37.** (a) 3.49×10^{-3} J/K (b) $1.29 \times$
10^{-3} J/K **39.** The heat capacity begins at zero and increases linearly (electrons). Then after a
few degrees it increases as T^3 (lattice) and approaches the value 3 (lattice saturated),

whereupon it nearly levels off, increasing linearly and slowly (electrons) until it reaches the value 4.5 (electrons saturated), whereupon it levels off for good.

Chapter 23

1. (a) Outer orbits overlap with the electron orbits of neighboring atoms, whereas inner orbits may not. (b) Outer orbits have greater overlap and therefore greater splitting. **3.** The same as Figure 23.2a, but the band is full rather than half full, and μ moves up to a position between the two bands. **5.** (a) $\mu_{Pt} = 6 \times 10^{-4}$ m^2/(V s), $\mu_{Cu} = 4 \times 10^{-3}$ m^2/(V s) (b) Most conduction electrons in conductors are stuck well below the Fermi level and cannot respond at all to imposed electric fields. The average mobility is heavily weighted by these non-mobile electrons. **7.** 2.1×10^{29}/m^3 and 1.8×10^{29}/m^3. They are 2.1×10^3 and 1.8×10^3 times larger, respectively. **9.** (a) Set $n_e = n_h$ and solve for μ. The required conditions are $\varepsilon_v < \mu < \varepsilon_c$, and $|\varepsilon - \mu| \gg kT$. (b) 0.013 eV **11.** (a) $n_e = n_h = 4.9 \times 10^{16}$/m^3 (b) $n_e = n_h = 1.0 \times 10^8$/m^3 (c) $n_e = n_h = 1.1 \times 10^{21}$/m^3 **13.** (a) 1.0×10^{16}/m^3 (b) 1.0×10^{15}/m^3 **15.** (a) 6.7×10^{15}, 5.9×10^{19}, 2.5×10^{13}, 7.7×10^{23}/m^3 (b) 2.0×10^{-4}, $5.5, 3.5 \times 10^{-6}$, 1.0×10^6A/(V m) **17.** 1.05 eV **19.** 9 K **21.** 6.4 eV **23.** (a) 0.256 eV (b) 1.25×10^{21}/m^3 (c) 1.7×10^{11}/m^3 (d) 0.460 eV, 1.25×10^{21}/m^3, 1.9×10^{19}/m^3 (e) 590 K **25.** (a) 0.256 eV (b) 1.25×10^{21}/m^3 (c) 1.8×10^{16}/m^3 (d) 0.4 eV (in actual fact it is 0.394 eV, because the material is not quite purely intrinsic. We must have (no. of holes in valence band)+(no. of donor electrons)=(no. of electrons in conduction band). Then we would use the law of mass action for the product $n_h n_e$. 5.7×10^{21}/m^3, 4.4×10^{21}/m^3 (e) 440 K **27.** (a) 0.405 eV (b) 0.202 eV (c) 162 K
29. 20 K (Only 10^{-5} of the donors are ionized, so the Fermi level lies almost on the donor levels.) **31.** (a) 0.7 eV (b) 8.0×10^{21}/m^3, 5.5×10^9/m^3 (c) 5.5×10^9/m^3, 8.0×10^{21}/m^3 (d) 2.0×10^{22}/m^3, 1.9×10^{11}/m^3 **33.** (a) 1.3×10^{11}/m^3 (b) 1.3×10^{11} /m^3 (c) 1.1×10^5 m/s (d) 1.15×10^{-3} A/m^2

Chapter 24

1. (a) $P_1/P_0 = e^{-\beta(\varepsilon_1-\varepsilon_0)} = e^{-50} = 10^{-21.7}$ (b) 1160 K, 1.16 K **3.** 2.4×10^{-3}
5. $(M, S) = (\pm 6\mu, 0); (\pm 4\mu, k \ln 6); (\pm 2\mu, k \ln 15); (0, k \ln 20)$ **7.** 5.4 cm/s **9.** The volume and temperature both decrease. The spin entropy increases. The entropy in the kinetic and other degrees of freedom decreases. **11.** (a) 4.6×10^{-3} (b) 3.5×10^7 **13.** (a) 1.2×10^{-7} (b) 8.6×10^6 times more **15.** 8×10^{-4} K **17.** (a) 0.555×10^{22} (b) 0.550×10^{22} (c) 1.46×10^{-4} J/K (d) 0.012 K **19.** 5.3×10^{-5} K **21.** 0.9999999999 **23.** (a) 11.0×10^{-10}m (b) Spacing $= 3.5 \times 10^{-10}$m; vibrations are three times larger. **25.** The normal fluid component slows down and stops, but the superfluid component keeps on going without friction (i.e., it flows through the normal component). **27.** 0.65 **33.** 2050 km, 5.8 km **35.** (a) $x = 0.885(M/M_s)^{1/3}$ and $x = 0.356(M/M_s)^{1/3}$ (b) 600 km, 4.1 km (c) $1.2M_s$, $4.7M_s$

Index

Printed in the United States
By Bookmasters